Human Genes and Genomes

Science, Health, Society

Human Genes and Genomes

Science, Health, Society

Leon E. Rosenberg and
Diane Drobnis Rosenberg

ELSEVIER

AMSTERDAM • BOSTON • HEIDELBERG • LONDON
NEW YORK • OXFORD • PARIS • SAN DIEGO
SAN FRANCISCO • SINGAPORE • SYDNEY • TOKYO
Academic Press is an imprint of Elsevier

Academic Press is an imprint of Elsevier
32 Jamestown Road, London NW1 7BY, UK
225 Wyman Street, Waltham, MA 02451, USA
525 B Street, Suite 1800, San Diego, CA 92101-4495, USA

First edition 2012

British Library Cataloguing-in-Publication Data
A catalogue record for this book is available from the British Library

Library of Congress Cataloging-in-Publication Data
A catalog record for this book is available from the Library of Congress

ISBN : 978-0-12-385212-0

For information on all Academic Press publications
visit our website at elsevierdirect.com

Typeset by TNQ Books and Journals Pvt Ltd.
www.tnq.co.in

Dedicated to our children
Robert, Diana, David, and Alexa
And to our grandchildren
Suzannah, Sam, Eli, Celia, and Abraham

CONTENTS

Leon E. Rosenberg has been a professor in the Department of Molecular Biology and the Woodrow Wilson School of Public and International Affairs at Princeton University since 1998. From September 1991 to January 1998, he was Chief Scientific Officer of the Bristol-Myers Squibb Company.

Prior to joining Bristol-Myers Squibb, Dr Rosenberg was dean of the Yale University School of Medicine, a position he had held since 1984. During his 26-year affiliation with Yale, he worked as a research geneticist, teacher, clinician, and administrator. In 1965, he was appointed assistant professor of medicine at Yale. He was named professor of human genetics, pediatrics, and medicine in 1972, the same year he helped establish the Department of Human Genetics at Yale and became its first chairman. A specialist in inherited metabolic disorders in children, Dr Rosenberg and his colleagues conducted pioneering laboratory investigations into the molecular basis of several inherited disorders of amino acid and organic acid metabolism.

Dr. Rosenberg received his BA and MD degrees, both summa cum laude, from the University of Wisconsin. He completed his internship and 1 year of residency training in internal medicine at Columbia-Presbyterian Medical Center in New York City. He then moved to Bethesda, Maryland, to begin a 6-year association as an investigator with the metabolism service of the National Cancer Institute.

Dr Rosenberg's honors include election to the National Academy of Sciences and to the Institute of Medicine, recipient of the Borden Award from the American Academy of Pediatrics and of the Kober Medal from the Association of American Physicians, and honorary Doctor of Science degrees from the University of Wisconsin and the Mt Sinai School of Medicine. He was the medalist for the Australian Society for Medical Research in 2002. He is a past president of the American Society of Human Genetics, the Association of American Physicians, the Funding First Initiative of the Mary Lasker Trust, and the Association of Patient Oriented Research.

Diane Drobnis Rosenberg is the Chief Operating Officer of Meadowgate Farm Alpacas, LLC, in Lawrenceville, NJ. She has been active on numerous boards, including the Foundation for Health in Aging Board for the American Geriatrics Society, the Friends Board of the Institute for Advanced Study, and SAVE, a private Princeton animal shelter.

At companies such as McGraw-Hill, Elsevier Science Publishing, and Williams & Wilkins, Ms Drobnis Rosenberg acquired and published life sciences books and texts as well as medical journals. In 1990, she started a consulting firm that worked with medical societies on business arrangements for their journals.

Ms Drobnis Rosenberg was born in Washington, DC in 1944 and grew up in Evanston, Illinois. She is an Alumna of Wellesley College, where she majored in Biology, and also the University of Pennsylvania, where she studied Molecular Biology, having been awarded a NASA traineeship.

FOREWORD

The Rosenbergs, the authors of *Human Genes and Genomes: Science,Health and Society,* and I, share a deep admiration for Archibald Garrod (1857-1936), physician-scientist and student of human biology. I will explain, briefly, why Garrod and his ideas, deserve mention here and why he is so admired.

In 1908, Garrod held the prestigious Croonian Lectureship under the auspices of the Royal College of Physicians, London. The lectures themselves were delivered on four separate days and published separately in the journal *Lancet.* They were then published together as a book in 1908, with additional text, under the title *Inborn Errors of Metabolism*. Garrod used the Croonian Lectures to present his novel concept of inborn errors affecting metabolism. He illustrated it with his own findings gleaned from the careful study of patients harbouring one of four rare charter inborn errors: albinism, cystinuria, pentosuria, and alkaptonuria; the latter condition had already been used by Garrod, in an article published in 1902, to introduce the new theme of human "chemical individuality". At the time, Garrod's Croonian Lectures were salient primarily among biochemists; they were controversial for geneticists, because Mendelism was in abeyance among the biometricians; and they were almost irrelevant for the physicians in his audience. Today, the lectures are recognized as landmarks in the ever-evolving knowledge of human biology, biochemistry, genetics, and medicine.

Metabolism functions somewhere near the core of life itself in autonomous agents such as ourselves. Garrod saw metabolism as a dynamic process involving a normal chemical species (metabolites) distributed in and flowing through normal pathways which could be made variant by Mendelian inheritance. No small gift of understanding! Garrod's insight was huge, yet at the time he would have known little or nothing about enzymes and genes for which even the names were still mysteries awaiting revelation at some distant future. Nonetheless, his insights were correct, and they were given coherence by later discoveries. With hindsight, we can see that Garrod had qualities to fit an aphorism from antiquity: the fox knows many things but the hedgehog knows one big thing (Archilocus 680-645 B.C.). Garrod's unifying vision of human biology (the hedgehog in him) also encompassed its particularities (the fox in him). Accordingly, one could say that Garrod was a hybrid 'hedgefox' by nature.

A second, slightly enlarged, edition of *Inborn Errors of Metabolism* appeared in 1923, but Garrod's hope that many new examples of these variant phenotypes would have been discovered did not materialize until 1935 when he was told about the discovery of phenylketonuria, regarded today as a classic example of a Mendelian inborn error of human metabolism.

Meanwhile, Garrod had been enlarging his awareness of variance at the interface between heredity and experience. In his famous essay of 1902 on the incidence of alkaptonuria, Garrod had begun to broaden the spectrum of illnesses, apparently explained in part by heredity, to include idiosyncratic reactions to food and drugs, as examples. He encapsulated the corresponding inborn susceptibility in the term "diathesis". Garrod saw diathesis (or heredity) as a contributor to cause even when there was no overt pattern of Mendelian inheritance for the ailment. This prescient view was illustrated with examples of various complex disorders showing familial tendencies. Under the title *Inborn Factors in Disease,* Garrod gathered these subtle ideas together and published them in his second great book in 1931. This book was republished with extensive annotations in 1989 by Scriver and Childs, with a Foreword by Lederberg, under the title *Garrod's Inborn Factors in Disease* (Oxford University Press).

I see *"Inborn Factors"* as the companion to *"Inborn Errors"*, the two books serving as forerunners to medical genetics as we know it today. This explains our admiration for Garrod and his ideas. Unfortunately, the prematurity of these insights hindered their wider recognition and a paradigm shift in thinking was too long delayed.

Whereas inborn errors are relatively rare events in the population, we now know that "inborn factors" make enormous contributions to the corresponding burden of common diseases in human society. If Garrod gave us new ways to think about the causes of illnesses, we were also given better ways to answer an abiding question in health care: **Why does this person have this ailment now?** Garrod brought genetics into medical thinking. In his day, these were esoteric ways of knowing; today they are pragmatic and mainstream. To know about DNA and genes and about biochemical networks and genomes, has acquired resonance and relevance. To be aware of the genetic landscape that shapes us as autonomous organisms is not an esoteric exercise; it becomes a facet of human culture when it is a way of knowing the biological infrastructure of our personal and collective health and our being. While such knowledge is transforming, it is also being translated into useful tactics such as diagnostic tests, therapies, and practices to neutralize or even eliminate the disadaptive effects of our diseases. These are some of the whys and the whats of knowing more about human genetics, why the Rosenbergs have written this book, and why Dr. Rosenberg teaches his course at Princeton.

When we are better informed with greater knowledge, we may be able to answer, with greater wisdom, three questions posed by the poet T.S. Eliot:

Where is the life we have lost in living?
Where is the wisdom we have lost in knowledge?
Where is the knowledge we have lost in information?

(Choruses from the *Rock, Part* I, lines 15-17.)

After pondering these unsettling questions, the reader may find encouragement in four other often-cited lines of poetry by the same poet:

We shall not cease from exploration
And the end of all our exploring
Will be to arrive where we started
And know the place for the first time.

(*Little Gidding*: Lines 239-242, from *Four Quartets*).

While modern genetics and ancient heredity will not cease to permeate culture, there will always be room for the poetry.

I will end with a thought about the culture of science. Richard Feynman proposed that science is driven by a continual dialogue between sceptical inquiry and a sense of inexplicable mystery. May you, reader and student, always enjoy sceptical inquiry, yet never lose sight of mystery and its excitement.

Charles R. Scriver MDCM FRS
Alva Professor Emeritus of Human Genetics
McGill University

PREFACE

The decision to write this textbook was made in the earliest days of January, 2010, though it had been in our minds and lives for 36 years. This decision came about collaboratively as follows. LER was engaged in his perennial task of updating his one semester course for undergraduates at Princeton University, entitled "Genes, Health, and Society." This year's effort was unusually strenuous, aimed at complete integration of information about "modern" genomics with "classical" genetics. As he lamented that, in his view, none of the textbooks in the field were optimal for his course (too long or too short; too detailed or too brief; aimed at science majors rather than non-majors; for medical or graduate students, not undergraduates), his wife of 32 years, DDR, repeated advice she had given him before and throughout their marriage, "Why not write your own book, then, with my assistance?" And so we have.

AUDIENCE: This book is designed for use in a one-semester course particularly, though not exclusively, for college or university undergraduates. It should also be of use to some medical, dental, nursing, or graduate students, to prospective genetic counselors, and to interested members of the general public. It presupposes no prior exposure to the subject of human genes and genomes, aiming instead to present the principles of the field and to follow these with a discussion of its many applications and implications. If our aim is true, this book will be neither above the head of the non-major in science nor below the foot of those concentrating on the biological, chemical, or physical sciences.

THESIS: The syllabus for the Princeton course that inspired this textbook begins with two questions: "Why should an educated person living in the twenty-first century know something about human genes and genomes?" and "What should they know?" Here is our answer: The educated person should be conversant with human genes and genomes because it is scientifically exciting, esthetically elegant and beautiful, and increasingly relevant to every day life. We doubt that anyone will contest the view that genetics was the most thrilling of the sciences during the second half of the twentieth century and continues its dazzling pace in the twenty-first. Each seminal discovery (always rare) and the many important contributions that have followed them have changed the way we think about living organisms and have expanded the core of human knowledge. They have literally transformed our view of biological molecules, cells, organ, individuals, populations, and the evolution of the human species. For those fortunate enough to experience this scientific journey, it has been like taking a strenuous upward climb in the autumn, in that the shapes and colors of each panoramic view become more breathtaking the higher one climbs. Thus, it behooves an educated person to know something about genetics for the same reason that they should be aware of such other core subjects as mathematics, philosophy, history, literature, and music.

Another answer to the "why" question is the growing relevance of the science of genetics and genomics to health, medicine, and society. Our individual uniqueness begins with our genetic blueprint. Our health depends on the way our genes interact with our environment(s). Our myriad medical ailments result, in whole or in part, from genetic mutations. Our attitudes regarding the occasional explosive interfaces between science and religion are often shaped by the implications of discoveries in genetics and genomics to bioethics, morality, and law.

As for what an educated person should know, we believe that an understanding of the language of genetics and genomics is required for intelligent, critical evaluation of what one hears from a physician or politician, of what one reads in the newspaper or on the web, or of the personal decisions one makes about such things as marriage and reproduction. Complex language is what differentiates humans from all other species. Knowing something of the specialized language of genetics and genomics will, likewise, facilitate communication between individuals and among groups.

EDUCATIONAL CHALLENGES: Human Genetics is a complex subject and becoming more so. This fact presents challenges for those who teach this subject as well as to their students.

Teachers need to:

- recognize that some of their students have little grounding in the biological sciences, yet have an interest in human genetics as great as that of students familiar with the subject;
- present the concepts that the subject is built on in such a way that its applications and implications follow in an orderly and coherent way;
- include the mathematical underpinnings of such subjects as risk assessment, linkage, and genotype frequency;
- be aware of the religious and political forces that make some topics in human genetics charged and controversial;
- use examples from the clinic, the media, and personal experiences to make the subject live and breathe.

Students need to:

- understand that learning the language and concepts of human genetics is required before its more accessible applications and implications can be understood and debated;
- remember that much of the information they have gained about genetics at the dinner table or from the media is mythologic or just plain wrong;
- embrace the numerical aspects of the field because they are both its underpinnings and its means of problem-solving;
- distinguish among such critical words as "cause", "risk", "probability", and "susceptibility".

ORGANIZATION: This book is composed of 19 chapters arranged in a logical progression in which each successive chapter builds on the one(s) before. There are three main parts: core concepts; genetic disorders; and genes in populations and individuals. To enrich the text, the book contains 74 photographs, 170 line drawings, and 61 tables. Further, there are 21 boxed entries, which are entitled "intersections", "amplification", or "implications"; they are designed to highlight special topics.

EDUCATIONAL TOOLS: Each chapter is followed by a set of review questions and exercises that allow students to test their understanding. Answers to all of these questions are found at the end of the book. The book's end also contains the following: a list of classic and contemporary readings; a comprehensive glossary; and a thorough, well organized index.

In addition to these embedded tools, there are two varieties of web-based resources. The first is the following list of websites containing much useful information about general or specific topics in the field of human genetics and genomics.

General Information

American Society of Human Genetics

www.ashg.org

American College of Medical Genetics

www.acmg.org

Genetics and Public Policy Center

www.DNApolicy.org

National Cancer Institute

www.cancer.gov

National Institutes of Health

www.nih.gov

National Society of Genetic Counsellors

www.nsgc.org

Online Mendelian Inheritance in Man

www.omim.org

Specific Topics

American Society of Gene Therapy

www.asgt.org

Bioethics Blog

http://blog.bioethics.net

Council for Responsible Genetics

www.councilforresponsiblegenetics.org

Innocence Project

www.innocenceproject.org

International Society for Stem Cell Research

www.isscr.org

National Library of Medicine

www.medlineplus.gov

National Organization for Rare Disorders

www.rarediseases.org

The second source is the companion website for the book provided by Elsevier – http://www.elsevierdirect.com/9780123852120. This site contains all images in electronic format for convenience in teaching, direct links to the above referenced sites of interest and additional information about the book.

CONTENT: The following notes provide a "thumbnail" sketch of the content of each chapter, and include a few, representative illustrations.

Chapter 1 frames the field by raising commonly asked questions about such topics as eye-color, hair-color, and height. It is aimed at drawing the student into the subject.

Chapter 2 presents a complete list of all "core concepts" which lead off each subsequent chapter. It is a distillate of the entire book, useful to many but not all readers.

Chapter 3 explains what we mean when we say that genetics is the science of heredity and individual variation. It focuses on the unity of living things, and on the historical landmarks of the field from Mendel to the present day.

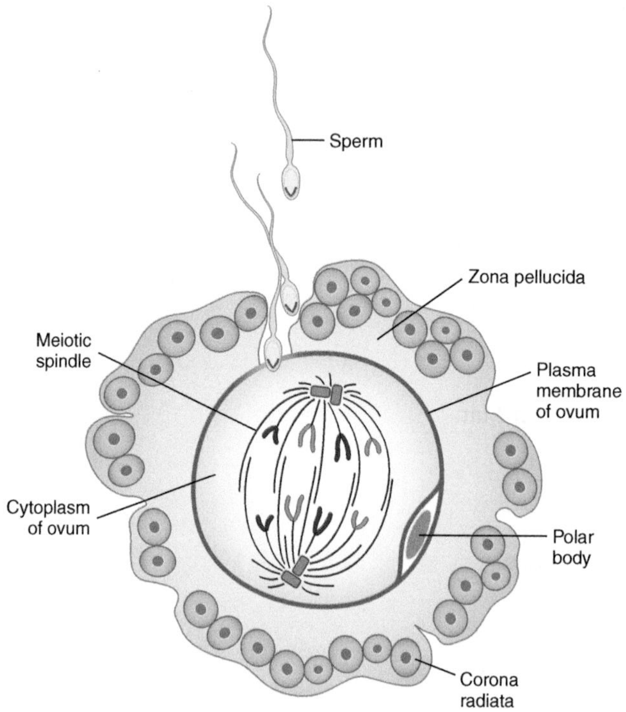

FIGURE 4.11
Fertilization of an egg by a sperm. As a single sperm penetrates the corona radiata and zona pellucida, a series of chemical reactions facilitate penetration of the plasma membrane and prevent other sperm from gaining entry. After fertilization has occurred, meiosis II in the egg is completed and the male and female pronuclei form.

Chapter 4 describes where genes are found (in chromosomes) and what they program (cell division, fertilization, embryonic and fetal development, cellular differentiation, and cell death). It makes clear what we mean by the phrase, "chromosomal basis of heredity."

Chapter 5 tells the story of how genes are transmitted from one generation to the next. It highlights the life and work of the first geneticist, Gregor Mendel, and of the modes of inheritance that bear his name.

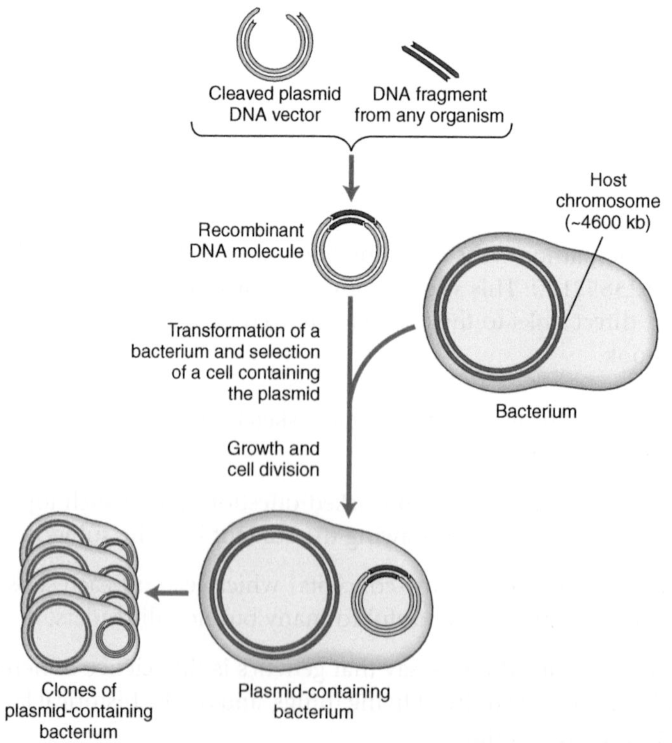

FIGURE 6.21
Creating recombinant DNA molecules with plasmid vectors. A DNA fragment, prepared by digestion with *Eco*R1, is joined to an *Eco*R1 cleaved plasmid; the recombinant plasmid then transforms a bacterial cell, where the plasmid is replicated as the bacteria proliferate; the recombinant plasmids are isolated and the cloned donor fragments of DNA pooled.

Chapter 6 moves to the molecular level. It deals with DNA from many vantages: how we came to know that it is the "stuff" of genes; how it is constructed; how it replicates; how we distinguish between a gene and a genome; how scientists came to sequence the entire human genome and decipher some of its secrets.

Chapter 7 addresses the function of genes and genomes. It presents the central dogma of molecular biology by which information in DNA is converted to its functional elements—proteins. To amplify this central theme, such subjects as differential gene expression and gene silencing are presented.

Chapter 8 defines the term, mutation, as a heritable change in the nucleotide sequence of DNA. It underscores that mutations of many kinds exist and produce positive and negative phenotypes, as well as being the driving force for biological evolution.

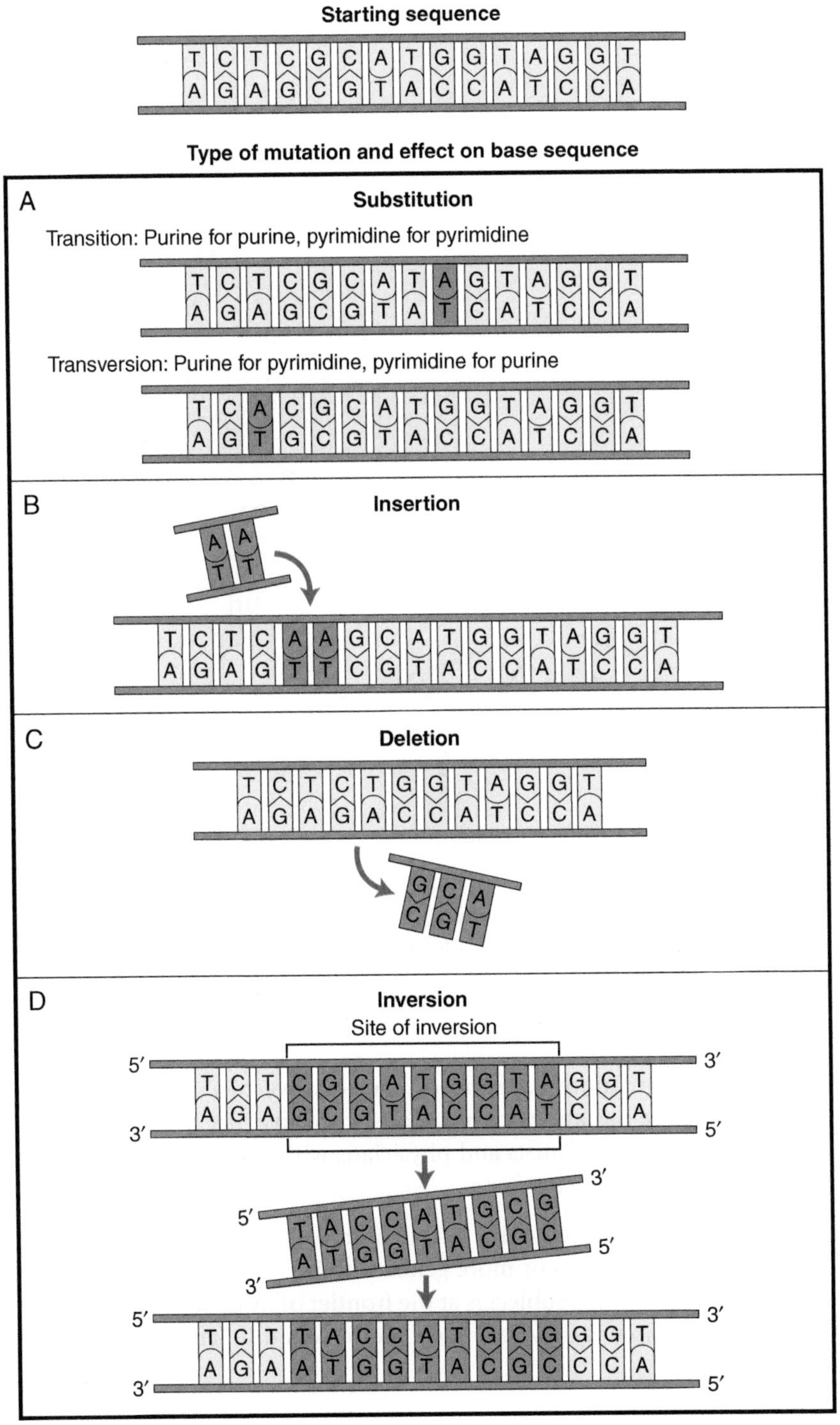

FIGURE 8.1 Classes of mutations affecting DNA structure. See text for further definition of substitution, insertion, deletion, and inversion.

Chapter 9 lays out the incontrovertible evidence for biological evolution. It explains what is meant by "decent from a common ancestor," and construction of life trees. It argues that creationism and intelligent design are supernatural constructs, not evidence-based science.

Chapter 10 asks why no two human beings are genetically identical, and underscores how far we are from answering the profound question of individual uniqueness.

Chapter 11 introduces the broad subject of genetic disorders, classifies them, and then focuses on abnormalities of chromosome number or structure. It underscores that most chromosomal defects occur during meiosis and that only a small minority of such defects are compatible with life, and even fewer, with good health.

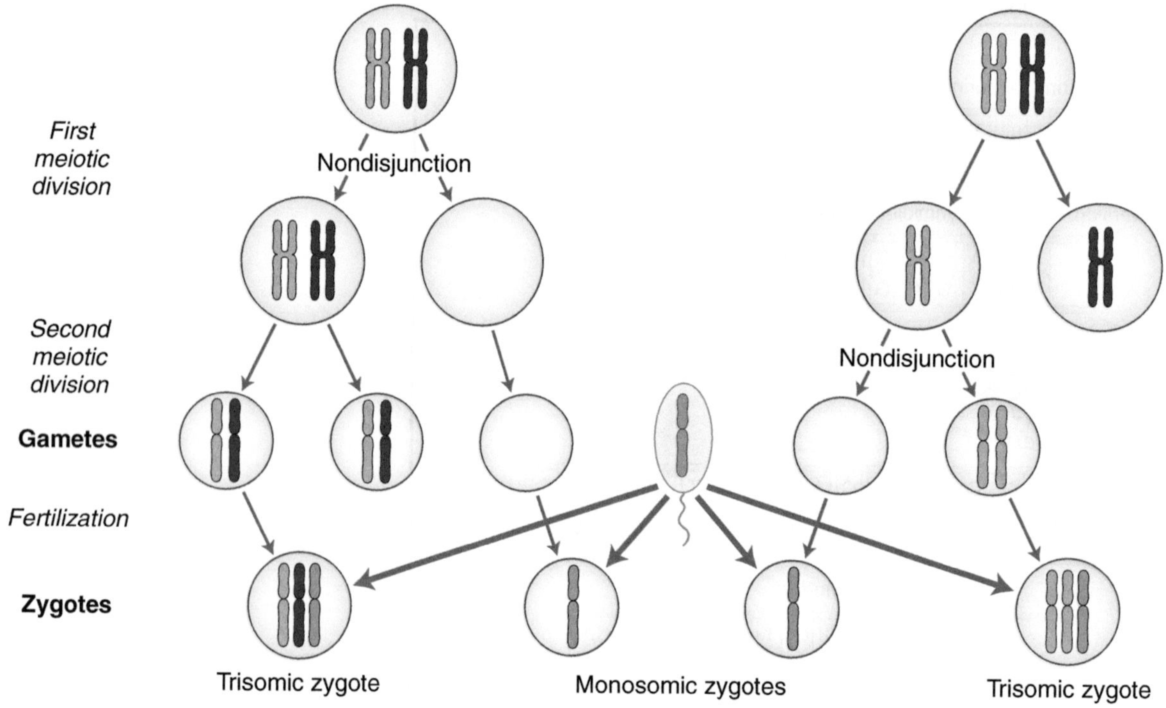

FIGURE 11.4
Consequences of non-disjunction occurring in either meiosis I or meiosis II. A single pair of chromosomes is depicted—the paternal homologue is in pink, the maternal homologue in red. As discussed in the text, the homologues present in the two trisomic zygotes are not identical.

Chapter 12 states that single-gene mutations have already been discovered at more than 2,000 loci (out of a total of 21,000). These mutations have variously positive, negative, or neutral phenotypic effects. They have provided scientists and physicians with invaluable information about biological processes and disease mechanisms.

Chapter 13 introduces the fundamental theme that most common traits and most common disorders result from the interactions of two or more genes with the external environment (multifactorial traits). It points out that this subject is at the frontier of human genetic research and that genomic and epigenomic approaches have merely offered a tiny "crack in the door" toward its understanding.

Chapter 14 states that exploration of the human genome has recently revealed a new class of genetic disorders – namely that due to previously unrecognized architectural changes in the genome typified by extra gene copies, missing gene copies, misplaced chromosomal segments, and other changes referred to as "copy number variations."

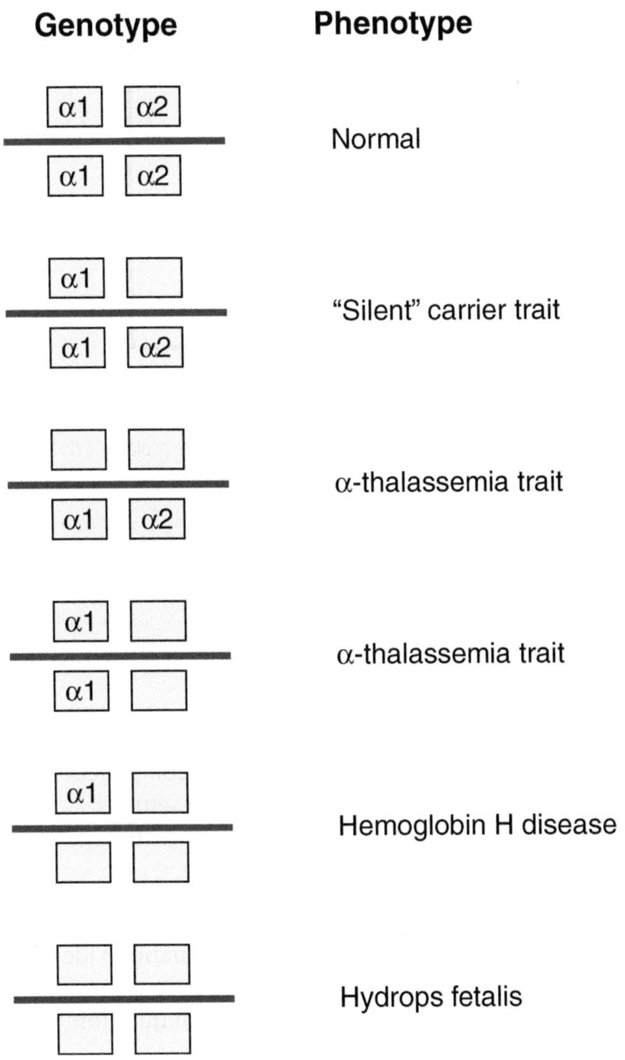

FIGURE 14.4
Genotypes and phenotypes in α-thalassemia. Top: normal people have four copies of α-globin genes (2A1, 2A2). Bottom: in the absence of all four copies (shown as empty boxes), infants succumb *in utero* or at birth with hydrops fetalis (meaning fluid accumulation in the fetus). Four intermediate genotypes reflect progressive loss of A1 or A2 copies. See text for details.

Chapter 15 states that about 5% of all live-born humans have major or minor congenital abnormalities (also called birth defects). It touches on the huge subject of how genes control the prenatal development of multicellular organisms. It points out that birth defects are generally caused by chromosomal, single-gene, or multifactorial abnormalities, but that a small fraction result from teratogens in the environment of pregnant women.

Chapter 16 discusses the concept that cancer is a genetic disease, usually caused by mutations in somatic cells rather than in germ cells. It defines the phenotypic features of the cancer (malignant) cell, and points out that viruses, chemicals, and other environmental insults can lead to the appearance of "cancer genes" which lead cells to grow out of control.

Gene class	Genotype	Representation	Gene product	Cell proliferation
Protooncogene (dominant)	Homozygous normal		Normal structure and amount	Controlled
	Heterozygous		Normal structure and amount Abnormal structure or excessive amount	Uncontrolled
Tumor suppressor (recessive)	Homozygous normal		Normal structure and amount	Controlled
	Heterozygous		Normal structure and amount Absent or inactive	Controlled
	Homozygous mutant		Absent or inactive	Uncontrolled

FIGURE 16.6
The two major forms of cancer-causing genes. An oncogene acts dominantly; mutation of one allele alters the expression of its gene product. A tumor suppressor gene acts recessively; mutation of both alleles are required before abnormal gene expression is produced. Mutation of one allele only does not disturb function.

Chapter 17 brings the presentation of genetic disorders to a close by discussing the various means of detecting and treating such conditions. It underscores that detection may be carried out throughout the human life cycle and that many treatment modalities exist – with variable effectiveness. It summarizes the state of our information about the potential of gene therapy and of the use of embryonic stem cells.

Chapter 18 shifts the focus from individuals and families to populations – including the entire human species. It introduces the mathematical cornerstone of population genetics, mentions its exceptions, and shows how it is used in clinical and counseling settings.
It concludes by discussing human origins, migrations, and genetic variability within and among ethnic groups and races.

Chapter 19 uses particular examples to forecast the view that "personalized" genetics and genomics is more than a slogan. It posits that the means are, or soon will be, at hand to identify susceptibility to many conditions, to explain the individual efficacy and adverse effects of pharmaceuticals, and to offer progressively more information about the central question: why am I different from every other human being on planet Earth.

ACKNOWLEDGEMENTS

Recognizing individuals and groups who have, in one way or another, contributed to this project is both necessary and unsatisfying. It is necessary because this textbook would not have been completed without them. It is unsatisfying because merely listing them conveys no sense of the gratitude we feel for their contributions—particularly for those people who don't even know that they have made a difference. With this disclaimer, we recognize the following.

Author of the Foreword. We can think of no one more appropriate to write the foreword to this book than Charles Scriver, a distinguished human geneticist, and our compatriot, friend, and confidant of many decades' duration.

Authors of prior books. We benefited greatly from the efforts of others who have authored outstanding textbooks in the field of genetics in general, and human genetics in particular. These include T. Gelehrter, F. Collins, and D. Ginsburg, *Principles of Medical Genetics;* D. Hartl and E. Jones, *Genetics: Analysis of Genes and Genomes*; L. Hartwell, L. Hood, M. Goldberg, A. Reynolds, L. Silver, and R. Veres, *Genetics: From Genes to Genomes*; R. Lewis, *Human Genetics: Concepts and Applications*: E. Mange and A. Mange, *Human Genetics;* R. Nussbaum, R. McGinnis, and H. Willard, *Thompson and Thompson's Genetics in Medicine.*

Readers and reviewers. Early on, several people read drafts of chapters and offered criticism and encouragement: Arianna D'Angelo (who was also instrumental in preparing the glossary), Daniel Drobnis, and Steve Hanks. As the book neared completion, several colleagues graciously read the entire book and helped us see things that were unclear or in error. We therefore owe much to Mike Dougherty, Margretta Seashore, and Hunt Willard.

Secretarial expertise. We are grateful to Amanda Denzer-King for her organizational and technical expertise regarding a number of the book's components (text, figures, tables, captions, and additional readings). Gail Huber and Ellen Brindle-Clark also provided valuable secretarial help at critical moments.

Elsevier Publishing Company. Several of our publisher's staff tolerated the authors' constant prodding with (generally) good humor—and did some prodding of their own: first and most constant, Christine Minihane (acquisitions editor), Ceil Nuyianes (art manager), Marie Dean (artist), Jessika Bella-Mura (developmental editor), Rogue Shindler (project manager), Caroline Johnson (production manager), and Sue Armitage (copy editor). Together, they and their many associates underscore the meaning of the word "globalization" in the world of scientific publishing. We thank them, wherever they are.

Colleagues. We could write a decent-sized book of its own to recognize the literally hundreds of colleagues in the world of human genetics who have interacted with one or both of us over the years and who have, indirectly or directly, led us toward this project. They include scientific collaborators, university leaders, academic comrades, and teaching assistants. We know who they are, and are certain that many of them could identify their anything-but-trivial contributions.

Students. This book would not have been written without the stimulus provided by the few thousand medical students at the Yale School of Medicine and, more recently, the several hundred undergraduates at Princeton University who were taught by LER. These latter students have caused the compass of LER's career to point away from laboratory research, administration, and public advocacy and toward what it means to be a responsible, committed, full-time educator.

PART 1

Introduction and Core Concepts

CHAPTER 1

Framing the Field

CHAPTER OUTLINE

Have you ever asked yourself questions such as these?

- Why are some people short, and others tall, and why are some children of tall parents short (Figure 1.1)?
- Why is red hair so much more common in the Scottish or Irish than in the Chinese or Japanese (Figure 1.2)?
- Is it true that a blue-eyed child must have two blue-eyed parents (Figure 1.3)?
- What does being a "carrier" for cystic fibrosis mean?
- Should all women be tested for genes that predispose to breast cancer?
- How does DNA "fingerprinting" exonerate people charged with murder?
- Is direct-to-consumer genetic testing valid and useful?

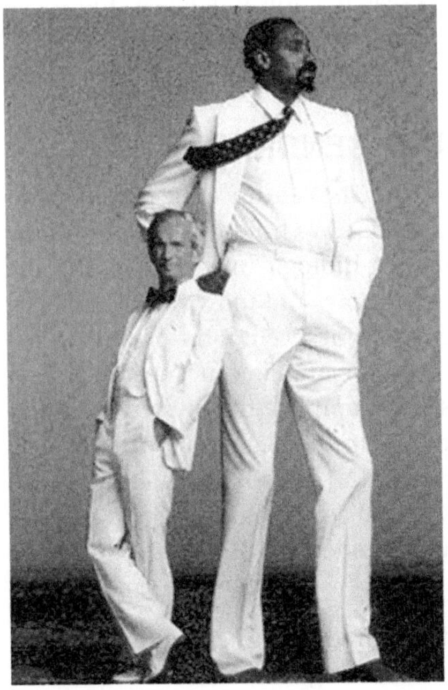

FIGURE 1.1
Willie Shoemaker, a famous horse racing jockey, and Wilt Chamberlain, a famous NBA basketball player. Even though they dressed alike, their differences in height, weight, and skin color are discernable. Willy's weight was under 100 lbs, and he was only 4 feet 11 inches tall, while Wilt's weight was 275 lbs and he was 7 feet 1 inch tall. Multiple genes working together with the environment are responsible for all these differences, as well as some that are not as visible, such as susceptibility to disease.

Human Genes and Genomes. DOI: 10.1016/B978-0-12-385212-0.00001-9

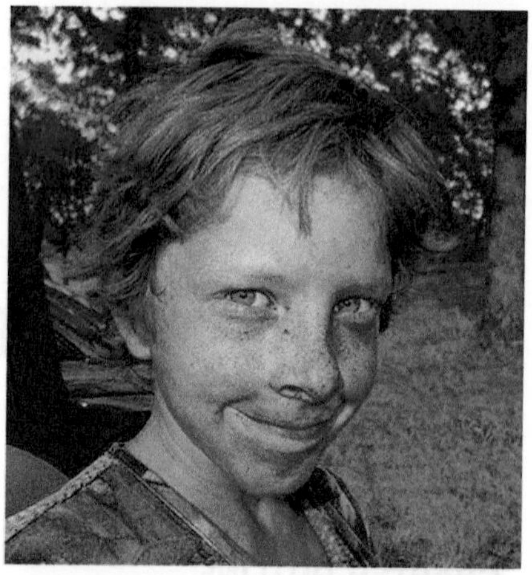

FIGURE 1.2
This young woman has a relatively common variant of the gene that codes for a protein called the melanocortin 1 receptor. This causes her to have red hair, fair skin, and freckles.

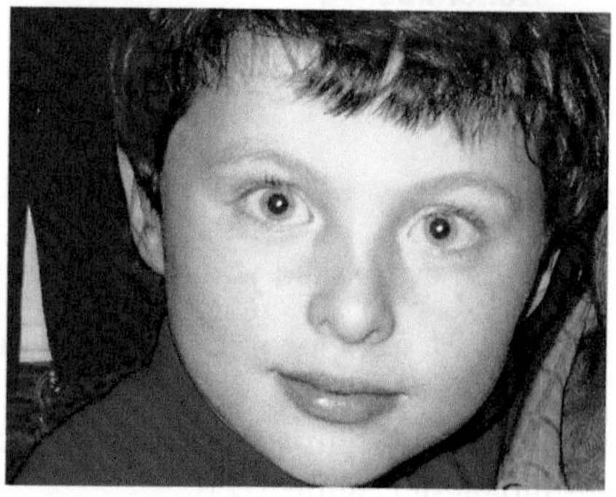

FIGURE 1.3
A healthy blue-eyed boy whose parents both have blue eyes. Some blue-eyed people have brown-eyed or green-eyed parents.

- What does the Genetic Information Nondiscrimination Act (GINA) prohibit?
- Why is the discipline of genetics so embroiled with societal controversies about stem cell research or the teaching of evolution?

These questions will be answered in this text. The science behind these answers is detailed, complex, and has important implications for laboratory research, clinical practice, and social issues.

RATIONALE FOR STUDYING GENETICS AND GENOMICS

In our view, it is incumbent on an educated person living in the 21st century to be familiar with the principles of science in general, and of genetics, in particular. We hold this view for many reasons. First, the study of genetics and genomics explores our biologic universe today in the same excitement-capturing way that astrophysics explored our physical universe yesterday. Rather than seeing stars and planets, geneticists see genes, genomes, cells, and organisms. Rather than landing men on the moon, geneticists now "walk" the human genome in search of landmarks. Rather than planning trips to Mars, geneticists contemplate the mystery of individual uniqueness, in scientific, not just philosophical, terms. In the nearly 60 years since Watson and Crick proposed the double helical structure of **DNA** (deoxyribonucleic acid) (Figures 1.4, 1.5), the molecule of heredity, wave after wave of discoveries have made genetics the most thrilling field in the sciences, perhaps in all human endeavors. And that is why an educated person must understand it.

Second, the study of the genomes of *Homo sapiens* and hundreds of other species has provided powerful—indeed incontrovertible—evidence for biological evolution from

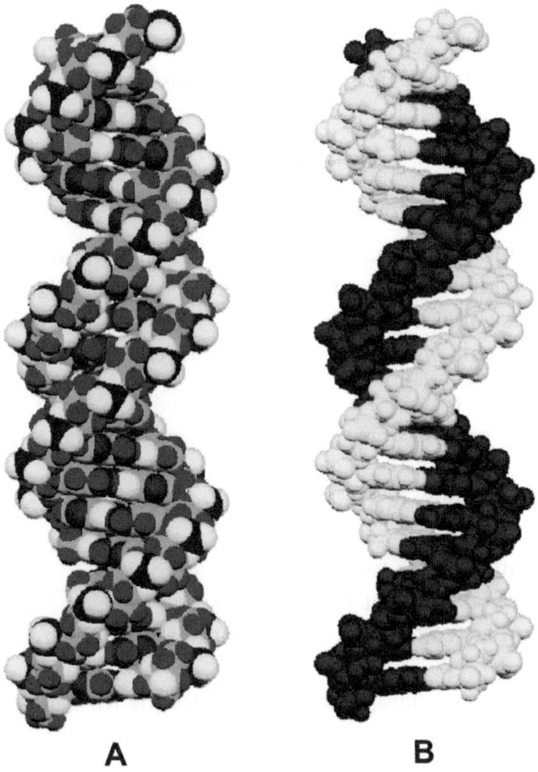

FIGURE 1.4
Ball-and-stick models of the DNA double helix. The structural elegance is obvious; the details will be discussed in Chapter 6.

a common ancestor and for the mechanism of natural selection. It grounds the study of human origin and early migration, and reminds us that evolution is not "done," but continues in us.

Third, there is much more than scientific excitement here. There is growing relevance to health, disease, and disability. Much of who we are, where we came from, and where we are

FIGURE 1.5
Looking down through the center of a model of DNA's double helix. The details will be discussed in Chapter 6. The colors match those in Figures 1.4A.

going comes down to our genes and genomes. They define our biologic boundaries, or, to paraphrase Shakespeare, what our flesh is heir to. We will, increasingly, be able to use genetic information to make health-related decisions. Thus, awareness is a prerequisite for living a healthy life.

WHY STUDY HUMAN GENETICS AND GENOMICS NOW?

The previous paragraphs offer a rationale for understanding genetics. But why now, and why *human* genetics? Without doubt, much that we know has been learned from bacteria, plants, yeast, flies, worms, and mice. Learning from these organisms will continue. Today, however, we can apply this fundamental information to humans and ask about humanness in all its complexity:

- about the relationship between the size of the human brain and the fraction of our genome expressed in it;
- about how our immune system wards off pathogenic organisms—even ones that we've not been exposed to before;
- about how normal traits are transmitted from parents to children;
- about why some traits cluster in certain populations, but not in others;
- about thousands of rare and common disorders resulting from deleterious mutations in our genes;
- about the value of and rationale for premarital, prenatal, and neonatal testing for genetic disorders;
- about genetic influences on beneficial and adverse responses to medications;
- about the benefits and risks of direct-to-consumer information concerning our own genome;
- about a near future when it will be possible to have one's complete genome sequence determined for $1,000 or less using ever more sophisticated automated gene sequencing machines (Figure 1.6).

Every one of these topics has moved out of the laboratory, the clinic, and the classroom to the public square. Not a day goes by without some mention of genes and genomes in print or on the Internet; in news media or on blogs; in movies or on television. Distinguishing fact from fiction (or fantasy), or telling the difference between understanding and misunderstanding, requires a frame of reference and some knowledge of terms and concepts.

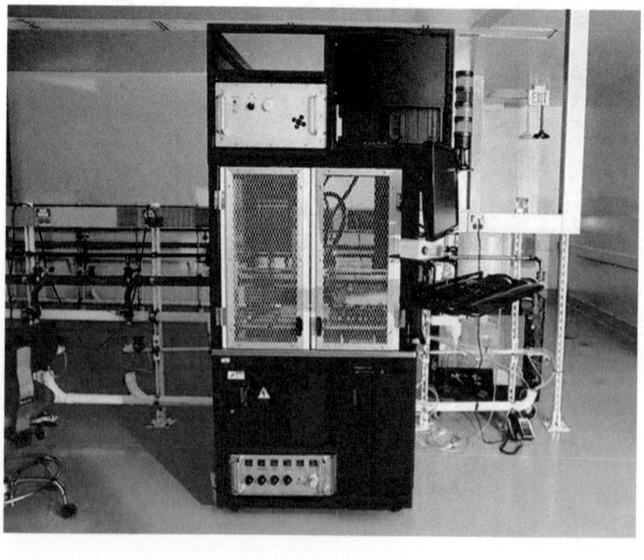

FIGURE 1.6
Automatic DNA sequencing machine. Its output must be linked to high-speed computer equipment. Machines like this one have lowered the cost of complete DNA sequencing in humans by more than 100,000 fold already, and will reduce the cost further.

INTERSECTIONS: BIOETHICS

The Unfortunate Misuse of Human Genetics

The smallest genetic differences among people, real or imagined, can sometimes lead to tragic actions. The conflict between the Tutsi and Hutu people of the country of Rwanda in central Africa illustrates this. Recent genetic studies indicate that a close ethnic kinship exists between these two tribes, generations of gene flow obliterating any clear-cut physical or genetic differences between them.

When the Belgian government and its colonial power seized control of Rwanda in 1916, they created a strict system of racial classification, dividing the previously unified population and putting one group (Tutsi) above the other (Hutu). Conducting a census, the colonialists classified the people throughout Rwanda based on a phenotypic (observable characteristics) scheme. This classification was based on nothing more than the size of the nose, color of the eyes, and height. People with a longer nose, lighter eyes, and greater height were said to be Tutsi. They were given positions of greater authority and opportunity, as well as favorable land possession. The result of splitting the population of Tutsis and Hutus in this way was a conflict between the different groups, particularly in 1994, that has lasted into the 21st century, resulting in the deaths of hundreds of thousands of people. That the horror of this recent genocide should have been based on scientifically spurious discrimination is both cruel and ironic

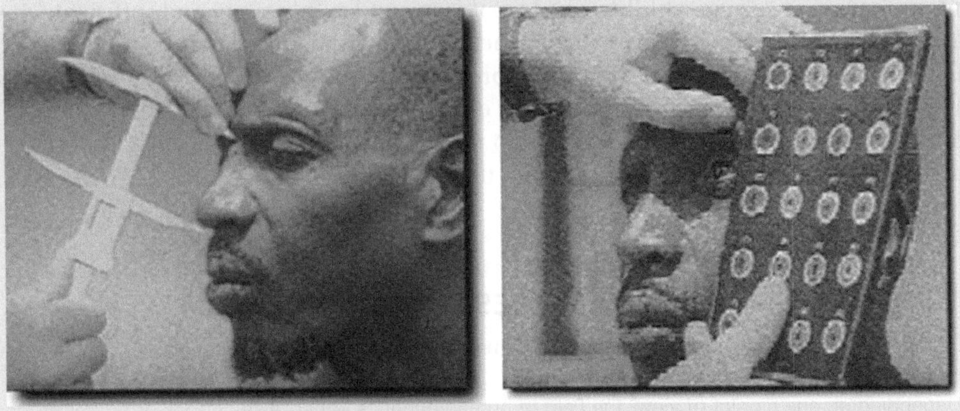

The size of the nose and the color of the eyes were factors that determined whether a person was classified as Hutu, Tutsi or Twa.

DIVERGENT PUBLIC REACTIONS TO THE FIELD

An unusually wide array of public reactions has greeted the sometimes breathless pace of discovery in genetics and genomics. Most people take hope from the field in the form of scientific discoveries that tell us about the world we live in, in understanding the maladies that beset us throughout our lives, and in development of a personalized health care system.

A minority of the population sees the field in darker hues. Some doubt that scientists will describe their work candidly and without "hype," or that scientists will conduct their work within accepted ethical boundaries. Other people have fears, not just doubts—they fear that employers, insurers, and law enforcement personnel who have access to genetic information will intrude upon their privacy; they fear that novel reproductive technologies such as **in vitro fertilization (IVF)** and **pre-implantation genetic diagnosis (PGD)** may be the opening wedge toward the end of sexual reproduction. Still others see in neonatal screening and genetic counseling the specter of neo-eugenics (manipulating human mating and reproduction to "improve" the human species) and runaway genetic determinism. Finally, some people are unalterably opposed to the field for moral and religious reasons. They may reject prenatal diagnosis and embryonic stem cell use, for example, and, by extension, everything else about the applications of genetics to individual choices.

It is not difficult to account for the widely divergent views just mentioned—hopes, doubts, fears, and opposition. Fundamentally, the divergence depends on one's view of science. If one believes that science can improve the world we inhabit through understanding it, then advances in genetics and genomics will be perceived in a positive, even exuberant, way. But the more muted one's belief in the power to make things better, the greater the likelihood that hopes will be transmuted to other elements. Lack of trust in science and scientists may provoke doubts. Frank mistrust will generate fears. Those who desire supernatural (rather than natural) explanations for human events may be inclined to see genetics as an attack on their religious, moral, and/or ethical precepts and, therefore, seek to see the field impeded, even halted.

ANTICIPATED RETURNS ON EDUCATIONAL INVESTMENT

This book will allow the student to comprehend the language, the direction, and the uncertainties in the field of genetics and genomics. It will help distinguish between such critical words as "cause" and "susceptibility," and what is meant by "necessary and sufficient." It will remind students of those instances when geneticists halted their own work on **recombinant DNA technology**, **germ line gene therapy**, **reproductive cloning**, and more, because they were concerned that the benefits in the work were exceeded by the risks. It will allow students to interpret and criticize what they encounter in the media and, thereby, be able to develop an informed position on a wide range of topics that will affect their lives.

REVIEW QUESTIONS AND EXERCISES

1. Discuss briefly four reasons why the field of human genetics has been at the forefront of science and medicine during the past half century.
2. Depending on one's point of view, the field of human genetics either raises hopes or provokes fears. Discuss briefly three kinds of hopes and three kinds of fears.
3. As each new genetic technology has been developed, it has run into intense opposition from a segment of society. Describe briefly three such technologies and the origin of the opposition engendered.
4. An article in a major US newspaper was entitled "Gene Screen: Will We Vote Against a Candidate's DNA?"
   a. Is this headline mere attention grabbing? Why, or why not?
   b. What societal changes would have to occur before mandating disclosure of a candidate's genome sequence would be appropriate—if ever?

CHAPTER 2

Introducing the Core Concepts

CHAPTER OUTLINE

To participate in discussions about genes, genomes, health, and society intelligently, we must understand the principles and language of the field.

- When we say that a **mutation** in gene "A" causes disease "B," we must understand what a mutation is, how mutations occur, and why and under what circumstances they may cause disease.
- When the *Wall Street Journal* publishes an article entitled, "Gene May Foil AIDS Virus," we must understand the relationship between genes and the environment and realize that some mutations can be beneficial, while others are deleterious.
- When for-profit companies market direct-to-consumer analysis of your genome using **SNPs** (single-nucleotide polymorphisms, pronounced "snips"), we must know the difference between **genes** and **genomes** and understand what SNPs are and gauge their importance.
- When several groups claim to have "**cloned**" human beings, we must ask, "What's the scientific evidence?" and "Should human cloning even be attempted?"
- When people say that geneticists are "playing God," we must ask how anyone who believes that God is omnipresent and omnipotent can imagine any human masquerading as God.

The next 17 chapters will discuss, in some detail, the core concepts that attempt to offer a basis for considering answers to the above questions. To set the stage for this discussion and to allow the student to appreciate each concept's and each chapter's relationship to one another, the

Human Genes and Genomes. DOI: 10.1016/B978-0-12-385212-0.00002-0

concepts will simply be stated here. You will note that some chapters have more than one core concept. Taken together, the core concepts provide a concise summary of this book.

CHAPTER 3: GENETIC VARIABILITY

Genetics is the science of **heredity** and **inherited variation**. It considers the biological information all living organisms require to grow, develop, reproduce, and die. It considers, as well, how traits are transmitted from one generation to the next. The chemical nature of this information is the same in organisms as simple as microbes and as complex as man.

CHAPTER 4: GROWTH, DEVELOPMENT, AND REPRODUCTION

All organisms package their genetic or genomic information in tiny cellular structures called **chromosomes**. The relationship between the words *genome, chromosome,* and *gene* may be thought of as a set of Russian dolls: a genome is composed of a set of chromosomes; a chromosome is composed of a set of genes; a gene is composed of a set of nucleic acids. In humans, and other **eukaryotes** (organisms composed of one or more cells containing a visible nucleus and organelles) which reproduce sexually, chromosomes are paired, one inherited from each parent. Human cells contain 23 pairs of chromosomes, for a total of 46. These **diploid** cells—cells that contain two copies of every chromosome—transmit their chromosomes during two kinds of cell division: **mitosis** and **meiosis**. In mitotic cell division, which occurs in all cells of the body, each "**daughter cell**" receives the identical chromosome complement of its parent cell. In meiosis, which occurs only in germ cells (reproductive cells), the diploid number (46) is halved, producing **haploid gametes** (cells that will fuse during fertilization), each with 23 chromosomes. Inherited variation between any two people is largely a function of two events that occur during meiosis: **random chromosome separation** and **crossing over** between members of each chromosome pair. Reproduction proceeds through the union of **gametes**, an **oocyte** (egg) from the female and a **sperm** from the male. The fertilized egg, or **zygote**, is **diploid**, as are all other body cells. Prenatal and postnatal development occurs through mitosis, which is how the single-cell zygote develops into a 40-billion cell newborn and a 100-trillion-cell adult.

CHAPTER 5: TRANSMISSION OF GENES

Traits are transmitted from one generation to the next by **genes**. The Austrian monk, Gregor Mendel, discovered the basic rules of this transmission—that is, **gene segregation** and **assortment**—in the mid-nineteenth century by studying pea plants. Trait transmission in human pedigrees follows the same rules. Traits attributable to single genes (referred to as "**Mendelian traits**") are inherited as **dominants** or **recessives**. They may be **autosomal** or **sex-linked**. Traits attributable to multiple genes and those demonstrating **imprinting** or maternal inheritance display more complex patterns.

CHAPTER 6: STRUCTURE OF GENES, CHROMOSOMES, AND GENOMES

Genes are the units of inheritance. They are composed of **DNA (deoxyribonucleic acid)**. Genes store information in linear DNA molecules composed of a four-letter chemical alphabet—composed of the bases **G (guanine)**, **C (cytosine)**, **A (adenine)**, and **T (thymine)**, which form the unique part of **nucleotides**, and which give each gene its unique function. Each gene has a unique sequence of these nucleotides, which, in turn, underlies its unique function. The DNA molecule is a double strand of nucleotides carrying **complementary G-C** or **A-T base pairs**. The complementarity of double-stranded DNA is the key to understanding how DNA functions in inheritance.

Within the cells of an organism, DNA molecules carrying genes are assembled into chromosomes: **organelles** composed of DNA and associated proteins. The sum total of genes and chromosomes in each cell is its **genome**. The nuclear human genome consists of 24 distinct kinds of chromosomes, composed of 6 billion nucleotides and about 21,000 genes. Only about 2% of the DNA in the human genome is composed of genes that encode proteins. The remainder is comprised of nucleotide sequences with variably understood regulatory and evolutionary significance.

Genomics, the study of whole genomes, was made possible by determining the complete DNA sequence of humans—and that of about 400 other organisms. This triumph of modern genetics, facilitated by employing old technologies (such as gene cloning, restriction mapping, gel electrophoresis) and new ones (such as high throughput DNA sequencers and power computational tools)—is beginning to reveal the kinds and extent of variation in DNA structure not previously anticipated and not yet well understood. Already, however, the study of remarkably common **single nucleotide polymorphisms (SNPs)** and **copy number variations (CNVs)** is providing us with information about mechanisms of susceptibility to many human traits and disorders.

CHAPTER 7: EXPRESSION OF GENES AND GENOMES

The central dogma of gene action is abbreviated: **DNA→RNA→Protein**. Genes transmit their information by being **transcribed**, i.e., their DNA codes for a complementary RNA referred to as **messenger RNA (mRNA)**. In turn, mRNAs are edited, leave the nucleus, and are translated into **proteins**—the molecules largely responsible for the functions cells carry out. Translation occurs according to a universal **genetic code**: the sequence of nucleotides in mRNA, read as **triplets** (called **codons**) that specify the sequence of **amino acids** in proteins. This fundamental pathway of gene expression is modulated such that only a subset of genes is active in any one tissue—thereby explaining **differentiation**, that is, the structural and functional differences between, for example, a brain cell and a heart cell. Gene expression may be modified in a number of ways: by changes in DNA primary structure such as mutations; by chemical (**epigenetic**) modification of DNA; and by interference with mRNA structure or function produced by small RNA species, called siRNAs and miRNAs.

CHAPTER 8: MUTATION

A **mutation** is a heritable change in the nucleotide sequence or arrangement of DNA. In the positive, evolutionary sense, mutations are responsible for the **selective advantage** that one species gains over another. In the negative sense, mutations cause or increase susceptibility to thousands of human disorders. Mutations can occur at three different levels: **genome**, **chromosome**, and **gene**. Genome mutations result from **missegregration** (failure of chromosomes to properly segregate) during meiosis and produce changes in chromosome number. Chromosome mutations are caused by rearrangements in the structure of a chromosome, such as **translocations** or **deletions**. Gene mutations alter the base sequence of a gene. Mutations occur in **germ cells**, **somatic cells**, and **mitochondria.** Germ-line mutations affect DNA in all cells and are transmitted to offspring. Mitochondrial mutations affect mitochondrial DNA, all of which is inherited from the mother's egg; accordingly, they are inherited maternally. Somatic mutations affect cells in a single tissue and are not transmitted to the next generation. Mutations act, generally, by perturbing the function of proteins and regulatory processes, and they may do so from the earliest to the latest stages of life.

DNA is not static. Rather, its sequence changes at a slow but measurable rate all the time. Mutations are of central importance to all life forms, including humans. Without mutations, our species would not have evolved over several billion years. Without mutations, individuals would not differ from one another, even within the spectrum that is considered "normal." Without mutations, humankind also would not be faced with a myriad of spontaneous abortions, inherited disorders, or cancers. Therefore, we must comprehend both the nature of mutations and the effects they produce.

CHAPTER 9: BIOLOGICAL EVOLUTION

Biological evolution is the central organizing principle of modern biology and has revolutionized our understanding of life on Earth. It has provided a scientific explanation for why there are so many different kinds of organisms on our planet and how all these organisms became part of a continuous lineage—including humans. DNA is central to understanding biological evolution because all traits in virtually all organisms originate from DNA and are transmitted from generation to generation through it. New mutations in DNA result in the variation of traits that enable the organism to adapt more successfully to its environment. Through increased **reproductive fitness**, such traits will increase in frequency in a population from one generation to the next, as will the frequency of the DNA changes responsible for them. The process by which **genotypes** best suited to survive and reproduce in a given environment, and so gradually increase the overall ability of the population to thrive, is called **natural selection**.

CHAPTER 10: HUMAN INDIVIDUALITY

Each human being has his or her unique genome. This includes monozygotic twins whose DNA sequences are identical but whose packaging of DNA differs slightly. This genetic uniqueness expresses itself in the form of physical, chemical, and behavioral individuality. In the past, an understanding of this individuality was barely hinted at by the existence of rare, **single-gene disorders** and by estimating the frequency of **polymorphic protein variants** in a population of healthy people. As we enter the genomic era, genome-wide studies, including complete DNA sequences on a small but increasing number of people, have already attested to considerable differences in **single nucleotide polymorphisms** (**SNPs**), small **insertions** and **deletions**, **copy number variations** (CNVs), and in **heterozygosity** for rare disease genes. As such studies are conducted on more individuals and groups, they will lead to greater understanding of normal traits and common disorders. Ultimately, this will get us closer to finding out how each of us differs genetically from anyone else.

CHAPTER 11: CHROMOSOME ABNORMALITIES

The 46 chromosomes found in the nucleus of humans cells—22 pairs of autosomes and one pair of sex chromosomes—are best examined during mitotic **metaphase**. Metaphase chromosomes can be visualized under a microscope by staining them with various dyes. Even more definitive identification employs **fluorescence *in situ* hybridization** (**FISH**). Chromosomes may be examined *in situ* using light microscopy on intact cells. More often, a **karyotype** prepared by cutting out and mounting individual chromosomes from a metaphase spread is studied. In a typical karyotype, chromosomes are displayed in homologous pairs according to size and to the position of the centromere.

Abnormalities in chromosome number (**aneuploidy**) or structure are found in 0.7% of newborns, but this figure is a gross underestimate of the significance of such abnormalities to human health: approximately 75% of all human conceptuses are aborted spontaneously due to such chromosomal defects. Cytogenetic testing (the study of chromosomes) is indicated under the following circumstances:

- advanced maternal age;
- previous infertility;
- still births;
- neonatal deaths;
- birth defects;
- chromosome abnormalities in first-degree relatives.

Most kinds of aneuploidy—**polyploidy** (extra sets of all chromosomes), **trisomy** (extra copy of one chromosome), or **monosomy** (loss of a single chromosome)—are found in aborted fetuses. Those aneuploid states compatible with full-term gestation include: trisomy for chromosomes 13, 18, and 21; different sex chromosome trisomies (XXX, XXY, XYY) and **quadrisomies** (XXXX; XXYY); and monosomy for the X chromosome. A variety of structural rearrangements are also observed cytogenetically. These include **insertions**, **deletions**, **inversions**, and **translocations**. The clinical consequences of such aneuploidies or structural rearrangements, most of which result from **non-disjunction** or chromosome instability occurring during meiosis, depend on several factors:

- which chromosome (or chromosomes) is involved;
- which genes have been perturbed;
- how much chromosomal material has been added or subtracted;
- what compensatory mechanisms exist (such as X chromosome inactivation);
- whether genomic imprinting is involved.

CHAPTER 12: SINGLE-GENE DEFECTS

Mutations of single genes have been documented at more than 10% of the estimated 21,000 genetic loci in the human genome that code for proteins. These single-gene mutations have a considerable effect on child health: they occur in 0.4% of newborns; they are responsible for 5% of hospitalizations; and they cause 8% of deaths. The study of these disorders—once called "**inborn errors of metabolism**" and now generally referred to as "**inherited metabolic diseases**"—has been of value in several ways: by elucidating normal biochemical pathways of **anabolism** and **catabolism**; by defining the biochemical mechanisms of myriad disorders and the nature of the gene mutations that cause them; and by using this information to develop diagnostic tests and therapeutic strategies.

This value has been extracted by studying disorders caused by single mutations at four logically constructed, ascending levels: the clinical phenotype; the metabolic pathway or specific reaction; the protein affected; and the genetic locus perturbed. Although each disorder is unique, reflecting as it does the locus and its product, some generalities deserve mentioning:

- Most disorders are genetically **heterogeneous**, that is, each results from a wide variety of different mutations (**missense**, **nonsense**, **frameshift**, and **splicing**) which interfere with the function of a single protein. A few are caused by trinucleotide repeats.
- They reflect modification of the structure and function of one or more of the kinds of proteins found in the human body: enzymatic, structural, regulatory, circulating, and membrane. Organ dysfunction follows from differential gene expression.
- Some are inherited as dominant traits, others as recessives; some are autosomal, others sex-linked.
- Some, like red hair, are benign traits; others are uniformly fatal during childhood. Still others are compatible with extended life but impair organ function in serious ways.
- Whereas some conditions are found with near equal prevalence in all ethnic groups, most show ethnic clustering.

CHAPTER 13: MULTIFACTORIAL TRAITS

Most common human phenotypes reflect interactions between genes and the environment and are termed "**multifactorial**." Some multifactorial phenotypes are ubiquitous physiologic traits such as height, weight, and mathematical aptitude. Others are a multitude of common disorders encountered at birth (cleft lip, neural tube defects), in children (asthma, juvenile diabetes mellitus), or in adults (high blood pressure, coronary artery disease, schizophrenia). A small number, like eye color and finger-tip ridge count, are called polygenic traits because they result from the action of two or more genes with little or no environmental contribution.

Multifactorial traits run in families, but not according to Mendelian modes of inheritance. Some, like height, are quantitative traits in that they vary continuously over a range of measurement and display a normal (Gaussian) distribution curve. Others, such as cleft lip or schizophrenia, are qualitative traits with two classes of people—affected or unaffected, although even in affected individuals there may be a range of phenotypes. A variety of mathematical tools are employed to determine that genes play a part in quantitative or qualitative multifactorial conditions and to estimate the magnitude of that genetic contribution. These tools include **empiric study**, **intrafamilial correlation**, and **twin concordance**.

Identifying the particular genes involved and determining how they interact with one another and with the environment has proven to be much more difficult. Although there is not a single instance in which complete understanding of a multifactorial trait's biological basis has been produced, progress is being made. Such progress uses **candidate gene studies** (seeking mutations in genes whose function is relevant to the trait), **linkage analysis** (examining pedigrees for linkage between the trait and one or more genetic markers), and **genome-wide association studies** (in which the frequency of single nucleotide polymorphisms is compared in controls and those with a particular trait or disorder). As complete genome sequencing becomes technically feasible and financially affordable, more comprehensive understanding is expected.

CHAPTER 14: DISORDERS OF VARIABLE GENOMIC ARCHITECTURE

The human genome has, until very recently, been thought of in classical terms:

- Chromosomes and genes occur in pairs—one from each parent
- A pair of alleles at each locus produces dominant, recessive, and sex-linked traits
- Mutations produce heritable changes in DNA that often have deleterious results.

The Human Genome Project has taught us, however, that these classic understandings are over-simplified and incomplete: the function of chromosomal DNA is sometimes affected by epigenetic chemical modification; some genes are duplicated and others have multiple alleles; some Mendelian disorders result from mutations at two different loci, or even three. From these observations has come the current view of the genome:

- that only 2% of it codes for proteins;
- that about half is present in single copy sequences;
- that almost half of it consists of a variety of repeating sequences;
- that variation in copy number affects 10–20% of the genome.

It is already safe to conclude that the human genome is dynamic—not fixed; that it is capable of great variability which is occurring continually. Although any two humans are 99.6% identical in genomic DNA sequence, the other 0.4% of non-identity is already beginning to provide a glimpse of what makes each human genetically unique.

CHAPTER 15: BIRTH DEFECTS

Birth defects occur in 5–7% of newborns. About half of these defects affect major organs, including the brain, heart, and limbs; the remainder are minor (crooked fifth finger, missing fingernails, etc). These birth defects reflect but a small minority of the lethal embryonic errors that occur during human gestation and which end in spontaneous abortion. Five critical cellular events must occur during successful embryonic and fetal development: **proliferation**, **differentiation**, **migration**, **communication**, and **apoptosis**. These events—some sequential, some concurrent, all critically timed—take place during the embryonic (weeks 1–8) and fetal (weeks 9–38) periods of gestation.

This program is controlled by a series of genes acting at precise intervals in precise fashion. Some of these genes (maternal-effect) are encoded by the mother's genome and synthesize products transferred, first, to the oocyte, and then to the zygote and early embryo. The remainder of the genes controlling development are the embryo's own, and they control such critical processes as:

- **segmentation** (dividing the embryo into parts from head to tail);
- **pattern formation** (fate of cells in each segment);
- **cell signaling** (chemical communication between and within cells);
- **apoptosis** (programmed cell death essential for tissue remodeling).

When this developmental program works perfectly, a single-cell zygote ultimately becomes a 40-billion cell neonate. But a myriad of accidents lead to the birth of children with defects. Each of the three major classes of genetic disorders (chromosomal, single gene, and multifactorial) underlies a fraction of birth defects. The nature, severity, and outcome of any particular defect depend on what part of the genome is affected and to what degree.

CHAPTER 16: THE GENETICS OF CANCER

Cancer is a large group of diseases that result when cells divide out of control and acquire the ability to spread beyond their prescribed borders. Cancer is a genetic disease because it is caused by mutations, but it is rarely inherited because these mutations occur, in most cases, in somatic cells rather than in germ cells. Cancer affects one in three people during their lifetime and is the second leading cause of death in the United States. In women, the most common and most lethal cancers originate in the lung, breast, and colon. In men, cancers of the lung, prostate, and colon predominate. In sharp contrast to cancer, also called a **malignant tumor**, a **benign tumor** is defined as a collection of cells that grows out of control but does not spread into surrounding tissues.

Fundamentally, cancer results from dysfunction of the repeating pattern of mitotic cell growth and division, called the **cell cycle**. This cycle (composed of the four phases G_1, S, G_2, and M) is tightly controlled by the concerted action of hundreds of genes. Through their products, these genes provide the components which make it possible for a single cell to grow, replicate its DNA, prepare to and then divide into two identical daughter cells, and, when necessary, die. Two large families of proteins—the **cyclins** and the **cyclin-dependent kinases**—regulate the transition from one phase of the cell cycle to the next (for instance, from G_1 to S). In addition to these two large classes of proteins, many others—called **checkpoint proteins**—scan and arrest the cell cycle so that any damage that has occurred to the genome (intrinsically or extrinsically) may be repaired. Cancer is the result of some kind of failure of this intricate set of cellular checks and balances to work perfectly.

Cancer cells differ phenotypically from normal cells in many particular ways. Their proliferation is not controlled by such usual phenomena as contact with a neighboring cell, chemical signaling between adjacent cells, and failure to respond to cell-death signals. Their genomes are unstable, as evidenced by a variety of chromosome abnormalities. They demonstrate immortality, meaning that they no longer observe the usual limits on the number of cell divisions—a process in which **telomeres** (chromosome ends) and the enzyme that replenishes their length, telomerase, play a critical role. They develop the ability to invade surrounding tissue and to **metastasize** by, in part, stimulating the formation of new blood vessels.

Among genetic diseases, cancer is unique. It is caused not by a single-gene mutation but by a sequence of mutations at different loci in the progeny of a single cell. No one of these mutations by itself produces the cancer phenotype; rather, only their successive nature ultimately topples the cell from healthy to potentially lethal. Cancer is usually sporadic, meaning that it occurs in only one member of a family; this reflects the fact that cancer-predisposing germ-line mutations (which are heritable) are rare, while somatic ones are common. Cancer is the essence of a multifactorial trait in that environmental exposures (such as tobacco, X-rays, bacteria, viruses, dietary substituents) interact with gene mutations in the disorderly process, ultimately overwhelming the cell's ability to control its growth.

Mutations in two large classes of genes—**oncogenes** and **tumor suppressor genes**—are the best-studied examples of "cancer genes." Oncogenes are produced by mutations of their normal gene homologues, **proto-oncogenes**. Once formed, oncogenes are gain-of-function mutations that drive cell proliferation. Oncogenes behave dominantly in that mutation of one of two alleles is sufficient to perturb cellular function. Tumor suppressor genes, in contrast, normally function to inhibit cell division or activate apoptosis. They act recessively in that both alleles must be mutant before they are unable to carry out their usual "braking" function. Hundreds of different mutant oncogenes and mutant suppressor genes have been identified and characterized. Though they vary greatly in structure and function, it is their propensity to arrange themselves in a dysfunctional repertoire that ultimately leads to cancer.

CHAPTER 17: DETECTION AND TREATMENT OF GENETIC DISORDERS

Clinical genetics depends on applying the principles of human genetics and genomics to the detection and treatment of the many kinds of genetic disorders discussed previously. Here, we present a framework for thinking about the intertwined subjects of **detection** and **treatment**—starting with detection. There are three main reasons to detect (or diagnose) genetic disorders:

- first and foremost, to institute treatment for the affected person;
- second, to provide information to family members regarding their risk for developing the same condition;
- third, to seek answers to scientific questions posed by the disorder, such as its mode of inheritance, its precise cause, and its frequency in different ethnic populations.

Detection (or testing) is increasingly being done throughout the cycle of human life: premarital, preconceptual, preimplantation, prenatal, and during childhood or adulthood. In general, premarital through prenatal testing is carried out to avoid the birth of those with untreatable conditions, or to alert parents to their presence. In contrast, detection in newborns, children, and adults is performed with the ultimate goal of instituting treatment. An increasingly broad array of biological materials is used in detection systems: blood, urine, cells, proteins, RNA, and DNA. In addition to having to ascertain the analytical and clinical validity of any test employed, testing for genetic disorders sometimes raises complex ethical and social questions that must be discussed with those seeking testing. To assist in answering the range of questions posed, families are referred for **genetic counseling**—a communication process facilitated by a team of professionals, including physicians, nurses, and master degree-trained genetic counselors.

Some genetic disorders may be cured surgically (for instance, cleft lip, congenital heart defects, and localized cancers). Some are not treatable (such as Tay-Sachs and α-thalassemia). A larger subset may respond beneficially to a variety of therapeutic strategies and modalities, such as substance avoidance, dietary restriction, product replacement, cofactor supplementation, pharmaceutical administration, organ or adult **stem cell transplantation**, and gene therapy. All together, these modalities are beneficial for only a minority of patients with genetic disorders, thereby constituting a major challenge to science and clinical medicine.

Gene therapy and stem cell therapy have been the subject of much attention in recent years. Although *in vivo* and *ex vivo* methods of gene therapy have been worked on for 30 years, clinical progress has been disappointingly slow. Currently, it has proven to be beneficial in less than a handful of disorders. At present, there is no instance in which use of embryonic stem cells has been shown to produce clinical benefit. It seems likely but not certain that gene therapy and embryonic stem cell therapy will benefit patients with a variety of conditions in the years and decades ahead.

CHAPTER 18: POPULATION AND EVOLUTIONARY GENETICS

Population genetics is the quantitative study of the distribution of **genetic variation** in a population and of how the frequencies of its genotypes, alleles, and phenotypes are maintained or changed. It seeks answers to such practical questions as why the frequency of phenylketonuria (PKU) in Caucasians is so much greater than in Japanese, or why the frequency of the sickle cell allele varies markedly in people from different West African countries. The mathematical cornerstone of population genetics is the **Hardy-Weinberg law** or principle. The law has two parts. First, it states that in a large, randomly mating population with two alleles at a locus (for example, A and a), there is a simple relationship between these allele frequencies (frequency of $A = p$; frequency of $a = q$) and the genotype frequencies (p^2, $2pq$, or q^2) they define. Second, it holds that this relationship between allele and genotype frequencies, constructed simply on the binomial expansion of $(p + q)^2$, does not change from one generation to the next. When a population conforms to this two-part law, it is in **Hardy-Weinberg equilibrium**. In such populations, the law is of great value in showing why dominant traits do not increase in frequency from one generation to the next and why recessive traits do not decrease. Further, the law is regularly used in genetic counseling settings where estimates of genotype, allele, and carrier frequencies are calculated from limited phenotypic information in small families, such estimates then being employed to estimate specific genetic risk.

Hardy-Weinberg equilibrium is never fully realized in human populations because it is perturbed by one or more deviations. First, individuals do not usually mate randomly. Mating is more often **assortative** (mate choice depends on geographic proximity), **stratified** (within an ethnic subset), or **inbred** (among relatives or a small group). Second, allele frequencies do not remain constant for a number of reasons: random or chance events producing major changes in population size and composition (called "**genetic drift**"); migration of individuals from one population to another, followed by mating between the populations, referred to as **gene migration**; **new mutations** that occur at a low rate constantly; and **natural selection** in which some genotypes are better suited to reproduce and thrive (called "**fitness**") and therefore give rise to a disproportionate share of offspring. A particular form of such selective advantage occurs when gene–environment interaction leads to the situation in which the fitness of heterozygotes for a particular genetic condition exceeds that in either homozygote. This is referred to as **heterozygote advantage**, and has been best studied in the relationship between sickle cell anemia and malaria.

Such examination of single-gene frequencies and perturbations is now being complemented and supplemented by genome-wide studies employing SNPs and CNVs. These genomic approaches have revealed that most genetic variation occurs within a population rather than between two populations—adding additional complexity to the meaning of the word "**race**" and making it clear that such population categories as European, Asian, African, and Hispanic, while distinct in terms of their geographic origins, are in no way distinct genetically.

As we understand more about the structure of genes and genomes, that information informs our ideas about the evolution of populations. **Molecular evolution** is concerned with determining how the study of genomes, chromosomes, genes, and proteins helps us account for the evolution of our species—and other species as well. The study of molecular evolution employs many techniques (DNA hybridization, chromosome banding, amino acid sequences in proteins, and whole-genome sequencing), all aimed at providing more precise estimates of the timing of evolutionary events (molecular clocks), and of the relationship between our species and that of others near or distant from our own (ancient DNA).

CHAPTER 19: PERSONALIZED GENETICS AND GENOMICS

Each of us is unlike anyone else who is alive today or who has ever lived. This uniqueness is a product of our genes and our environment. Among the many goals of the study of human genes and genomes is to help understand this individual uniqueness. Although we have barely scratched the surface of such understanding, useful signposts are being constructed. We will discuss three of them: **DNA fingerprinting**, **pharmacogenetics**, and **direct-to-consumer genomic testing**.

DNA fingerprinting (or **profiling**) is based on the nearly 50% of our genome that is composed of a variety of repetitive DNA sequences. In so-called **microsatellites**, the repeating units are 2–10 bp in length; in **minisatellites** the repeats are 10–100 bp long. Microsatellites and minisatellites are extremely polymorphic, meaning that there are a multitude of alleles at these loci within the population. One form of these alleles is called **variable number of tandem repeats (VNTRs)**, and it is in their identification that DNA fingerprinting is possible. Such fingerprints are developed by obtaining a sample of DNA from viable biologic material, cutting it with a particular bacterial restriction enzyme, running the DNA out using polyacrylamide gel electophoresis, hybridizing it with a large set of VNTR probes, and developing the profile of fragments using **radioautography** or **fluorescence**. The extraordinary power of this approach is shown by estimating that if 24 different minisatellite probes are used, there is only a 1 in 700 trillion chance of identity between any two individuals (except for identical twins). The many uses of DNA fingerprinting include forensic identification of the guilty and the innocent; proof of paternity and maternity; identification of victims of catastrophic events; and confirmation of the identity of long deceased persons—famous and infamous.

For instance, we all take many medicines. Sometimes these medicines are efficacious, other times toxic. Optimally, we would like to know how to predict efficacy and toxicity for any medicine in any person. No profiling exists with which to do this. We know that genes form part of the basis for differences in response to medicines (a field called **pharmacogenetics**). In some instances, genes have been identified that control the absorption, distribution, and metabolism of pharmaceuticals. In other situations, specific genes regulate the target of the pharmaceutical. Certain enzyme systems, such as the **cytochrome P450** family, play a central role in metabolism of administered drugs, and their variation predicts accurately both efficacy and toxicity. In a slowly growing list of examples, we are able to understand the pharmacogenetic basis for response to such important medicines as opiate analgesics, antiplatelet agents, antibiotics, and antimalarials. Extending this small body of information will depend on use of genomic (as well as genetic) information, and on major modifications in the systems by which pharmaceutical companies make medicines and practicing physicians use them.

A third approach toward understanding genetic uniqueness is already leading to a new industry—**direct-to-consumer genetic testing (DTCGT).** It depends on SNP profiling of an individual's DNA and comparing it to profiles from many thousands of other people's. The most robust of the small companies providing DTCGT now provides information on susceptibility to more than 120 serious and harmless traits, response to nearly 20 medicines, carrier status for more than 20 single-gene disorders, and ancestry. It is almost certain that DTCGT will burgeon in coming years in the form of more tests for more conditions, more medicines, and more disorders.

CHAPTER 3

Genetics

CHAPTER OUTLINE

CORE CONCEPT

Genetics is the science of heredity and inherited variation. It considers the biological information all living organisms require to grow, develop, reproduce, and die. It considers, as well, how traits are transmitted from one generation to the next. The fundamental nature of this information is the same in organisms as simple as microbes and as complex as humans.

HEREDITY AND VARIATION

What do we mean when we say, "genetics is the science of heredity?" We mean that units of inheritance, called **genes**, constitute the blueprint on which each organism and each individual constructs its developmental plan. We mean that genes are responsible for transmitting physical, chemical, and behavioral traits from parents to offspring. We mean that the study of genetics attempts to explain why "like begets like"—that is, why humans give birth to humans only, and mice, to mice only. We mean that genetics seeks to account for the enormous amount of biological information that allows an organism to reproduce, grow, mature, and die. We mean that genetics considers the basis for the remarkable diversity of living things (Figure 3.1).

HISTORY

Ancient History

Our species, *Homo sapiens,* first appeared about 200,000 years ago. It seems likely that humans first became curious about heredity nearly as long ago, but there is no record of exactly when. However, there is a record, beginning 5,000–7,000 years BCE, of human understanding of certain aspects of heredity as seen through cultivation of plants:

Human Genes and Genomes. DOI: 10.1016/B978-0-12-385212-0.00003-2

FIGURE 3.1
Informational content in DNA generates great diversity among living things, as shown using the following examples (clockwise from upper left): helical-shaped bacteria (*Campylobacter jejuni*); banana tree (*Musa sapienta*); red-footed booby (*Sula sula*); African leopard (*Panthera pardus*); humans (*Homo sapiens*).

Mayans were cultivating maize; the Chinese were growing rice; and the Assyrians were caring for the date palm tree (Figure 3.2). All these early peoples were employing genetics as they crossed individual plants to produce hybrids with the characteristics they sought.

A few millennia later, the Hebrews and Greeks demonstrated their awareness of genetic principles. The Jewish Talmud says that a woman whose brother and son have bled to death following circumcision should not circumcise her next-born son(s), demonstrating an early understanding of the hereditary nature of hemophilia. The Hippocratic School of medicine believed that semen was composed of "humors" responsible for inherited traits.

AMPLIFICATION: ANCIENT HISTORY

Aristotle's Understanding of Heredity

Aristotle (circa 384–322 BCE) was a Greek philosopher, known as "the father of biology." He taught in Athens at the Lyceum, or Peripatetic school, from 335 BCE until his death. He was greatly interested in the natural world and wondered why it is that animal and plant species almost always produce offspring very similar to other members of the same general group. His views of heredity, which were widely held for almost 2,000 years, recognized that both the mother and father contributed biological material to the creation of the offspring.

Aristotle felt that a child is the product of its parents' comingled blood. He believed, as did his predecessor Pythagoras, that semen was a fluid collected from the entire male body. This fluid, combined with the mother's menstrual blood, caused the child to grow. The father provided the miniature individual and the female provided the supportive environment for growth, in his theory.

Aristotle observed that parents who had suffered some kind of mutilation or loss of body parts produced a whole offspring, so he refuted the idea that there was a direct transfer of body parts from parent to offspring. He instead believed that the offspring is gradually generated from an undifferentiated mass by the addition of parts. He concluded that inheritance involved the potential of producing certain characteristics rather than the absolute production of the characteristics themselves.

Aristotle also had a systematic bent. He classified everything in nature from lower to higher, starting with non-living and ending with humans. He referred to this hierarchy as the "steps of nature." His classes were as follows:

1. Non-living beings
2. Lower plants
3. Higher plants
4. Sponges
5. Jellyfish
6. Shellfish
7. Insects
8. Crustacea cephalopoda
9. Ovipara
10. Whales
11. Ovoviparous quadrapeds
12. Humankind

Aristotle, detail from the fresco *The School of Athens* by Raphael.

Little changed in the understanding of genetics until the 17th and 18th centuries. In 1651, William Harvey (who gained fame for his correct interpretation of the circulatory system) proposed his *ex ovo omnia* theory, meaning everything comes from the egg. In his view, human life began with oogenesis (the production of the egg), and substances in the gametes would produce adult structures.

When Anton van Leeuwenhoek developed his microscope in the 1660s, he was able to see single-celled organisms for the first time. In 1677, he first described the **spermatozoa** of insects, dogs, and humans. Some men of his time, called "**preformationists**," believed that a **homunculus**, a tiny, fully formed human, resided in the head of a sperm (Figure 3.3). Leeuwenhoek, however, felt that each sperm was the seed of an individual creature and would give rise to the next generation if properly nourished in the womb.

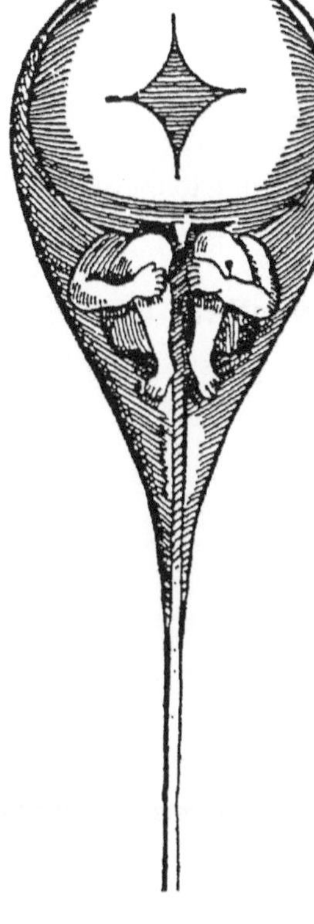

FIGURE 3.3
The homunculus. More than 200 years after Leeuwenhoek invented the microscope, many microscopists clung to the misperception that a fully formed miniature fetus was contained in the head of a sperm.

FIGURE 3.4
Gregor Mendel. This Austrian monk made the conceptual leap in 1866 that opened the modern era of genetic science.

Other scientists during this period contributed to the advancement of knowledge of hereditary variation. Robert Hooke first used the term "**cell**," based on microscopic studies of cork. Johannes Purkinje was the first to describe a cell's nucleus. Rudolph Virchow proposed the cell theory that said all cells come from pre-existing cells.

However, it was not until the 1850s and the 1860s that Gregor Mendel's (Figure 3.4) experiments with pea plants revolutionized scientific thought about heredity, even though it took 40 years for his work to be rediscovered and much longer to be appreciated. Mendel worked with pea plants and determined that units of inheritance determined their observable traits. In other words, what we can observe with our eyes when we look at an organism is determined by something within the organism that we cannot see. These units, which he called "**elements**" and we call genes or alleles, come in pairs. These pairs separate (**segregate**) during seed formation and recombine in the fertilized plant. He went on to propose that some traits determined by these units are **dominant** and others **recessive** and that two unrelated traits would assort independently—that is, would segregate uninfluenced by the other pair. It is staggering to consider that, in the few sentences we have just written, Mendel prefigured what geneticists have been proving and amplifying for the last century and a half.

Twentieth Century History

Forty years later, William Bateson came upon Mendel's work, or, if you will, his buried treasure, and discussed it with his colleague, Archibald Garrod. Bateson was the first to use the word "**gene**" for Mendel's element and recognized that the four human disorders Garrod was studying behaved like Mendel's recessives. Once Mendel's principles had been resurrected, the field of **genetics** was born. During the first half of the 20th century many key observations were made (Table 3.1), some of the most notable being that the behavior of chromosomes during meiosis explained **Mendel's laws** (Walter Sutton); that "**inborn errors**" caused metabolic diseases in humans (Archibald Garrod); that **sex-linked inheritance** had its own rules and could be studied in drosophila (the fruit fly) (Thomas Hunt Morgan); that genes were composed of **DNA** (Frederick Griffith and Oswald Avery); that bacteria are useful organisms with which to study genetics (Salvatore Luria, Max Delbruck, Joshua Lederberg); that genes may be transposed from one location to another with observable effects (Barbara McClintock); that genetic mutations could be caused by X-rays (Herman Muller); and that genes produced their effects by directing the synthesis of proteins (George Beadle and Edward Tatum).

TABLE 3.1 Pioneering Discoveries in Genetics

Year	Contribution	Scientist(s)
1866	Gene segregation and assortment	Mendel
1902	Chromosome behavior during meiosis	Sutton
1908	Inborn errors of metabolism	Garrod
1910	X-linkage	Morgan
1929	Transforming principle	Griffith
1941	One gene, one enzyme	Beadle/Tatum
1944	Genes are composed of DNA	Avery/MacLeod/McCarty
1946	X-ray mutagenesis	Muller
1948	Molecular pathology of hemoglobin S	Pauling
1952	Enzymatic defect in glycogenoses	Cori/Cori
1953	DNA structure: a double helix	Watson/Crick
1956	46 chromosomes in human cells	Tjio/Levan
1959	Enzymatic synthesis of DNA	Kornberg/Ochoa
1959	Aneuploidy in Down syndrome	LeJeune
1960	The Philadelphia chromosome in leukemia	Howell/Hungerford
1961	X chromosome inactivation	Lyon
1963	Triplet genetic code	Nirenberg/Holley/Khorana
1970	Restriction enzymes	Nathans/Smith
1972	Recombinant DNA	Cohen/Boyer
1976	Oncogenes	Varmus/Bishop
1977	DNA sequencing methodology	Sanger/Gilbert
1980	Restriction fragment length polymorphism	Botstein
1986	Positional cloning	Orkin/Kunkel
1998	RNA interference	Fire/Mello
2003	Human genome sequence	Many contributors
2007	Homologous recombination	Capecchi/Smithies
2009	Telomeres and telomerase	Blackburn/Greider

In retrospect, it is easy to see that during the first half of the 20th century the discipline of genetics was only gathering force for the explosion of information that characterized the second half of the 20th century and continues until today. Ushered in by James Watson and Francis Crick's remarkable proposal that DNA is a **double helix**, no other scientific field moved with the pace of genetics. It captured the attention of scientists working with organisms as diverse as bacteria and humans, and every creature in between. Its basic tenets were universal, explaining the structure and function of genes. It helped explain inherited human traits and disorders in biochemical and molecular terms. It established the existence of previously unrecognized kinds of molecules, such as **cell surface receptors**, **oncogenes** (genes that help turn a normal cell into a tumor cell), and **prions** (infectious agents composed entirely of protein). It led to the study of **genomes**, which, in turn, is propelling us toward a greater understanding of evolution, variation, and individuality.

CONSERVATION OF GENETIC INFORMATION

Everything we know says that the structure and function of genes and their products has been conserved throughout evolution. DNA is, for all organisms other than some viruses, the genetic material. The genetic code, by which the language of coding genes is translated into the language of proteins, is universal. DNA and protein sequences performing a specific function in bacteria have regularly been retained throughout the animal kingdom, indicating evolutionary descent from a common ancestor. In Figure 3.5, such conservation is shown for part of a single protein from six species. Finally, gene mutation produces beneficial and deleterious phenotypes in all organisms and is a fundamental driving force for evolution by natural selection.

H. sapiens	---MGDVEKGKKIFIMKCSQCHTVEKGGKHKT	GPNLHGLFGRKTGQAPGYSYTAANKNKGIIW
M. musculus	---MGDVEKGKKIFVQKCAQCHTVEKGGKHKT	GPNLHGLFGRKTGQAAGFSYTDANKNKGITW
D. melanogaster	---AGDVEKGKKLFVQRCAQCHTVEAGGKHKV	GPNLHGLIGRKTGQAAGFAYTDANKAKGITW
C. elegans	---AGDYEKGKKVYKQRCLQCHVVDS-TATKT	GPTLHGVIGRTSGTVSGFDYSAANKNKGVVW
A. thaliana	----GDAKKGANLFKTRCAQCHTLKAGEGNKI	GPELHGLFGRKTGSVAGYSYTDANKQKGIEW
S. cerevisiae	---PGSAKKGATLFKTRCQQCHTIEEGGPNKV	GPNLHGIFGRHSGQVKGYSYTDANINKNVKW

H. sapiens	GEDTLMEYLENPKKYIPGTKMIFVGIKKKEER
M. musculus	GEDTLMEYLENPKKYIPGTKMIFAGIKKKGER
D. melanogaster	NEDTLFEYLENPKKYIPGTKMIFAGLKKPNER
C. elegans	TKETLFEYLLNPKKYIPGTKMVFAGLKKADER
A. thaliana	KDDTLFEYLENPKKYIPGTKMAFGGLKKPKDR
S. cerevisiae	DEDSMSEYLTNPKKYIPGTKMAFAGLKKEKDR

☐ Indicates similar sequence
☐ Indicates identical sequence

FIGURE 3.5
Conservation of amino acid sequences in the cytochrome C protein. Equivalent portions of the protein in six organisms are shown. Yellow blocks indicate identical sequences; green blocks indicate similar sequences. Note the remarkable similarity in sequences among these widely different species. The six organisms (from top to bottom) are: human, mouse, fruit fly, roundworm, mustard plant, and yeast. The single letters (PGS...) are the accepted symbols for the 20 amino acids.

HUMAN VARIATION

As technology has advanced such that variation can be studied at **phenotypic**, **genotypic**, and **genomic** levels, the magnitude of our understanding of human variation has expanded remarkably. The study of single-gene defects indicated that rare mutations in 1/500 to 1/1,000,000 people could produce disorders. The study of chromosomal abnormalities showed that less than 1% of live-born humans have microscopically definable changes.

The examination of **protein polymorphisms** revealed that normal variation in amino acid sequence occurs in about one-third of proteins studied. The study of single nucleotide polymorphisms showed that any two humans differ at ~0.1% of their unique DNA sequences. Recent discovery of copy number variants has startled the genetics community by showing that each of us, in terms of our entire genome sequence, differs from anyone else by as much as 0.5%.

We can say with confidence that as sequencing entire human genomes from large populations becomes feasible, we will move toward a more full understanding of human variation and genetic diversity. This will occur during the next few decades and, with it, a wonderful view of what makes each of us unique will emerge.

REVIEW QUESTIONS AND EXERCISES

1. More than 2,000 years ago, Aristotle wrote and thought about why "like begets like."
   a. What does the expression "like begets like" mean?
   b. How close was Aristotle to understanding human reproduction?
   c. Why did it take two millennia to answer the question Aristotle asked?
   d. Describe the basis for "like begets like" using today's terms and concepts.
2. Gregor Mendel is rightly called the father of the science of genetics.
   a. What fundamental concepts did he discover?

b. Why was he able to make his remarkable discoveries almost a century before we knew that DNA was the genetic material?
c. Name three scientists who can rightly be called Mendel's scientific sons and explain what each did.
d. What is the modern term for Mendel's "elements"?

3. Why is it appropriate to refer to genes as a biologic blueprint? What structures does this blueprint use to "build" an organism such as a human being?

CHAPTER 4

Growth, Development, and Reproduction

CHAPTER OUTLINE

CORE CONCEPT

All organisms package their genetic information in tiny cellular structures called **chromosomes**. In humans, and other **eukaryotes** (organisms composed of one or more cells containing a visible nucleus and organelles) which reproduce sexually, chromosomes are paired, one inherited from each parent. Human cells contain 23 pairs of chromosomes for a total of 46. These **diploid** cells—cells that contain two copies of every chromosome—transmit their chromosomes during two kinds of cell division: **mitosis** and **meiosis**. In mitotic cell division, which occurs in all cells of the body, each "**daughter cell**" receives the identical chromosome complement of its parent cell. In meiosis, which occurs only in germ cells (reproductive cells), the diploid number (46) is halved, producing **haploid gametes** (cells that will fuse during fertilization), each with 23 chromosomes. Inherited variation between any two people is largely a function of two events that occur during meiosis**: random chromosome separation** and **crossing over** between members of each chromosome pair. Reproduction proceeds through the union of **gametes**, an **oocyte** (egg) from the female, a **sperm** from the male. The fertilized egg, or **zygote**, is **diploid**, as are all other body cells. Prenatal and postnatal development occurs through mitosis, which is how the single-cell zygote develops into a 40-billion cell newborn and a 100-trillion-cell adult.

Chapter 3 emphasized that the science of genetics concerns itself fundamentally with the storage and transmission of biological information. This chapter begins to explore how this information is packaged and what it accomplishes after it is transmitted. Central to this subject is the chromosomal basis of heredity.

Human Genes and Genomes. DOI: 10.1016/B978-0-12-385212-0.00004-4

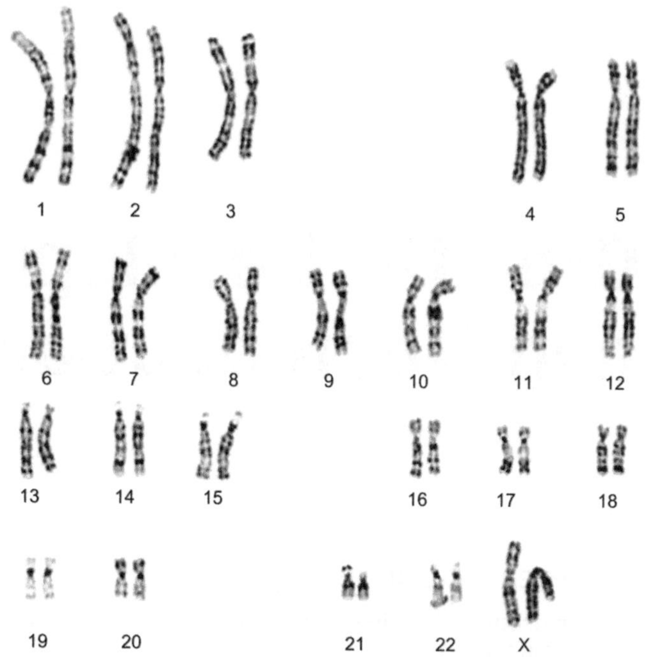

FIGURE 4.1
Karyotype of a human female. Metaphase pairs are arranged in order of decreasing size and numbered accordingly. In females, there are 22 pairs of autosomes (1—22) and a pair of similar sex chromosomes (XX). Male karyotypes look much the same except that the sex chromosomes consist of one X and one Y. Homologous chromosomes share a characteristic pattern of dark and light bands.

CHROMOSOMES: NUMBER AND NATURE

All living organisms package their genetic information into tiny cellular structures called **chromosomes**. Bacteria and other organisms that lack a nuclear membrane (**prokaryotes**) have 1 chromosome; some yeast have 4; peas have 14; humans have 46. These 46 chromosomes are found in the cell's nucleus. Each chromosome is a single linear molecule of DNA containing a unique subset of genes arranged linearly. The 46 chromosomes are composed of 23 pairs (Figure 4.1).

Members of a pair (referred to as **homologous chromosomes** or **homologues**) carry matching information inherited, respectively, from the mother and father. Each paired homologue carries the same genes in the same order, but the versions of those genes at any locus (location) may be the same or different. These versions of a gene are called **alleles**. A person has only two alleles at any locus, but the number of allelic versions may be small or large. The 22 pairs of visibly similar chromosomes are called **autosomes**. The remaining pair, the **sex chromosomes**, is visibly similar in females (XX), but not in males, where one of the two sex chromosomes, called the **Y**, is much smaller than its partner, called the **X** (Figure 4.2).

All cells of the body not involved in reproduction—also known as **somatic** cells—contain the complement of 46 chromosomes just described. Because they contain two copies (2N) of each of the 23 chromosomes, they are referred to as **diploid** cells. In contrast, **germ cells**—sperm and egg (also called an **oocyte**) cells—contain 23 chromosomes, one of each pair (N). They are called **haploid** cells. At the time of fertilization, a haploid oocyte (N) and a haploid sperm (N) fuse to produce a diploid **zygote** (2N). All cell divisions thereafter retain the diploid chromosome number.

CELL DIVISION

The zygote consists of a single cell. A full-term newborn has about 40 billion cells; an adult human, about 100 trillion cells. The remarkable process responsible for this dramatic proliferation of cells, and the one that enables the equally amazing events that, ultimately, make

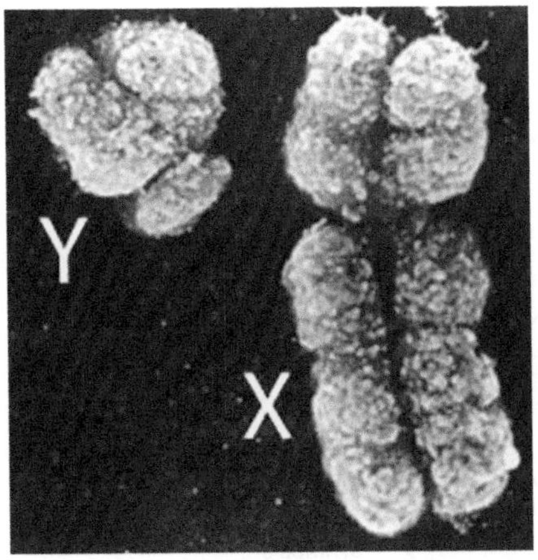

FIGURE 4.2
Electron micrograph of human Y (left) and X (right) chromosomes. Females have two X chromosomes, males an X and a Y.

cells of one tissue (such as brain) differ from that of any other (such as skeletal muscle), is cell division.

There are two kinds of cell division, **mitosis** and **meiosis**. Mitotic division makes possible the growth, differentiation, and regeneration of the body. Each mitotic division produces two daughter cells, each with the identical chromosome number (46) and gene complement of the parent cell. In contrast, meiosis occurs only in specialized cells (called **germ cells** or **germ line cells**) in the **ovary** and **testis**, and culminates in the formation of haploid gametes, the oocyte and the sperm. As we will see later in this chapter, meiosis accomplishes more than halving the chromosome number in gametes; through the processes of **random segregation** of chromosomes and **recombination of chromosome** regions (also called "**crossing over**"), it assures the uniqueness of each **zygote**.

When something goes awry in these division processes, abnormalities of chromosome number or structure can occur and have profound clinical consequences. Many spontaneous abortions, birth defects, and cancers are caused by a state of genetic imbalance resulting from meiotic or mitotic errors in the distribution of chromosomes to daughter cells. We will discuss such **mutations** later in the book.

Mitotic Division

Mitotic division consists of the separation of chromosomes and the distribution of **cytoplasmic contents** (everything in the cell except the nucleus) from one parent cell to two identical daughter cells. In a cell not undergoing mitotic division, the chromosomes are not condensed, and are, therefore, not visible by light microscopy. They become visible as the cell prepares to divide. The entire series of events from one cell division to the next is called the **cell cycle**.

THE CELL CYCLE (FIGURE 4.3)

Interphase, G_1 and G_0

An average adult cell spends most of its life in the period between two successive mitotic divisions, called **interphase**. Interphase encompasses three of the four phases of the cell cycle (G_1, S, and G_2). Immediately after dividing, the cell enters **G_1** (G standing for "**gap**"), a period when no new chromosomal synthesis occurs. During this period, the cell performs its normal biochemical functions as part of the organ to which it belongs. Cells that will never divide, such as red blood cells or neurons, remain in this quiescent state once they are fully differentiated. In this case, the G_1 phase can also be referred to as the **G_0** (**G "zero"**)

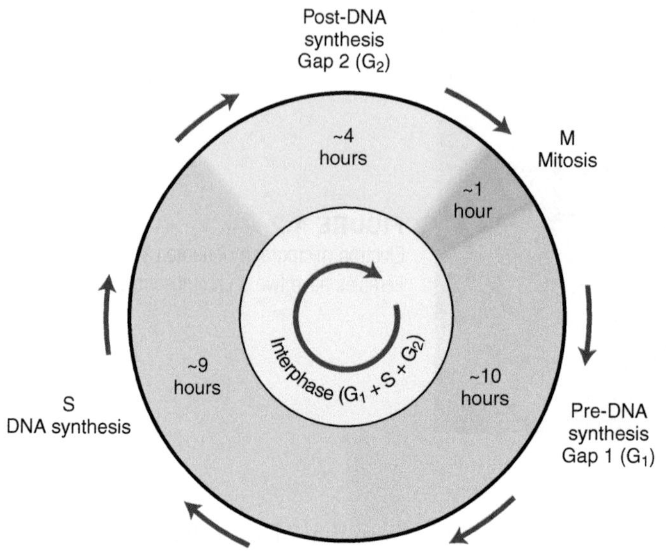

FIGURE 4.3
The cell cycle. Chromosomes replicate to form chromatids during S phase (synthesis). Sister chromatids segregate to daughter cells during M (mitosis). The intervals between S and M are called gap phases (G_1 and G_2), during which most cell growth and a variety of synthetic processes occur. In humans, this cycle is accomplished within 24 hours—in only one of which does cell division occur.

phase. Some cells, such as liver cells that normally exist in G_0, will return from G_0 to G_1 to recover from liver damage. They need to return to the cell cycle in order to make new cells. During G_1, the cell begins synthesizing the molecules it will need to become two separate cells. Some cells, such as bone marrow cells, will have a short G_1 period of approximately 10 hours. Other cells, such as the rapidly dividing embryonic cells, will never enter G_0 and will skip G_1 altogether.

The S Period

During the G_1 period, each cell contains one diploid copy of the genome. At the end of this period the cell enters the **S period**, during which the cell gradually increases in size, doubling its total mass. During S, each homologue replicates to become a two-part chromosome consisting of two sister **chromatids**. These two sister chromatids are held together at the **centromere**, a specialized region near the center of the chromosome. Individual chromosomal segments have their own characteristic replication time during this period of about 9 hours. By the end of this period, the cell contains two full copies of its chromosomal material. In other words, at this point the cell has become, temporarily, doubly diploid.

The G_2 Period

This is a short period following the S period and before mitosis (M). During G_2, the cell continues to grow, and more proteins and other molecules are manufactured. These molecules are necessary to form the new structures of the two daughter cells, particularly the new cell membranes that will enclose each cell.

Mitosis (M)

M, the shortest of the phases of the cell cycle, is the interval during which cell division occurs. In human cells, the total duration of the cell cycle is about 24 hours. M occupies only 1–2 hours of this total. M is conventionally described as having four

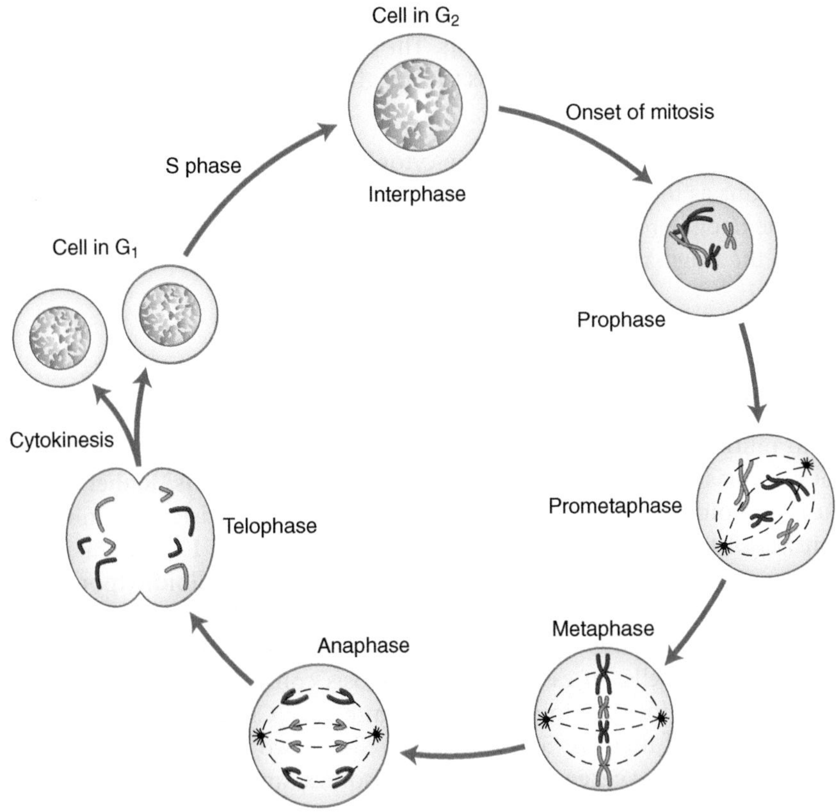

FIGURE 4.4
Mitotic cell division. Two chromosome pairs are shown. Note that each homologous chromosome has already been duplicated before mitosis starts and consists of two sister chromatids. In this figure, red designates maternally inherited chromosomes and green designates paternally inherited chromosomes. The details of the various phases and their relationship to the cell cycle are discussed in the text. The net result: two daughter cells identical to the parent cell.

phases—prophase, metaphase, anaphase, and telophase—that flow continuously from one to the next. Each phase is characterized by specific events (Figure 4.4).

1. *Prophase.* The cell enters M with its chromosomes already having been replicated during S. At the beginning of **prophase**, each chromosome consists of two identical strands called **chromatids.** The two chromatids are held together at a particular region, called the **centromere**. As prophase ends, the chromosomes condense further and the nuclear envelope breaks down.
2. *Metaphase.* At the start of **metaphase**, the **mitotic spindle** forms, which will be responsible for separating the chromosomes into two complete sets. The spindle is composed of a series of protein fibers whose poles (called **centrioles**) connect to the chromosomal centromeres. Following this attachment, each chromosome is moved to a location near the center of the cell, equidistant from the poles. This positioning of all the chromosomes makes up the metaphase plate, as shown in Figure 4.4. At this point, the chromosomes are maximally condensed and easiest to see.
3. *Anaphase.* Two critical events occur during **anaphase.** First, the proteins holding the centromeres together are degraded and the two centromeres separate. Second, the two strands of each chromosome—that is, the **sister chromatids**—move toward opposite poles of the spindle because the spindle fibers shorten and contract. As anaphase ends, the chromosomes lie in two groups at opposite poles of the spindle. Each group contains the number of chromosomes (**2N**) characteristic of the species.

4. *Telophase.* During **telophase**, the nuclear membrane forms around each group of chromosomes and the spindle disappears. The condensed chromosomes de-condense and become invisible as distinct entities. As the daughter nuclei form, each nucleus resumes its interphase appearance.

To complete the process of cell division, the cytoplasm splits by a process known as **cytokinesis**, which begins as the chromosomes near the spindle poles and ends with the formation of two complete daughter cells, each with its own nucleus containing all of the genetic information of the original cell.

Meiotic Division

In contrast to mitotic division, which occurs in all body cells, **meiotic division**, or **meiosis**, is a type of cell division found only in male and female **germ cells**. Rather than producing two identical, diploid daughter cells from one round of cell division, as mitosis does, meiosis consists of two rounds of cell division (called "**meiosis I**" and "**meiosis II**"), ultimately yielding gametes each of which is haploid rather than diploid. Upon completion of meiosis I (also called "**reduction division**"), each daughter cell contains half of the diploid number of chromosomes (thus is called "**haploid**"). In meiosis II, each haploid cell divides into two more haploid cells. In essence, then, meiosis constitutes the means by which one diploid cell divides into four haploid gametes (Figure 4.5). Critical for the organism, **genetic diversity** is a product of meiosis. Such diversity reflects two properties, each occurring during meiosis I: **random segregation** of chromosomes, and **recombination** between members of a chromosome pair. What follows are more detailed descriptions of meiosis I and meiosis II.

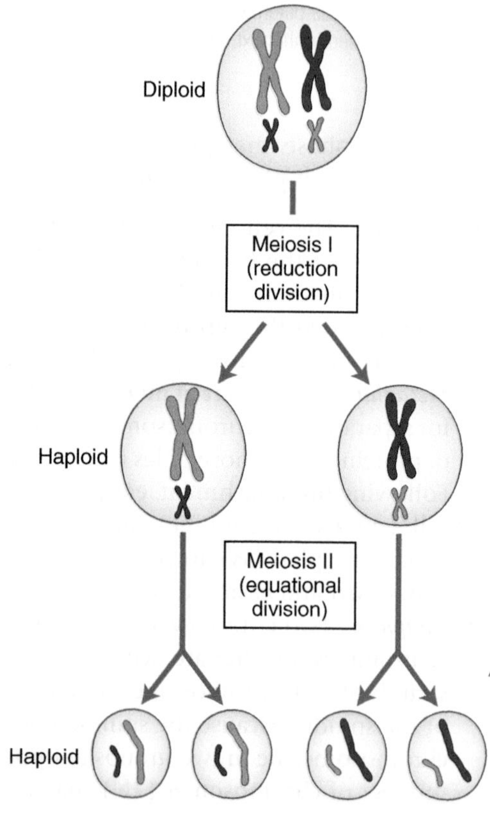

FIGURE 4.5
A simplified representation of meiosis, which occurs only in germ cells. Two different chromosome pairs are shown, one long and one short, and each consists of two sister chromatids. The colors indicate parent of origin. Note the essential character of meiotic cell division: one round of chromosome replication before meiosis I (reduction division), followed by two rounds of chromosome segregation, leading to the haploid number of chromosomes in each of four gametes.

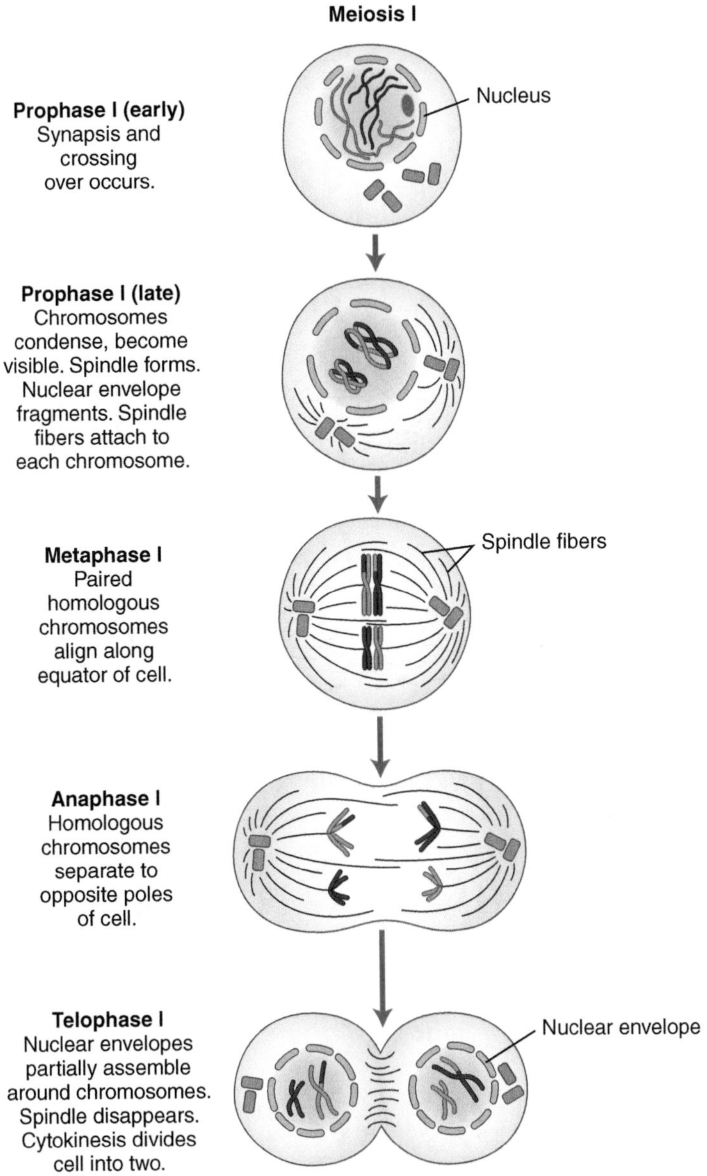
Meiosis I
Prophase I (early)
Synapsis and crossing over occurs.
Nucleus
Prophase I (late)
Chromosomes condense, become visible. Spindle forms. Nuclear envelope fragments. Spindle fibers attach to each chromosome.
Metaphase I
Paired homologous chromosomes align along equator of cell.
Spindle fibers
Anaphase I
Homologous chromosomes separate to opposite poles of cell.
Telophase I
Nuclear envelopes partially assemble around chromosomes. Spindle disappears. Cytokinesis divides cell into two.
Nuclear envelope

FIGURE 4.6
Detailed view of meiosis I. The critical features are: recombination (or crossing over, the exchange of genetic material between homologues) during prophase; random separation of homologous chromosomes during anaphase (2N to N); and cytokinesis (cell division) during telophase, producing two non-identical daughter cells.

MEIOSIS I (FIGURE 4.6)

The same phases discussed in mitosis (prophase, metaphase, anaphase, and telophase) occur here, but with important differences, particularly in prophase I.

Prophase I

Prophase I can be divided into several stages, as follows. **Leptotene** begins with the already replicated chromosomes becoming visible as thin threads. The two sister chromatids are so closely aligned that they are not distinguishable. During **zygotene**, homologous chromosomes begin to align along their entire length by a process called **synapsis** that is necessarily precise. Each pair of chromosomes is held together by a ribbon-like protein and

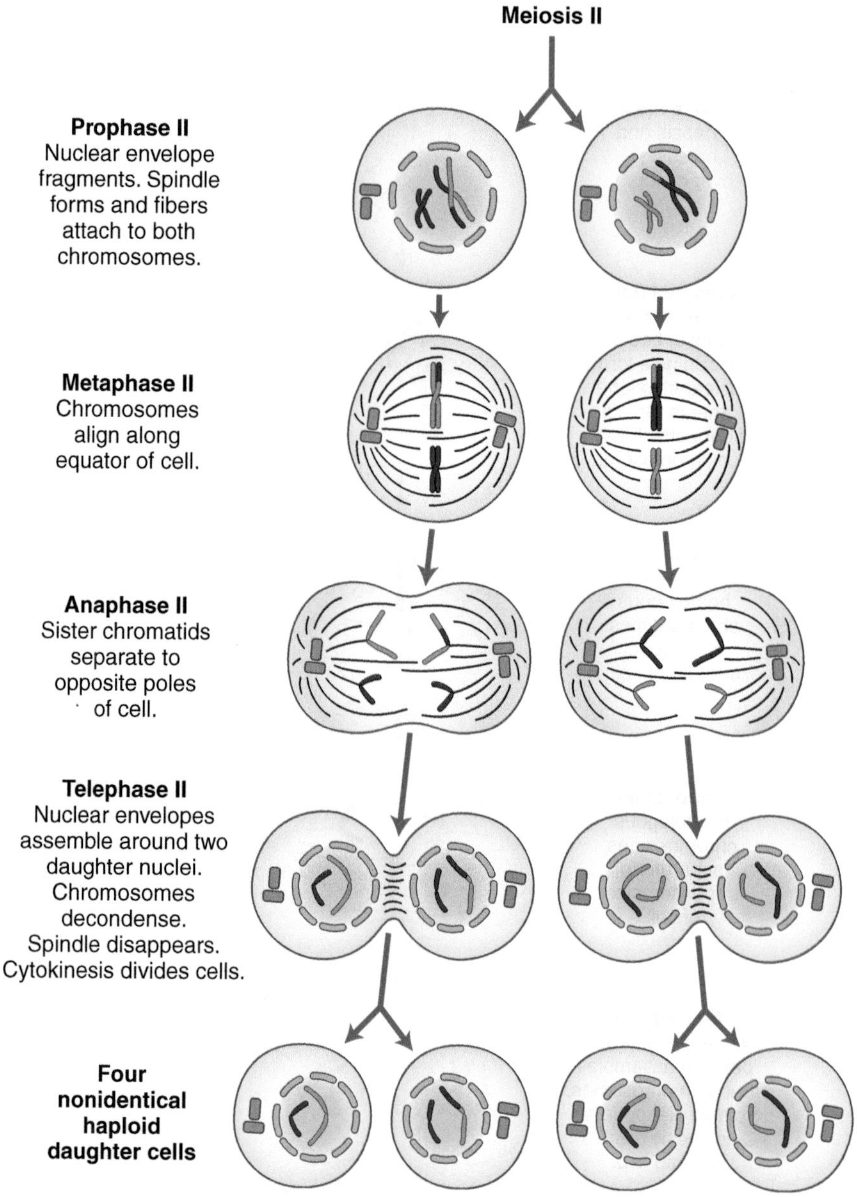
Meiosis II
Prophase II
Nuclear envelope fragments. Spindle forms and fibers attach to both chromosomes.
Metaphase II
Chromosomes align along equator of cell.
Anaphase II
Sister chromatids separate to opposite poles of cell.
Telephase II
Nuclear envelopes assemble around two daughter nuclei. Chromosomes decondense. Spindle disappears. Cytokinesis divides cells.
Four nonidentical haploid daughter cells

FIGURE 4.7
Detailed view of meiosis II. The critical features are: each daughter cell is haploid (N) as prophase begins; sister chromatids separate during anaphase; and cell division during telophase yields four gametes genetically non-identical to each other, because of crossing over and random homologue segregation during meiosis I.

forms the **synaptonemal complex**. Then, during **pachytene**, the pairs of chromosomes become condensed and coiled. During this stage, crossing over (or recombination) between members of a chromosome pair occurs. In this process, the two chromosomes literally trade similar parts of their structure, called homologous regions, with each other. After recombination has occurred, the synaptonemal complex begins to break down, and the two components (known as the **homologues**) begin to separate from each other. This phase is called **diplotene**. Eventually, the two homologues are held together only at points called **chiasmata** (crosses), which likely reflect points of crossing over. On average, two to three crossovers are noted per pair of chromosomes. Finally, during **diakinesis**, the chromosomes are maximally condensed.

Metaphase I

As with mitosis, metaphase I begins after the disappearance of the nuclear membrane. A spindle forms, and the paired chromosomes align themselves along the equatorial plane.

Anaphase I

The two members of each homologous pair and their respective centromeres are drawn to opposite poles by a process called **disjunction**. Accordingly, the chromosome number is halved because the maternal and paternal homologues for a particular chromosome have been separated. Crucial for the formation of the gametes, the movement of one type of chromosome (e.g., #4) is independent of any other type (e.g., #15). Said another way, the maternal homologue of one pair of chromosomes will go to one pole, but the maternal homologue for another pair may go to the other pole, and so on, for all 23 pairs. This **random segregation** means that the possible combinations of the 23 chromosome pairs that can be found in the gametes are 2^{23} (or more than 8 million). The magnitude of the variation that can be transmitted from parent to child is actually much greater than this because of crossovers. Thus each chromatid contains segments from each parent, perhaps three to five in number, of alternating maternal and paternal derivation.

Telophase I

The two haploid sets of chromosomes have been collected at opposite poles of the cell, and the nuclear envelopes reform.

MEIOSIS II (FIGURE 4.7)

The second meiotic division is similar in all ways to mitosis, except that its end result is the formation of 4 gametes, each containing 23 chromosomes and each containing its own unique set of chromosomal material.

AMPLIFICATION: HISTORICAL DEVELOPMENTS

The First Observation of Meiosis and Mitosis

Meiosis was discovered and described for the first time in 1876 by the distinguished German biologist Oscar Hertwig (1849–1922) in sea urchin eggs. Edouard Van Benden (1846–1910) described it at the level of the chromosomes in Ascaris worm eggs in 1883. It was only in 1890, however, that the German biologist, August Weismann (1834–1914), realized that two cell divisions were necessary to transform one diploid cell into four haploid cells and thereby maintain the diploid number of chromosomes after fertilization. In 1911, the American geneticist Thomas Hunt Morgan (1866–1945) observed crossovers during meiosis in *Drosophila melanogaster*, which provided the first evidence that genes from parents are "mixed" during meiotic division.

Meiosis probably evolved about 1.4 billion years ago. Germ cells that undergo meiosis are found in the five major eukaryotic super groups, *opisthokonts*, *amoebozoa*, *rhizaria*, *archaeplastida*, and *chromalveolates*. The only super group that does not exhibit meiosis in all its organisms is *excavata*. *Euglenoid* is an example of a genus in this group in which meiosis is non-existent.

The English biologist Walther Flemming, in 1879, was the first person to describe how chromosomes moved during mitotic cell division. Although cells that had undergone division had been observed by Carl Nageli almost 40 years earlier, he had studied dead cells in salamander embryos and felt that what he was seeing reflected abnormalities of such non-living cells. Flemming, observing cell division in living salamander embryos where cells divided at fixed intervals, described the whole process of mitosis. He saw the doubling of chromosomes and their subsequent equal distribution to the two daughter cells. He coined the terms prophase, metaphase, and anaphase that are still in use today.

(Continued)

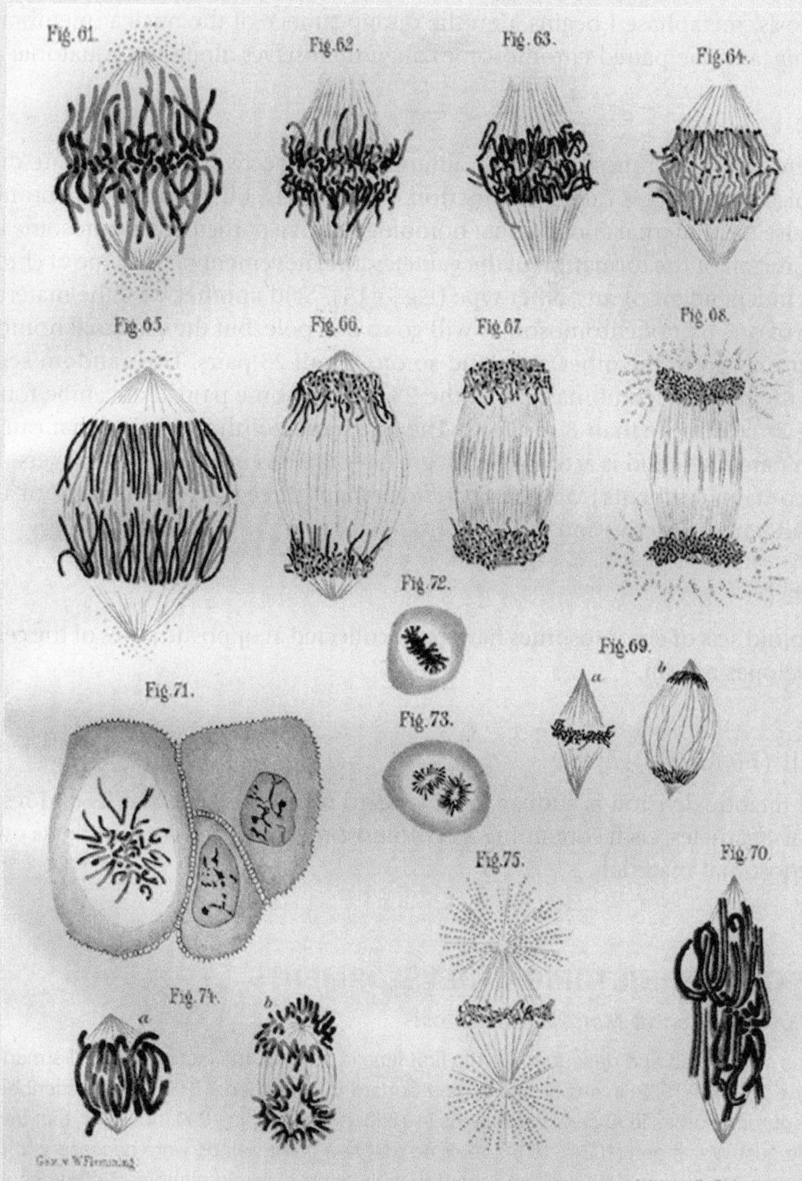

Drawing of mitosis by Walter Flemming.

This illustration is one of more than 100 drawings from Flemming's *Cell Substance, Nucleus, and Cell Division* (1882).

(*Continued*)

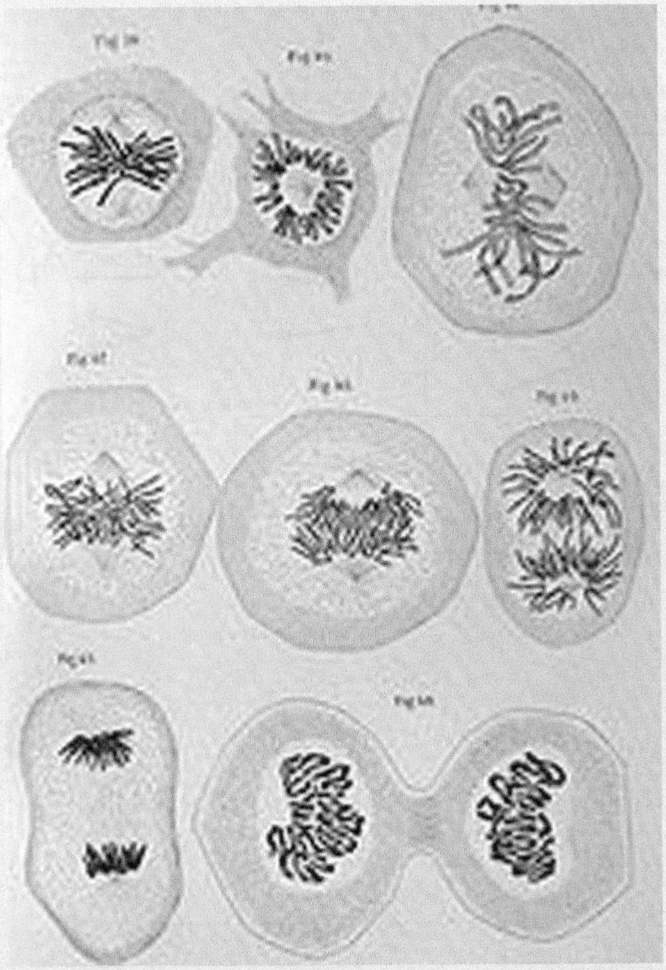

Illustration from O. Hertwig's text concerning the developmental history of humans and vertebrates, showing the cell nuclei of a species of salamander.

Gametogenesis

The meiotic events responsible for the formation of male and female gametes are both similar and different. They are similar in their net result—the formation of haploid germ cells capable of combining into a zygote—but they differ in several important ways. First, the meiotic events take place at different times in the life cycles of males and females (Figures 4.8, 4.9). Meiosis I begins in the ovary during fetal life; in the testis meiosis I begins only at puberty. In the male, meiosis II takes place almost immediately after meiosis I, the entire process being completed in 64 days. In contrast, meiosis II in females occurs only after fertilization, meaning that cells destined to become **oocytes** (eggs) remain suspended in prophase I for many years. The duration of this suspension differs for each oocyte because only one is released from the ovary during the adult menstrual cycle. Thus, oocytes "ripe" for fertilization may reside in the ovary for 30 or more years.

Second, each meiotic cycle in males produces four identical sperm, each capable of fertilizing an egg. In females, only one of the four gametes is capable of combining with a sperm. As shown in Figure 4.9, spermatogenesis begins with cells (**spermatogonia**) that line the **seminiferous tubules**. After many rounds of mitosis, primary **spermatocytes** are differentiated.

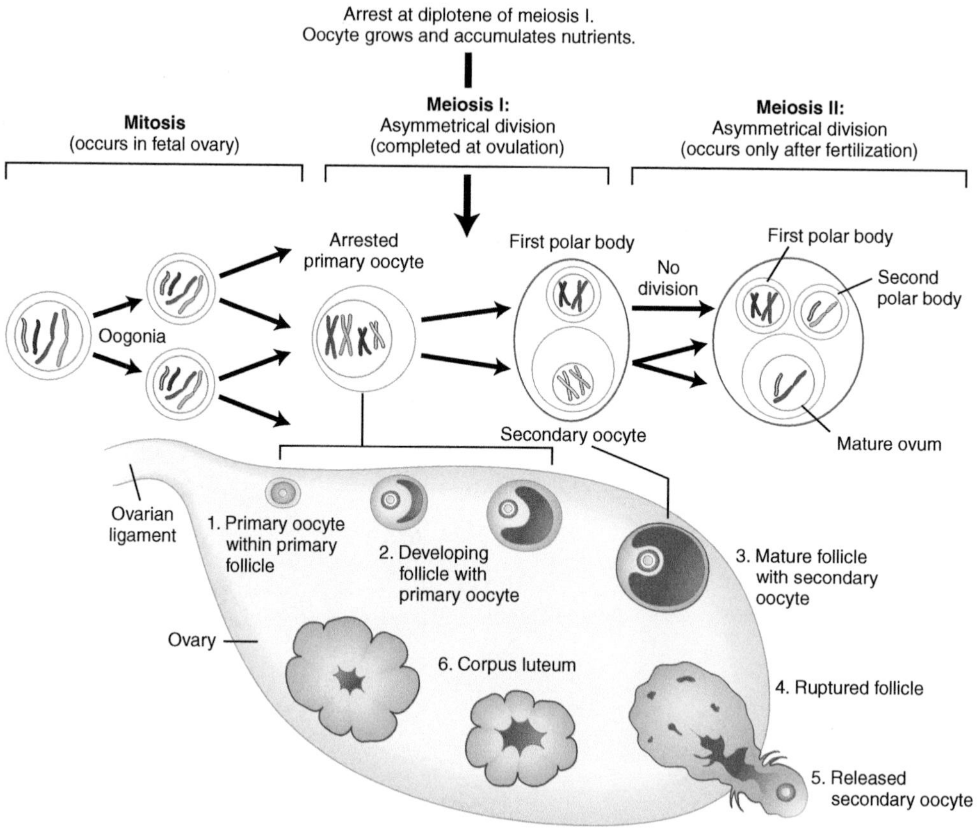

FIGURE 4.8
Human oogenesis. Egg formation begins in the fetal ovary and arrests during prophase of meiosis I. It completes meiosis I after ovulation and completes meiosis II only after fertilization. The ovarian germ cells (oogonia) produce, first, a large number of primary oocytes, and, after meiosis I, secondary oocytes. Following meiosis II, an egg or ovum is formed.

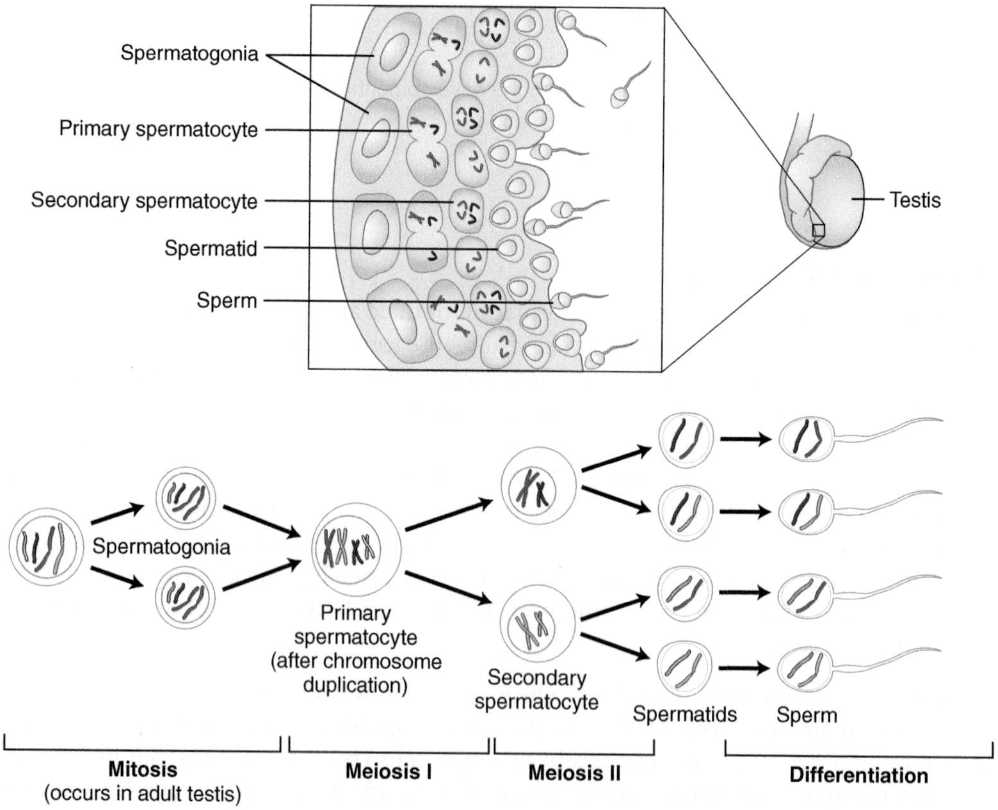

FIGURE 4.9
Human spermatogenesis. Sperm form continuously in the testis after puberty. Germ cell formation begins in spermatogonia, which are cells located at the periphery of seminiferous tubules. Meiosis I occurs in primary spermatocytes yielding spermatids, and, then, tail-bearing sperm. The meiotic divisions occur in cells progressively closer to the lumen, or open space, inside the seminiferous tubule.

These undergo meiosis I to become haploid secondary spermatocytes and become **spermatids** after meiosis II.

Events are quite different in females (Figure 4.8). In the fetal ovary, **oogenesis** begins with cells in the ovarian cortex (**oogonia**) that are the product of many mitotic cycles, starting with primordial germ cells. Each oogonium becomes the central cell in a growing follicle and develops into the primary **oocyte**, as early as the third month of prenatal development. Most primary oocytes begin meiosis by entering prophase I. By the time of birth, primary oocytes have completed prophase and are called secondary oocytes. They complete meiosis I, however, only at the time of ovulation. Meiosis II begins soon thereafter, but is completed if, and only if, fertilization occurs. During oogenesis, one of the two haploid products of meiosis I contains most of the parent cell's cytoplasm and organelles; the other, much smaller cell, is called the **first polar body**, and usually undergoes dissolution. This pattern repeats itself during meiosis II, yielding the fertilized ovum and the second polar body. Meiosis I and II, then, produce a single, viable gamete.

Third, there is a remarkable difference between the number of sperm formed and the number of eggs formed. Males produce about 200 million sperm per ejaculate, and an estimated trillion during a lifetime. A female has several million oocytes in her ovaries at the time of birth, but most of these degenerate. Only about 400 oocytes eventually mature and are ovulated, one per month from menarche to menopause.

Fourth, the ovum is much larger than the sperm and contributes virtually all the cytoplasmic components to the zygote (Figure 4.10). Among the most significant results of this "unequal" cytoplasmic inheritance is the distribution of **mitochondria**. These organelles, with their small, circular genome, are maternal in origin, and are transmitted to offspring accordingly. As will be discussed in Chapter 5, this maternal inheritance has important clinical implications.

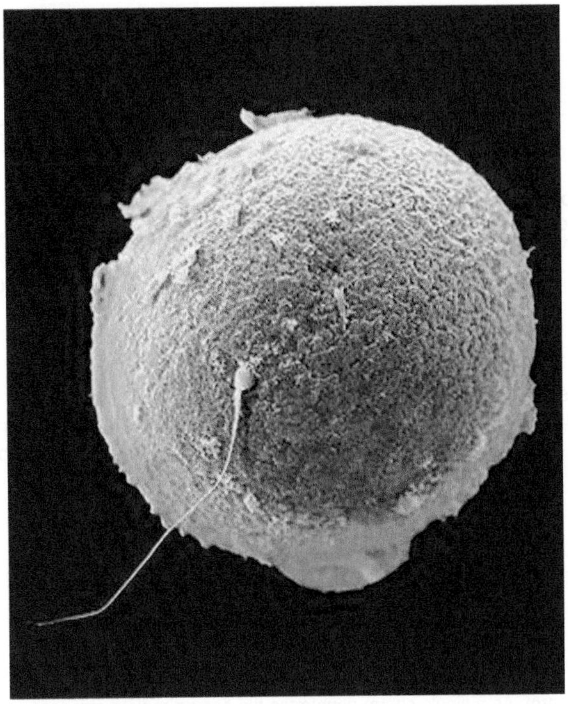

FIGURE 4.10
Micrograph of a sperm beginning to penetrate an egg. Note the huge disparity in size between sperm and egg.

PRENATAL DEVELOPMENT

Fertilization

During sexual intercourse, hundreds of millions of sperm are ejaculated into the vagina. Thereafter, time is of the essence: sperm are viable for about 3 days; the ovum must be fertilized within 12–24 hours after ovulation. During this critical interval, a number of things must happen: sperm must reach the egg; a single sperm must penetrate the egg; and the haploid genomes of the sperm and egg, respectively, must unite to form the diploid zygote.

The passage of sperm from the **vagina**, through the **uterus**, and into the **Fallopian tube** is aided by: chemical activation of sperm via female signals; secretion of a substance by the oocyte that attracts sperm; contractions of female pelvic musculature; and the moving tails of the sperm. Even so, only a few hundred sperm find their way to the oocyte. Fertilization, which generally occurs in the Fallopian tube, is initiated when a sperm contacts the cells surrounding the oocyte (called the *corona radiata*) (Figure 4.11). A structure at the head of the sperm, the *acrosome,* then bursts, releasing enzymes that break down barriers in the cell layers that protect the oocyte, and permitting contact between the outer membranes of the sperm and oocyte. This interaction is followed by a remarkable sequence: chemical and physical signals move across the oocyte surface, prohibiting interaction with other sperm; the head of the sperm (but not its tail) enters the oocyte; the nuclear membranes of the sperm and egg dissolve; the two sets of chromosomes (called **pronuclei**) approach each other; and, after a round of replication of the chromosomes, the nuclear membrane forms and the zygote is created.

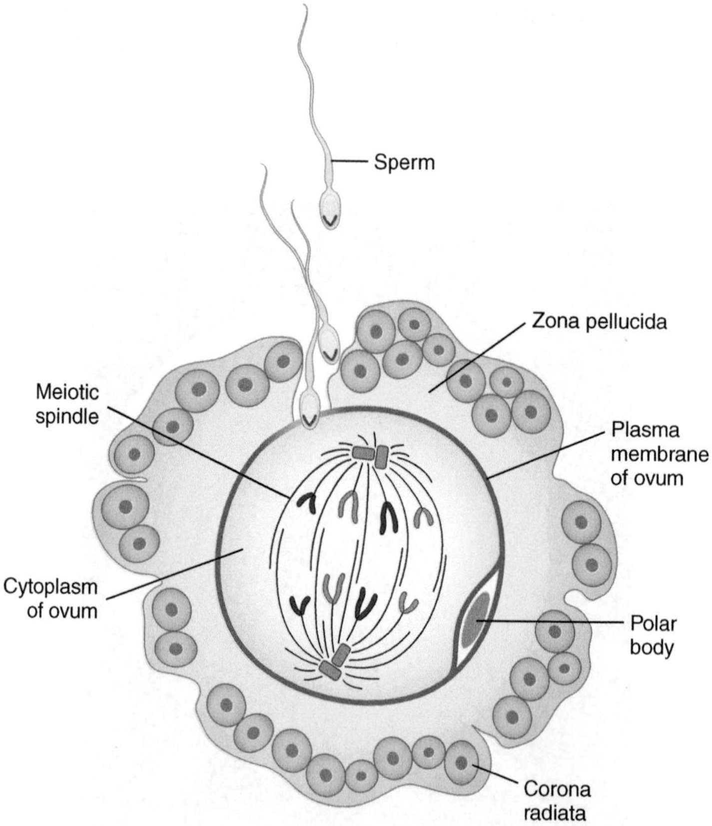

FIGURE 4.11

Fertilization of an egg by a sperm. As a single sperm penetrates the corona radiata and zona pellucida, a series of chemical reactions facilitate penetration of the plasma membrane and prevent other sperm from gaining entry. After fertilization has occurred, meiosis II in the egg is completed, and the male and female pronuclei form.

Gender Determination

The zygote's sex is defined by its sex chromosome complement. All oocytes have an X chromosome (because female diploid cells are XX). Sperm may carry either an X chromosome or a Y, in keeping with the XY genotype of the diploid male. Thus, fertilization of an X-carrying oocyte by an X-carrying sperm will produce an XX zygote, that is, a female; fertilization by a Y-carrying sperm will produce an XY zygote, a male.

This common knowledge has an uncommon explanation. Although the X chromosome contains 30 times as many genes as the Y does, it is a gene on the Y chromosome that determines maleness or femaleness. Up to about 6 weeks of embryonic life, XX and XY embryos have the same collection of undifferentiated cells in a region called the **gonadal** or **germinal ridge** (Figure 4.12). In the presence of a Y chromosome containing the **SRY gene** (for "sex regulating region of the Y"), this gene's product, called **TDF** (testis determining factor), directs the primitive gonad to become a testis. In an XX embryo that lacks this gene, an ovary forms instead. Secondary sex characteristics follow from this primary sex determination. (So, rather than lamenting the tiny size of the Y and the paucity of its genes, we should note the Y's huge contribution to human existence.)

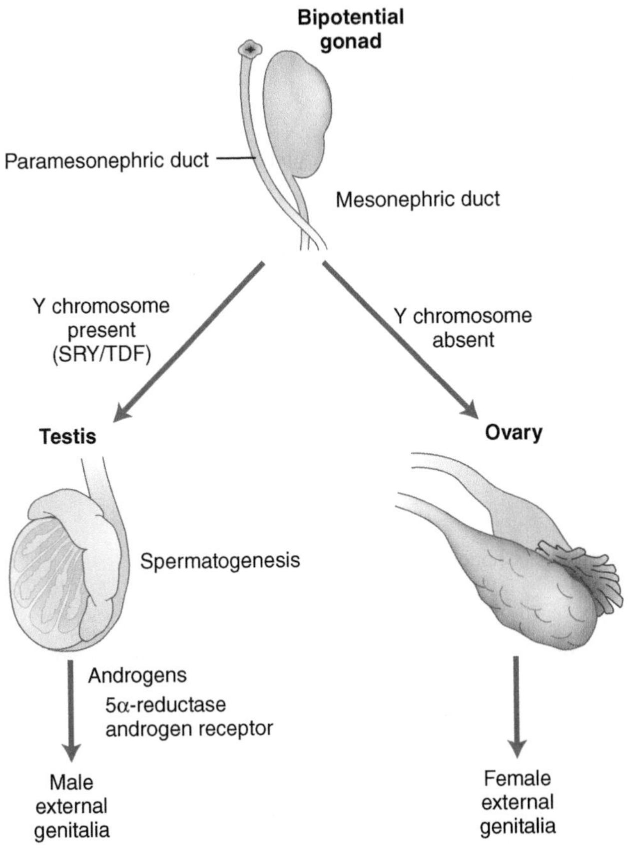

FIGURE 4.12
Sex differentiation in male and female gonads. At 6 weeks' gestation, both genders have the same undifferentiated germinal ridge. In the presence of SRY (sex regulating region of the Y), encoded on the Y chromosome, the primitive gonad develops into the testis. In the absence of SRY, an ovary forms instead. Secondary sex characteristics follow this primary sexual differentiation.

Twinning

Twins and other multiples (triplets, quadruplets, and so on) are not uncommon products of conception. **Identical (monozygotic** or **MZ) twins** occur in about 3 per 1,000 births. They form from a single fertilized ovum that splits, and, therefore, have identical genomes. They are called monozygotic because they are derived from the same fertilized egg. MZ twins may be further subdivided at the time the early embryo splits into two parts (Figure 4.13). If splitting occurs before day 5, the twins will each have their own **chorion** (the outermost sac that encloses the embryo) and their own **amnion** (the innermost sac that contains the embryo). This is the situation in about one-third of MZ twins. In nearly two-thirds of cases the early embryo splits between day 5 and day 9, resulting in two embryos with separate amnions but with a single chorion. In approximately 1% of twins, the early embryo splits soon thereafter, yielding twins with a single amnion and chorion. In order that a complete individual will develop no matter what kind of MZ twinning occurs, it must be true that these early embryonic cells have the same developmental potential.

Fraternal (dizygous or DZ) twins occur in a larger fraction of live births (6–20 per 1,000). DZ twins result when two sperm fertilize two eggs. This situation may arise when one egg is ovulated from each ovary at the same time or when two eggs are ovulated from one ovary. Genetically, these DZ twins are no more similar than siblings born at different times.

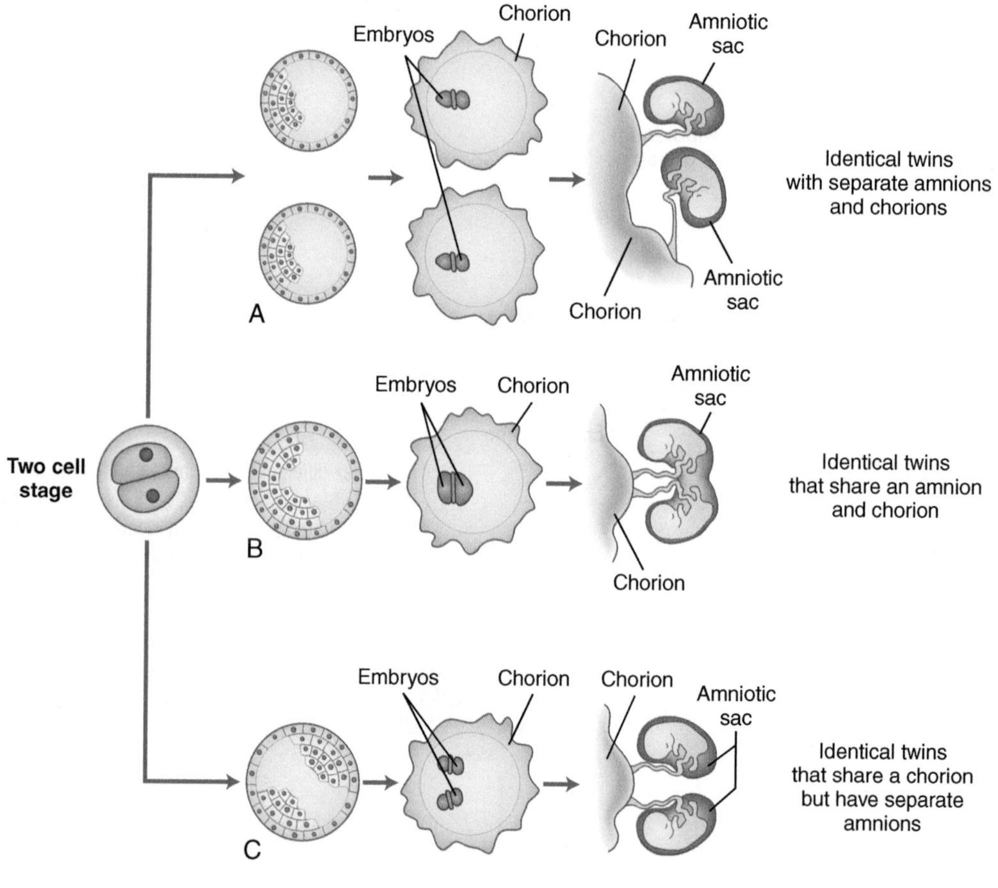

FIGURE 4.13
Types of identical (MZ) twins. If the blastocyst splits before day 5, shown in A, each twin has its own chorion and amnion. This occurs in about one-third of cases. If splitting occurs after day 5 but before day 9 (shown in C) an event responsible for nearly two-thirds of cases, the twins share a chorion but have their own amnion. In the small remainder of cases, shown in B, the twins share a common amnion and chorion.

Cleavage and Implantation (Figure 4.14)

About 24 hours after fertilization, which takes place in the Fallopian tube, the zygote undergoes a series of mitotic cleavages, doubling the number of cells from two to four to eight. Each of these cells is called a **blastomere**. By day 4, the cluster of 16 cells is called the

morula. By about day 5, the **blastocyst** is forming. The blastocyst is composed of a fluid-filled center and a lining of cells, called the **inner cell mass**. The inner cell mass will become the **embryo**. By day 7, usually, the blastocyst penetrates the uterine lining (the **endometrium**) and becomes the embryo. Soon thereafter, the embryo becomes fully implanted. During implantation, the "pregnancy hormone," **human chorionic gonadotropin** (HCG), is secreted by the blastocyst's outermost cells (called the **trophoblast**). HCG stimulates progesterone synthesis by the mother's ovary, which prevents menstruation for the duration of the pregnancy.

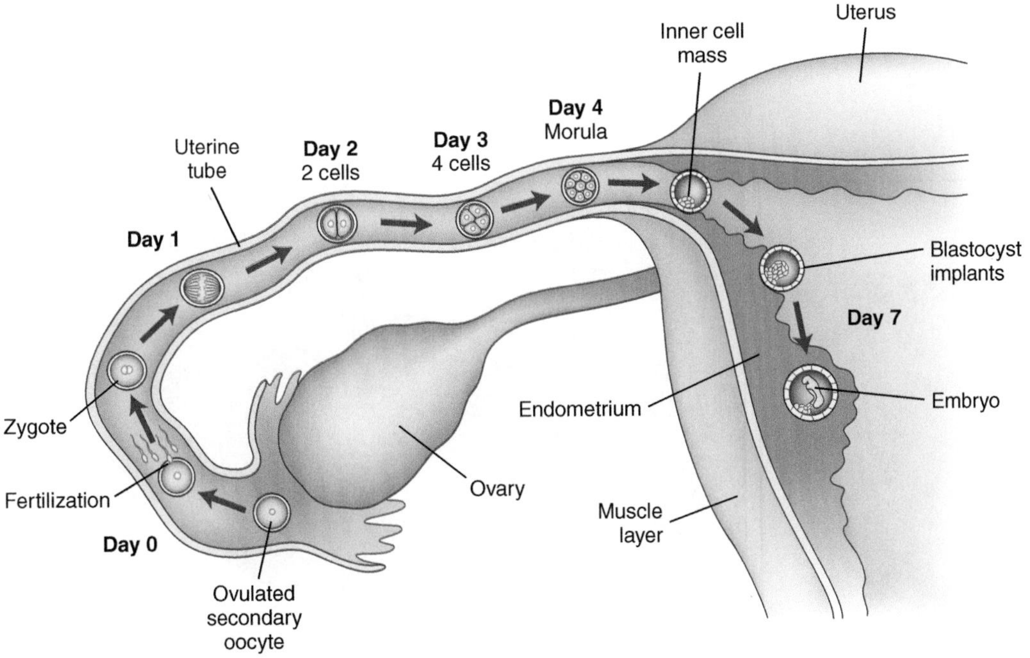

FIGURE 4.14
From ovulation to implantation. Following ovulation, fertilization occurs, in the Fallopian tube, usually within 24 hours, yielding the zygote. A series of events, highlighted by cell cleavages and movement toward the uterus, then ensues for 5–6 days. On or about day 7, the blastocyst implants in the uterine wall and the embryonic period of development begins.

Figure 4.15 depicts this early human development using photomicrographs. Structures of particular note are the **polar bodies** (4.15A), the blastocyst (4.15 G), and the "hatching" embryo (4.15H). Hatching is a process whereby the early embryo loses its protective zona pellucida, which is necessary for implantation in the endometrium.

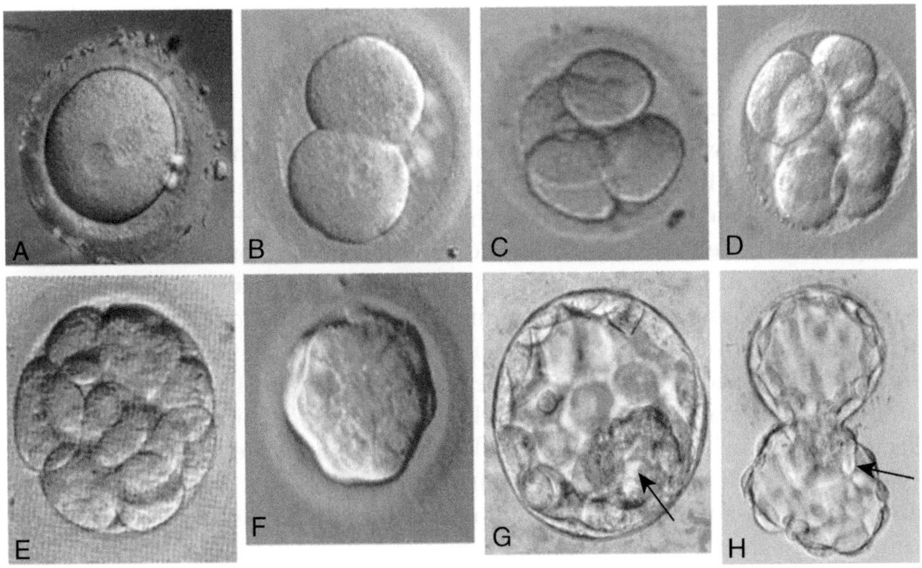

FIGURE 4.15
Human development from fertilization to blastocyst formation. (A) The fertilized egg on day zero; note two pronuclei at the center and two polar bodies at the periphery. (B) A two-cell embryo, 1 day after fertilization. (C) A four-cell embryo on day 2. (D) An eight-cell embryo on day 3. (E) A 16-cell embryo. (F) The 16-cell embryo undergoes compaction to become the morula. (G) The blastocyst, with its inner cell mass (arrow), begins to form on day 5. (H) The embryo (arrow) hatches from the zona pellucida.

The Embryonic Period

Weeks 2 through 8 of pregnancy are called the embryonic period, which is extraordinary in many ways. At the cellular level, genes control the division (**proliferation**), acquisition of specific function (**differentiation**), movement (**migration**), and programmed death (**apoptosis**) of cells. At the organ level, genetic control of the pattern and rate of growth and of the appearance of each body organ (**morphogenesis**) takes place. At the organismal level, the genetic blueprint prescribes the size, shape, and location of all of the body's parts. This exquisite process is a compelling area of study on one hand and a constant source of wonder on the other hand, particularly taking into account the large number of genetic and environmental insults that may interfere with the myriad events taking place. Some of the ways this process can be disrupted will be detailed in Chapter 14.

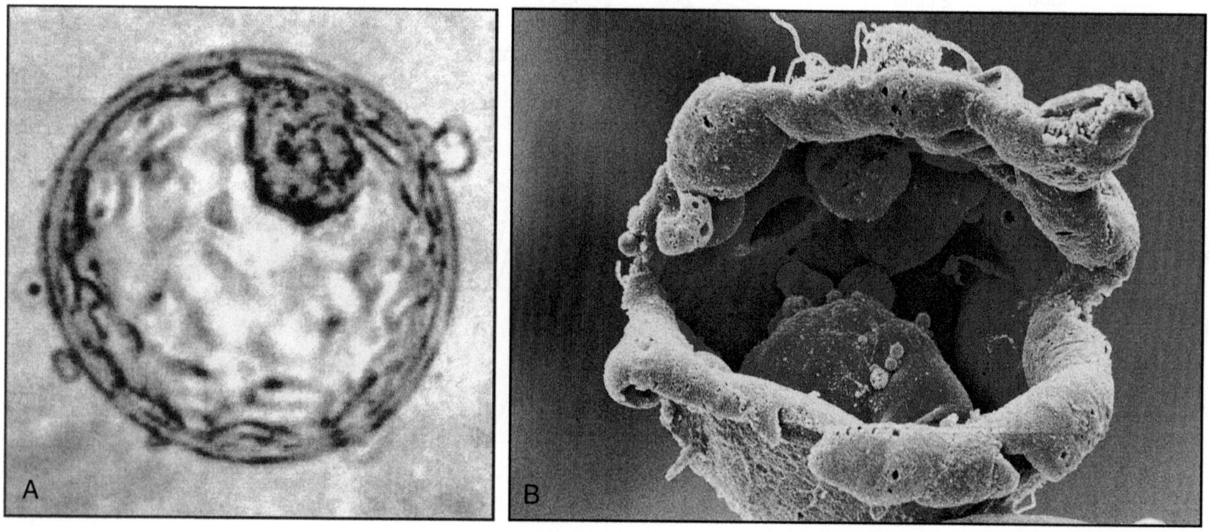

FIGURE 4.16
(A) A 7-day old human blastocyst. The berry-like collection of cells protruding into the center of the fluid-filled structure is the inner cell mass (ICM). Embryonic stem cells (ESCs) make up the ICM. The blastocyst is delimited from its external surroundings by a plasma membrane. (B) Scanning electron micrograph of a human blastocyst. The plasma membrane is shown in beige color, the cellular interior, in red. The largest portion of the inner cell mass (ICM) is discernible as the half moon-shaped structure at 6 o'clock.

STEM CELLS (FIGURE 4.16)

The inner cell mass of the blastocyst is composed of two kinds of cells: those that will become the mature organism (the **epiblast**), and those that will develop into the **placenta**, the **chorion**, and the **amniotic membranes**. The cells that will develop into the completed embryo are called **embryonic stem cells (ESCs)**. These undifferentiated cells have two critical properties (Figure 4.17): they can renew themselves, and they can differentiate into cell types characteristic of each tissue and organ. ESCs can undergo mitotic division to yield two identical daughter cells. Alternatively, they may divide into two unlike daughter cells: another ESC and a **progenitor cell**, defined as a cell that has begun to differentiate. Progenitor cells may—depending on the signals they receive—develop variously into brain, heart, muscle, and all other cells that define each organ particularly (Figure 4.18). ESCs are termed **pluripotent** because they can develop into all of the body's differentiated cell types. Progenitor cells are termed **multipotent** because they can develop into many different kinds of cells, but only of one organ or organ system.

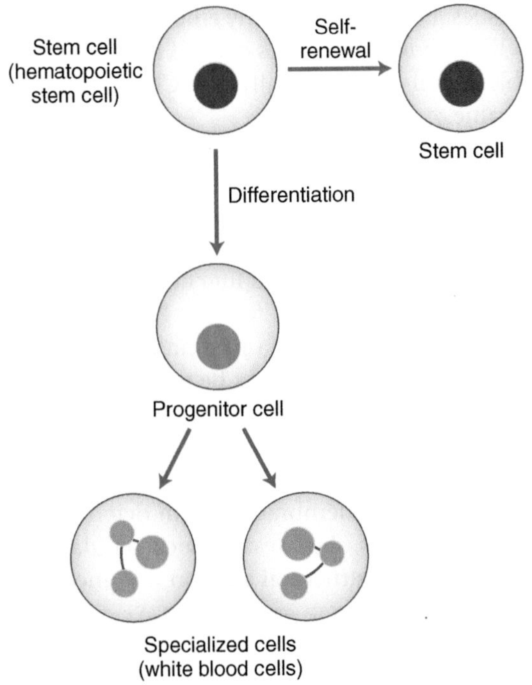

FIGURE 4.17
Features of stem cells. Pictured here is a diagram of a stem cell found in the bone marrow. It can give rise to all of the body's blood cells (red cells, white cells, and platelets). It is distinguished from other cells by two properties: self-renewal and differentiation. Self-renewal is the property of dividing and giving rise to undifferentiated cells like itself. Differentiation begins with formation of progenitor cells and ends with formation of highly specialized cells unique to the origin of the stem cell.

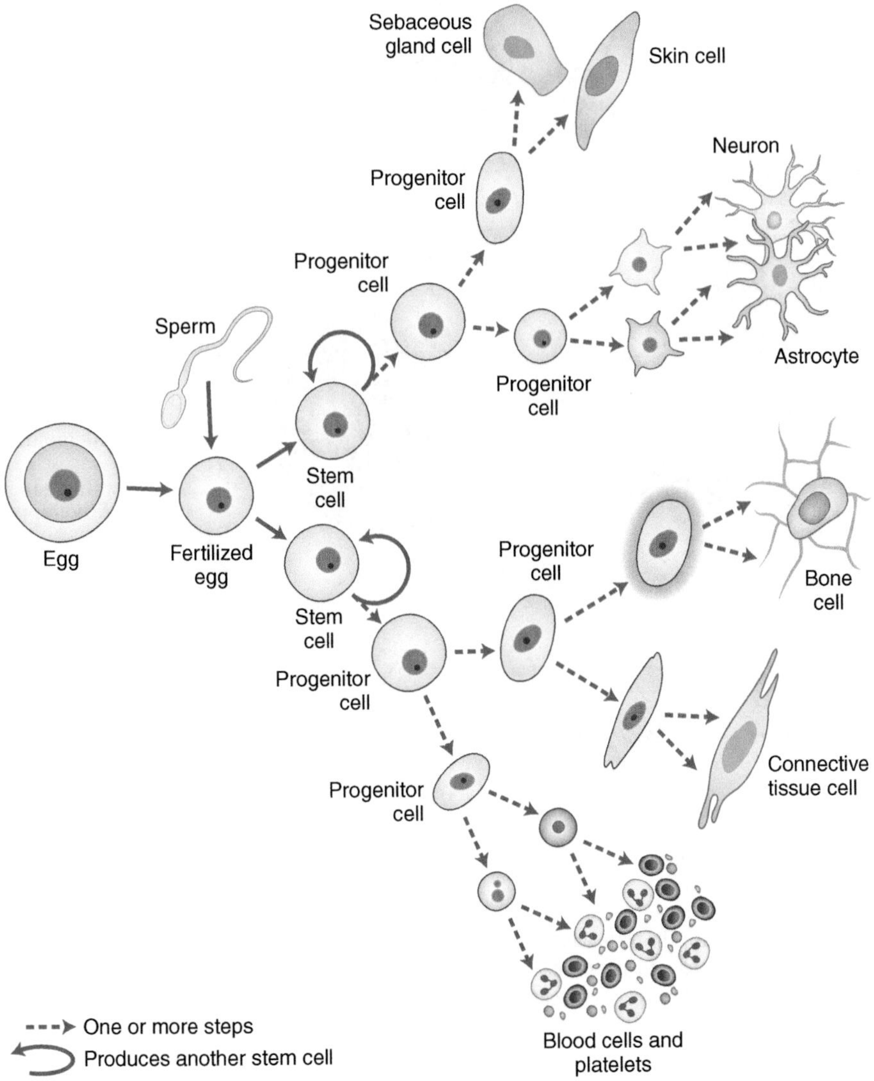

FIGURE 4.18
Pathways from embryonic stem cells (ESCs) to differentiated cells. ESCs are formed through cleavages of the fertilized egg. They are capable of self-renewal and formation of progenitor cells. Some progenitor cells are fated to become differentiated bone cells; others, connective tissue or bone cells; others, nerve cells; and so on. This is why ESCs are termed pluripotent, and adult stem cells (ASCs), obtained from differentiated tissues, multipotent.

Multipotent stem cells are found in most adult tissues, not just in the early embryo. These cells are called **adult stem cells (ASCs)** to distinguish them from ESCs. ASCs derived from the bone marrow have shown themselves to be remarkably useful therapeutically (see Chapter 17), raising hopes that ESCs, too, may have clinical value.

DIFFERENTIATED CELLS

The human body contains about 250 different kinds of differentiated cells. Each organ or tissue has its unique collection of cell types that carry out the specialized function of that body part. Nonetheless, differentiated cells share many common anatomic features Figure 4.19. Each contains a nucleus in which the chromosomes reside. Each has a **plasma membrane**, enclosing it, and separating it from other cells or the extracellular spaces between cells. Each has a **cytoplasm** containing a variety of substructures, known as **organelles**, capable of carrying out a plethora of functions. Among these cytoplasmic organelles, some deserve special mention: **ribosomes**, the site of protein synthesis; the **endoplasmic reticulum**, a membrane network in which proteins are made; the **Golgi apparatus**, a membrane system determining the fate and direction of the transport of proteins; **lysosomes**, the sites at which cellular substituents are degraded; the **cytoskeleton**, which provides structural support and motility;

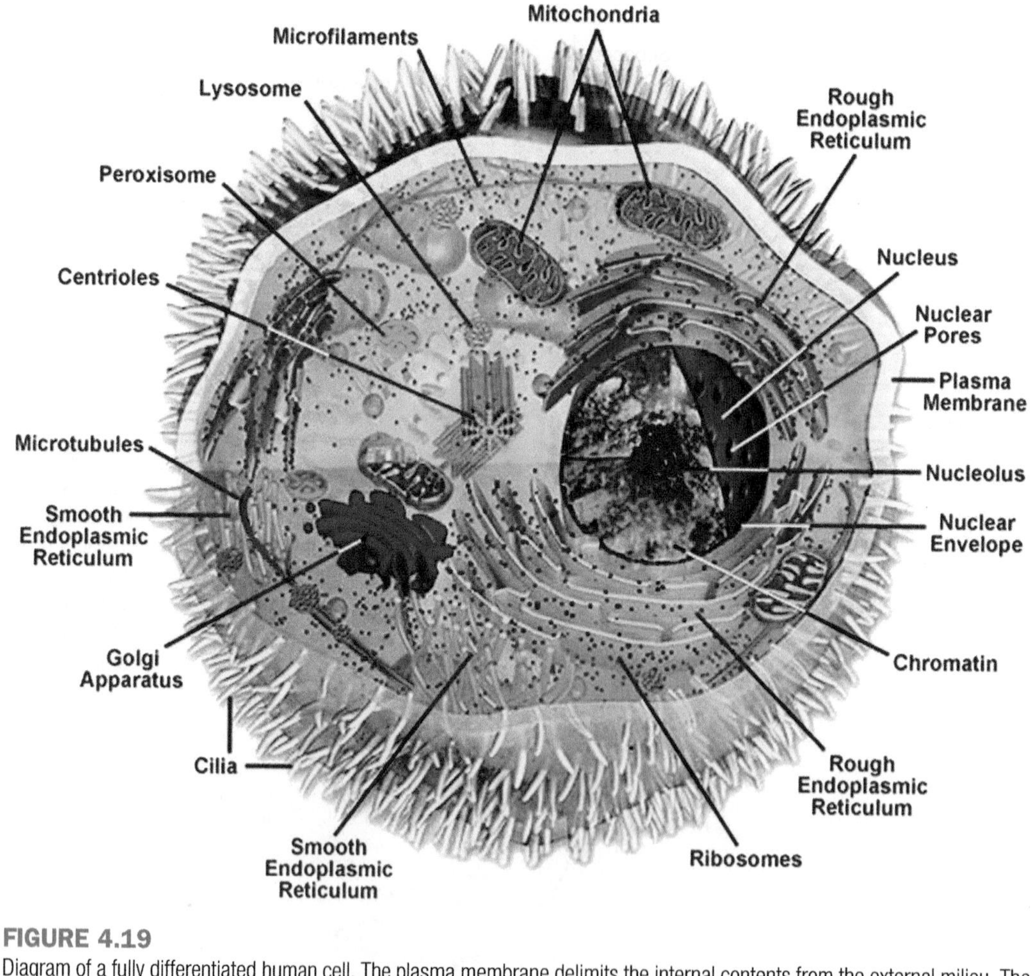

FIGURE 4.19
Diagram of a fully differentiated human cell. The plasma membrane delimits the internal contents from the external milieu. The nucleus, enclosed within a nuclear envelope, contains the cell's chromosomes and nucleolus where ribosomes are assembled before export to the nucleus. As shown, the cytoplasm is composed of a variety of organelles, each with its own structure and function. See text for additional details.

and **mitochondria**, the energy generating packets, which have their own complement of genes and enzymes.

PROGRAMMED CELL DEATH

Just as a sculpture is made through addition and preservation of desired features and removal of undesirable ones, so does human development. Programmed cell death (or **apoptosis**) occurs contemporaneously with cellular proliferation and **organogenesis** (literally meaning "the birth of organs"). It occurs wherever cells have been damaged and where tissues need to be remodeled. Thus apoptosis is critical to such events as the separation of digits in the hands and feet, formation of the anus and the heart's ventricular septum, and removal of cells damaged by environmental insults. During apoptosis, likened to cell suicide, a complex cellular program is triggered, leading to the degrading of the genetic material, condensing of the nucleus, and **blebbing** (irregular bulging) of the plasma membrane. These changes are followed by destruction of these "marked" cells by tissue **phagocytes** (white blood cells that ingest them). When apoptosis is impaired, a number of abnormal events may ensue, such as **syndactyly** (fusion of adjoining digits) and congenital heart defects. In addition, defective apoptosis is a critical event in the genesis of cancer.

ORGANOGENESIS

Once implanted, the embryo undergoes **gastrulation**, the arranging of the cells into a recognizable structure consisting of three cellular compartments, termed germ layers. These layers are called the **ectoderm**, **mesoderm**, and **endoderm** (Figure 4.20). The outer or **ectodermal** layer gives rise to the central and peripheral nervous system, to the skin, and to certain glands. The inner, **endodermal**, layer gives rise to the layer of cells lining the respiratory and intestinal tracts, and to portions of the liver and pancreas. The middle, **mesodermal**, layer will become such structures as the reproductive organs, kidneys, cardiac and skeletal muscle, and connective tissue.

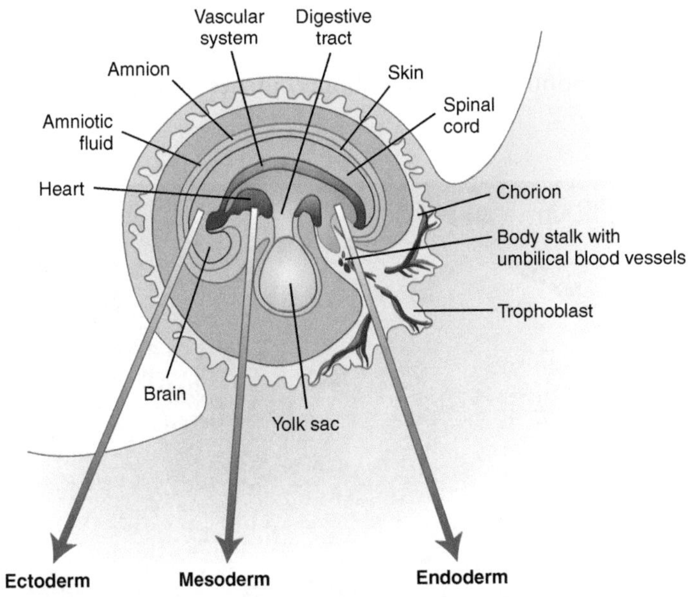

FIGURE 4.20
A human gastrula (a 14-day embryo) drawn to emphasize three embryonic cell layers: ectoderm,mesoderm, and endoderm. The tissue cells that each of these layers is fated to become are listed below the appropriate names and are discussed further in the text. The supporting structures are also identified: yolk sac, amnion, chorion, and trophoblast.

To summarize, during the first 8 weeks, the major developmental events are the initiation of the nervous system, the establishment of the basic body plan, and the appearance of each of the internal organs. By this time the embryo has all of its organs, including the sexual ones, and all the types of cells necessary for full development (Figure 4.21).

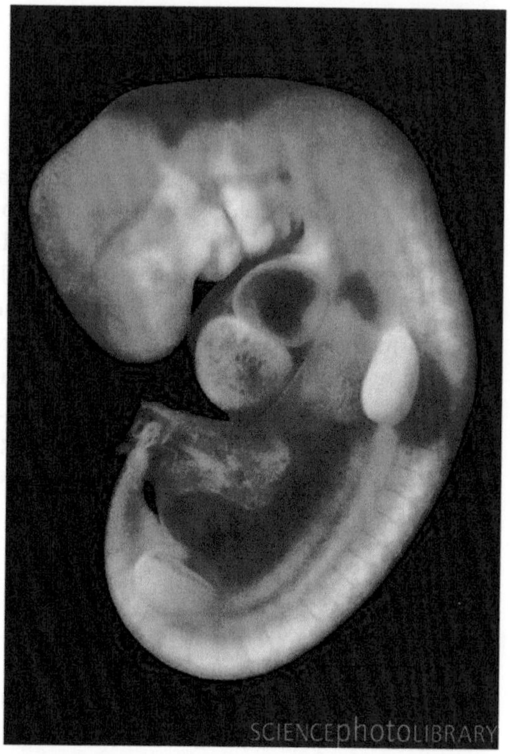

FIGURE 4.21
First-trimester human embryo. The 2-month-old (56 days) embryo has all the tissues and organ systems it will have at birth. At this stage it would be difficult to distinguish the human embryo from that of other simpler organisms.

The Fetal Period

From weeks 8 through 40, the developing human—now called a **fetus**—undergoes growth and maturation. Its appearance and capacity change in small and large ways: facial features assume those more like a newborn; cartilage becomes bone; the fetus moves, kicks, and puts its fingers in its mouth; it urinates and defecates; it takes amniotic fluid into its lungs and expels it.

During the second trimester, change continues and accelerates: hair and nails appear; vocal cords are present; the typical knees-to-chest position is assumed; blood vessels are apparent under the skin. By the end of the sixth month, the fetus is about 9 inches long and weighs about 1 pound (Figure 4.22).

FIGURE 4.22
A human fetus at age 24 weeks. By this age he is swallowing and breathing amniotic fluid and is shown here in direct contact with the amniotic membrane.

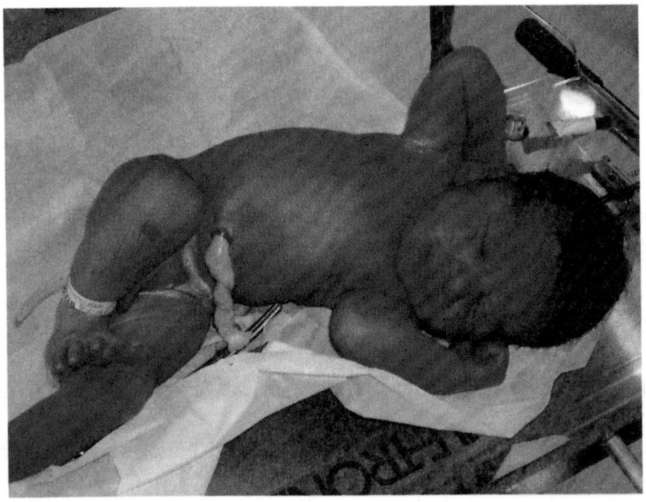

FIGURE 4.23
A full-term, healthy newborn boy, who has completed the remarkable journey from gametes, through embryonic and fetal development, to his own progressively independent life.

The third trimester features formation of neural networks in the brain, maturation of intestinal and respiratory systems, and appearance of subcutaneous fat. The stage is now set for leaving the protected environment of the womb for the progressively less protected environment beyond. By then, marvelously, the single cell zygote has become a 40-billion-cell newborn, made up of more than 250 cell types (Figure 4.23).

REVIEW QUESTIONS AND EXERCISES

1. Choose the phrase in the right column that best matches the term in the left column.

a. sex chromosomes	1. haploid germ cells
b. cytokinesis	2. pairing of homologous chromosomes
c. synapsis	3. division of the cytoplasm
d. anaphase	4. process that yields two identical cells from one diploid cell
e. interphase	5. testicular cells that undergo meiosis
f. chromatid	6. an array of chromosomes in a cell
g. mitosis	7. one of two identical halves of a replicated chromosome
h. autosomes	8. the parts of the cell cycle other than M
i. karyotype	9. chromosomes that do not differ between the sexes
j. centromere	10. process that yields four haploid cells from one diploid cell
k. gametes	11. chromosomes that differ between the sexes
l. polar body	12. connection between sister chromatids
m. meiosis	13. the time during mitosis when sister chromatids separate
n. spermatocyte	14. cell produced during meiosis that fails to become a gamete

2. Describe briefly the two meiotic events that result in each of us being genetically unique.

3. Distinguish between each of the following pairs of terms:
   a. Genome and chromosome
   b. Gene and allele
   c. Prokaryote and eukaryote
   d. Aneuploidy and polyploidy.

4. Identify three major differences between male and female meiosis.

5. In our species when a cell enters meiosis each chromosome is composed of two chromatids.
   a. How many chromosomes are there per cell at the end of meiosis I? How many chromatids?

b. How many chromosomes are there per cell at the end meiosis II? How many chromatids?
c. When is the diploid number of chromosomes restored?

6. At the cellular level, what is the fundamental difference between identical and non-identical twins?

7. What is the fundamental purpose of meiosis in terms of reproduction of a eukaryotic species such as ours?

8. What is the probability that all of your chromosomes have come to you from your father's father and your mother's mother? Would you be male or female? Disregard crossing over.

9. Some people in our society believe that all forms of reproductive genetic technologies are morally indefensible and should be banned. You have been asked to address such a group.
   a. Describe briefly five such technologies.
   b. Make the case that all of these technologies do not carry the same "ethical burden" as defined by your listeners.
   c. Discuss briefly how these technologies may be of value to couples desiring healthy children of their own.

CHAPTER 5

Transmission of Genes

CHAPTER OUTLINE

CORE CONCEPT

Traits are transmitted from one generation to the next by **genes**. The Austrian monk, Gregor Mendel, discovered the basic rules of this transmission—that is, **gene segregation** and **assortment**—in the mid-nineteenth century by studying pea plants. Trait transmission in human pedigrees follows the same rules. Traits attributable to single genes (referred to as "**Mendelian traits**") are inherited as **dominants** or **recessives**. They may be **autosomal** or **sex-linked**. Traits attributable to multiple genes and those demonstrating imprinting or maternal inheritance display more complex patterns.

Thus far, we have defined genetics as the science of heredity and have discussed the role of chromosomes and genes in reproduction, development, and cell structure. Now we must turn to another extraordinary question: How are traits transmitted from one generation to the next? The simple answer is, by genes. To understand the profundity of this answer, we must delve deeper into the concepts laid out in this chapter.

Human Genes and Genomes. DOI: 10.1016/B978-0-12-385212-0.00005-6

FAMILIAL SIMILARITIES

As humans, we are eternally fascinated with ourselves: Where did we come from and where are we going? What is the meaning of the word "family?" The very word "gene" comes from the Greek word that means, "giving birth to." No matter how little people know about genetics, they know they tend to look and act more like their parents, their grandparents, and their siblings than like strangers. They notice familial traits such as facial features, hair distribution, intelligence, and mannerisms. This starts in the nursery: people will often say, "She looks just like her father." Harmless traits such as freckles, hair color, cowlicks, baldness, broad thumbs, and crooked pinky fingers tend to run in families.

Certain disorders run in families, too. Before we had any knowledge of genetics, humans noticed that disorders such as hemophilia or gout or cancer or diabetes appeared more often in closely related people. Genetics offers the opportunity to confront and understand similarities and differences among family members as no other paradigm can. For instance, why does a child resemble a grandparent but not either parent? Or, why do brothers with the same parents look remarkably dissimilar? This kind of question is part of the field's attraction for people, its fascination. Think of it: a mutation in just one gene out of our 21,000 genes can destroy the ability to speak, and a defect in another single gene determines whether we will contract Huntington disease.

MENDELIAN INHERITANCE

It is rare for a whole body of science to carry the name of a single person. Yet this is the case with genetic transmission. It began with Gregor Mendel and his pioneering work in plants. He followed each principle of the scientific canon: observation, formulation of a hypothesis, experimentation, interpretation, and publication. He is widely acclaimed as the first experimental geneticist and as a worthy contemporary of Charles Darwin and Alfred Russel Wallace.

Gregor Mendel's Life (1822–1884)

Born to a farmer whose father was a gardener, Mendel learned early how to care for plants and trees. He left home at age 10 to attend a school for able students. Abbot Cyril Napp of the Augustinian monastery at Brunn, Austria (now Brno, in the Czech Republic) recognized Mendel's scholarly promise and acted as his mentor as Mendel became priest and monk. Abbot Napp sent Mendel to the University of Vienna, where he followed a diverse curriculum: mathematics, physics, chemistry, paleontology, botany, and plant physiology.

After his return to the monastery, Mendel became aware of Napp's interest in animal breeding. In the 1840s, Napp had asked three questions about sheep breeding: What is inherited? How is it inherited? What is the role of chance in heredity? Between 1857 and 1863, Mendel worked in a monastery's garden (Figure 5.1). There, he sought answers to these questions using the garden pea (*Pisum sativum*) as his experimental organism. In the supportive and intellectually rich environment of the monastery, Mendel conducted experiments on more than 29,000 pea plants. By 1865, when he presented his work entitled "Experiments in Plant Hybridization" to the Natural History Society of Brunn, he was in possession of a German translation of Darwin's *On the Origin of Species*, and mentioned the term "evolution of organic forms" in his discourse. Mendel's work was published in 1866 in the Proceedings of the Natural History Society of Brunn.

Two years later he was named Abbot of the monastery, and apparently had little time for scientific experimentation thereafter. Mendel's seminal work was rejected or unnoticed until the dawn of the 20th century, when his findings were confirmed and their significance noted.

FIGURE 5.1
Mendel's garden at the monastery in Brno (then Brunn).

Mendel's Experiments

ORGANISM AND DESIGN

Two misconceptions prevailed when Mendel began his work: that one parent (the male) transmits most inherited characteristics to offspring; and that parental traits become mixed (or blended) in the offspring. Mendel demolished both ideas. He did so by choosing a suitable organism and by carrying out a large enough number of carefully controlled experiments so that results were quantifiable and verifiable. Mendel chose to work with the garden pea, *Pisum sativum*, because it had several favorable characteristics: a short growing season, enabling data collection and repetition; plant anatomy that clearly separated eggs (female) from pollen (male), and that lent itself to artificial fertilization and to essentially no "false impregnation" (Figure 5.2); and a series of "characters" (traits) that were of two discrete varieties (such as yellow or green peas; tall or short plants; round or wrinkled peas).

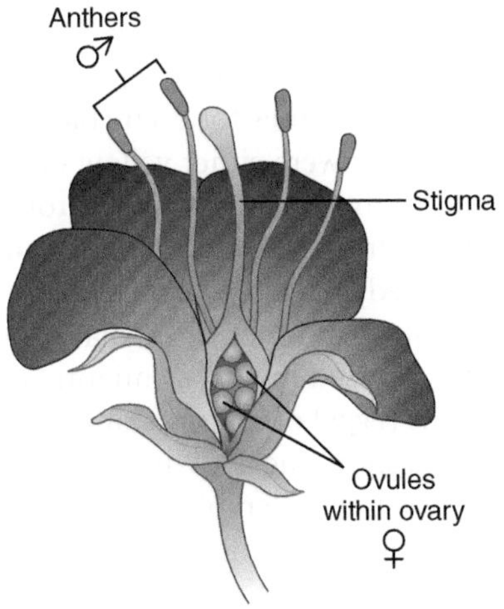

FIGURE 5.2
Anatomy of the *Pisum sativum* flower. Pollen is produced by the (male) anthers. It is deposited on the stigma, which is connected to the (female) ovary. The ovary, which becomes the pea pod, contains ovules, the immature seeds that become peas.

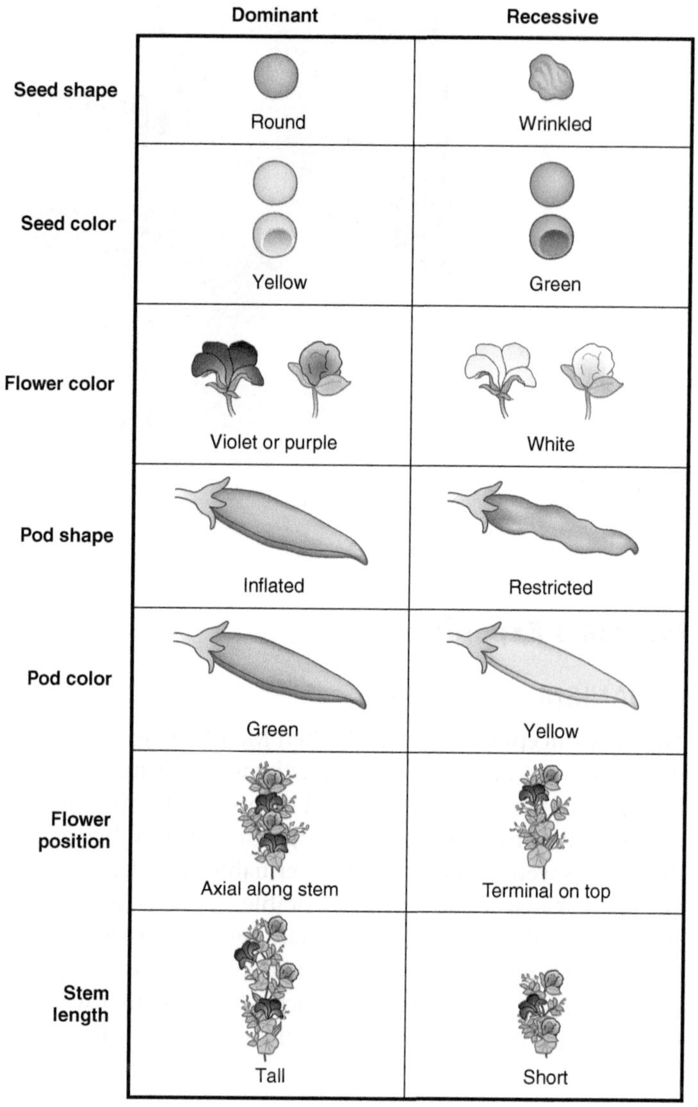

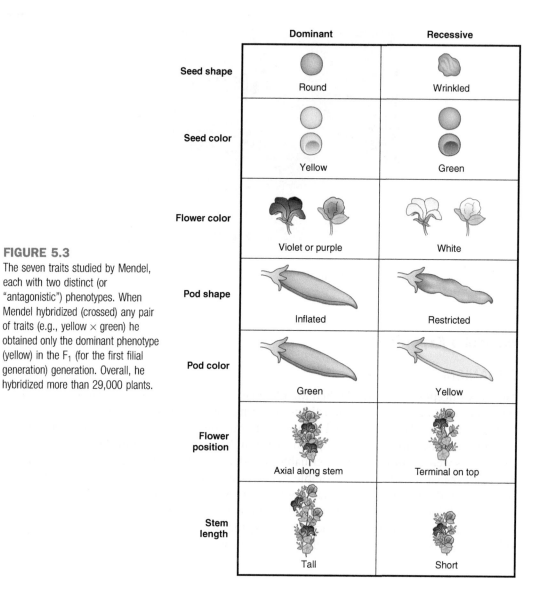

FIGURE 5.3
The seven traits studied by Mendel, each with two distinct (or "antagonistic") phenotypes. When Mendel hybridized (crossed) any pair of traits (e.g., yellow × green) he obtained only the dominant phenotype (yellow) in the F_1 (for the first filial generation) generation. Overall, he hybridized more than 29,000 plants.

Mendel chose seven "differentiating characters" for study (Figure 5.3). For each character, two easily discernible forms (he called them "antagonistic") were available. Over as many as eight generations, he developed pure-breeding lines for each character. For example, his **pure-breeding plants** referable to seed color were either yellow or green; for seed shape, either round or wrinkled; for flower color, either purple or white. From those plants that bred true, he took plants of one differentiating character (for example, round) and **cross-fertilized** them with their opposite number (wrinkled). He then collected and analyzed the offspring (**hybrids**) of these dissimilar parents. It is noteworthy that he also carried out **reciprocal crosses**, that is, ones in which he reversed the differentiating traits of the male and female plants. Thus, he could fertilize seeds (eggs) of a short plant with pollen of a tall one, and seeds (eggs) of a tall plant with pollen of a short one. He wrote that it was "immaterial to the form of the hybrid" which parental type was the pollen plant and which the seed. These results demolished the idea that one sex contributes more to inheritance than the other.

THE LAW OF GENE SEGREGATION

Mendel carried out a large series of experiments, called **monohybrid crosses**, over several years of the sort described in Figure 5.4. He did this with each pair of phenotypes shown in

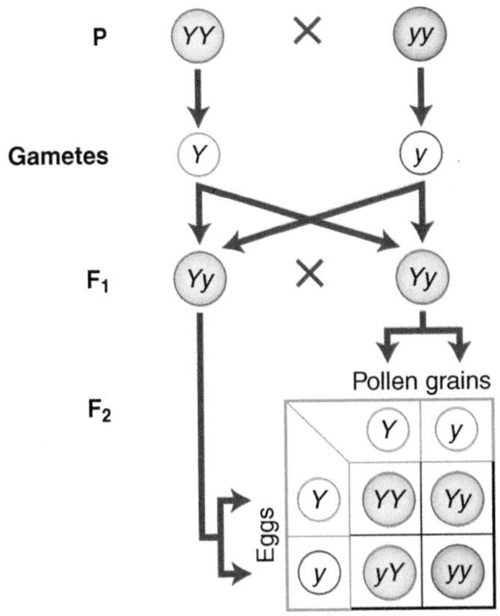

FIGURE 5.4
Visual summary of a typical monohybrid cross, conducted by Mendel. In the parental cross (P) he used true-breeding plants producing yellow or green peas. Yellow plants produced only *Y* gametes; green plants, only *y* gametes. In the cross, F_1, all peas had *Yy* or its *yY* equivalent genotype and all were yellow phenotypically. He then self-crossed the *Yy* plants and obtained, in the F_2, yellow and green peas at a ratio of three yellow to one green (as accounted for in the Punnett square at the bottom). This kind of experiment led to his principle of gene segregation and to his use, for the first time, of the words "dominant" and "recessive."

Figure 5.3, but we shall use seed (pea) color as an example. Crossing, in the **parental generation (P_1)**, true-breeding plants yielding yellow peas with true-breeding plants yielding green ones, Mendel observed hybrid progeny in which all the peas were yellow (the **F_1** generation). Plants of this F_1 generation were allowed to self-fertilize, and the peas of the next generation (F_2) were counted and scored. Of more than 8,000 peas collected, 6,022 were yellow and 2,001 were green—an almost perfect ratio of 3 yellow to 1 green. Using each of the other six characters, Mendel obtained the same result—self-fertilization of the single character observed in the F_1 yielded both parental characters in the F_2 at a ratio of 3 : 1.

These findings were incompatible with the idea of blending. Each parental character was recovered intact in the F_2, rather than being "lost" in the F_1. Mendel reasoned that the yellow peas in the P_1 were not identical to the yellow peas in the F_1 because the P_1 yellows were true breeding and the F_1 yellows were not. He proposed that the trait which appeared in the F_1 was dominant and that the trait which disappeared in the F_1 but reappeared in the F_2 was recessive. But what accounted for the reproducible 3 : 1 ratio?

Mendel proposed—in an astonishingly prescient way—that each plant carried two copies of a unit of inheritance for each trait, one inherited from the male, one from the female. He proposed further that each unit comes in alternative forms that give rise to the differentiating characteristics he studied (yellow–green, round–wrinkled, etc.). Today, we call his "**units**" "**genes**" and his "**alternative forms**" "**alleles**." He went on to propose that the two alleles found in cells of a mature plant segregate (separate) during germ cell formation and reunite, one from each parent, at fertilization. Mendel set out to find laws of inheritance. This was his first: the law of **gene segregation**.

The law explains the 3 : 1 ratio in the F_2 as follows (Figure 5.4), using the visually accessible **Punnett square** (a diagram that is used to predict an outcome of a particular cross or breeding experiment). The true-breeding yellow pea plants (P_1) have two copies of the dominant allele, denoted *Y*; the plants yielding only green peas have two copies of the recessive allele, denoted *y*. (Capital letters generally depict the dominant allele, small letters the recessive one.) Gametes of these P_1 plants (*YY* and *yy*, referred to as "**homozygotes**") are either *Y* or *y*. At fertilization, all zygotes are *Yy* (**heterozygotes**). Because *Y* is dominant, all plants are yellow. When these plants are self-fertilized, the male and female each produce gametes that are either *Y* or *y*. In the F_2, then, 1/4 of the progeny are *YY*, 1/4 are *Yy*, 1/4 are *yY*, and 1/4 are *yy*. Given that *Y* is dominant, and that *Yy* and *yY* are equivalent, the ratio between yellow and green peas is 3 : 1.

THE LAW OF INDEPENDENT ASSORTMENT

Mendel's work with single characteristics demonstrated that alleles segregate during reproduction. He went on to study the simultaneous inheritance of two seemingly unrelated characteristics. How would two pairs of alleles segregate in a dihybrid cross? he asked. To answer this question, he employed, for instance, peas differing in color and shape (Figure 5.5). First, he obtained true-breeding yellow round peas (*YYRR*) and true-breeding green wrinkled ones (*yyrr*). He then crossed these parental (P_1) varieties and found that all the F_1 were yellow and round (as expected for doubly heterozygous plants—*YyRr*—whose phenotype expresses the dominant alleles). When he self-crossed (fertilized the plant with its own pollen) the F_1 plants, four phenotypes were evident in the F_2: yellow round, yellow wrinkled, green round, and green wrinkled. That is, there were two parental types and two "new ones" produced in the dihybrid cross (recombinants). In one experiment, typical of many, he found 315 yellow round peas, 101 yellow wrinkled, 108 green round, and 32 green wrinkled.

From these data he proposed that segregation of one pair of alleles had no effect on the segregation of another pair—that gene pairs assort independently. Said another way, the presence of an allele for one gene has no effect on the presence of an allele for a second gene. From the Punnett square in Figure 5.5, the proof for this thesis is at hand. In the gametes of the F_1 self -cross (female and male), four genotypes are expected (*YR, Yr, yR, yr*)—identical in female and male. Sixteen genotypes, then, are predicted in the F_2. Of these, 9/16 are yellow round, 3/16 yellow wrinkled, 3/16 green round, and 1/16 green wrinkled. Note that this conforms almost exactly to the observed numbers and phenotypes of the peas in the typical

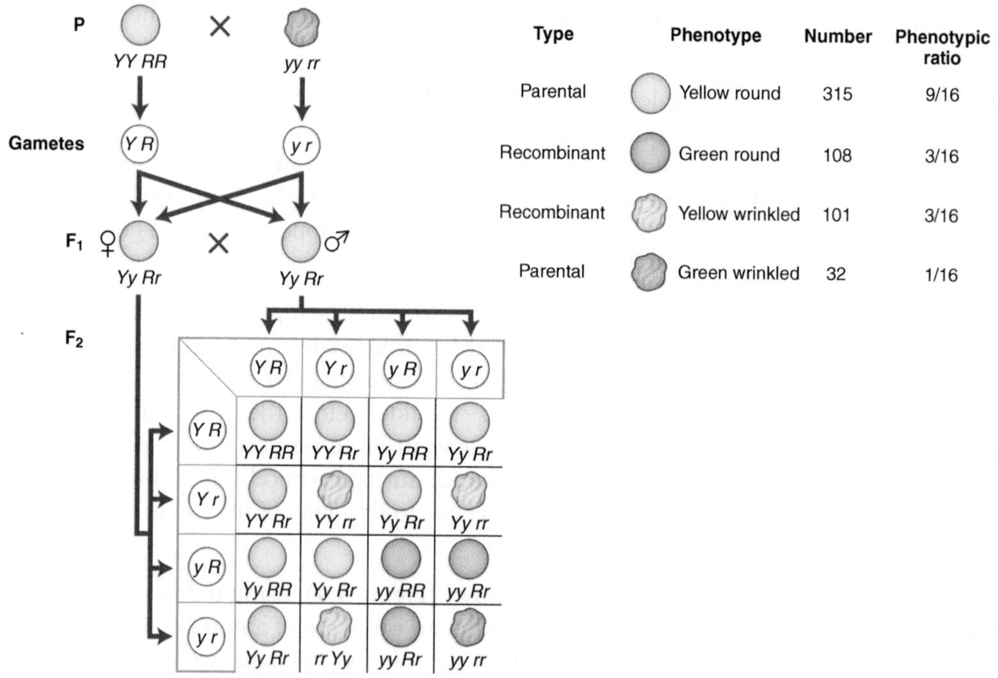

FIGURE 5.5
A typical dihybrid cross leading to the principle of independent assortment. Two different phenotypes were studied simultaneously (e.g., yellow/round and green/wrinkled). As noted, the experimental design (P, F_1, F_2) was the same as used in the mono-hybrid cross shown in Figure 5.4. In the F_1, all peas displayed both dominant characters—yellow and round. In the F_2, the four possible phenotypes were recovered at the ratios shown on the right. The gene for yellow and the gene for round acted independently of one another; that is, they assorted.

experiment noted above. As important, the ratio of yellow to green, or round to wrinkled, was 12 : 4 (or 3 : 1), exactly as in the monohybrid crosses discussed earlier.

RESTATEMENT AND SIGNIFICANCE

Mendel's two laws—of segregation and of assortment–are stated individually in Table 5.1. The first says that alleles segregate during gamete formation; the second says that segregation of one pair of alleles occurs independently of another pair. These laws (or principles) have been confirmed countless times in the past 150 years. They hold for all organisms that are diploid. Mendel's generalities have been followed by hosts of particularities and mechanisms, even involving the "characters" he studied. We now know that green peas result from a defect in breaking down the green pigment, chlorophyll, to its yellow product. We know, too, that wrinkled peas result from a failure to synthesize the starch amylopectin because of a deficiency of starch-branching enzyme 1 (SBE1).

Yet we must also acknowledge the irony associated with the response to his 1866 paper. It met with disappointment initially, and then with neglect. No one validated his work—in plants or other organisms—for more than 30 years. Only after chromosomes had been seen in cells did scientists begin to recognize that Mendel's laws conformed to the visible segregation of chromosomes during cell division. Finally, his work was confirmed and acknowledged independently by Carl Corren, Hugo de Vries, and Erich von Tschermak in 1900. Shortly thereafter, Bateson pointed out to Garrod that Mendel's law of gene segregation could explain Garrod's studies on "inborn errors" in humans. That Mendel's experiments are still described in textbooks and classrooms everywhere attests to the appropriateness of calling him "the father of genetics."

TABLE 5.1 Mendel's Laws

Law	Definition
Gene segregation	The two alleles for each trait segregate (separate) during gamete formation such that each gamete is equally likely to possess either allele. At fertilization, one allele from each parent unites in the zygote.
Independent assortment	During gamete formation, segregation of the members of any pair of alleles is independent of segregation for any other pair of alleles.

AMPLIFICATION: LANDMARK SCIENCE

The Rediscovery of Mendel's Work

You may ask yourself, "Why did Mendel's work go unrecognized for almost 35 years?" Considering he was an obscure monk who went into monastery management after he completed his prescient work, a better question might be, "How was his work rediscovered and by whom?"

Gregor Mendel first communicated his work on pea plants when he read his paper, Experiments on Plant Hybridization, at two meetings of the Natural History Society of Brunn, in 1865. His work was well received, but its true importance for inheritance was not recognized. It was seen only as relevant to plant hybridization. For a long 35 years its significance went unnoticed.

Three men, Hugo de Vries, Carl Erich Correns, and Erich von Tschermak, all working independently on different plant hybrids, have been credited with the rediscovery of Mendel's work, although even they may not have understood entirely the work's significance. A fourth man, Walter Sutton, was the first to relate Mendel's laws to what he observed under the microscope when observing cell division in the sperm of male grasshoppers.

(Continued)

Hugo de Vries

In the 1890s, De Vries, a Dutch botanist, conducted a series of experiments hybridizing varieties of multiple plant species. He published a paper about his work in the late 1890s that, at first, failed to mention Mendel's very similar work, though he had read this work and had altered some of his own terminology to correspond to Mendel's. When his paper was published in January 1900, Correns, criticizing the originality of de Vries' work, forced de Vries to acknowledge that Mendel's work had predated his own by 30 years.

Carl Erich Correns

Correns, a German, was a student of Karl Wilhelm von Nageli, a renowned botanist with whom Mendel corresponded about his work with peas. Nageli had failed to understand how significant Mendel's work was, but Correns, after searching the literature, found that Mendel had reached the same conclusions as he had while working with hybrid pea plants. Thus, Correns' work provided further proof for Mendel's theories. Correns published his first paper on January 25, 1900, which cited both Darwin and Mendel, yet he did not fully recognize the relevance of Mendel's work to Darwin's ideas.

Erich von Tschermak

Erich von Tschermak, a young Austrian, also worked with pea plants and had results similar to those of Mendel's. As he wrote up his results, he discovered a copy of Mendel's work from the library at the University of Vienna and properly acknowledged it. When his paper was published in June 1900, he was also credited with the rediscovery of Mendel's work. As it happened, his grandfather, Eduard Fenzi, had taught Mendel botany during his student days in Vienna.

Walter Sutton

It was left for Walter Sutton, a graduate student at Columbia University who was studying the sperm cells in male grasshoppers in 1902, to finally understand the parallels between Mendel's laws and what he was seeing when observing meiosis. He concluded that Mendel's principle of segregation of alleles could be accounted for by the segregation of homologous chromosomes at meiosis. He also was able to account for Mendel's second principle of independent assortment when he proposed that the several different "factors" described by Mendel (now called the genes) are carried individually on the chromosomes.

Mendelian Inheritance in Humans

The language of human genetics is filled with words derived from Mendel: homozygote, heterozygote, gene, allele, genotype, phenotype. We do not, of course, carry out experimental crosses and produce thousands of scorable progeny in people. Instead, we observe families, their chance matings, and the pedigrees they beget. These observations have made it possible to identify, characterize, and catalogue thousands of human traits resulting from the actions of single genes of the kind Mendel used in coming to the principle of gene segregation. The most widely used database of this kind is called **Online Mendelian Inheritance in Man (OMIM)**. It currently lists some 9,000 documented conditions: some harmless traits; others disorders of varying severity. Thus, nearly half of the estimated 21,000 human genes (see Chapter 6) have been shown to follow Mendelian inheritance.

TERMS

We have already introduced most of the terms found in Table 5.2, but understanding them is crucial to discussing the inheritance of single-gene traits in man. Some amplification is necessary. Genotype refers either to the pair of alleles at a single locus or to a person's entire set of genes. Phenotype is the observable expression of a genotype, in other words, a trait. Such expression may be biochemical, cellular, morphological, or clinical. Genotype and phenotype may describe normal or abnormal traits or conditions. The terms homozygous and heterozygous are easily understood, and refer to genotypes for loci on the autosomes (chromosomes 1 through 22). Hemizygous refers to genes located on the X chromosome in males, who have only one X.

TABLE 5.2 Language of Mendelian Traits

Term	Definition
Locus	The position occupied by a gene on a chromosome
Genotype	The pair of alleles at a locus; the complete genetic constitution of an individual
Phenotype	The observed biochemical or morphological expression of a genotype
Homozygote	Having two identical alleles at a locus
Heterozygote	Having two different alleles at a locus
Hemizygote	Having only one, instead of two, alleles at a locus
Dominant	An allele or trait expressed in heterozygotes
Recessive	An allele or trait only expressed in homozygotes

PEDIGREES

Seeking traits resulting from the action of a single gene in humans is much more difficult than working with plants, flies, or mice, for example. Human families are small, and experimental mutagenesis is forbidden. Despite these limitations, great progress has been made in identifying traits reflecting the action of single genes. This is so because humans are interested in their families, and in their family trees. Constructing and analyzing pedigrees is a critical part of the study of any human trait or disorder. The larger the number of people and generations in a pedigree, and the more certain the description of individuals in it, the more information it will reveal.

Pedigrees (as constructed by geneticists) are charts made using standard symbols (Figure 5.6). The extended family is called a "**kindred**." Males are depicted by squares (□), females by circles (○), those of unknown or undesignated sex by diamonds (◇). A horizontal line between a male and a female denotes a mating. A vertical line connecting a mating line to a second horizontal line below represents parents and their offspring (referred to as **sibs** or a **sibship**). Roman numerals denote generations, usually by birth order in their sibship (subscripts). A diagonal arrow identifies the person in the pedigree who first brought the family to the attention of the geneticist. A double horizontal line denotes a mating between relatives (**consanguinity**). Filled-

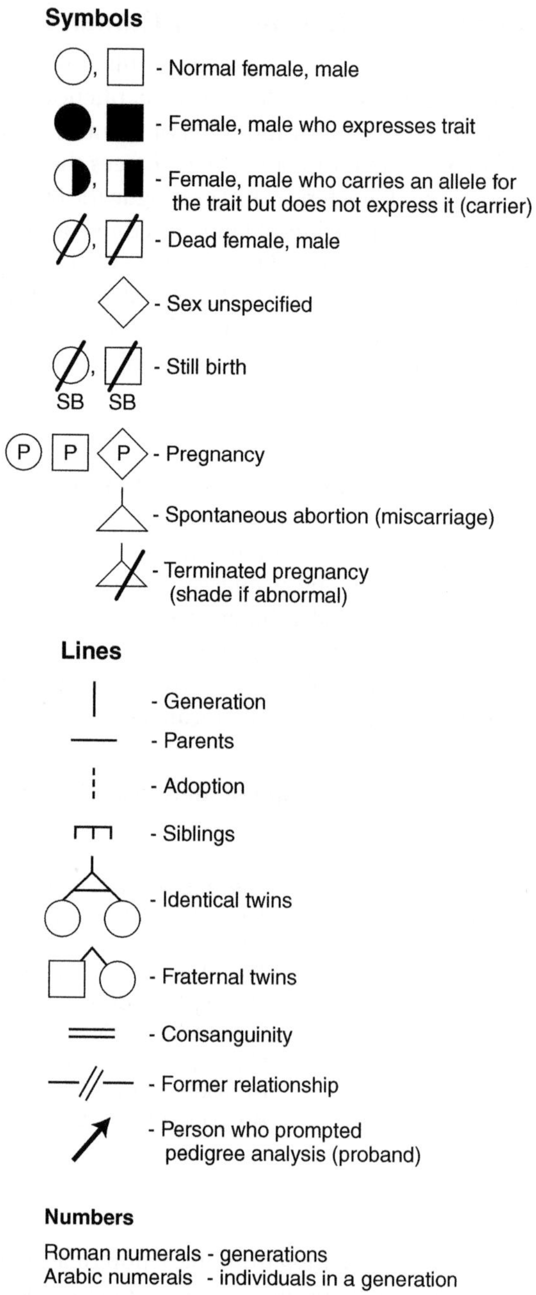

FIGURE 5.6
Symbols generally used in constructing human pedigrees. These drawings of shapes and lines are of great value in genetic counseling. Details are provided in the text.

in symbols identify individuals with the trait or disorder being analyzed; open symbols, those unaffected; half-filled symbols, those who are heterozygous carriers.

Relatives are classified as **first degree** (parents, sibs, offspring), **second degree** (grandparents, grandchildren, uncles, aunts, nephews, nieces, half sibs), **third degree** (first cousins), and beyond, depending on the number of steps in the pedigree between any two individuals. Figure 5.7 illustrates the way these symbols are employed to construct a pedigree. Once a pedigree has been drawn, one may ask: Is the trait likely inherited? Is a single locus, that is, one point on a chromosome, involved? Is the phenotype dominant or recessive? Is the gene on an autosome or, alternatively, on a sex chromosome? For example, the condition depicted in Figure 5.8A is clearly familial, but it is not genetic. It happens to be an extended family with malaria. On the other hand, the condition shown in Figure 5.8B is not familial (that is, it is sporadic), but it is genetic. It happens to be a rare family with albinism, which is inherited as

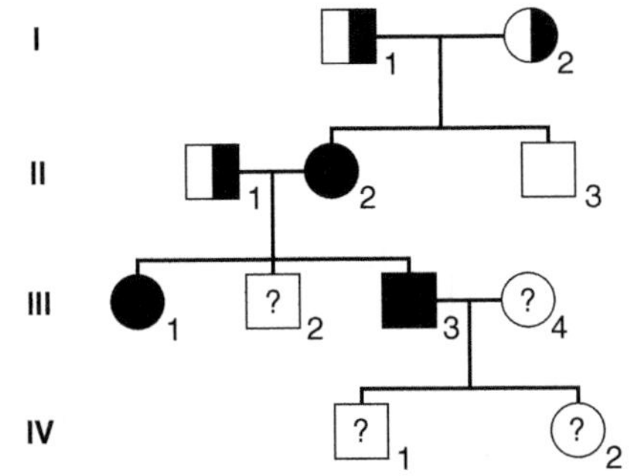

FIGURE 5.7
A stylized pedigree drawn to illustrate use of the symbols shown in Figure 5.6 and discussed in the text. This family could depict the common trait of redheadedness, inherited as an autosomal recessive (the filled-in symbols). Because II2 is affected, I1 and I2 each must be carriers. Likewise, II1 must be a carrier because two of his children (III1 and III3) are redheads. What deduction can you draw about the genotypes of III2, III4, IV1, and IV2?

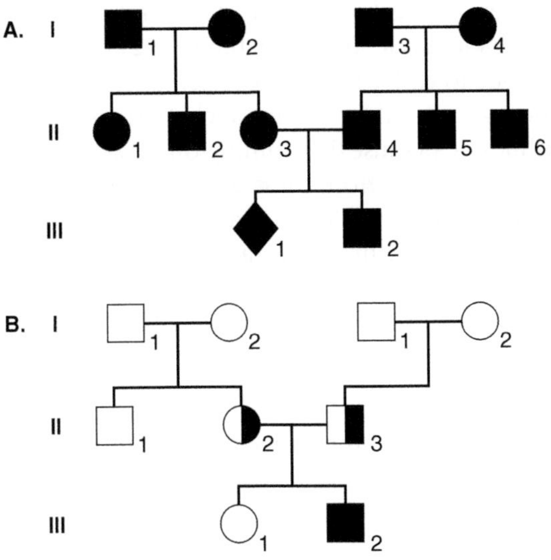

FIGURE 5.8
(A) A family reminding us that "familial" is not always synonymous with "genetic" or inherited. The filled-in symbols denote individuals from an African kindred, who have had malaria. (B). A family reminding us that a recessively inherited disorder e.g., albinism (filled-in symbol) may not be familial.Thus, in this kindred, albinism is found only in one person (II2).

an autosomal recessive trait. With these caveats in mind, let us examine the "signatures" of classic single-gene patterns of inheritance. Four will be discussed: **autosomal dominant**, **autosomal recessive**, **X-linked dominant**, and **X-linked recessive**.

Autosomal Traits

AUTOSOMAL DOMINANT TRAITS

There are nearly 4,000 autosomal dominant traits in the OMIM database. Many are harmless, such as crooked fifth fingers, or dimples, or freckles. Others, including **Huntington disease** and **neurofibromatosis,** are serious and progressive. Whether harmless or harmful, the pedigree pattern is the same (Figure 5.9). Its characteristic features are:

1. Males and females are equally likely to be affected;
2. Each affected offspring has an affected parent;
3. On average, half of the offspring of an affected parent are affected;
4. Offspring of unaffected individuals are unaffected.

The pedigree pattern (describing a family with Huntington disease) follows simply from Mendel's law of gene segregation. It is "vertical," meaning that affected individuals are seen in

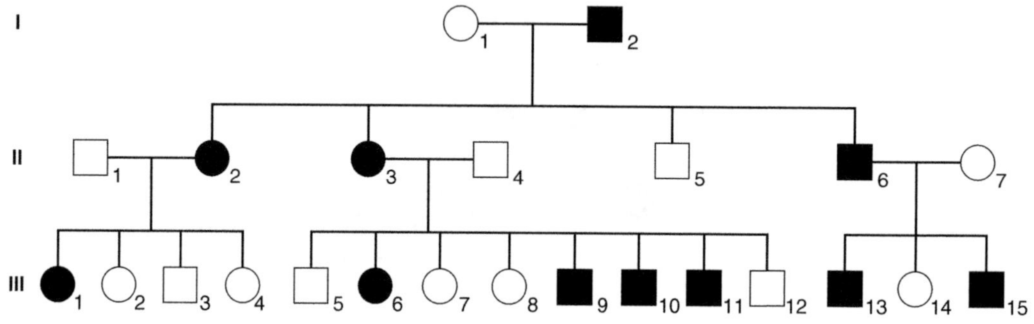

FIGURE 5.9
Pedigree of a three-generation family with an autosomal dominant trait, Huntington disease. Filled-in symbols denote affected individuals of genotype *Hh*. Unaffected individuals (open symbols) have the *hh* genotype. Note that each affected person has an affected parent, that about equal numbers of males and females are affected, and that affected males or females can have affected male or female offspring.

successive generations. Affected individuals are heterozygous (*Hh*) for the dominant allele responsible for the disorder. Thus, each affected person has a 50% probability (1/2) of transmitting either the dominant allele or the normal one to each offspring. Offspring receiving the dominant allele will be affected. Offspring homozygous for the recessive allele (*hh*) are unaffected and cannot transmit the disorder to their children. Because the Huntington disease gene is rare, few *HH* homozygotes exist. Probability estimates follow from these understandings. For example, a grandson of a person with Huntington disease has a 25% probability (1/4) of being affected: 1/2 that his parent will have received the disease gene from his grandfather; 1/2 that his parent, if affected, will transmit the Huntington gene to him. According to a basic rule of statistics called the product rule, the probability of two independent events occurring together is the product of the probability of each event; thus, $1/2 \times 1/2 = 1/4$.

The great 20th century folk singer and workers' rights advocate, Woody Guthrie (Figure 5.10), had Huntington's disease, which shortened and diminished his creative life. Fortunately, neither of his two children are affected.

FIGURE 5.10
Photo of Woody Guthrie (left), the famous folksinger and songwriter who died of Huntington disease in 1967. Arlo, his son (right), who is an equally celebrated musician, did not inherit the disorder.

AUTOSOMAL RECESSIVE TRAITS

The OMIM database contains nearly 4,000 traits inherited as autosomal recessives. As with dominant traits, some are of no medical consequence, such as red hair or inability to taste a bitter substance. Many others produce clinical and chemical abnormalities requiring diagnosis and treatment, such as cystic fibrosis or phenylketonuria or sickle cell anemia. Autosomal recessive traits, as illustrated in Figure 5.11, have a characteristic pedigree pattern with the following features:

1. Males and females are equally likely to be affected;
2. Parents and children of affected individuals are usually unaffected;
3. On average, 1/4 of sibs in a sibship containing an affected individual are affected;
4. Parents of affected individuals are more likely to be related.

The genetics accounts for these features. Assume that the pedigree in Figure 5.11 is from a family with phenylketonuria. Because the gene is located on an autosome, the trait will affect both genders. Because the trait is recessive, only individuals receiving a recessive allele (*p*) from each heterozygous parent (often called **carriers**) with *Pp* genotype will be affected homozygotes (*pp*). Because most autosomal recessive traits are rare, homozygotes will transmit a single, recessive allele to each offspring, who generally receive a normal allele (*P*) from the other parent, producing heterozygotes (*Pp*). Because each heterozygous parent has a 1/2 chance of transmitting the recessive allele to each offspring, the likelihood that two carriers will have an affected offspring is $1/2 \times 1/2$, or 1/4 for each pregnancy. Because relatives (first cousins, for example) are much more likely than an unrelated couple to carry a recessive allele inherited from a common ancestor, **consanguinity** (meaning mating between close relatives) is often seen in rare, autosomal, recessive traits. Figure 5.12 depicts this pedigree's critical features using the Punnett square. It illustrates the genotypes and phenotypes, and the expected ratio of each.

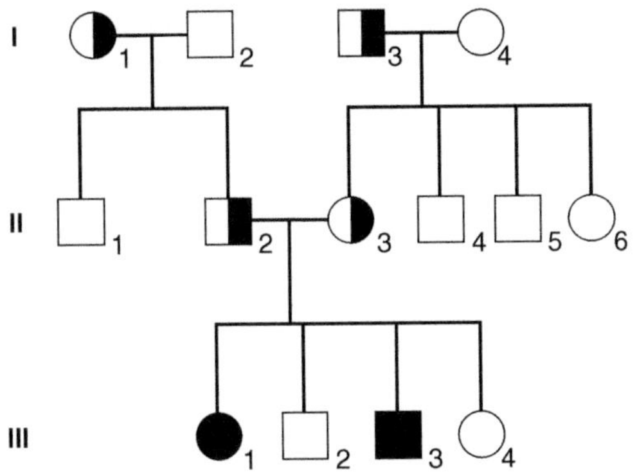

FIGURE 5.11
Pedigree of a three-generation family with an autosomal recessive trait, phenylketonuria (PKU), a disorder of phenylalanine metabolism. Filled-in symbols denote those affected (*pp*); open symbols, those unaffected (PP); and half filled-in symbols, those who are carriers (*Pp*). Note that affected individuals appear only in generation III, and may be male or female.

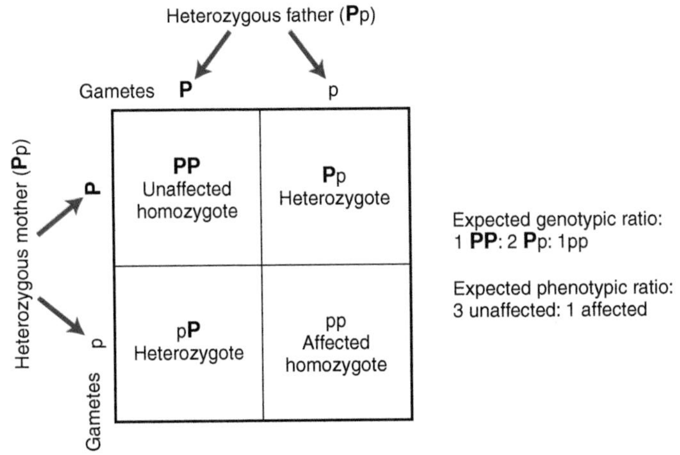

FIGURE 5.12
Punnett square showing expected genotypes and phenotypes for affected individuals in Figure 5.13 and for their parents.

X-LINKED TRAITS

Hemophilia, Duchenne muscular dystrophy, and red/green color blindness are only three of the more than 1,000 known **X-linked traits** in humans. X-linkage is a phenomenon not of autosomes, but of the sex chromosomes. Because females have two X chromosomes, and males only one, X-linked traits have their own particular pattern of inheritance. For any mutant allele on the X chromosome, females, having two Xs, have three possible genotypes: homozygous normal, heterozygous, and homozygous affected. In contrast, males have only two genotypes under this circumstance—hemizygous normal or hemizygous affected—because they only have one X chromosome. This unequal distribution of X chromosomes plays a critical role in phenotypic differences between males and females for X-linked traits. But it is not the only critical determinant. A second one has to do with X chromosome inactivation.

X CHROMOSOME INACTIVATION AND DOSAGE COMPENSATION

Given that mammalian females have two X chromosomes and males one, it follows that females would be expected to have twice as much activity as males for any gene on the X chromosome. This must have been disadvantageous evolutionarily, because humans and most other mammals have evolved a means of dosage compensation (that is, of correcting for the expected effect of two X chromosomes versus one). Early in embryonic development, perhaps as early as the 16-cell stage, one of the two X chromosomes in each somatic cell of a female is inactivated by chemical modification of the DNA (see Chapter 6). Such inactivation is random, meaning that in any cell, initial inactivation may affect either the X chromosome transmitted by the father or the X from the mother.

The pattern of X inactivation is then fixed and is passed on to all daughter cells throughout mitotic cell division and development. Thus, females are **mosaics** regarding X-linked gene expression—some cells express alleles transmitted by the paternally derived X, but not the maternally derived one; others just the opposite (Figure 5.13). This biologic mosaicism has important impacts on phenotypic expression for X-linked traits in heterozygous females (see the following section).

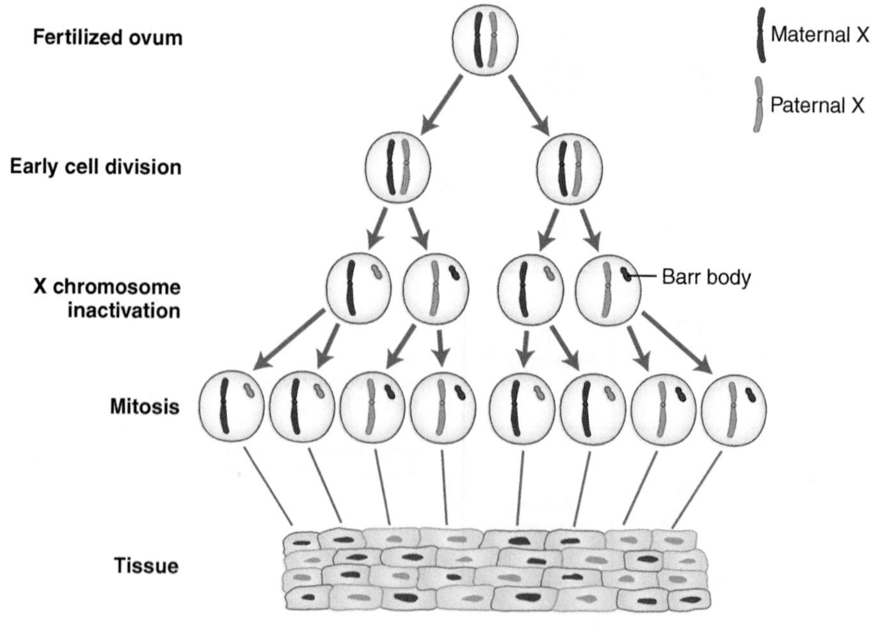

FIGURE 5.13
X chromosome inactivation. For the first several embryonic cell cleavages, both X chromosomes (red and green) are active. Then one X in each cell, derived from the father or mother, is randomly inactivated (or silenced). Thereafter, that same X chromosome is inactive in all cell progeny. Only a small number of genes on the X escape inactivation. The mosaic pattern shown in the resulting cells is depicted at the bottom.

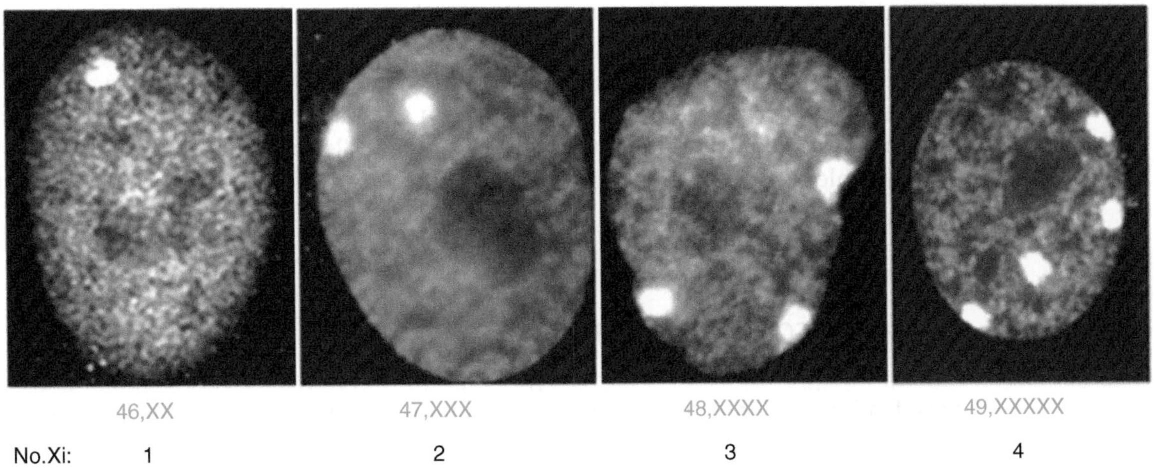

FIGURE 5.14
Visual counterpart of X chromosome inactivation. In fluorescence micrographs of interphase cells from females, the inactivated X chromosome is seen as a densely fluorescent, intranuclear inclusion called the Barr body (after the cytologist who first observed it). At the left, cells from a normal XX female have one Barr body. As the number of X chromosomes in a female's cells increases to three, four, or five, the number of Barr bodies increases correspondingly to two, three, and four.

The inactivated X chromosome can be identified morphologically as well as functionally. As seen in Figure 5.14, each normal female cell has a white-staining, rounded body near the inner surface of the nuclear membrane. This is the inactivated X chromosome, and it is called the **Barr body**. As expected, male cells have no such body, because their single X chromosome does not undergo inactivation.

X-LINKED DOMINANT TRAITS

The pedigree shown in Figure 5.15 is typical for a form of vitamin D-resistant rickets produced by a dominant allele on the X chromosome. Only a few such X-linked dominant traits are known in humans, and each is rare, but their features are distinct:

1. Females are twice as likely to be affected as males;
2. Affected males have no affected sons and no unaffected daughters;
3. Affected females have equal numbers of affected males and females;
4. Unaffected males have neither affected sons nor affected daughters.

The 2 : 1 ratio of affected females to males follows from the fact that females transmit an X chromosome to all their progeny, while males transmit their X to only half. Given that an

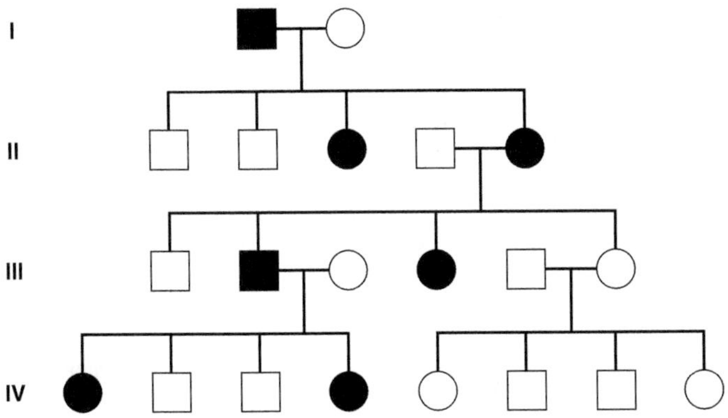

FIGURE 5.15
Pedigree of a four-generation family with an X-linked dominant trait, an inherited form of rickets. Note that affected males have no affected sons and no unaffected daughters, and that affected females have affected sons and daughters. Typically, there are about twice as many affected females as males, and phenotypic severity is greater in males because of X chromosome inactivation in females.

affected female is presumably heterozygous for the mutant allele, she has a 1/2 (or 50%) chance of passing it on to each daughter and to each son. Affected males are the key to distinguishing an X-linked dominant trait from an autosomal dominant one. Because an affected male transmits a Y chromosome to each son (and not an X), male-to-male transmission of an X-linked dominant trait does not occur. Reciprocally, each daughter of an affected male receives his mutation-bearing X chromosome; thus all daughters of an affected male are affected.

X-LINKED RECESSIVE TRAITS

Hemophilia A is the most widely recognized X-linked recessive disorder. It has affected the royal families of Europe because Queen Victoria was a carrier, and a number of her male descendants were affected. Figure 5.16 depicts a typical pedigree for an X-linked recessive trait that we'll assume was hemophilia A. The features are distinctive:

1. Affected males far outnumber affected females;
2. Half of the brothers of an affected male are affected, but none of his sisters are;
3. An affected male has no affected sons;
4. Some female carriers may express the phenotype with variable severity.

Each of these features has a logical explanation. If we use *H* for the dominant allele and *h* for the recessive one, females have three possible genotypes (*HH*, *Hh*, *hh*), hemizygous males only two (*H*–,*h*–). Because the disorder is recessive and rare, affected females would have to be *hh*: such homozygosity is an extraordinarily rare event. Thus, the vast majority of affected individuals are males, who inherit the mutation from their carrier mothers. Because a carrier female (*Hh*) will transmit either X chromosome to half of her daughters and half of her sons, each son has a 1 in 2 chance (50%) of being affected; each daughter, a 1/2 chance of being a carrier. Because males do not transmit their X chromosome to sons, no male-to-male transmission of the affected phenotype will be seen. The phenotypic variability observed in *Hh* females follows from the random inactivation of X chromosomes discussed previously. By probability alone, one would expect that half of the cells of a female carrier would express the normal X and half the mutant one. But the random inactivation pattern is set very early in embryonic life, making it possible, even probable, that the mosaic pattern will be shifted in favor of either normal cells or mutant ones.

The "royal" hemophilia pedigree is shown in Figure 5.17. Because, in the 19th century, Queen Victoria of England had no affected antecedents, it is assumed that the hemophilia mutation appeared first in her germ cells. Royalty was trumped by genetics insofar as hemophilia was concerned.

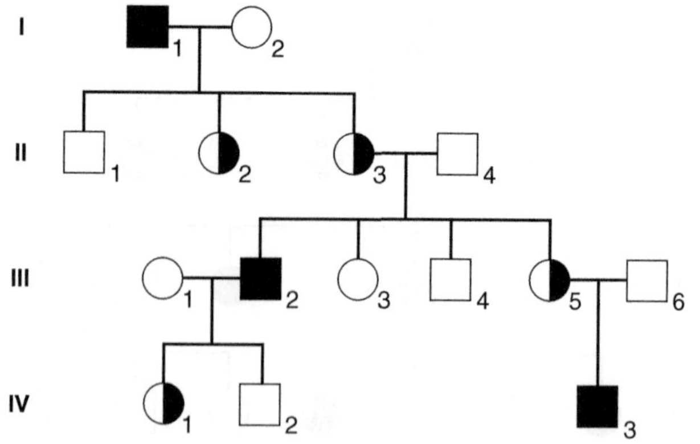

FIGURE 5.16
Pedigree of a four-generation family with an X-linked recessive trait, hemophilia A. The pattern of those affected tends to appear diagonal (rather than vertical or horizontal). Note that all affected individuals are male, that the mutant allele is transmitted by (usually unaffected) carrier females, and that affected males have no affected sons.

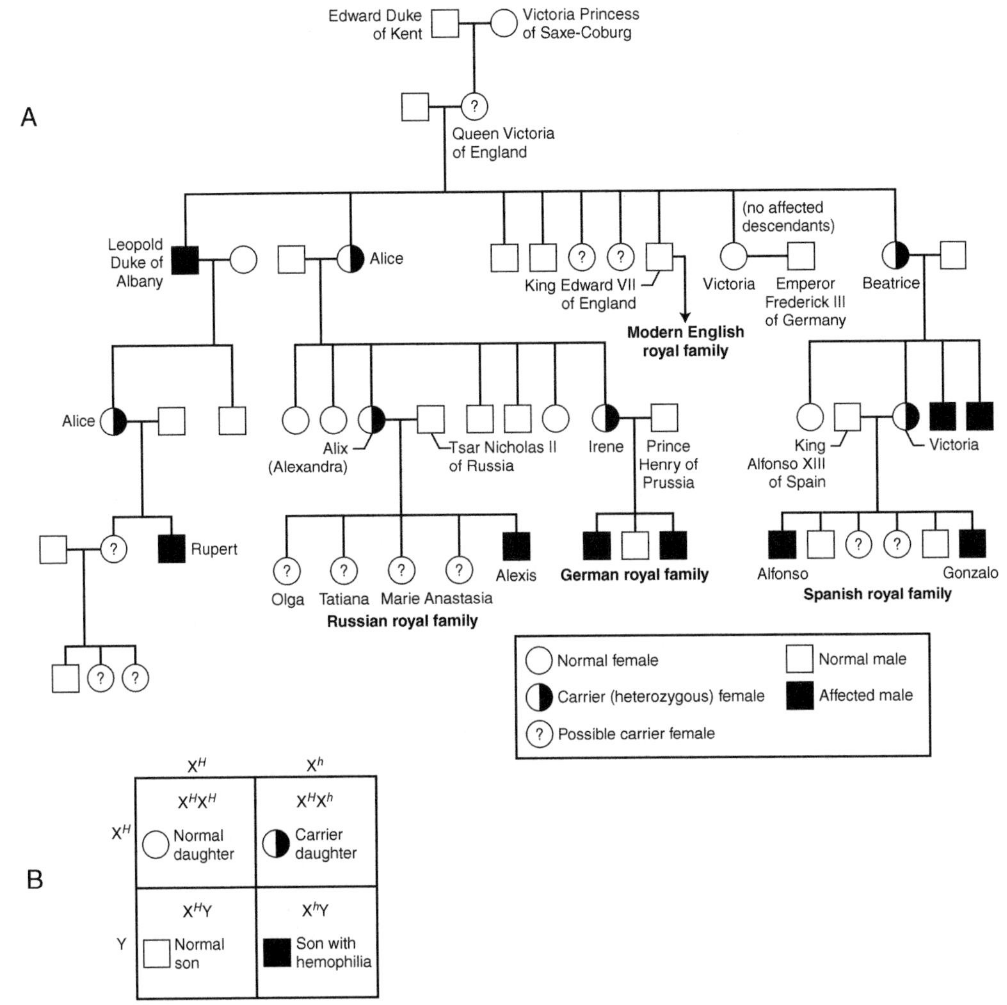

FIGURE 5.17
Hemophilia A in the royal families of Europe. (A) The six-generation pedigree (partial only) shows that Queen Victoria was a carrier; that Leopold was affected and that Alice and Beatrice were carriers in generation III; and that from there, hemophilia was transmitted to the royal families of Russia, Germany, and Spain—but, ironically, not of England. (B) The Punnett square shows how a carrier female (top) and a normal male (left side) can have an affected son.

APPLYING PRINCIPLES OF GENE SEGREGATION TO GENETIC COUNSELING

Clinical geneticists and genetic counselors use Mendel's law of gene segregation every day. Among the most common and most important questions asked by parents are: Will our baby have such and such condition? Will our baby be a carrier or be unaffected? Figure 5.18 depicts such a case study. Whitney (II2) is African American and has a brother, Scott (II3), with sickle cell anemia. Whitney is unaffected, as are both of her parents. Whitney, who is pregnant, wants to know whether the fetus she is carrying will have sickle cell anemia or whether he will be a carrier (often called sickle cell trait.) No one in her husband James' (II1) family has sickle cell anemia. You answer Whitney's questions as follows: sickle cell anemia is inherited as an autosomal recessive trait, meaning that her brother is genotypically ss; he inherited one s allele from his mother and one from his father, thus both parents are *Ss*. Because Whitney doesn't have the disorder, there is a 2/3 chance that she is a carrier (from the Punnett square). If she is a carrier, the probability of transmitting the mutant allele is 1/2; therefore, the net probability that she will pass on the mutant allele for sickle cell anemia is $2/3 \times 1/2 = 1/3$. This is more

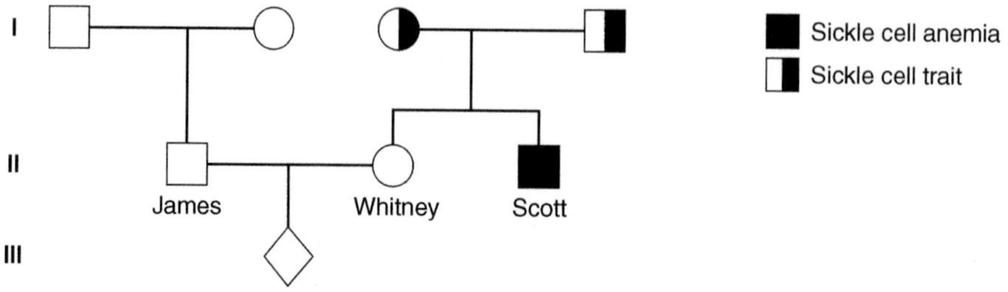

FIGURE 5.18
Estimating probabilities using Mendel's law of gene segregation. Question asked: What is the likelihood that the fetus of Whitney and James, each African American, will have sickle cell anemia, like Whitney's brother Scott? Neither of Whitney's or James' parents have sickle cell anemia. To answer the question, one must first determine the probability that Whitney is a carrier, then the probability that she will transmit the mutation to her unborn fetus, then the probability that this fetus will inherit the mutation from James, and, finally, the probability that the unborn will be affected.

Question: Likelihood that III ◇ will have sickle cell anemia.

Solutions:
1) Estimate probability that Whitney has sickle cell trait: $^{2}/_{3}$ (because she is unaffected).
2) Estimate probability that if she has the trait, Whitney will transmit it to her fetus: $^{1}/_{2}$.
3) Estimate probability that James has the trait: $^{1}/_{10}$ (because about 10% of African Americans have the sickle cell trait.
4) Estimate probability that if he has the trait, James will transmit it to his fetus: $^{1}/_{2}$.
5) Total probability (from the product rule): $^{2}/_{3} \times {}^{1}/_{2} \times {}^{1}/_{10} \times {}^{1}/_{2} = {}^{1}/_{60}$.

Perspective: A risk of $^{1}/_{60}$ is small, but it is more than 10 times the prevalence of $^{1}/_{700}$ in the African American community. Both the probability and the perspective are part of accurate and useful genetic counseling.

than three times the frequency of the carrier state for sickle cell anemia in the African American population (1/10). In the absence of sickle cell anemia in her husband's family, the likelihood that they will have a child with sickle cell anemia is $2/3 \times 1/2 \times 1/10 \times 1/2 = 1/60$, compared to about 1/700 in the African American population. This information can be improved upon considerably by testing Whitney and her husband for sickle cell trait. If either of them is not a carrier, they will not have a child with sickle cell anemia. If both are carriers, their chance of having an affected baby is 1/4. Analogously, families are counseled for autosomal dominant and X-linked traits.

EXCEPTIONS TO EXPECTED MENDELIAN PATTERNS

Mendel carefully selected traits that he could cross, self-cross and back-cross (crossing a hybrid with one of its parents)—always looking for either/or phenotypes (yellow : green; round : wrinkled; etc.). Human geneticists have no such luxury. We study an organism remarkably more complex than the garden pea. It should come as no surprise, then, that we encounter many situations in which the pedigree patterns predicted in the previous pages are not followed perfectly. This does not mean that the laws of segregation and assortment no longer pertain—they most assuredly do. This does mean that we need to identify (and hopefully understand) a growing number of exceptions or deviations from the idealized patterns of inheritance just presented. We will comment on a number of these situations briefly.

Penetrance

In classic Mendelian segregation, phenotype always follows genotype. In clinical situations, however, this is often not the case. Pedigrees may contain individuals who have the same genotype as someone with a particular phenotype, but who lack the phenotype themselves. **Penetrance** is defined as the fraction of individuals of a given genotype who have the expected phenotype. Huntington disease, for example, shows nearly 100% penetrance, meaning that virtually everyone with this dominantly inherited mutation develops the disorder. Many conditions show incomplete or reduced penetrance. In this situation, epitomized by a dominantly inherited form of polydactyly (extra fingers or toes), we find individuals with 10 fingers and 10 toes who have a parent and a child each with more than the normal complement of digits. This is what we mean by the term **incomplete penetrance**, or skipped generation. Some human traits may be 70 or 80% penetrant (that is, 70–80 people out of 100 with the genotype have the phenotype.)

Expressivity

In contrast to penetrance, which is an all-or-none phenomenon, **expressivity** refers to the magnitude (or severity) of phenotypic expression. We speak of **variable expressivity** when referring to situations in which phenotypic differences occur in genotypically identical people. In polydactyly, for instance, some affected people will have six fingers on each hand, and others may have six fingers on one hand, but five fingers and a rudimentary sixth on the other hand. Variable expressivity is often encountered in metabolic disorders inherited as autosomal recessive traits. Accordingly, two or more sibs with identical mutations can have modestly different or dramatically different clinical manifestations.

Reduced penetrance and variable expressivity must have molecular bases, such as modifying genes, effects of age, different environments, and chance. We have little precise understanding of these phenomena currently, but are beginning to glimpse the mechanisms underlying such remarkable and important complexity, as will be mentioned now.

Epistasis and Modifier Genes

Certain traits result from the action of two genes, rather than one. In such situations, one gene may have a large influence on a phenotype (the major gene), the other a small effect. The latter is called a **modifier gene**. Tail-length in mice is a well-studied example of such major gene/modifier gene interaction. Closely related conceptually is **epistasis**, defined as gene interaction in which the effects of alleles of one gene hide the effects of another gene. Epistasis has been well studied in plants and dogs; there is a rare and complex example in humans affecting ABO blood groups. Surely, many more kinds of modifier and epistatic gene effects in humans will be added to the current limited repertoire.

Incomplete Dominants and Recessives

Mendel's dominant and recessive factors were absolute, all-or-none. For yellow/green peas, for instance, genotypes *YY* and *Yy* produced yellow peas, genotype *yy* green ones. No orange peas, or any other intermediate color, were observed. There are well-characterized conditions in humans and other species where dominance and recessivity are not absolute. For example, in sickle cell anemia, classified as an autosomal recessive trait, *Ss* carriers may develop clinical abnormalities under rare circumstances: athletic exhaustion, high altitude flying, battlefield dehydration. So it would be fair to call sickle cell disease an incomplete recessive or an incomplete dominant.

Even clearer is a rare but important disorder of cholesterol metabolism, called **familial hypercholesterolemia (FH)**. Three genotypes exist, which we will denote *FF, Ff, ff*. *FF* individuals have normal serum cholesterol concentrations and no clinical phenotype. *Ff* individuals have moderately elevated serum cholesterol concentrations and are prone to heart attacks in midlife and to cholesterol deposits in their tendons. Individuals that are *ff* have dramatically elevated serum cholesterol and usually die of heart attacks before age 20. Each genotype, then, has its own phenotype. Should we call FH incompletely dominant or incompletely recessive?

Genomic Imprinting

Mendel's law of gene segregation is built on the idea that both alleles of a given gene—one inherited from the mother, the other from the father—are expressed equally in each offspring. Said another way, a mutant allele of an autosomal gene is equally likely to be transmitted from either gender to an offspring of either gender. This axiom, however fundamental, is not without its exceptions. We now know that for a few hundred autosomal human genes (and an unknown number in other mammals) only one allele is expressed in offspring, and the other one is silenced by chemical modification. This is referred to as **genomic imprinting**. Its molecular mechanism will be discussed in Chapters 6 and 7, but it must be mentioned here for its effect on pedigrees.

For any imprinted gene, the silencing of either the paternal or maternal allele is fixed and occurs during germ cell development. At some loci the paternal allele is imprinted, at others the maternal allele is imprinted. This silencing of one or the other allele has no clinical consequences unless there is a mutation or chromosomal deletion affecting the non-imprinted allele. For example, if gene *A* is imprinted such that its maternal allele (A^m) is inactivated and A^p, the paternal allele, is active, a mutation knocking out A^p would result in the complete absence of expression of *A* in the offspring. There are well-known instances of precisely this kind in humans. They will be discussed subsequently.

Why does genomic imprinting occur in mammals? The answer is, we don't know. Some have proposed that imprinting is the result of competition for resources between mother and fetus. This thesis is supported by the fact that many imprinted genes are involved in apportioning growth-affecting nutrients between mother and fetus. But many other imprinted genes have no obvious relationship to such hypothesized maternal–fetal competition.

Uniparental Disomy

How would you explain the extremely rare situation in which a child with an autosomal recessive condition had only one carrier parent? Similarly, how would you account for male-to-male transmission for an X-linked trait? The answer: an error in meiosis such that both chromosomes are inherited from one parent (let's say the father), rather than one homologue from the father, the other from the mother. This is called **uniparental disomy**, and its implications are clear. If the father just mentioned is a carrier for a recessive disorder and contributes both copies of the relevant chromosome to his offspring, the child will be affected even if the mother is homozygous normal. Uniparental disomy needs to be excluded when pedigrees seem to defy the law of segregation. (So does a much more common occurrence: **non-paternity**, meaning that the husband is not the father.)

MITOCHONDRIAL INHERITANCE

Humans have about 21,000 nuclear genes. They constitute the protein-encoding portion of our nuclear genome. However, we have another genome composed of only 37 genes. It exists in our mitochondria, and its mutations produce characteristic and unusual pedigree patterns.

The Mitochondrial Genome

As discussed in Chapter 4, our cells contain organelles called **mitochondria**. These organelles carry out hundreds of reactions, each catalyzed by different proteins. **Mitochondrial genes**, however, code for only a tiny fraction of these mitochondrial proteins. The mitochondrial genome is 16,500 bases long. It is contained in a single, circular chromosome. The mitochondrial genome codes for 37 products: 13 **polypeptides** that are subunits of proteins generating cellular energy; 22 different kinds of **ribonucleic acids (tRNAs)** that are necessary for protein synthesis (see Chapter 7); and 2 **ribosomal genes**, also required for protein synthesis. Most cells contain hundreds of mitochondria, and most mitochondria contain many circular genomes. Hundreds of rearrangements and mutations of the mitochondrial genome have been described. While the clinical consequences of these mutations vary widely, they all produce a characteristic pedigree pattern based on the fact that we inherit all of our mitochondria from our mother, none from our father.

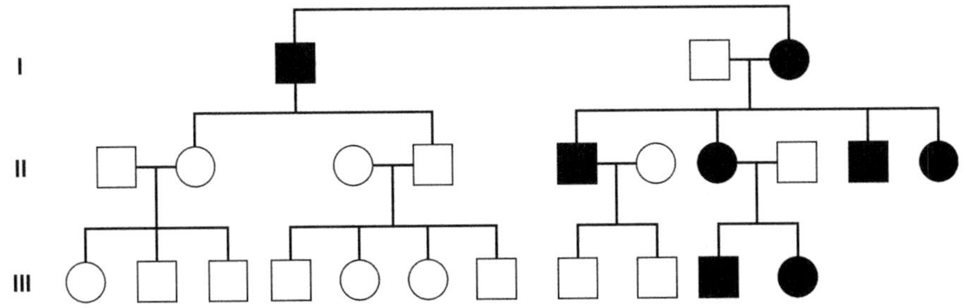

FIGURE 5.19
Pedigree of a three-generation family with an hereditary form of blindness caused by a mutation in mitochondrial DNA. Note that a brother and sister in generation I are affected; that none of the children or grandchildren of I1 are affected; and that all of the children of I3 are affected, as are both children (III10 and III11) of II7.

Mitochondrial Segregation

During meiotic and mitotic cell division, mitochondrial DNA replicates (as does nuclear DNA), and the newly formed mitochondria with their replicated mitochondrial genomes assort randomly in daughter cells, as opposed to the rigorously controlled 1:1 segregation of nuclear chromosomes discussed in Chapter 4. This is called **replicative segregation**, but it is not the only special feature of mitochondrial inheritance. The second feature relates to the fact that each cell contains many copies of the mitochondrial genome. When a mutation occurs in mitochondrial DNA, it is initially present in only one of the mitochondrial DNA molecules in a cell. When cells divide, a cell containing a mixture of normal and mutant mitochondria may transmit its mitochondria such that its daughter cells receive only normal or only mutant mitochondria. This is called **homoplasmy**. Alternatively, daughter cells may receive variable fractions of normal mitochondria and of mutant mitochondria by chance alone. This situation, referred to as **heteroplasmy**, has great bearing on the phenotypic expression of mitochondrial gene mutations. It may result in variable expression of such mutations in different tissues and variable transmission of mutations from one generation to the next.

Maternal Inheritance

The last defining characteristic of the genetics of mitochondria is **maternal inheritance**. Mitochondria found in sperm either don't enter the oocyte or are rapidly eliminated from the zygote. Thus, only mitochondria found in the oocyte (which may number in the thousands to hundreds of thousands) propagate during early rounds of cell division and embryogenesis. Because mitochondrial DNA is inherited from the mother only, mutations of mitochondrial DNA display a characteristic pedigree pattern (Figure 5.19), assuming that the affected females are homoplasmic:

- All male and female offspring of an affected female will be affected;
- None of the male and female offspring of an affected male will be affected.

If the affected woman is heteroplasmic, the mitochondrial DNA molecules transmitted to her offspring will be both mutant and normal. Accordingly, the phenotype in her offspring will be variable, ranging from severely affected to essentially unaffected.

REVIEW QUESTIONS AND EXERCISES

1. Choose the phrase in the right column that best matches the term in the left column.

a. recessive	1. the unit of inheritance
b. alleles	2. observable characteristics
c. F_1	3. having two different alleles at a gene locus
d. gene	4. the alleles a person has
e. heterozygote	5. the allele that is not expressed phenotypically in a heterozygote
f. genotype	6. the separation of the two alleles of a gene into different gametes
g. dominant	7. having two identical alleles at a gene locus
h. phenotype	8. alternate forms of a gene
i. segregation	9. alleles of one gene segregate into gametes randomly with respect to the alleles of a second gene
j. homozygote	10. the alleles expressed phenotypically in the heterozygote
k. independent assortment	11. offspring of the parental (P1) generation

2. When flipping a coin, what is the probability that it will land heads? The probability that the next flip will also be heads? The probability of flipping 3 heads in a row? Should you be surprised if 10 flips come up 7 heads and 3 tails?

3. Clinodactyly (short fifth fingers) is usually inherited as an autosomal dominant trait. Sam has this condition, as does his grandson, Felix. Felix's father (and Sam's son), Phil, is unaffected. No one else in the family has clinodactyly. How do you account for this situation?

4. Why are there so few Y-linked traits? What are the characteristic pedigree features of a condition inherited as a Y-linked trait? Name a Y-linked trait.

5. Consider the following pedigree: Esther and Amos are married; each has been deaf since birth. They have five children. Three (Ann, Joseph, and Daniel) are deaf. Two non-identical twins (Bob and Abigail) are unaffected. Joseph married Sally, who is unaffected. They have four children; two affected (Clyde and Laura), two unaffected (John and Thomas).
   a. Draw this family's pedigree using standard symbols. Identify those affected with deafness and those unaffected.
   b. What is the most likely mode of inheritance of this form of deafness? Why do you come to this conclusion?
   c. Which, if any, modes of inheritance can be excluded in this family? Why?
   d. What if all of Esther's and Amos's children and grandchildren had been unaffected? Explain this very different pedigree pattern.

6. Gaucher's disease, an autosomal recessive trait, is much more common in Ashkenazi Jews than in the general population. Charles, a Jewish man, has a sister with Gaucher's. His parents, grandparents, and three other sibs are unaffected. Charles's wife, Rachel, remembers that the brother of her paternal grandfather had Gaucher's.
   a. Draw this family's pedigree.
   b. What is the probability that the first child of Charles and Rachel will have Gaucher's disease?
   c. What is the probability that their first and second child will each be affected?

7. A man named David suffers from an inherited defect in tooth enamel resulting in brown teeth. His wife, Sonia, has normal teeth. Each of their three sons (Paul, George, and Ringo) have normal teeth but each of their two daughters (Joni and Judy) have brown teeth. Joni marries a man with normal teeth and they have one affected son and one unaffected daughter.
   a. Draw this pedigree using standard symbols.
   b. What is the most likely mode of inheritance of brown teeth?
   c. What other mode(s) of inheritance are possible?
   d. What mode of inheritance can be excluded from consideration?

8. Dosage compensation is a fundamental property of mammalian species.
   a. What is dosage compensation?
   b. Which human chromosome displays dosage compensation?
   c. What is the mechanism for dosage compensation in humans?
   d. How does dosage compensation manifest itself phenotypically?

9. Polydactyly (extra fingers and toes) is a rare condition inherited as an autosomal dominant trait. A man with six fingers on each hand and six toes on each foot and his unaffected wife seek genetic counseling. Their only son has six fingers on each hand, but five toes on each foot.

a. What is the probability that their next child will have polydactyly?

b. They tell you that their next child will be unaffected since their first child is affected. What is your response?

c. What explains the phenotypic difference between father and son?

d. What is the likelihood that this couple will have four affected sons in a row?

10. A rare genetic disease that results in an inability to speak normally runs in the Brown family. The patriarch, George, is affected, as is his wife, Lucille. They have four children: two sons and a daughter who are affected; one son, Michael, who speaks normally. Michael is a widower with one unaffected son. Michael's sister is married to an unaffected man, and they have a newborn daughter.

a. Draw the Brown family pedigree.

b. What modes of inheritance can be ruled out?

c. What is the most likely mode of inheritance of this speech defect?

d. What is the probability that George and Lucille's granddaughter will be affected, assuming that her father is unaffected?

11. Sarah and Ben have seven children (Bert, Ernie, Elmo, Charlie, Lucy, Christopher, and Sue). Sarah and each of her children suffer from an allergic response to caffeine characterized by hives and facial swelling. Ben doesn't have this problem. The seven children grow up and some have children of their own. Each of Lucy's three sons is affected, as are each of Sue's two daughters. Neither of Bert's two sons nor of Ernie's two sons is affected. The other three children (Elmo, Charlie, and Christopher) have no progeny yet.

a. Draw the pedigree, identifying those affected and those unaffected.

b. What is the most likely mode of inheritance of this condition? Why?

c. If Elmo marries an unaffected woman, will any of their children be affected? Explain.

12a. Enumerate the characteristics of *Pisum sativum* (the garden pea) that made it a suitable organism for Mendel's discoveries of the fundamental principles of genetics.

12b. Using these characteristics, why would it be impossible to carry out similar experiments in humans?

CHAPTER 6

Structure of Genes, Chromosomes, and Genomes

CHAPTER OUTLINE

In earlier chapters, we have stressed that genetics is the science of heredity. From a conceptual sense, we have talked about how biological information makes possible such fundamental processes as reproduction and hereditary transmission. Now we will move to the molecular level and answer such questions as:

1. What is the composition of a gene?
2. How are genes passed on from one generation to the next?
3. What is the relationship between genes, chromosomes, and genomes?

GENE COMPOSITION

CORE CONCEPT

Genes are the units of inheritance. They are composed of **DNA (deoxyribonucleic acid)**. Genes store information in linear DNA molecules composed of four different nucleotides – each containing a nucleic acid base, deoxyibase, and phosphate. The four bases are **G (guanine)**, **C (cytosine)**, **A (adenine)**, and **T (thymine)**. Each gene has a unique sequence of these nucleotides, which, in turn, underlies its unique function. The DNA molecule is a double strand of nucleotides carrying **complementary G–C or A–T base pairs**. The complementarity of double-stranded DNA is the key to understanding how DNA functions in inheritance.

Human Genes and Genomes. DOI: 10.1016/B978-0-12-385212-0.00006-8

That genes are composed of DNA (deoxyribonucleic acid) was not obvious as recently as 70 years ago. The first strong evidence that DNA, not protein, is the genetic substance came from the work of Fredrick Griffith in 1928. His remarkable experiment showed that genetic information from dead bacterial cells could be transmitted to live cells. He worked with two kinds of pneumococcal bacteria: an avirulent R form and a virulent S form (Figure 6.1). As expected, mice lived after being injected with the R form and died after receiving the S form. When the S form was heat-killed before mouse injection, the animals survived; however, when they were injected with a mixture of live R form and heat-killed S form, they died. Further, bacteria cultured from the blood of the latter mice were of the living S form. This meant that something (called by him the "**transforming principle**") from the heat-killed S bacteria had transformed the living R bacteria to S. (Transformation is the ability of a substance to change the genetic characteristics of an organism.)

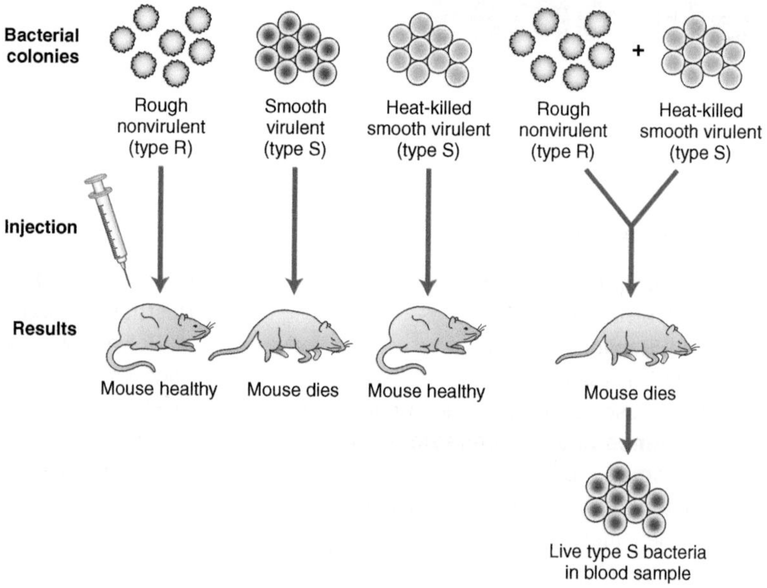

FIGURE 6.1
Demonstration of bacterial transformation. Griffith showed that a substance in heat-killed, virulent (type S) bacteria transformed non-virulent (type R) bacteria into virulent ones (see text for additional details).

That this "something" was DNA was proved by two quite different kinds of experiments. In one, Oswald Avery and his colleagues spent 15 years (1929–1944) purifying the transforming principle. They determined it chemically to be DNA. They then treated the principle with various enzymes (Figure 6.2), to determine whether a molecule other than DNA could cause transformation. Enzymes that broke down RNA, protein, or polysaccharide had no effect on the transforming principle, but a DNA-degrading enzyme totally destroyed its activity. These investigators concluded that the transforming principle was DNA.

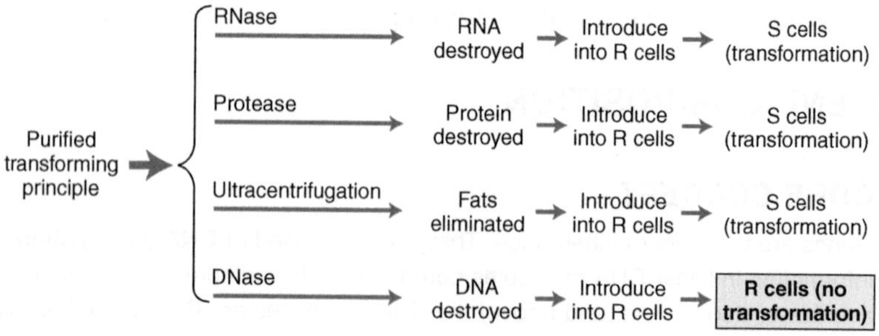

FIGURE 6.2
Avery, MacLeod, and McCarty showed that the transforming principle was DNA, not protein. They treated the purified transforming principle with several different enzymes *in vitro* and found that only exposure to DNase (an enzyme that degrades DNA) destroyed the ability to transform type R bacteria to type S.

Many scientists remained skeptical, still clinging to the idea that the transforming principle was a protein. Alfred Hershey and Martha Chase settled this matter in 1952 using bacteriophage, a virus that infects bacteria. In their famous experiment (Figure 6.3) they labeled phage

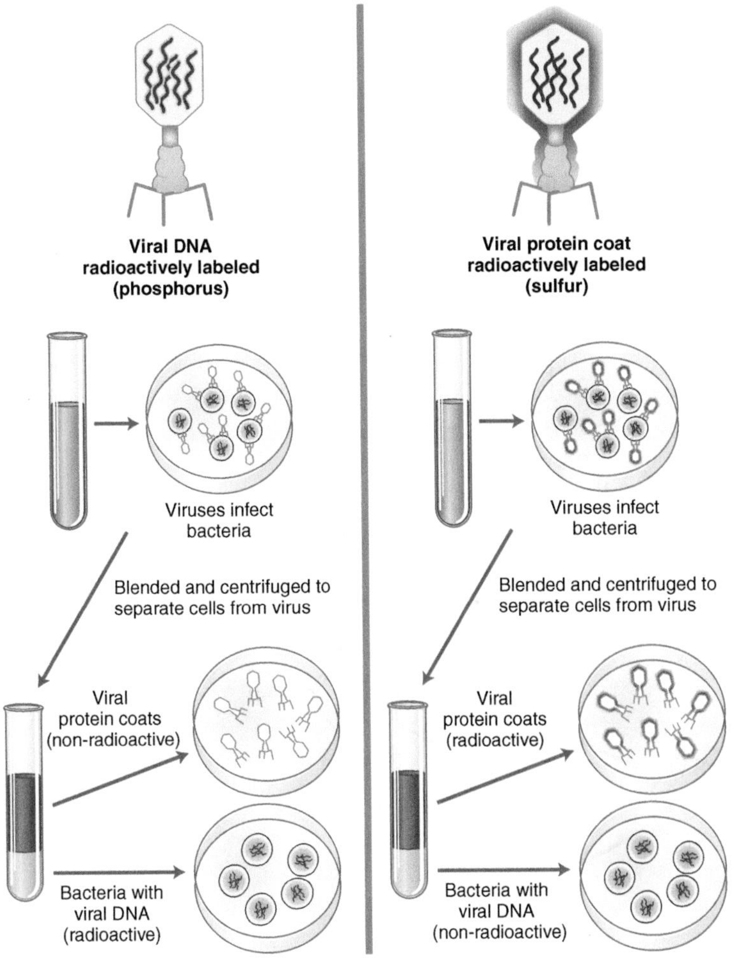

FIGURE 6.3
The "blender experiment" of Hershey and Chase. Bacteria were infected with viral particles labeled either with P^{32}-labeled DNA (red; top left) or S^{35}-labeled proteins (orange, top right). When the bacteria were then separated from the viral "ghosts," the P^{32} fractionated with the bacteria, the S^{35} did not, confirming that genes are made of DNA.

proteins with radioactive sulfur (S^{35}) and phage DNA with radioactive phosphorus (P^{32}). They found that the labeled DNA partitioned with the phage-infected bacteria, but the labeled protein did not. It remained with the non-infectious membrane "ghosts." Even more important, and central to the proof, was the fact that the labeled DNA was found in the next generation of phage particles (those released from the phage-infected bacteria). This meant that the DNA must have been encoding the information necessary for the production of a new phage. They concluded that phage genes are DNA, a conclusion that supported Avery's work and removed any doubt. DNA was accepted as the genetic material.

DNA Structure

DNA is a long **polymer**—a molecule made up of repeating subunits—consisting of **nucleotides**. Each nucleotide is made up of a **deoxyribose sugar**, a **nitrogenous base**, and a **phosphate group** (Figure 6.4). Four different bases are found in DNA: two **purines** (**adenine** and **guanine**); and two

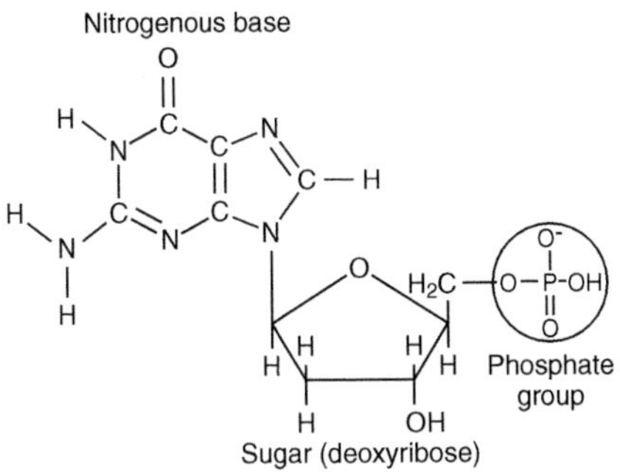

FIGURE 6.4
A nucleotide consists of a nitrogenous base (e.g., guanine), a deoxyribose sugar, and a phosphate group.

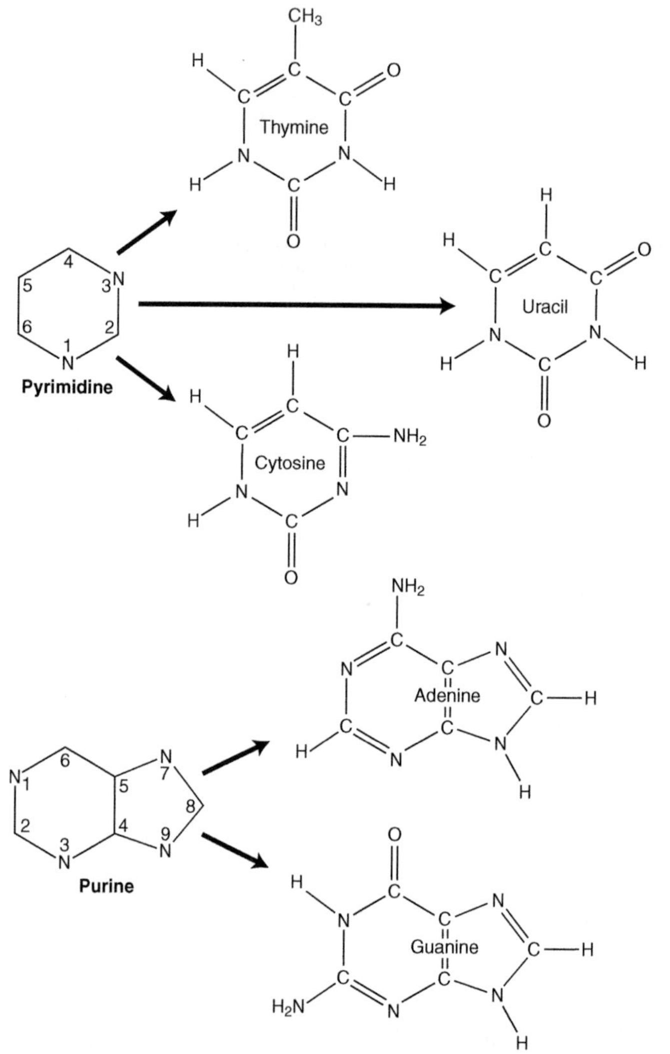

FIGURE 6.5
The four nitrogenous bases found in DNA. Guanine (G) and Adenine (A) are purines, each containing the same nine-membered ring structure. Cytosine (C) and Thymine (T) are pyrimidines, each with the same six-membered ring. The third pyrimidine, uracil, is found in RNA. The side chains distinguish these molecules from one another. The five bases are composed of carbon (C), nitrogen (N), hydrogen (H), and oxygen (O).

pyrimidines (**cytosine** and **thymine**) (Figure 6.5). A purine is a nitrogenous base that has two carbon-nitrogen rings, while a pyrimidine has only one carbon-nitrogen ring. A third pyrimidine, uracil (U) is found only in RNA and will be discussed later. Nucleotides are connected to one another by phosphodiester bonds between the sugar and phosphate substituents (Figure 6.6). A DNA sequence is "read" according to its bases: A (adenine), G (guanine), C (cytosine), and T (thymine). Its genetic information comes from the different sequences of As, Ts, Gs, and Cs. These polynucleotide polymers are many millions of nucleotides long.

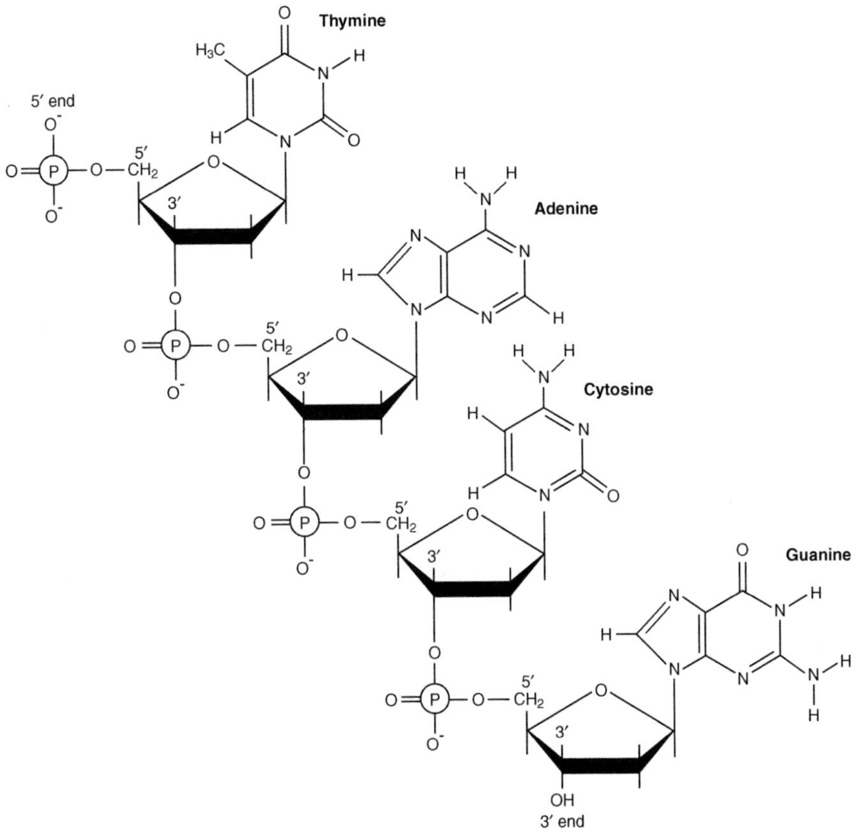

FIGURE 6.6
Nucleotides linked in a chain. A single DNA strand consists of a chain of nucleotides formed by bonds (called phospho-diester bonds) between the phosphate and the sugar molecule, making a backbone on which the bases (T, A, C, G) are assembled.

Watson and Crick's remarkable contribution was in proposing that DNA was a **double helix** of two nucleotide strands (Figure 6.7). They knew from prior work that the ratio of A to T in DNA was about one, as was the ratio of G to C. After reviewing the X-ray crystallography studies done by Rosalind Franklin, they deduced that DNA's double helix

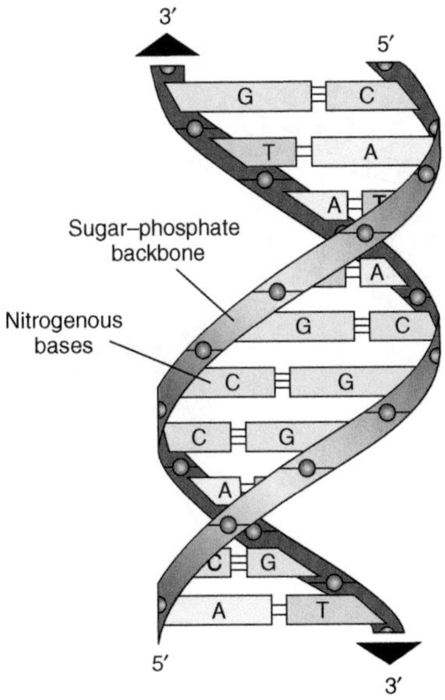

FIGURE 6.7
The DNA double helix. Two DNA chains wind around an axis with the sugar–phosphate backbones on the outside (pink) and pairs of bases meeting in the middle. The two chains are antiparallel, i.e., one runs from 5′ to 3′ downward; the other 5′ to 3′ upward.

consisted of two **antiparallel** strands—two strands running in opposite directions—in which As on one strand bonded to Ts on the other; similarly, Gs bonded to Cs. This complementary base pairing provided the means both for holding the DNA together and for allowing the two chains to separate during DNA replication, i.e., the process that exactly duplicates DNA during cell division, when genetic information is transmitted. Simply put, DNA can be thought of as a right-handed spiral staircase in which its two sets of stairs run in opposite directions. These sets of stairs are held together by chemical bonds between the "stairs:" A of one stair to T of its complementary stair, and so on. Because of this **complementarity**, knowledge of the sequence of stairs on one set makes obvious the sequence of the other.

DNA Replication

During each cycle of cell division—meiotic or mitotic—all the genetic information in a cell must be copied and transmitted to its progeny cells. This is accomplished through the process of DNA replication (Figure 6.8). Replication can be divided into two parts, **initiation** and **elongation,** with each part requiring the action of many enzymes. Initiation begins by unwinding the double-helical DNA to expose the polynucleotide chain and

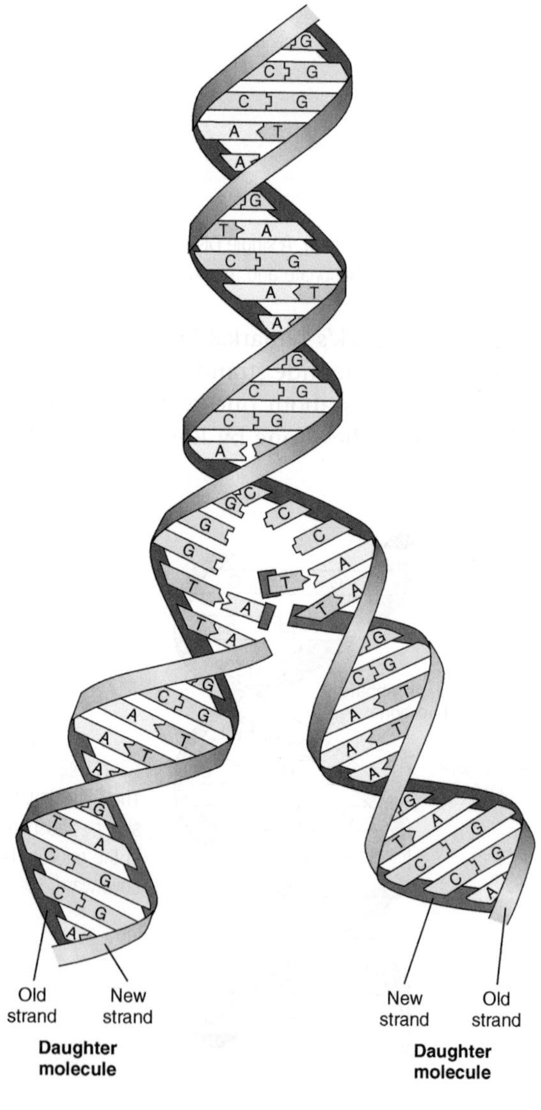

FIGURE 6.8
The widely accepted model of DNA replication. As the double helix unwinds, each strand serves as a template for the synthesis of a new strand by complementary base pairing. This results in a single double helix becoming two identical daughter double helices.

dissociate one strand from its partner. This is accomplished through creation of a replication fork and a replication "bubble" catalyzed by the enzyme DNA helicase. Then other enzymes, notably DNA polymerase, catalyze the synthesis of a new DNA strand complementary to the one it copies. Elongation then extends this copying process until its completion. DNA replication is semi-conservative, meaning that during copying, one strand of the "new" DNA is conserved from the parent molecule while the other is newly synthesized (Figure 6.9).

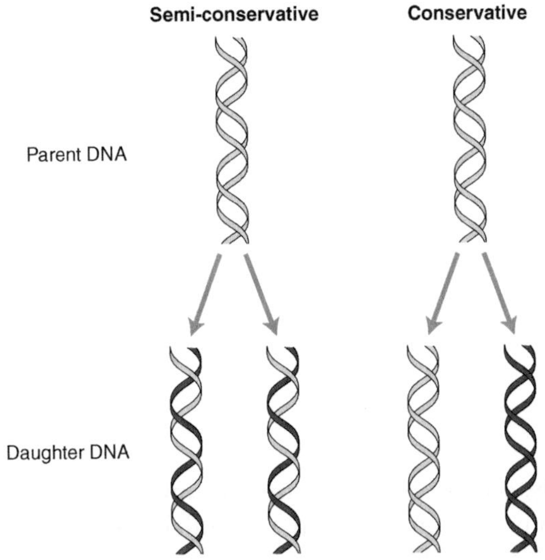

FIGURE 6.9
Semi-conservative replication of DNA. Parent DNA is copied such that one strand is conserved from the parent molecule (pink), and the other strand is newly synthesized (red). This type of replication is distinguished from conservative replication, shown at the right.

DNA Packaging

CORE CONCEPT

Within the cells of an organism, DNA molecules carrying genes are assembled into **chromosomes**: organelles composed of DNA and associated proteins. The sum total of genes and chromosomes in each cell is its **genome.** The human genome consists of two sets of 23 chromosomes, composed of 6 billion nucleotides and about 21,000 protein-coding genes. Only about 2% of the DNA in the human genome is composed of genes. The remainder comprises nucleotide sequences with variably understood regulatory and evolutionary significance.

Within the nucleus of all eukaryotic cells, genes are assembled into chromosomes. Humans have 46 chromosomes (22 pairs of **autosomes**, and one pair of sex chromosomes) (Figure 6.10). Each chromosome is a single molecule of DNA containing hundreds to thousands of genes. Chromosomes also have components other than DNA. These components include a variety of proteins: **histones**, which are fundamentally important to DNA packaging, and **non-histone** associated proteins. "**Chromatin**" is the word used to describe the complex of DNA and the proteins that make up a chromosome. By weight, chromatin is made up of equal amounts of DNA, histones, and associated proteins. Human cells have thousands of different DNA-associated proteins, each one responsible for a specific function. For example, some make up the structural scaffold of the chromosome; others are enzymes or transcription factors required for replication, separation, transport, regulation of gene expression, and other functions.

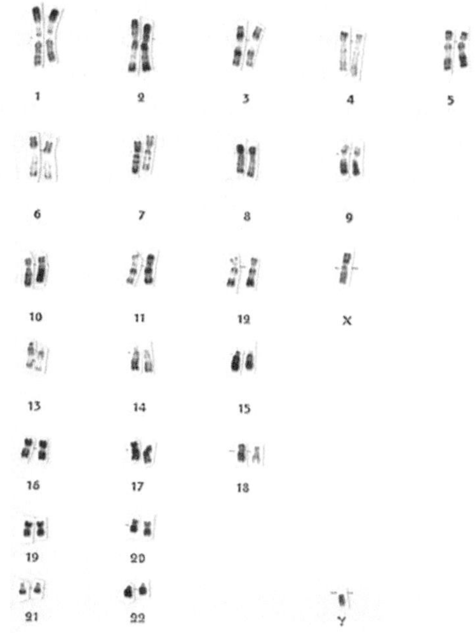

FIGURE 6.10
A normal human karyotype. The metaphase chromosomes are paired, sorted by size, and stained with Giemsa to yield the banded pattern. Twenty-two pairs of homologous autosomal chromosomes and the single X and single Y identify the donor as a male.

When looked at linearly, even the smallest chromosome is far too long to fit inside a cell's nucleus; it must be compacted, and this is the essential role of histones (Figure 6.11). First, DNA is wound around histones (like thread around a spool), forming **nucleosomes**. Next, the nucleosomes are coiled together into particles visible under the electron microscope. Five distinct histones make up the **nucleosome spools** (H1, H2A, H2B, H3, and H4). Higher-order packaging then takes place, ultimately leading to the compacted metaphase chromosome as it is usually depicted.

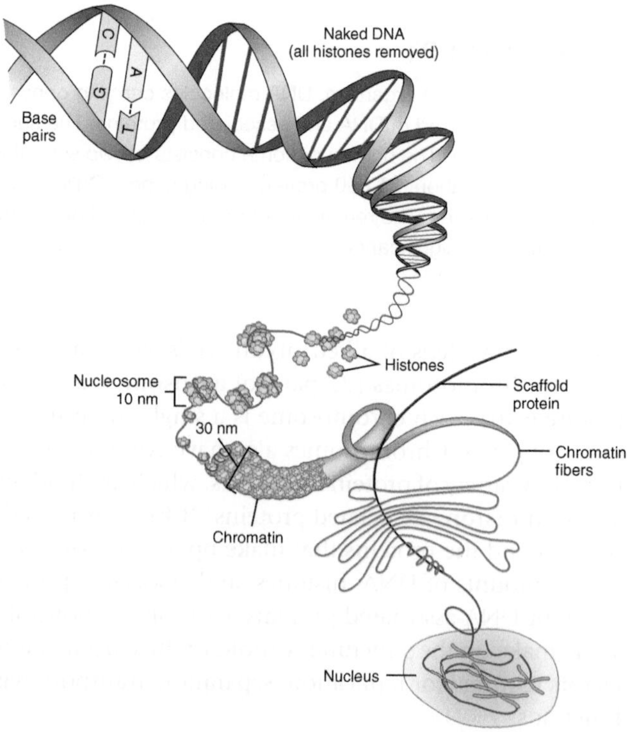

FIGURE 6.11
Super-coiling of DNA. The naked DNA, shown at the top, coils around octamers of four different histone subunits, forming nucleosomes. The nucleosomes, connected by linkers, are coiled together into fibers that are further compacted in the nucleus.

Chromosome Structure and Content

Karyotypes, the visual depiction of all the chromosomes in a cell, are prepared from cells just before they undergo cell division—that is, during metaphase. At this stage chromosomes are super-coiled enough to be visible under the light microscope. At all other stages in the cell cycle, chromosomes are uncoiled, linear structures, not visible except by electron microscopy.

A karyotype displays chromosomes in pairs, according to size (Figure 6.12A). The largest human chromosome (number 1) is nearly 250 million nucleotides long and contains about 2,200 genes (Figure 6.12B). The smallest chromosome (number 21) is about 40 million nucleotides in length and contains about 400 genes. The sex chromosomes (X and Y) are generally depicted in the lower-right corner of a karyotype.

As noted in Figure 6.12, decreasing chromosome length (from 1 through 21) does not correspond closely to the number of genes on any particular chromosome. This striking difference in "**gene density**" has important physiologic correlates that we will come back to later.

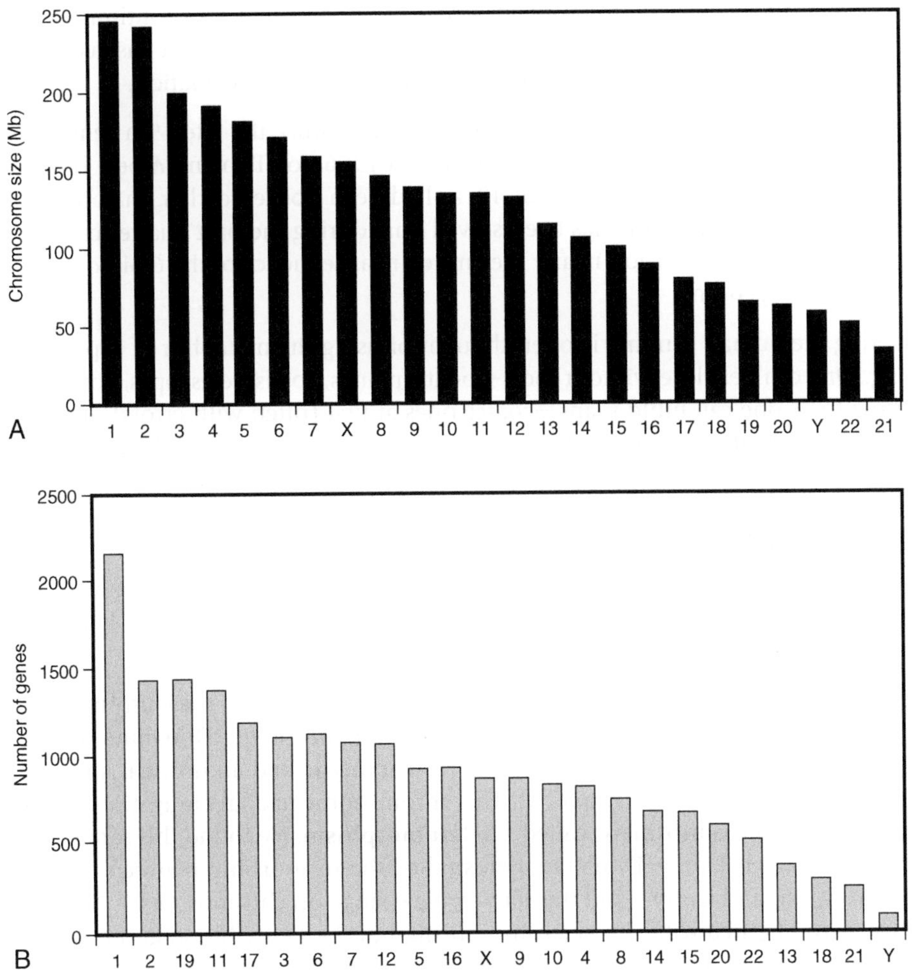

FIGURE 6.12
Size and gene content of the 24 human chromosomes. In (A), chromosomes (numbered along the abscissa) are ordered from left to right by size (in Mb, i.e., millions of base pairs). In (B), chromosomes are ordered by gene content (in number of genes per chromosome). Note that the abscissas of A and B do not correspond.

FROM GENES AND CHROMOSOMES TO GENOMES

CORE CONCEPT

Genomics, the study of whole genomes, was made possible by determining the complete DNA sequence of humans and that of about 800 other organisms. This triumph of modern genetics, made possible by employing old technologies (such as gene cloning, restriction mapping, gel electrophoresis) and new ones (such as high throughput DNA sequencers and powerful computational tools) is beginning to reveal the kinds and extent of variation in DNA structure not previously anticipated and not yet well understood. Already, however, the study of remarkably common **single nucleotide polymorphisms (SNPs)** and **copy number variations (CNVs)** is providing us with information about mechanisms of susceptibility to many human traits and disorders.

History of Discovery about Genes and Genomes

To appreciate the elegant relationship between genes and genomes, one must understand something about the history of this information. This history has taken a century and a half to unfold.

Mendel and Garrod pioneered the study of single genes, and that area continues to advance. We now recognize Mendelian traits for nearly half of our 21,000 genes. It has taken 100 years, by studying one gene at a time, to acquire our current level of information.

Chromosome structure, too, has been an object of study since the late 19th century, when Boveri first identified them under the microscope. In 1956, Joe Tsio and Albert Levan showed that humans had 46 chromosomes—not 48, as had been posited earlier. In contrast, the structure of the human genome has been solved only during the past quarter-century. Today, we can write with great precision nearly the entire linear sequence of the 6 billion nucleotides that make up our genome.

Deciphering the human genome is often thought of using the metaphor of exploration and mapping. One can't explore without tools—boats, planes, roads, telescopes, compasses, etc. With these tools one can build maps—coarse ones at first (often with errors), increasingly sophisticated ones later. This metaphor bears great relevance to solving the structure of the human genome.

Linkage Mapping

Until the late 1960s, not a single human gene had been mapped to an autosome. By this time, a number of genes had been mapped to the X chromosome because of the characteristic pedigree pattern of X-linked traits discussed in Chapter 5.

The first gene mapped to an autosome, reported in 1968, was that for the Duffy blood group of antigens. This was accomplished by a medical student named Roger Donahue. He did this by studying his own family. Donahue noted that his **karyotype** had an unusual appearance of one copy of chromosome 1 (Figure 6.13). This chromosome had an elongation and thinning of a region next to the **centromere**, called a **heteromorphism**. He found this same variation in other members of his family. Donahue then studied the relationship between blood groups—ABO, Rh, and Duffy—and this karyotypic change on chromosome 1. No ABO or Rh alleles were co-inherited with the heteromorphism, but the Duffy blood group was. Duffy blood group a (one of its two alleles) always was found along with the heteromorphism

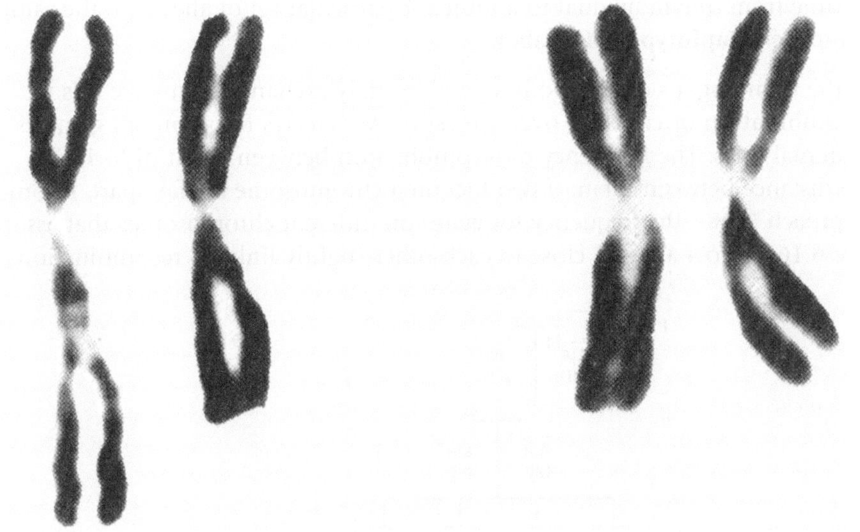

FIGURE 6.13
Chromosome 1 heteromorphism in the Donohue family. On the left, the Donohue chromosome 1, one of the two chromosomes is elongated and extended, the other is not. On the right, the usual chromosomes, neither chromosome is heteromorphic.

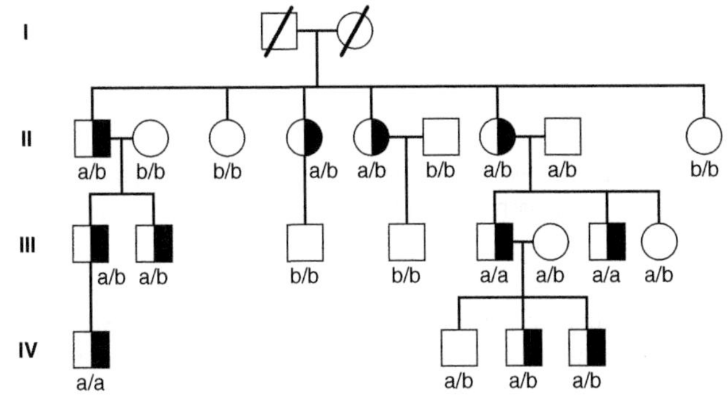

FIGURE 6.14
Demonstration of linkage of Duffy blood group to chromosome 1. In generations II, III, and IV of the Donohue pedigree, the chromosome with the heteromorphism (half-filled symbols) travelled with the Duffy group a allele in all cases. The Duffy a and b genotypes are shown under the symbols.

(Figure 6.14). Donahue concluded that the Duffy gene locus must be on chromosome 1, and that it was close to the heteromorphic region because no **recombinants** were found. In this case, recombinants would be created by crossing over between homologous chromosomes, as discussed in Chapter 4. If two loci are so near each other on a chromosome that recombination doesn't occur between them, geneticists say the loci are **linked**.

Today, the concept of **linkage** permeates the study of all mammalian genetics. Distinct from the independent assortment Mendel observed, linkage is defined as the co-inheritance of two or more non-allelic genes because their loci are in close proximity on the same chromosome. During meiosis, linked genes remain associated more often than the 50% probability expected for genes on different chromosomes. A **linkage map** depicts the relative positions of genetic markers on a chromosome; each chromosome has its own map. A **linkage marker** is a gene, nucleotide sequence, protein product, visual chromosome appearance, or disease useful in conducting linkage analysis (Figure 6.15). Genes are linearly arrayed on a chromosome, their

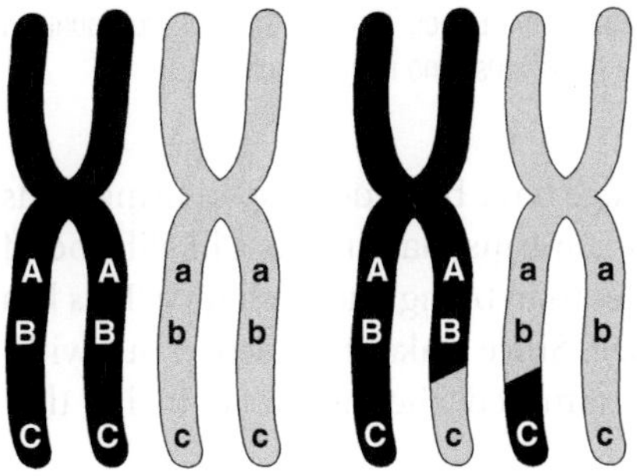

FIGURE 6.15
Stylized chromosomes displaying haplotypes and crossing over. In the black-colored chromosome on the left, alleles A, B, C of different genes comprise one haplotype, while alleles a, b, c on its homologue (light-colored chromosome) comprise another. On the right, meiotic crossing over is depicted: allele c is recombined onto the dark-colored chromosome, allele C to the light-colored one.

order being the same from one individual to another. A particular set of alleles on the same chromosome comprise a **haplotype** (ABC; abc).

During meiosis, the homologous chromosomes pair and may exchange segments; this is referred to as **recombination** or **crossing over**. Crossing over creates recombinant gametes different from parental ones. The frequency of recombination between a pair of loci is a function of the distance between them; if two loci on a chromosome are far apart, recombination may approach 50%—the frequency for genes on different chromosomes that assort randomly (Figure 6.16). If loci are very close to each other (tightly linked), recombination

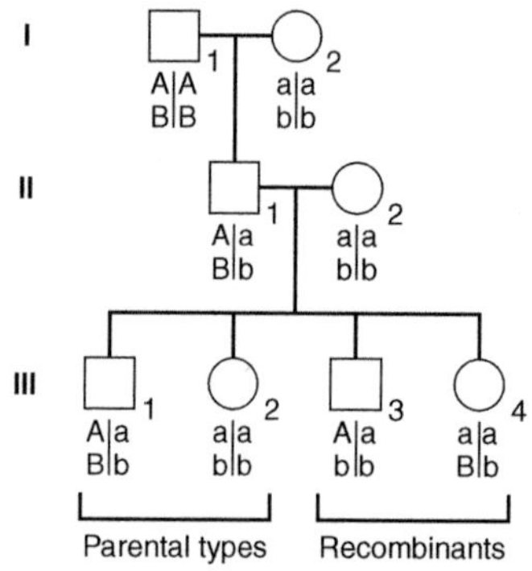

FIGURE 6.16
Absence of linkage between A and B loci. In generation III, half of offspring carry parental genotypes, half are recombinants (as would be expected for loci on different chromosomes or so far from each other on the same chromosome as to favor recombination between them).

between them will be very rare, even approaching zero (Figure 6.17). If the two loci are less tightly linked, recombination will occur at a frequency related to the distance between them (Figure 6.18). The frequency of recombination is usually expressed as Θ, and ranges from 0 (absolutely linked) to 0.5 (absolutely unlinked).

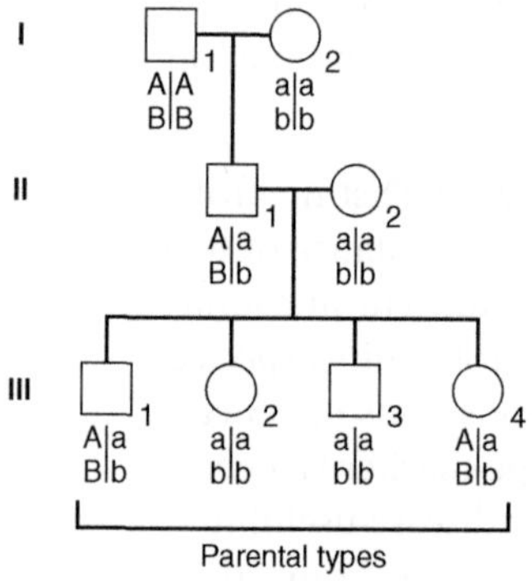

FIGURE 6.17
Complete linkage between A and B loci. All individuals in generation III are of parental genotypes; i.e., no recombinants are found.

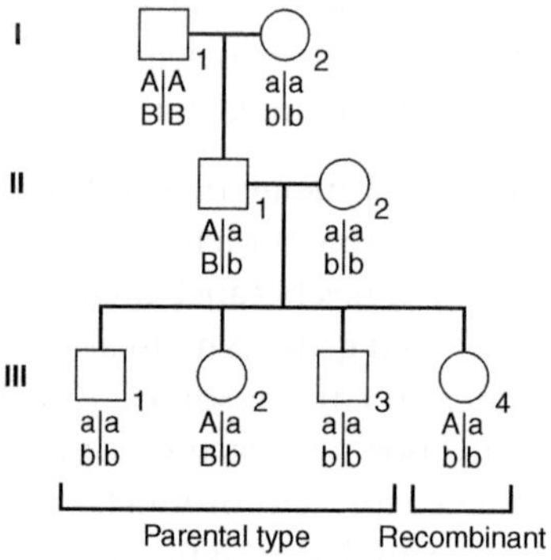

FIGURE 6.18
Intermediate linkage between A and B loci. Three of the four individuals in generation III have parental genotypes, one is a recombinant.

From these principles, more complex probabilities of linkage have been developed using odds ratios and logs of odd ratios (**LOD scores**). A LOD score of 3 means that there is a likelihood of more than 999/1,000 that the observed linkage is real rather than being due to chance. This has become the convention accepted as demonstrating linkage. Since linkage is carried out with pedigrees, the larger the pedigree or the more pedigrees examined, the more convincing the evidence is for or against linkage.

Recombinant DNA Technology

By the end of the 1970s only a few dozen genes had been mapped to autosomes because the number of usable markers was so few. With the discovery of **recombinant DNA technology**, all that changed. By recombinant DNA (rDNA), we mean a combination of DNAs from different origins, that is, different organisms (such as bacterial and human).

Recombinant DNA technology depends on five "tools." These are:

1. **Restriction enzymes** are bacterial enzymes that cut DNA, like a scissors, at specific sites, producing fragments of different sizes (Figure 6.19); some restriction enzymes recognize

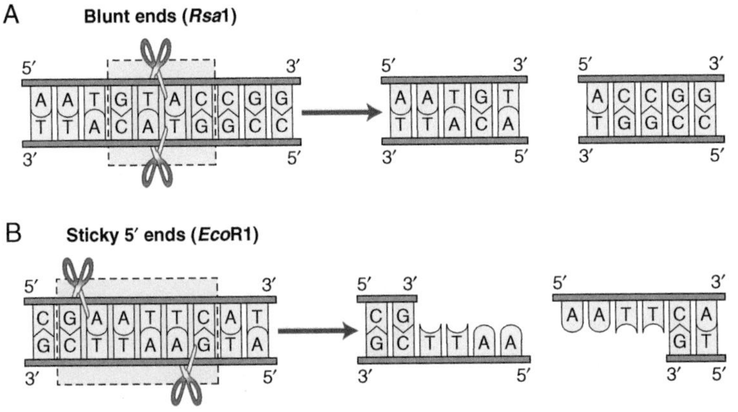

FIGURE 6.19
Restriction enzymes cut DNA at specific sites, producing fragments. Some enzymes (e.g., *Rsa*1) in (A) produce fragments with blunt ends, others (e.g. *Eco* R1) in (B) produce fragments with a 5′ overhang. Such enzymes evolved to protect bacteria from viruses that infect them.

a four-nucleotide sequence (e.g., AGGA), others a six-nucleotide one (GAATTC). These fragments can be run out by **electrophoresis** (where electrically charged molecules are allowed to migrate through a fluid or gel under the influence of an electric field, thereby being separated), producing a particular pattern of fragments arrayed by size, from largest to smallest (Figure 6.20).

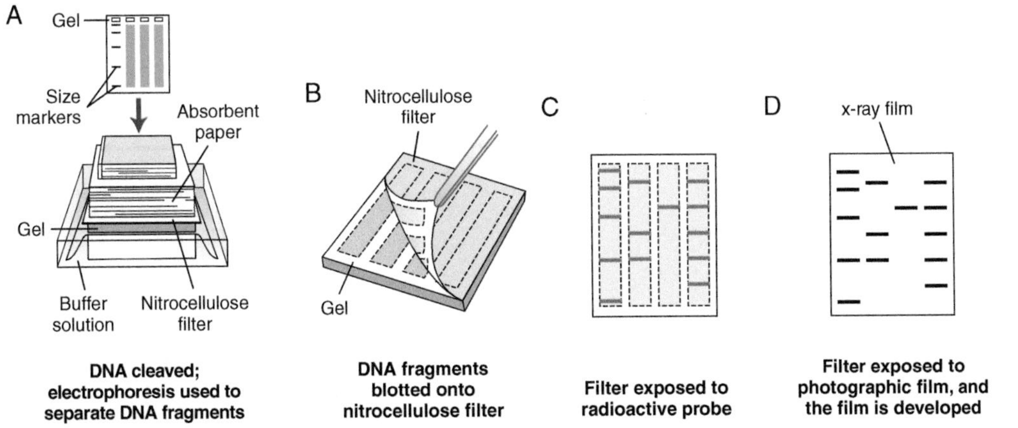

FIGURE 6.20
The Southern blot, used to identify DNA fragments. (A) DNA is cleaved and its fragments separated by electrophoresis; (B) fragments are blotted onto a nitrocellulose filter; (C) filter is exposed to radioactive probe; (D) filter is developed by exposure to photographic film. Bands assort with the largest fragments near the top of the film, the smallest ones near the bottom.

2. **Molecular cloning** is the process by which DNA fragments are spliced into viral or bacterial vectors, purified, and amplified (Figure 6.21).

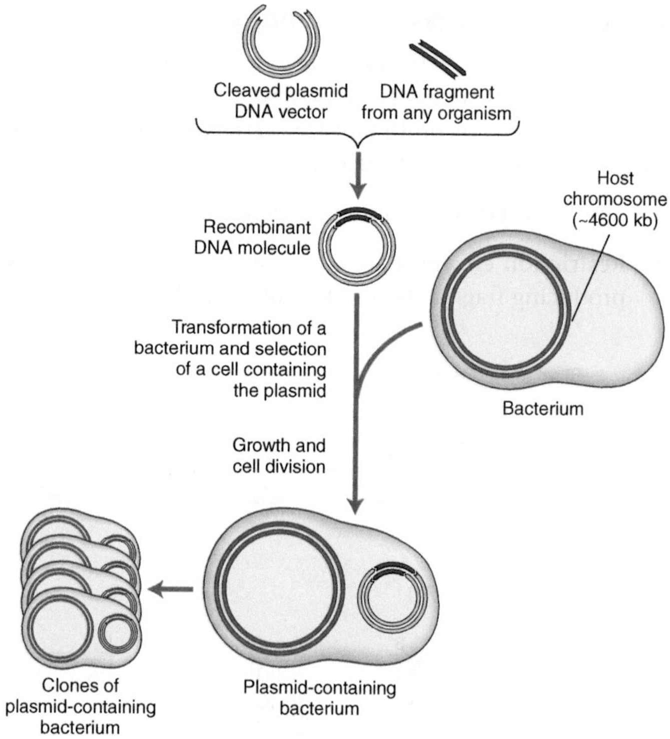

FIGURE 6.21
Creating recombinant DNA molecules with plasmid vectors. A DNA fragment, prepared by digestion with *Eco* R1, is joined to an *Eco* R1 cleaved plasmid; the recombinant plasmid then transforms a bacterial cell, where the plasmid is replicated as the bacteria proliferate; the recombinant plasmids are isolated and the cloned donor fragments of DNA pooled.

3. **Hybridization probes** are single-stranded, purified DNA sequences of varying length (25 to several thousand nucleotides) that are labeled with a radioactive isotope or fluorescent dye. Complementarity allows them to hybridize with, and thereby identify, corresponding sequences in cloned collections of DNA fragments.
4. **Polymerase chain reaction (PCR)** uses a specific bacterial polymerase to amplify a piece of DNA up to a billion-fold or more. This is a powerful means of obtaining sufficient DNA for a variety of purposes (Figure 6.22).

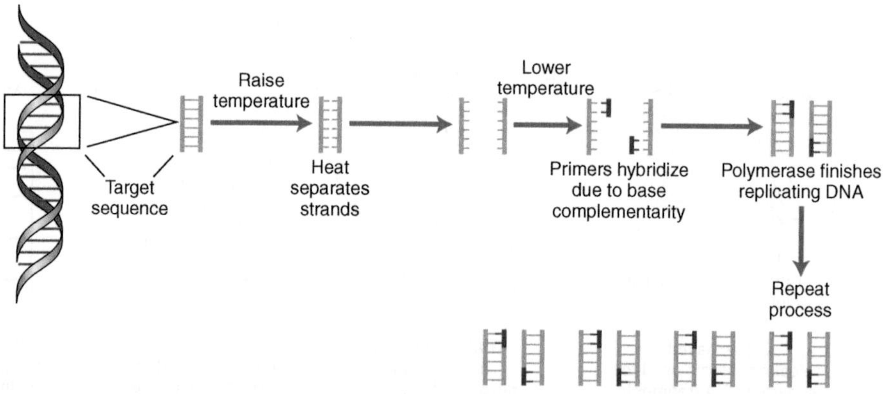

FIGURE 6.22
Amplification of a DNA sequence using the polymerase chain reaction (PCR). The isolated target sequence is mixed with a thermo-stable DNA polymerase (usually Taq 1). Amplification is accomplished by using specific primers and modifying the reaction temperature. Billions of copies of the target sequence are formed.

5. **DNA sequencing** reveals the order of base pairs in an isolated DNA molecule or fragment. As shown in Figure 6.23, the Sanger sequencing method deploys fluorescent-labeled analogues of A, T, G, and C that interrupt the synthesis of a DNA strand complementary to the template molecule. From the sequence of terminated fragments, the sequence of the original template can be determined—a once painstaking technique that has become fully automated.

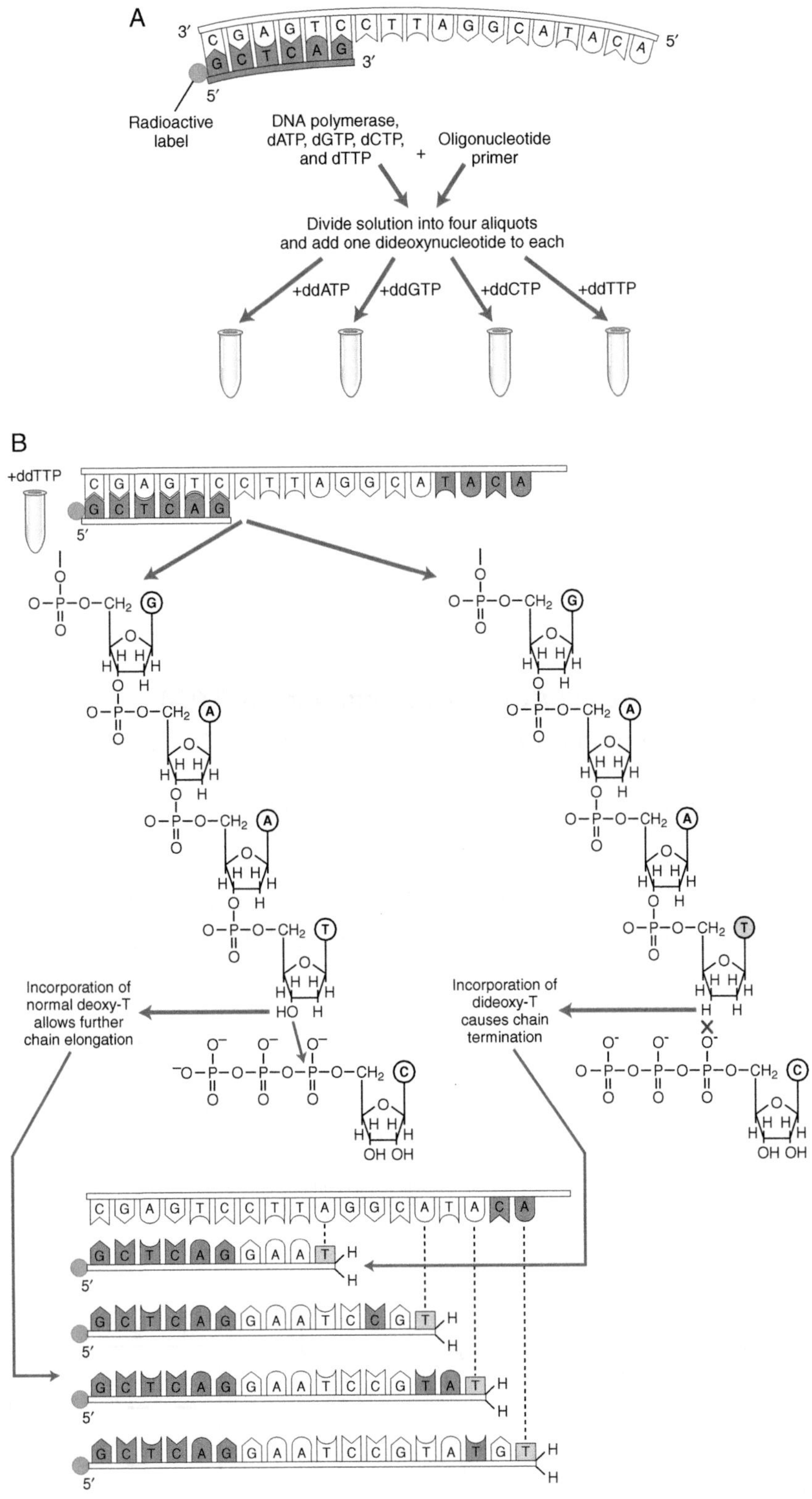

FIGURE 6.23
Sanger sequencing of DNA. (A) The DNA fragment to be sequenced is mixed with an fluorescently-labeled primer complementary to a portion of that sequence, then DNA polymerase and deoxynucleotides are added and the mixture is divided into four parts. (B) Into each part, a single chain-terminating, fluorescently-labeled dideoxy nucleotide is added. (C) After the dideoxy analogue has caused termination of the growing chain, the four aliquots are separated by electrophoresis according to size. The sequence is then read starting with the smallest fragment according to its dideoxy analogue.

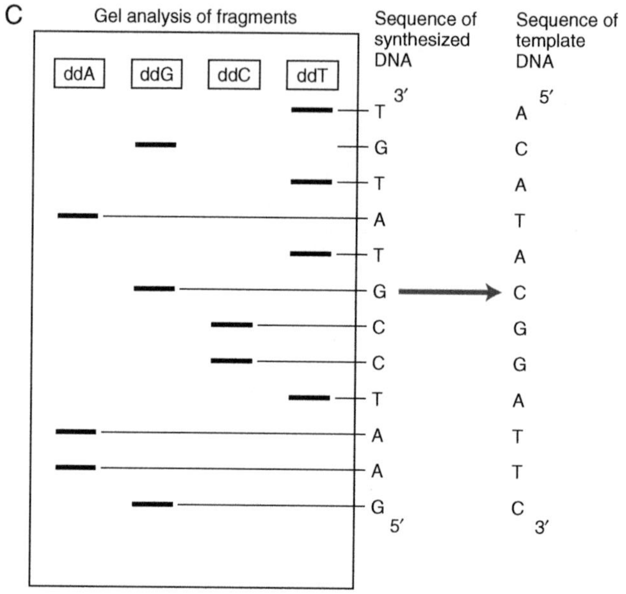

FIGURE 6.23
(Continued)

Restriction Fragment Length Polymorphisms (RFLPS)

When DNA from an organism (such as virus, fly, human) is cut by a particular restriction enzyme, a series of different-sized DNA fragments is produced. These fragments can be separated on a gel, thereby producing a particular pattern, ranging from very large fragments to very small ones. By hybridizing the fragments with a probe (i.e., a DNA sequence of many different types), a fragment length pattern is produced (Figure 6.24). By comparing such patterns in different individuals or families, the fragments can be used as alleles, that is, alternate forms of that particular **restriction fragment length polymorphism**, or **RFLP**.

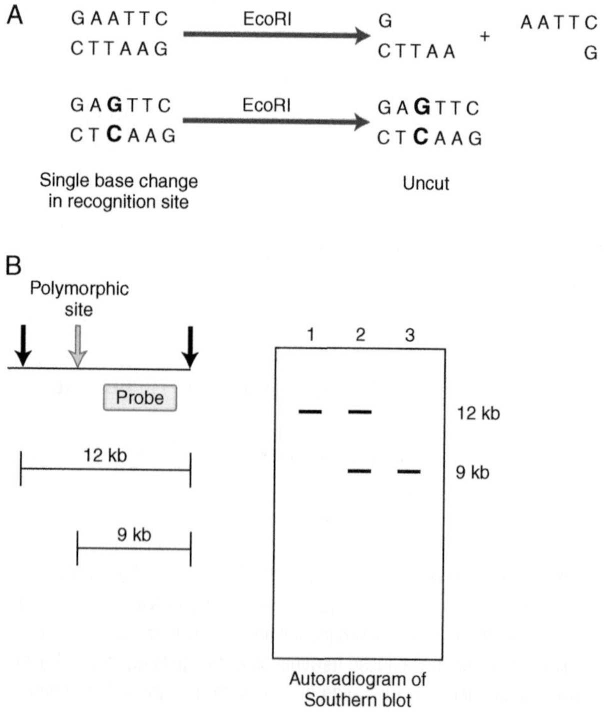

FIGURE 6.24
Restriction fragment length polymorphism (RFLP). (A) Bacterial restriction enzymes, e.g., *Eco* R1, can identify polymorphic sites in DNA that either are cut by the enzyme or not. The fragments, then, behave like a two-allele system, with two homozygotes and one heterozygote. (B) If electrophoresed DNA containing these fragments is probed with the isotopically labeled sequence of interest, a plot is obtained that distinguishes these genotypes from one another. In the example shown here, the solid arrows identify sites common to all genotypes, the colored arrow the polymorphic site. Two different-sized fragments are produced—12 kb and 9 kb. Homozygotes not having the polymorphic site (lane 1) have a single 12-kb band; heterozygotes (lane 2) have a 12-kb band and a 9-kb band; homozygotes having the polymorphic site (lane 3) have a single 9-kb band.

Combining this kind of pattern recognition with the mathematics of probability (i.e., LOD scores) provided linkage mapping with a powerful tool.

The Human Genome Project

We can now see that by the mid-1980s, the tools needed to solve the human genome were in place. These included rDNA technology, RFLPs and other new linkage markers, the use of positional cloning (see Chapter 7) to identify disease-causing genes, automated DNA sequencers capable of producing orders-of-magnitude more DNA sequence than obtainable manually, and bioinformatic power provided by high-speed computers.

In 1984 and 1985, several scientists and groups proposed that it was timely and important to sequence the entire human genome. Some prominent scientists embraced this view. Others did not, arguing that the effort was "big science" rather than traditional, small investigator-driven work, that it would siphon financial resources from more useful, hypothesis-driven science, and that it would be a tedious, "crank-turning" task. These skeptics argued further that, because most of the genome did not code for protein products, it was unnecessary to sequence all the DNA. To resolve this controversy, Congress directed the appointment of two committees of prominent scientists and policy makers—one by the National Academy of Sciences, the other by the Office of Technology Assessment. Both groups met for a year (1987–1988). Each concluded that sequencing the human genome would provide a huge amount of vital information relevant to the study of evolution, to a molecular view of health and disease, and to an understanding of gene regulation. Thus, in 1989, Congress launched the **Human Genome Project (HGP)**. The project was predicted to take 15 years and to cost about 3 billion dollars. To direct the project, a new center—the National Human Genome Research Institute (NHGRI)—would be established at NIH. The Department of Energy (DOE) would be engaged as well.

AMPLIFICATION: GENOMICS CONCEIVED

Inside the Committee Room: The Recommendation to Undertake the Human Genome Project

In 1986 the United States Congress directed the National Academy of Sciences (NAS) to address the following question: Should the United States embark on a project aimed at determining the complete sequence of human DNA? Accordingly, a "blue ribbon" committee of 1 woman and 15 men was appointed, chaired by Bruce Alberts, then a professor at the University of California, San Francisco (and later president of NAS). Each member of the committee had been elected to NAS; four were Nobel Laureates.

The deliberations of this group were as informative as they were fractious. Some members were convinced at the outset that the project had to be undertaken. They employed forceful arm-twisting and verbal bludgeoning to convince those several others who were undecided (among whom was this book's co-author, Dr Rosenberg). This central question was not the only contentious point of debate. Others included: whether genome mapping should precede sequencing or be conducted in parallel with it; whether all nucleotides should be sequenced or only those coding for proteins; whether the project should be carried out by a single center or by a consortium of many; and whether government responsibility should be vested in the National Institutes of Health (NIH) or the Department of Energy (DOE).

Alberts managed this group—some whose arrogance and bombast matched, and even exceeded, their accomplishments—with laudable even-handedness and political savvy. To his credit, only one member defected during the year-long deliberations, at the end of which the committee voted unanimously to recommend undertaking the project according to precisely defined phases and a prescribed administrative framework.

Now, some 25 years later, I still feel that this collegial effort was a vivid example of the value of honest, transparent peer review. I know that several others who were on the committee share this view. The Human Genome Project (HGP), whose importance was only guessed at initially, has already expanded our knowledge of human genetic variation and disease and will continue to do so well into the future.

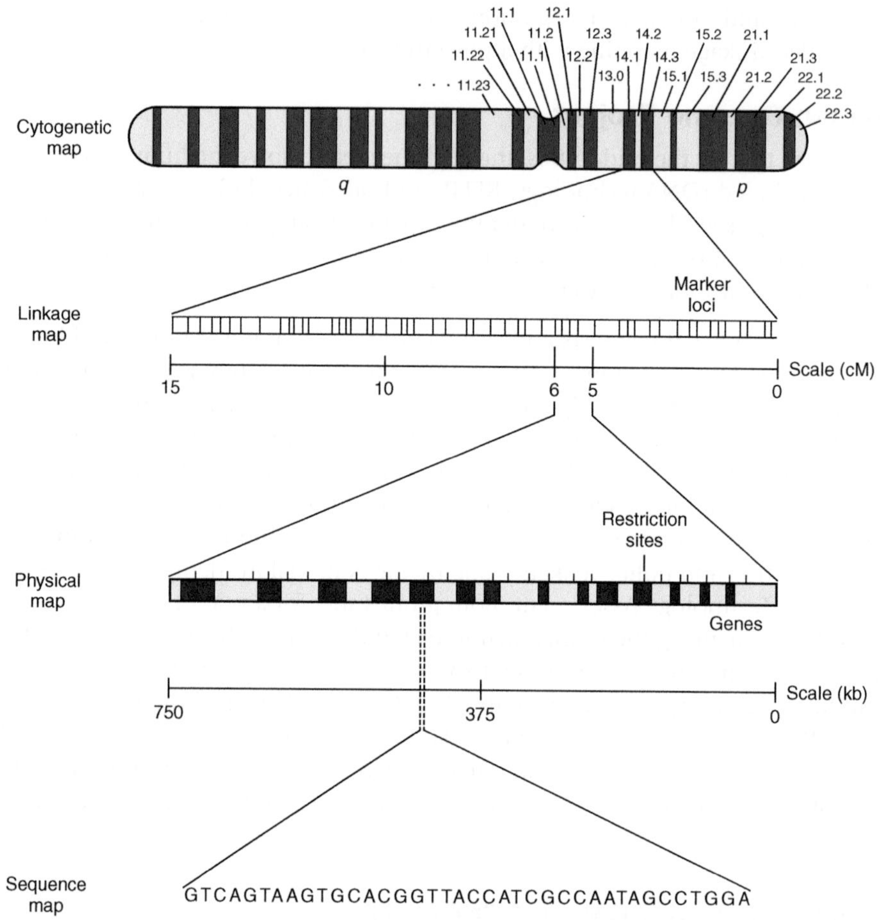

FIGURE 6.25

Phased strategy of the human genome project (HGP). (A) A cytogenetic map (top) is made using genes, genetic markers, telomeres, centromeres, and genetic disorders. It can locate sites 1 Mb apart. (B) A linkage map identifies millions of common polymorphisms (SNPs) and simple sequence repeats (SSR). It can locate sites hundreds of kb apart. (C) A physical map constructed of overlapping, cloned DNA fragments can locate sites tens of kb apart. (D) A sequence map (bottom) depicts the order of individual nucleotide bases.

STRATEGY

The HGP was to proceed in phases, with mapping preceding sequencing (Figure 6.25). First a dense cytogenetic map was to be prepared, placing linkage markers close to each other on every autosome and sex chromosome. Next, higher-resolution physical maps would be created for each small interval of each chromosome. These pieces of DNA would be placed into **bacterial artificial chromosomes (BACs)** using rDNA technology. Finally each piece of DNA would be sequenced, this output then being aligned using high-speed computer algorithms. To confirm and extend the information gained from sequences obtained from human DNA, sequences of other "model" organisms (such as bacteria, yeast, fruit flies, roundworms, and mice) would be obtained. From the outset, 3% of the funds appropriated for the project were set aside to consider and stay abreast of the ethical, legal, and social implications of the HGP (the ELSI Research Program).

PROGRESS

Between 1990 and 1995, great progress was made in cytogenetic and physical mapping. More than 75,000 highly informative markers (RFLPs, DNA sequences coding for proteins, simple repeats, translocations, and diseases) were placed along each chromosome at about 1 million base intervals. Then, more than 30,000 pieces of DNA were cloned, using the cytogenetic maps as starting points. These clones covered more than 99% of the genome.

Sequencing proceeded slowly, however, until the development of higher-speed, automated sequencers and even more powerful computers. Less than 1% of the human genome had been sequenced by 1995, but progress accelerated dramatically thereafter. The new technologies enabled a single sequencer to produce in 3 months what the entire scientific community had generated by 1995. This acceleration was produced, in major part, by intense competition between the government-sponsored consortium and an effort funded by a private company, **Celera**. In 1996, Celera published the first compete DNA sequence for any organism—the *Haemophilus influenza* species of bacteria. In short order, complete sequences for yeast, worm, fly, and plant sequences were solved, as was more and more of the human sequence. In 2001, a "rough draft" of the human genome sequence was published independently both by the government group and by Celera. This compilation represented more than 99.5% of the genome. The remainder was provided by 2003.

Interestingly, the two efforts used different sequencing strategies (Figure 6.26). The government-sponsored group sequenced DNA propagated in bacterial artificial chromosomes B(ACs) then used algorithms to join each of 220,000 pieces to decipher the entire genome. Celera deployed

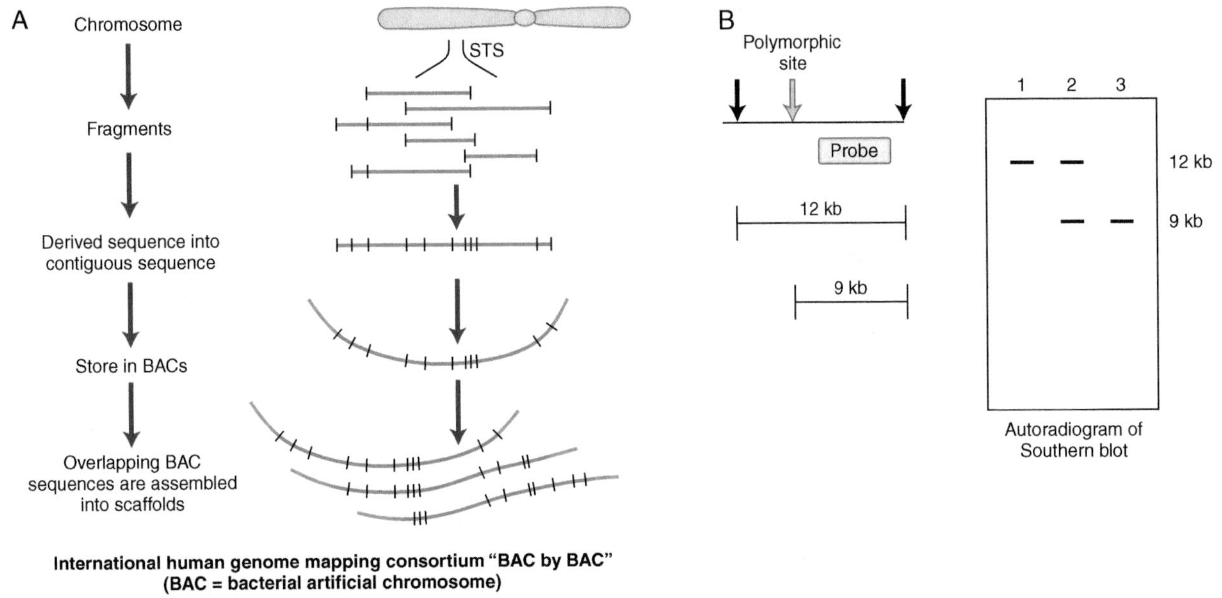

FIGURE 6.26
Two tactical approaches to sequencing the human genome. (A) The "BAC by BAC" approach used by the international consortium: the target genome is fragmented and cloned into bacterial artificial chromosomes (BACs); the fragments are organized into a physical map; the BAC clones are sequenced; the overlapping sequences are assembled. (B) The "shotgun" approach employed by Celera. The whole genome to be sequenced is "shot-gunned" (i.e., fragmented into different-sized pieces). These pieces are then sequenced directly and a genome-wide shotgun computer program is used to assemble the final sequence.

a "shotgun" approach in which the total genome was sheared enzymatically or by sonic means, the pieces sequenced, and then ordered with the aid of computer-driven informatic systems. Celera was aided measurably by the fact that all of the public consortium's sequence was in a database open to all. Celera's output was kept private, as would be expected for a for-profit venture. The timeline of the Human Genome Project is summarized in Table 6.1.

TABLE 6.1 Timeline of the Human Genome Project (HGP)

Year	Event
1985	Scientists raise the idea of sequencing the human genome
1987	Committees of the National Academy of Sciences (NAS) and the Department of Energy (DOE) recommend proceeding with the HGP
1989	Congress authorizes NIH and DOE to fund the HGP
1990	National Center for Human Genome Research (NCHGR) established at NIH (expanded to National Human Genome Research Institute (NHGRI) in 1992)
1994	Dense linkage map of human genome produced
1995	First complete genomic sequence published (of *H. influenzae*)
1998	Complete genomic sequences of yeast, roundworm, and fruit fly published
1999	Public consortium and private company (Celera) race to complete human genome sequence
2001	Rough drafts of complete human genome sequence presented independently by public consortium and by Celera
2003	Almost complete sequence of human genome published

FINDINGS AND INTERPRETATIONS

The HGP was deemed completed in 2003, but this is an overstatement. We still have much to learn about the fine details of the DNA structure, about the enormous amount of variation from individual to individual, and about the contribution of genome structure to gene, organ, and whole human function. Nonetheless, a great deal has been learned—much of it in the form of surprises. The major findings and revelations are as follows:

- The haploid human genome contains just over 3 billion nucleotides, divided unevenly into 24 linear molecules, one for each chromosome.
- The genome contains about 21,000 genes—that is, sequences that code for protein products. This gene complement is much less than what was anticipated when the HGP began.
- Approximately 2% of the genome consists of these gene-encoding sequences. The remainder, still being actively investigated, consists of a great variety of single-copy sequences and repeats.
- The non-genic single copy sequences are composed of non-coding RNAs, promoter and enhancer elements, and others yet uncharacterized. Altogether, these sequences make up about 50% of the genome.
- The other 50% of the genome is made up of a variety of repeats: **simple sequence repeats** (di-, tri-, and tetra-nucleotides); **centromeric repeats** of longer length; **segmental duplications** (10,000 to 300,000 base pair repeats that are found at more than one site in the genome); **pseudogenes**, inactive copies of RNA that have been reinserted into the genome; and **transposon-like repeats** similar in structure to those found in bacteria. **Transposons**, so-called movable elements, have importance that will be discussed in Chapter 7.
- Whereas our genome contains about 21,000 genes, our **proteome** (defined as the total inventory of proteins in a cell) ranges from 200,000 to even 1 million. This means that a single gene may encode information to make many proteins. It is now widely accepted that this inherent flexibility in gene structure and function is a central feature distinguishing humans from other simple (or complex) organisms.

These discoveries have led to a revised "world view" of the genome. In the old view, the genome was seen as a linear collection of individual protein-encoding genes acting alone, surrounded by a vast amount of "junk." In the new view, the genome is far more complex and dynamic: some genes are found within other genes; some genes code for RNAs—not proteins—and a portion of these RNAs regulate expression of other genes; some genes are more than 2 million bases long, others only 40,000 bases in length; some genes are regulated by sequences adjacent to or within them, while others are regulated by sequences far away. Looked at this way, the human genome is a highly complex, versatile master blueprint capable of being read in many ways and of designing a variety of structures and functions.

COMPARATIVE GENOMICS

Complete genome sequences from more than 800 organisms—from prokaryotic bacteria to eukaryotic *Homo sapiens*, have been presented. These genomes range in size from 700,000 base pairs in a microbial species to 640 billion base pairs in one variety of amoeba. Thus, there is no simple relationship between genome size and information content. Scanning the genomes of a representative group of "model" organisms reveals some interesting facts, not all understood by any means (Table 6.2).

TABLE 6.2 Comparative Genomics: Size and Complexity

Organism Type	Species	Genome Size (Mb)	Predicted Number of Genes	Number of Genes per Mb
Bacterium	*E. coli*	4.6	4,200	905
Yeast	*Saccharomyces cerevisiae*	12.1	5,800	483
Worm	*Caenorhabditis elegans*	100	19,100	197
Fly	*Drosophila melanogaster*	180	13,600	117
Mustard weed	*Arabidopsis thaliana*	125	25,000	221
Puffer fish	*Fugu rubripes*	380	38,000	118
Mouse	*Mus musculus*	3200	21,000	16
Human	*Homo sapiens*	3200	21,000	16

First, let us consider the matter of "**gene density**," defined as the number of genes per million base pairs. Bacterial gene density is about 900, compared to human gene density of 16. The puffer fish has a genome only about one-tenth the size of humans, yet it has more genes than we do. Thus, during evolution, there has been much greater expansion in genome size than gene number. It seems certain that such expansion provided adaptive advantage to one species over others, consistent with the fundamental ideas of natural selection.

Second, many genes have been conserved from the most simple species all the way to humans. This is taken as evidence that such genes control functions needed by all plant and animal species. Those genes lost or gained during evolutionary time must represent adaptive events. This view is strengthened further by the observation that human genes, for example, can be spliced into yeast and function as they do in human cells.

APPLICATIONS OF THE HGP TO HEALTH

With each passing month, the returns on the investment—financial and scientific—made in the HGP become greater. The HGP has accelerated the discovery of genes that cause or increase susceptibility to disease. It has expanded the repertoire of tests used in detecting disorders and identifying carriers. It has provided new targets for drug discovery and development. It has not, as some projected, revolutionized the practice of medicine, but that doesn't mean it may not do so in the next decades. Finally, it has raised important ethical, legal, and social questions that attest to the wisdom of staying abreast of these matters from the initiation of the HGP to the present. Each of these generalizations will be enlarged upon in subsequent chapters.

REVIEW QUESTIONS AND EXERCISES

1. Choose the phrase in the right column that best matches the term in the left column.

a. origin of replication	1. proved that genes are made of DNA
b. chromatin	2. a nitrogenous base composed of a single ring
c. Griffith	3. less than 2% codes for proteins
d. complementary bases	4. the sequence of bases at which unwinding of DNA occurs
e. linkage	5. non-covalent bonds that hold two strands of DNA together
f. bacteriophage	6. discovered transforming principle
g. pyrimidine	7. a nitrogenous base composed of two rings
h. genomic DNA	8. complex of DNA and proteins that make up a chromosome
i. Avery *et al.*	9. the co-inheritance of two genes because they are in close proximity on a chromosome
j. hydrogen bonds	10. two nitrogenous bases that pair by hydrogen bonds
k. telomere	11. the structure formed when DNA is wound around histones
l. nucleosome	12. the process of DNA duplication in which the new double helix contains one "old" strand and one "new" one
m. purine	13. a virus that infects bacteria
n. semi-conservative	14. the tips of chromosomes

2. The elements carbon, hydrogen, nitrogen, oxygen, phosphorus, and sulfur are common constituents of organic matter. Which of these elements is not found in DNA?
3. Distinguish between the following terms ending in "type."
   a. Haplotype
   b. Phenotype
   c. Karyotype
   d. Genotype.

4. What five experimental tools made the construction of recombinant DNA possible? Describe each briefly.
5. Why did the discovery of restriction fragment length polymorphisms (RFLP) accelerate construction of the human linkage map?
6. Answer the following questions about the human genome project (HGP).
   a. What was the goal of the HGP?
   b. Given that the cost of the HGP was 3 billion dollars, what was the approximate cost per nucleotide (assuming that all of the money was spent on sequencing human DNA)?
   c. If a human gene has homologues in chimps, mice, fruit flies, and bacteria, what does this tell us about the age and the function of that gene?
   d. To be strictly correct, each of us has two genomes, not one. What are these two?
   e. What does it mean to say that the human genome has a lower "gene density" than that of bacteria, yeast, or fruit flies?
   f. Now that the human genome has been completely sequenced, why do scientists still have to map and study genes?
7. It seems likely that within 10 years you will be able to carry an affordable card with your entire genome on it. List three ways that this information would be useful to you.
8. Prior to any autosomal gene mapping, a number of genes had been mapped to the X chromosome. Why was this possible?
9. Your Aunt Gertrude is disappointed to learn that humans have only the same number of genes as a puffer fish, and fewer genes than many plants. How do you cheer her up about the significance of gene numbers?

CHAPTER 7

Expression of Genes and Genomes

CHAPTER OUTLINE

CORE CONCEPT

The central dogma of gene action is abbreviated:

DNA → RNA → Protein.

Protein-coding genes transmit their information by being **transcribed**, i.e., their DNA codes for a complementary RNA referred to as **messenger RNA (mRNA)**. In turn, mRNAs are edited, leave the nucleus, and are translated into **proteins**—the molecules responsible for the functions cells carry out. Translation occurs according to a universal **genetic code**: the sequence of nucleotides in mRNA, read as triplets (called **codons**), that specify the sequence of **amino acids** in proteins. This fundamental pathway of gene expression is modulated such that only a subset of genes is active in any one tissue—thereby explaining **differentiation**; that is, the structural and functional differences between, for example, a brain cell and a heart cell. Gene expression may be modified in a number of ways: by changes in DNA primary structure; by chemical (epigenetic) modification of DNA; and by interference with mRNA structure or function produced by small RNA species, called siRNAs and miRNAs.

Human Genes and Genomes. DOI: 10.1016/B978-0-12-385212-0.00007-X

GENE EXPRESSION: HISTORY OF RESEARCH

The central dogma of gene expression is abbreviated: **DNA →RNA→Protein** (Figure 7.1).

Elucidating this unidirectional pathway composed of three classes of biological molecules has taken more than 100 years and is still undergoing refinement.

In 1908 Archibald Garrod, a British physician, recognized that "**inborn errors of metabolism**" were recessively inherited disorders. It took 30 more years (and the evidence that DNA is the genetic material) before Beadle and Tatum stated their "**one gene, one enzyme**" hypothesis. Their Nobel Prize-winning work, conducted with the fungus neurospora, said that a single gene codes for one, and only one, enzyme. This brilliant work joined the disciplines of genetics and biochemistry. Although additions, modifications, and corrections to the "one gene, one enzyme" formulation have occurred, the underlying concept remains powerfully explanatory.

The central dogma, first articulated by Francis Crick in the 1950s, states that the information encoded in DNA is transferred to proteins using **RNA** as an intermediate. Catalyzed by the enzyme **RNA polymerase**, one strand of DNA is used as a template to make a complementary, single-stranded RNA. This step is called "**transcription**," and is described in detail below. The RNA is processed to messenger RNA (mRNA) in the nucleus and is transported to the cytoplasm. The nucleotide sequence of the mRNA is used to define the order of amino acids in the protein. This step is called "**translation**" because the information in the nucleic acid "language" is translated into the language of amino acids and proteins.

In the 1950s, experiments with mutants of the *E. coli* bacterium revealed that the sequence of base pairs in DNA determines the sequence of amino acids in the protein in a **co-linear** way. This co-linearity adds a word-for-word relationship to the translation of nucleic acid "language" into its protein counterpart.

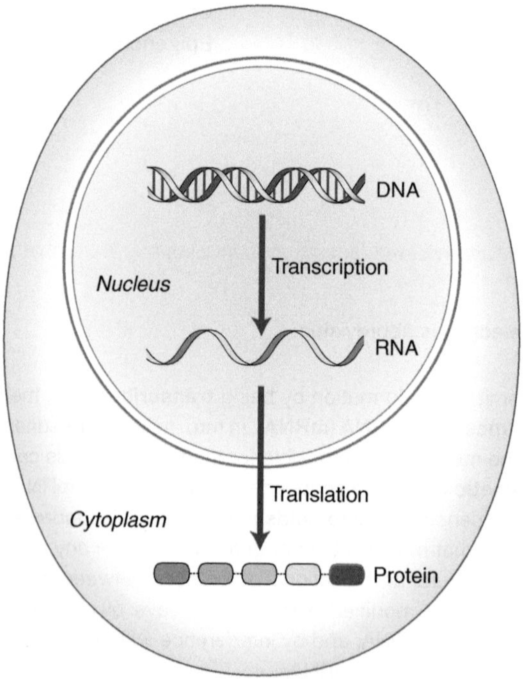

FIGURE 7.1

The central dogma of gene expression. Double-stranded DNA is transcribed into single-stranded RNA, which is translated into protein. Transcription occurs in the nucleus; translation, in the cytoplasm.

THE RIBONUCLEIC ACIDS (RNAs)

RNA Varieties

RNAs differ from DNA in important ways (Figure 7.2). In DNA, **deoxyribose** is the sugar molecule that makes up nucleotides. In ribonucleic acid (RNA), the sugar is **ribose**. Thymine in DNA is replaced by **uracil (U)** in RNA. Uracil is complementary to A. Further, most RNAs are single stranded, not double-stranded. Finally, RNAs are shorter in length than DNA, are generally more unstable and, in some cases, may act as enzymes.

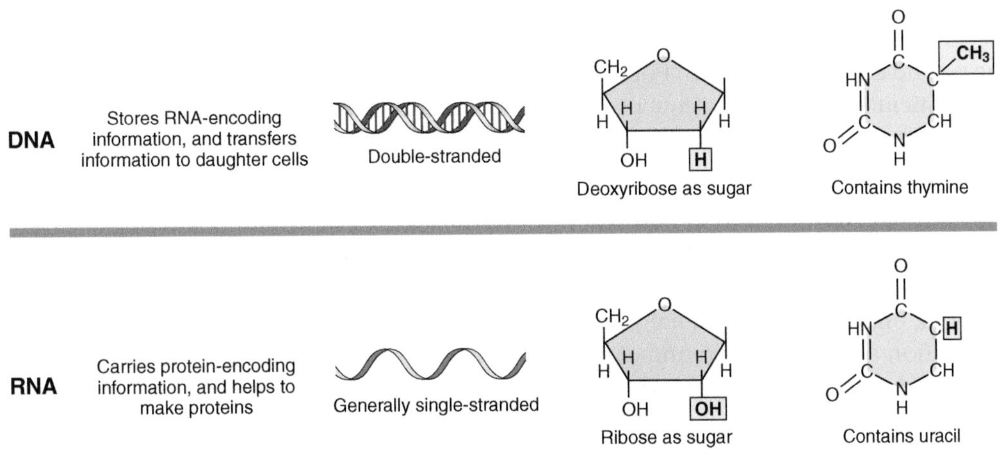

FIGURE 7.2
Chemical differences between DNA and RNA. DNA is double-stranded; RNA, generally single-stranded. In DNA, the sugar is deoxyribose; in RNA, it is ribose. Thymine, one of the two pyrimidines in DNA, is replaced by Uracil in RNA.

Four general categories of RNA molecules are recognized:

- **Messenger RNA (mRNA)** is synthesized from DNA and used as a template for protein synthesis.
- **Transfer RNAs (tRNAs)** are adaptor molecules that shuttle amino acids to their correct position in a growing polypeptide chain.
- **Ribosomal RNA (rRNA)** constitutes a portion of the ribosome; i.e., the cytoplasmic machine responsible for "reading" mRNA and synthesizing its co-linear polypeptide. Think of it as an assembly line, which rolls along the mRNA and attaches each amino acid brought to it by the tRNAs.
- **Micro RNAs (miRNAs)** and short interfering RNAs (siRNAs) are a recently described species. We now recognize hundreds of miRNAs, which are short in length and function, that regulate mRNA degradation or stability in the cytoplasm.

RNA Polymerases

RNA polymerases are large, multi-subunit complexes. They contact 70–90 base pairs of DNA in **promoter regions** used to initiate DNA transcription, during which DNA wraps around the polymerase. Eukaryotes have three different types of RNA polymerases:

- **RNA polymerase I (Pol I)**, used to produce the large ribosomal subunit;
- **RNA polymerase II (Pol II)**, used to produce the protein-encoding RNAs and the micro RNAs;
- **RNA polymerase III (Pol III)**, used to produce each of the tRNAs and the small ribosomal RNA subunit.

TRANSCRIPTION AND PROCESSING

Synthesis of Single-stranded RNA

Three steps characterize the process by which one DNA strand is used by RNA polymerase II (Pol II) to synthesize a complementary RNA version of that strand (Figure 7.3). The first step is **initiation**. Pol II binds to a 6- to 10-base pair (bp) sequence of nucleotides on the DNA to be copied. This sequence, called the **promoter region**, is usually about 35 bp upstream from the transcription start site. Promoters vary in sequence and in their ability to initiate transcription. After Pol II binds, it causes the DNA double helix to separate into the strand to be copied and its partner. RNA synthesis, then, begins at the **"start site"** in the **5′ to 3′ direction** (by convention 5′ to 3′ means left to right, or "downstream").

The second step is **elongation**. Pol II moves along the DNA strand and adds a ribonucleotide complementary to the base sequence of the DNA template strand. The DNA double helix reforms after it has been transcribed, and the 5′ end of the synthesized RNA protrudes from the transcription complex.

The third step, **termination**, occurs when Pol II reaches a specific termination sequence in the DNA. Then, the RNA molecule and the polymerase are released. The most common mechanism of chain termination occurs when Pol II reaches a particular DNA sequence that is able to fold back on itself, forming a hairpin loop. Other proteins, too, are needed to complete the termination event. Another round of transcription occurs before the first round ends, catalyzed either by the Pol II molecule used in the first round or by another molecule of Pol II.

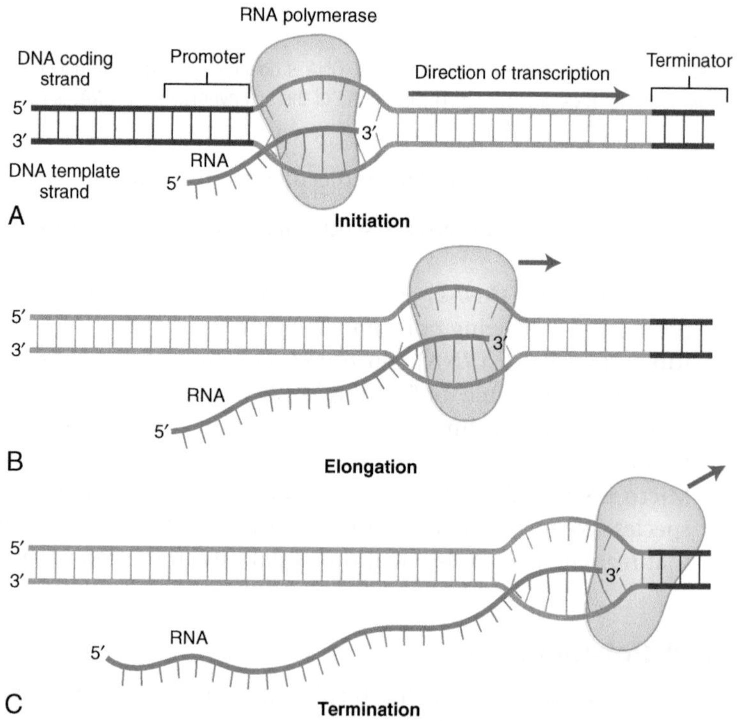

FIGURE 7.3
Transcription of DNA to RNA. Transcription occurs in three phases. (A) *Initiation*: RNA polymerase binds to DNA at the beginning of the gene to be copied. The polymerase recognizes and binds to promotor sequences adjacent to the gene, to be copied. (B) *Elongation*: the RNA polymerase moves along the chromosome, unwinding the double helix of DNA and producing a complementary single-stranded RNA copy. (C) *Termination*: signals in the DNA and in accessory proteins cause the polymerase to dissociate from the DNA, producing the primary RNA transcript.

MESSENGER RNA (mRNA): Primary Transcript

The molecule of RNA transcribed from its DNA template, called the **primary transcript**, contains all the information needed for encoding a protein, but it requires **"punctuation marks"** and, in eukaryotes, editing to make it function properly (Figure 7.4). The punctuation occurs at both the 5′ (or upstream) end of the transcript and at the other 3′ (or downstream) end. The 5′ end is modified by addition of a **substituent—methylguanosine**—called a **cap**.

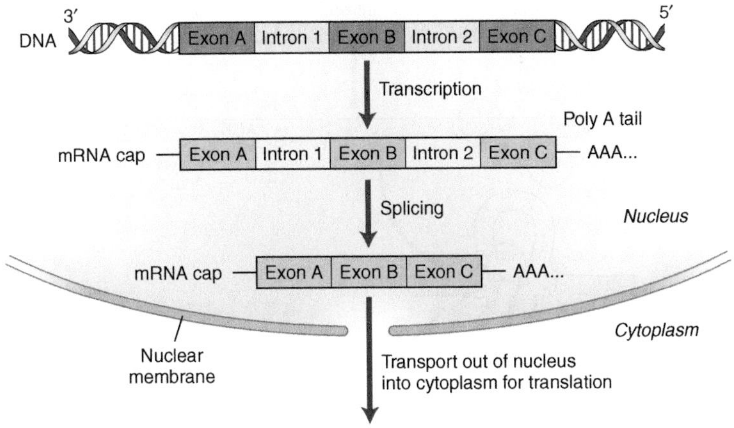

FIGURE 7.4
Editing the primary RNA transcript. First, the transcript takes on a 5′ cap and a 3′ poly A tail. Then the RNA is spliced to remove introns and join the exons to be included in mRNA. Finally, mRNA exits from the nucleus.

This cap is required for the mRNA to bind to the cytosolic protein-synthesizing machine, the ribosome, and thereby initiate protein synthesis. The 3′ end is altered after termination by addition of as many as 200 adenines, called the poly A tail. This tail plays a role in messenger RNA stability.

Processing

In prokaryotes, the length of the gene being transcribed matches that of the RNA transcript. That is not the case for most eukaryotic genes. Here, the DNA template was found (in work leading to a Nobel Prize) to be much longer than its corresponding mRNA. Scientists learned that the primary transcript—also called **pre-mRNA**—underwent extensive processing in the nucleus before being exported to the cytoplasm. Pre-mRNA contained sequences found in the DNA and the mRNA, called **exons**, and sequences found in the DNA but not in the mRNA, called intervening sequences, or **introns** (Figure 7.5).The primary transcript is then processed to remove the introns and join the exons to one another. This RNA processing—often called **splicing**—takes place before the primary transcript leaves its DNA template. In other words, transcription and splicing are coupled events whose efficiency and fidelity depend on numerous transcription factors, enzymes, and nucleoprotein complexes.

mRNA splicing takes place in nuclear particles called **spliceosomes**. They are composed of proteins and small RNAs. The spliceosome accomplishes three things:

- it identifies the upstream sequence in the transcript that demarcates the junction between exon and intron;
- similarly it identifies the downstream sequence that marks the intron/exon junction;
- it snips out the intron and rejoins the end of the upstream exon with the beginning of the downstream one. The upstream sequence is called the **donor site**, and the downstream sequence, **the acceptor site.**

When splicing has been completed, the mRNA travels from the nucleus to the cytoplasm where the translation apparatus resides.

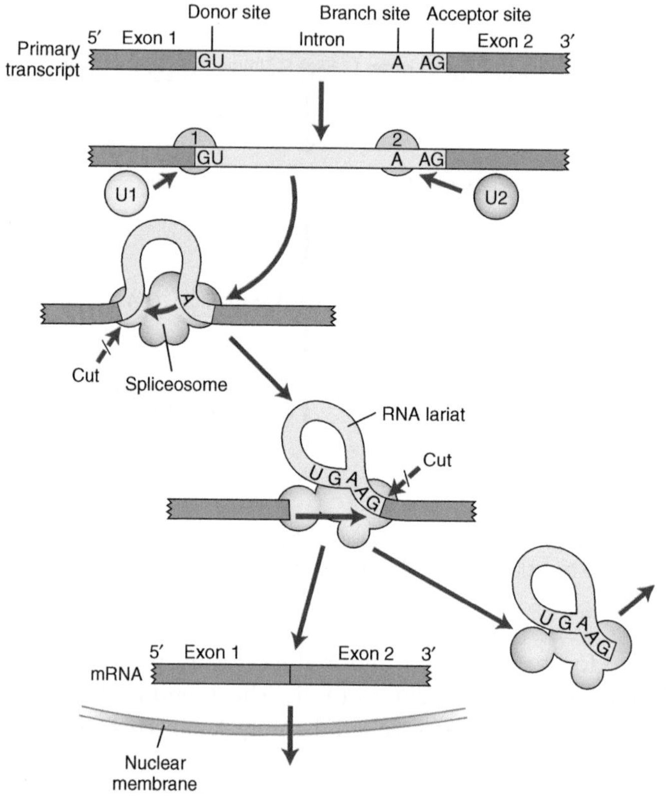

FIGURE 7.5
Details of RNA splicing. (A) Splicing begins with the binding of ribonucleoproteins (1, 2) to the splice donor and acceptor sites on the intron. (B) These sites are then brought together by other components of the spliceosome (U1 and U2). (C) The donor site is then cleaved, the RNA segment to be excised and the acceptor site assume a lariat configuration, and the lariat structure is released by cutting at the acceptor site. (D) The two exons are joined together.

Human Transcripts: Variability

RNA transcripts in humans range greatly in size and complexity. Some (globin, for example)) measure only 1.6 kb (thousands of bases) and are composed of three exons and two introns. Others, for example dystrophin, are 25 Mb long and contain 81 exons and 80 introns. Exons tend to be shorter than introns, the average exon length measuring 145 bases, the average intron length, 3,350 bases. Significantly, these averages are skewed greatly by exons and introns at the tails of the distribution curves for each molecular species.

THE GENETIC CODE

As information about gene and protein structure expanded in the 1950s, a critical question vexed molecular biologists: How could only 4 different nucleic acid bases (A, T, G, C) encode the information required to build proteins composed of 20 different amino acids (methionine, leucine, etc.)? Stated another way, how could a 4-letter alphabet instruct a 20-letter one? These questions were answered by discoveries leading to the **genetic code**.

Scientists reasoned that a 1 : 1 relationship between nucleotides and amino acids was not possible. **Codons** consisting of pairs of nucleotides, likewise, would be insufficient because 4 nucleotides in pairs can form only 16 possible combinations (4^2). **Triplets** of nucleotides would be enough, however, because 4^3 triplets would provide 64 possible codons—more than needed for 20 amino acids.

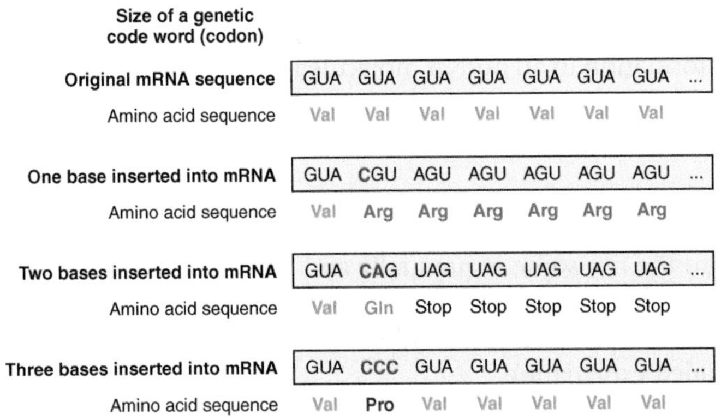

FIGURE 7.6
Evidence for a triplet code from bacterial mutagenesis (Crick, Brenner, and colleagues). Addition or deletion of one or two bases resulted in major alteration of the amino acid sequence in the encoded protein. Addition or subtraction of three bases retained the "reading frame," implying that the genetic code was a triplet one.

The notion that codons were triplets came from genetic experiments carried out by Crick, Brenner, and colleagues in 1961 (Figure 7.6). They made mutants in bacteriophage T4, and examined their effect(s) on a particular mRNA (called rII). All single base **insertions** and **deletions** and all **double mutants** (e.g., two nucleotide insertions or two deletions) interrupted the reading frame of the mRNA and changed the amino acid sequence of rII accordingly. When mutants were made that contained three insertions or three deletions, however, the reading frame was restored. These classic studies provided strong evidence for a triplet genetic code.

Unequivocal proof that the genetic code consists of triplet codons came from biochemical studies performed in 1964 by Nirenberg, Holley, Khorana and their colleagues. They conducted *in vitro* experiments in which protein synthesis was carried out in extracts of *E. coli*, charged with chemically synthesized, artificial mRNAs. The "eureka" moment occurred when an RNA composed only of U residues (polyuridylic acid) led to a polypeptide consisting only of phenylalanine. This meant that UUU was a codon for phenylalanine. Subsequent experiments with all kinds of artificial mRNAs led to the picture shown in Figure 7.7. The essential findings depicted in the figure are these:

- three codons (UGA, UAG, and UAA) are stop signals for translation;
- AUG is the initiator methionine codon and codes for internal methionines as well;
- the other 60 codons specify the remaining amino acids.

Some amino acids, phenylalanine, for example, have two codons (UUU and UUC). Other amino acids, e.g., leucine, have four codons (CUU, CUC, CUA, CUG). Different codons specifying a single amino acid are said to be synonymous. The code is called "**degenerate**" because more than one codon designates any single amino acid. In sum, the genetic code is virtually universal and consists of **non-overlapping, degenerate triplets.**

First letter	Second letter: U	C	A	G	Third letter
U	UUU Phe	UCU Ser	UAU Tyr	UGU Cys	U
U	UUC Phe	UCC Ser	UAC Tyr	UGC Cys	C
U	UUA Leu	UCA Ser	UAA Stop	UGA Stop	A
U	UUG Leu	UCG Ser	UAG Stop	UGG Trp	G
C	CUU Leu	CCU Pro	CAU His	CGU Arg	U
C	CUC Leu	CCC Pro	CAC His	CGC Arg	C
C	CUA Leu	CCA Pro	CAA Gln	CGA Arg	A
C	CUG Leu	CCG Pro	CAG Gln	CGG Arg	G
A	AUU Ile	ACU Thr	AAU Asn	AGU Ser	U
A	AUC Ile	ACC Thr	AAC Asn	AGC Ser	C
A	AUA Ile	ACA Thr	AAA Lys	AGA Arg	A
A	AUG Met	ACG Thr	AAG Lys	AGG Arg	G
G	GUU Val	GCU Ala	GAU Asp	GGU Gly	U
G	GUC Val	GCC Ala	GAC Asp	GGC Gly	C
G	GUA Val	GCA Ala	GAA Glu	GGA Gly	A
G	GUG Val	GCG Ala	GAG Glu	GGG Gly	G

FIGURE 7.7
The genetic code, read as triplets. Shown here is the relationship between the 64 RNA codons and the amino acids they code for. The amino acids are abbreviated as follows: Ala, alanine; Arg, arginine; Asn, asparagine; Asp, aspartate; Cys, cysteine; Gln, glutamine; Glu, glutamate; His, histidine; Ile, isoleucine; Leu, leucine; Lys, lysine; Met, methionine; Phe, phenylalanine; Pro, proline; Ser, serine; Thr, threonine; Trp, tryptophan; Tyr, tyrosine, Val, valine.

TRANSLATION: SYNTHESIS OF AN mRNA-DIRECTED POLYPEPTIDE

The series of events by which the information in an mRNA molecule is converted to a polymer of amino acids in a polypeptide is called **translation**. This process requires many components: mRNA, ribosomes, tRNA, aminoacyl tRNA synthetase, and protein factors facilitating initiation, elongation, and release.

The essential steps in translation can be summarized as follow: mRNA binds to the ribosome; aminoacylated tRNAs are brought to the ribosome sequentially and one at a time; peptide bonds are formed between each successively added amino acid; when the last amino acid has been added, the complete polypeptide is released from the ribosome. This intricate, biological assembly line warrants a more detailed explanation.

Initiation

Translation is begun by concerted interaction among the 5′ cap on the mRNA, a number of protein initiation factors, and a tRNA charged with methionine ($tRNA^{met}$) (Figure 7.8). This initiation complex then moves along the mRNA until the **anticodon** of $tRNA^{met}$ (UAC) recognizes its complementary codon (AUG) on the mRNA. This signals the beginning of polypeptide chain synthesis, an event associated with recruitment of the small ribosomal subunit and release of the initiation factors.

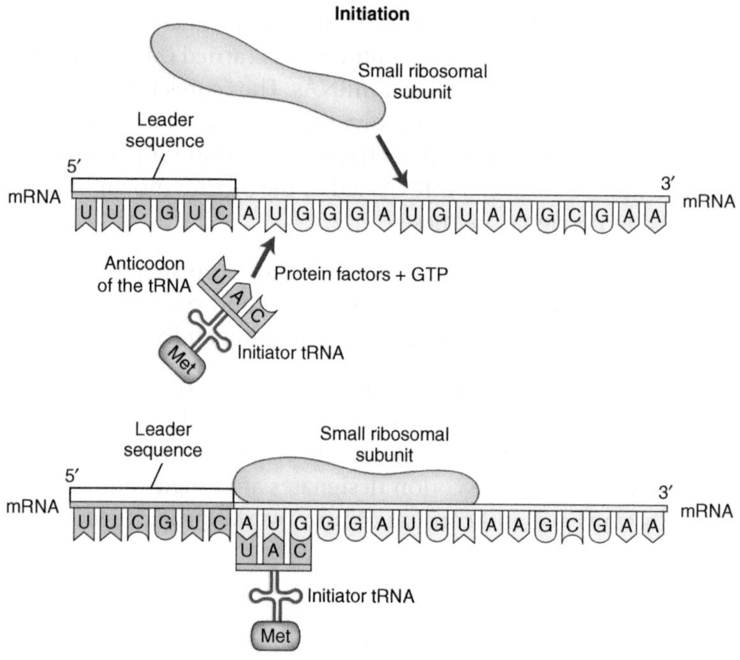

FIGURE 7.8
Initiation of translation. The small ribosomal subunit binds to the 5′ cap of the mRNA. It then migrates to the first AUG it encounters and interacts with Met-loaded tRNA, thereby starting 5′ to 3′ translation.

Elongation

Upon recruitment of an elongation factor and the large ribosomal subunit, polypeptide chain synthesis progresses as shown (Figure 7.9). This requires three events:

- bringing each aminoacylated tRNA into line;
- forming the next peptide bond in the growing amino acid chain;
- moving the ribosome to the next codon in the mRNA.

The ribosome provides a platform permitting each successive tRNA charged with its amino acid cargo to be put in its proper place. The first peptide bond joins the initiator methionine residue to whichever amino acid follows it in the mRNA-encoded sequence.

Subsequently, the smaller ribosomal subunit shifts in position, thereby allowing the next aminoacylated tRNA to discharge its amino acid onto the growing polypeptide chain. This sequence is repeated over and over until the polypeptide chain has been completed. Under optimal conditions, about 15 amino acids per second can be added to a polypeptide chain.

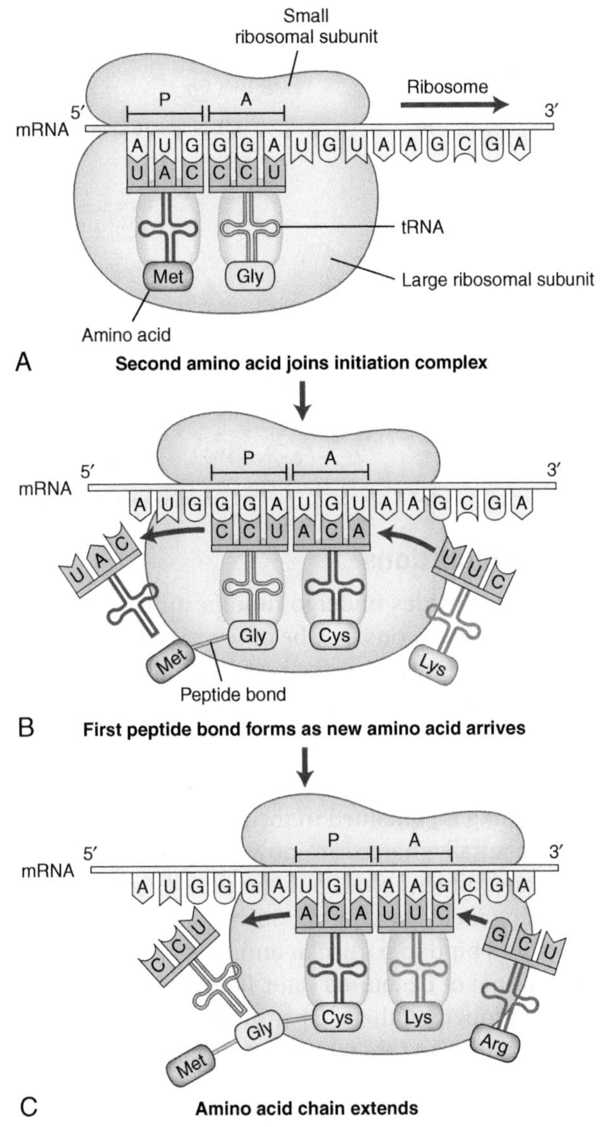

FIGURE 7.9
Elongation of polypeptide. (A) The large ribosomal subunit binds to the initiation complex; a tRNA carrying the second amino acid (glycine in this diagram) brings its cargo to the A site and forms bonds between its anticodon and the mRNA. (B) A peptide bond forms between the Met and Gly residues, and a third tRNA brings the next amino acid (e.g., cysteine) to the vacated A site. (C) The process repeats itself as the ribosome slides along the mRNA, adding one amino acid at a time.

Termination and Release

Termination of synthesis and release from the ribosome occur when a **stop codon** in mRNA is encountered, i.e., a codon that does not interact with a tRNA species (Figure 7.10). Upon recognition of the stop codon, a release factor interacts with the ribosome and the polypeptide chain. When this occurs, the ribosomal subunits are recycled to other mRNAs.

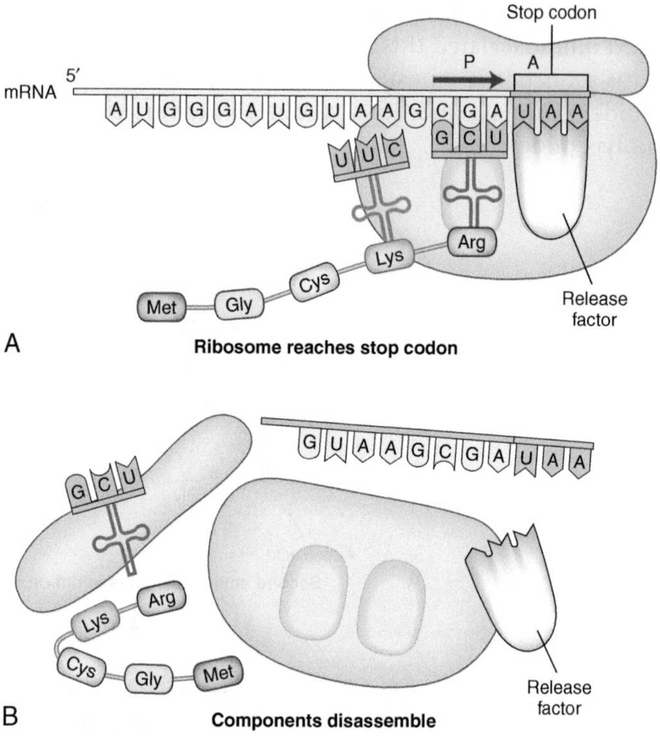

FIGURE 7.10
Termination of translation. (A) When the ribosome encounters a stop codon, a protein release factor binds to it, releasing the full-length polypeptide. (B) All the translational components disassemble.

Post-translational Modifications

Many newly synthesized **polypeptides** undergo modification before they assume their functional configuration. These modifications may be of several types: chemical change; folding; or formation of multi-subunit structures. In some cases, the first methionine is removed enzymatically. In others, the polypeptide is cleaved into two or more pieces. In still others, phosphate or other small chemical groups are added.

Polypeptides do not have functional capability until they fold. For the majority of proteins, the information required for folding is contained in the primary amino acid sequence. That is, the positively, negatively or neutrally charged amino acid residues dictate the formation of helices, pleated sheets, and other three-dimensional configurations (Figure 7.11). Other polypeptides require assistance in achieving their active forms. Such assistance is provided by proteins called **chaperones**. Some chaperones bind to particular amino acid residues, thereby preventing their aggregation into a dysfunctional or denatured state. Repeated cycles of binding and release give the polypeptide time to fold properly. Other chaperones form one- or two-chambered cylindrical structures that permit polypeptides to fold within a protected environment.

Some polypeptides, when properly folded, are physiologically active as single subunits. Other proteins are composed of two or more subunits. These multi-subunit proteins (called **quaternary structures**) consist of identical subunits in some cases and of non-identical subunits in others. For example, human hemoglobin is a tetramer (i.e., four subunits), two called alpha (α) and two called beta (β).

REGULATION OF GENE EXPRESSION

The gene was originally defined as a segment of DNA that encodes a protein. This central idea has stood the test of time, but it has been modified in multiple ways. Each of the central dogma's three "words" (DNA, RNA, protein) has been expanded upon and refined. The dogma is no longer a simple linear pathway. Rather, it has been shown to be a network of interacting webs producing a three-dimensional system of expression and control. A brief description of some of these modifications follows.

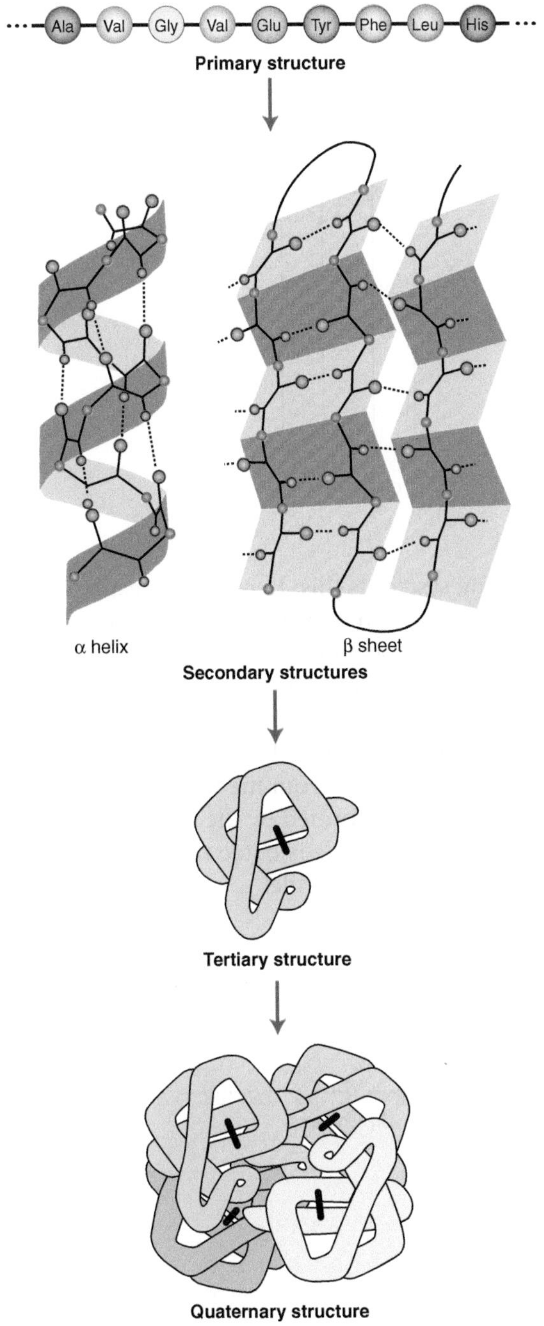

FIGURE 7.11
Four levels of complexity of structure. *Primary structure*: the sequence of amino acids in a polypeptide chain. *Secondary structure*: localized regions form structures such as pleated sheets and helices. *Tertiary structure*: the protein assumes its complete, three-dimensional form, incorporating its various sheets, helices, etc. *Quaternary structure*: some proteins are composed of more than one polypeptide subunit that are bound together chemically.

Somatic DNA Rearrangements

In rare instances, a single polypeptide chain is formed from multiple genes. In humans, the formation of **immunoglobulin (Ig)** antibodies exemplifies this mechanism. Five classes of Ig molecules exist, the most abundant being IgG. Antibodies are composed of four subunits: two **heavy chains** and two **light chains**, held together by bonds between sulfur atoms (Figure 7.12). Antibodies are made by **B cells**, a type of lymphocyte generated in the bone marrow. Each B cell produces only one type of antibody, whose specificity is produced by the sequence of amino acids in the so-called **variable regions** of the heavy and light chains (depicted as the regions above the fork in Figure 7.12). The remainder of the antibody consists of the "**constant region**," so named because they are the same in all immunoglobulin types.

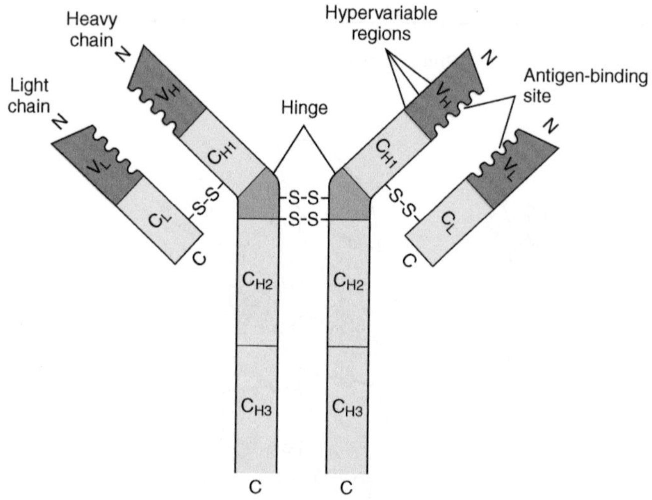

FIGURE 7.12
Antibody structure. Two heavy chains and two light chains, held together by disulfide (S—S) bonds, make up the general structure of antibodies. Heavy and light chains have variable domains (V_H and V_L) near their amino termini (orange) which associate to form the antigen-binding site. The remainder of each chain is composed of constant regions for the heavy (C_H) and light (C_L) chains.

The central question posed by this antibody-making system is this: How can we explain the ability of B cells to make 10^8 or 10^9 kinds of antibodies from only 10^4 kinds of DNA coding segments? The answer turns out to be extensive DNA splicing in the B cells (Figure 7.13). Germ cells contain a small number of clustered genes coding for the constant regions of light chains. On the same chromosome, but at some distance from this constant region cluster, is a much larger cluster of genes coding for the variable region of light chains. During B cell differentiation, one gene from the constant region is spliced to one gene from the variable region to make a single light-chain gene. Single heavy-chain genes are formed by the same kind of splicing mechanism. Thus, a single immunoglobulin polypeptide is made from multiple genes; this is a notable exception to the "one gene, one enzyme" rule.

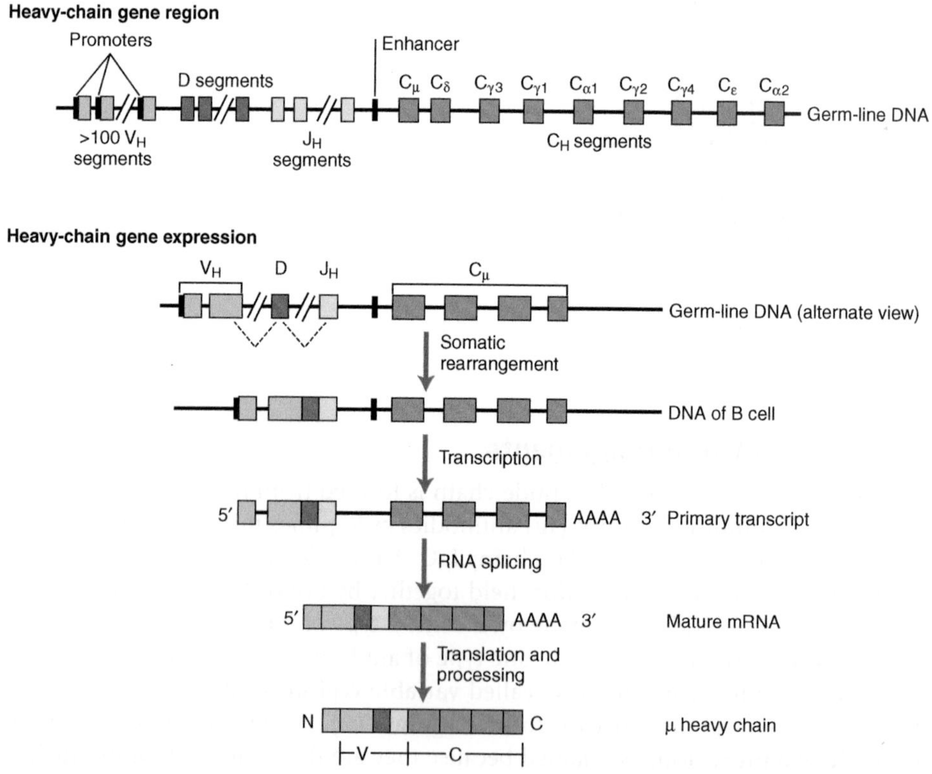

FIGURE 7.13
Antibody gene regions. The heavy-chain gene region is depicted here. Germ-line cells (top) contain more than 100 V_H segments, a small number of D and J segments, and 9 C_H segments. In B cells (below), somatic rearrangements bring together individual V_H, D, and J segments and splice them to the C_H segments. Transcription then leads to formation of the primary transcript which is edited to its mRNA and translated into a unique, heavy chain.

Making a complete antibody gene is more involved than simple splicing of constant and variable region genes because two other kinds of genes, J and D, are spliced into the complete antibody gene. J genes are also multiple, thereby providing for an even greater combinatorial repertoire of different antibody genes. In this way, the human immune system copes with the virtually unlimited number of antigens found in foods, bacteria, viruses, and other kinds of foreign substances or organisms.

Transcriptional Control

Gene expression is modulated in many ways: many genes are "turned on" in some cells and "turned off" in others; gene activity may vary by more than a hundred-fold in the same cell type; some genes are used in early development and not thereafter. Such fine-tuning is accomplished by a veritable orchestra of mechanisms, some intrinsic to DNA (called **cis-acting** elements), others extrinsic (called **trans-acting** elements).

There are two major types of cis-acting elements: **promoters** and **enhancers**. Promoters are short sequences of DNA located very near the site of initiation of transcription. Enhancers are DNA sequences of varying length and varying location. Some may be thousands of bases from the gene they regulate; others are located at the 5′ or 3′ end of the gene; still others, even in introns within the gene. Trans-acting elements, called **transcription factors**, are proteins that bind to promoter or enhancer regions, thereby accomplishing control of initiation or termination of transcription. Some transcription factors, called **activators**, increase the level of transcription. Others, called **repressors**, decrease the level of transcription. Transcription regulation often requires the formation of gene–protein complexes containing several transcription factors. Thousands of different transcription factors exist in human cells, each with a unique function or set of related functions.

Chromatin Remodeling

Chromosomal DNA (and its respective genes) is found in super-coiled nucleosomes (see Chapter 6). This conformation may prevent the binding of one or more transcription factors to the initiation site, thereby blocking gene transcription. To overcome this structural impediment, several different multi-protein complexes have been identified, called **chromatin-remodeling complexes (CRC)** (Figure 7.14). CRCs enable genes to be transcribed by energy-dependent mechanisms that are not well defined. They act to reposition the interaction between chromosomal DNA and nucleosomal proteins (principally histones) in such a way that access to DNA binding sites and, through them, initiations of transcription is accomplished.

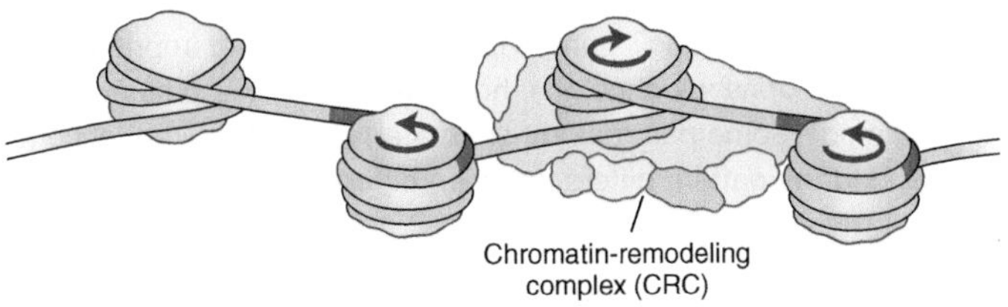

FIGURE 7.14
Chromatin remodeling complex (CRC). CRCs (green) are multiprotein structures capable of modifying histones or repositioning nucleosomes (yellow) along the DNA (pink). CRCs serve to enable chromatin to be transcribed.

DNA Methylation: Epigenetic Control

Gene expression may be modified by chemical modification of the DNA rather than by changes in the DNA sequence itself. Such modification is referred to as "**epigenetic**," to distinguish it from "genetic." Such epigenetic control is often exerted by methylation of cytosine residues in DNA (see Chapter 6) . After incorporation of these cytosines into DNA, an enzyme called DNA methylase adds a methyl group to the number 5 carbon atom. Cytosine methylation occurs preferentially in regions where many adjacent cytosine and guanine dinucleotides are clustered (so-called CpG islands).

Heavily methylated genes are often silenced or transcribed at a very low level. It is becoming clear that this kind of epigenetic change is more widespread than previously thought. Many genes on the inactive X chromosome found in mammalian female cells are highly methylated at their CpG islands, implying that methylation results in gene (and region) inactivation. More recently, hundreds of human genes have been shown to be **imprinted**, meaning that one of the homologues inherited from either the father or the mother has been silenced by **methylation**. X chromosome inactivation and imprinting have important correlates in human genetic disorders that will be discussed subsequently.

IMPLICATIONS: THE EPIGENOME AND EVOLUTION

Genomic Imprinting: Mom and Dad Work it Out

Each of us inherits 23 chromosomes from each of our parents. If the human species adhered strictly to Mendelian tenets, each of these chromosomes would be expressed equally; that is, it would make no difference whether any single autosome was inherited from the mother or father. Recently, however, it has been shown that as many as 1–2% of our genes behave such that expression depends on the parent they came from. In this subset, either the gene inherited from the mother or the one from the father is silenced by a chemical process in which methylation of the DNA inhibits (or silences) its expression. This phenomenon is called genomic imprinting; it gave rise to the idea that gene expression could be modified in ways other than by changing the sequence of nucleotides itself. Imprinting and its broader discipline, called epigenetics, has raised several questions.

First, what is the evidence for imprinting? In the 1980s, it became possible to produce mouse embryos that carried, respectively: one male and one female pronucleus (the haploid nucleus belonging to the sperm and the ovum); two male pronuclei; or two female pronuclei. Embryos with one male and one female proneucleus survived. All the embryos with two male or two female pronuclei died. This meant that something other than DNA distinguished between male and female pronuclei. Subsequently, evidence was accumulated from studies of the *insulin-like growth factor 2* gene (*Igf2*), which promotes fetal growth. If the gene was deleted on the chromosome contributed by the mother, the fetus developed to normal size; but if the *Igf2* gene inherited from the father was deleted, the mice were only about 40% of normal size. The simplest interpretation of these results was that the *Igf2* gene copy inherited from the mother is normally silenced, while that from the father is normally expressed.

Second, how extensive is this imprinting? Early on, it was thought that imprinting affected just a few genes. With the development of new technology, however, it is now estimated that as many as 500 autosomal loci are imprinted—some paternal, others maternal. Further, imprinting appears to be restricted to mammals, and the pattern differs between different species (for example, between mice and humans).

Third, why does imprinting exist? From an evolutionary perspective, imprinting would appear disadvantageous for survival. Two active copies of a gene would be expected to protect against the deleterious effect of random, recessive mutations. The leading theory to explain imprinting has been proposed by David Haig, an evolutionary biologist. He suggests a three-way tug-of-war among fetus, mother, and father. The fetus wants to extract as many nutrients from the mother's placenta as possible. The mother wants to allocate her resources evenly to all the children she may bear now or in the future, possibly with different fathers. The father is interested only in the survival of his own child. So the father passes on his *Igf2* gene in active form, the mother passes on hers in imprinted (silenced) form. The endgame of this evolutionary struggle would be silencing of growth-promoting genes in females and silencing of growth-inhibiting genes in males.

IMPLICATIONS: THE EPIGENOME AND EVOLUTION—Cont'd

Fourth, does imprinting affect tissues other than those regulating growth (such as the placenta to which the fetus is attached)? The answer is affirmative. When viable mice carrying two male or two female pronuclei were produced by mixing in some normal cells, those with two male pronuclei had large bodies and small brains; with two female pronuclei, it was the other way around. Subsequent work in mice indicated that the maternal genome controls growth of the cerebral cortex, the paternal genome that of the hypothalamus.

Doubtless, other questions and other answers about imprinting will emerge in the near future, and the field will impinge on such weighty matters as gene/environment interactions and on the expression of mutations in human disorders.

RNA Processing

Gene expression is controlled by events occurring after DNA transcription. One such event involves the processing of the primary RNA transcript to form mRNA. As discussed earlier in this chapter, eukaryotic genes contain exons and introns. The introns must be spliced out and the exons joined to form mRNA. An important means of gene regulation concerns **alternative splicing** of the primary transcript (Figure 7.15). This usually results in the formation of more than one kind of mRNA from a single gene and, thereby, more than one protein. In one case, all of a gene's exons are retained in the spliced RNA; in others, one or more exons are "skipped." Alternative splicing is a main reason for the wide disparity between the number of genes in the human genome (*circa* 21,000) and the number of proteins in the human cell, called the **proteome** (hundreds of thousands). Given that the human proteome is much larger than that of flies and worms, whereas the genomes of these species are of similar sizes, it follows that alternative splicing is a fundamental mechanism explaining evolutionary divergence and human complexity.

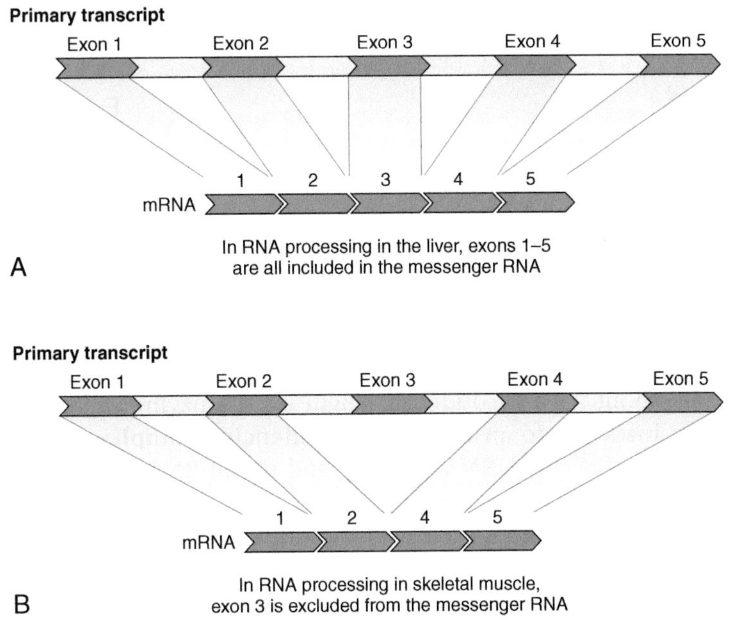

FIGURE 7.15
Alternative splicing of a primary transcript. (A) Exons 1–5 are included in the mRNA. (B) The same primary transcript is spliced such that exon 3 is omitted from the mRNA. Alternative splicing is a key means for generating several polypeptides from a single gene.

RNA Interference

In the past decade a totally new mechanism of gene regulation has been discovered. This involves interference with mRNA stability or translation by small, **interfering RNA** species hundreds, even thousands, in number. These RNAs have already been implicated in cancer and other human diseases. They may have great therapeutic promise as well. Such interfering RNAs are of two general types: **small interfering RNA (siRNA)** and **micro RNA (miRNA)**. Each of them is synthesized in double-stranded form (Figure 7.16).

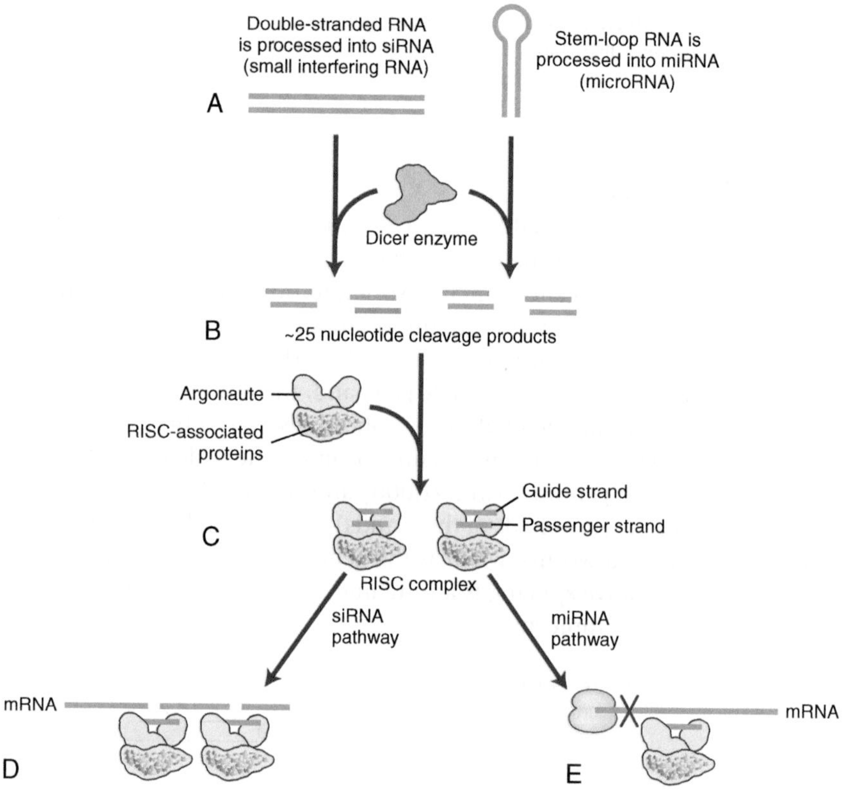

FIGURE 7.16
RNA interference. (A) Double-stranded siRNA and single-stranded miRNA are synthesized. (B) Processed siRNA and miRNA are cleaved into fragments of about 25 nucleotides by the enzyme, dicer. (C) RNA fragments are incorporated into the RNA-induced silencing complex (RISC). (D) siRNA causes mRNA cleavage. (E) miRNA causes mRNA translation to be blocked.

The double-stranded (ds) precursor of siRNA is produced by transcription of a region of DNA from both strands of the double helix. In contrast, the double-stranded precursor of miRNA is transcribed from a single strand of DNA in which a duplicated sequence is presented in inverted orientation, yielding a base-paired stem-loop molecule. These double-stranded RNAs are transported from the nucleus to the cytoplasm, where they are cut into fragments about 25 nucleotides in length by an enzyme called "**dicer**." These fragments are then loaded onto an **RNA-induced silencing complex (RISC)** composed of several proteins. Thereafter, the siRNA pathway and the miRNA path diverge. The siRNA–RISC complex binds to a single species of mRNA complementary in sequence to that of the siRNA. Then, the mRNA is cleaved and degraded by enzymatic activity of one of the RISC proteins. In contrast, the miRNA pathway involves less than a perfect match between the miRNA and its mRNA target, resulting in blocked translation of the mRNA. A key difference between siRNAs and miRNAs has to do with specificity: siRNAs are mRNA specific; miRNAs are not. Thus, a single kind of miRNA can interfere with translation of many—even hundreds—of mRNAs. There is still much to be learned about these interesting RNAs, but the excitement surrounding their importance—fundamental and applied—is warranted.

Translational and Post-translational Control

Inhibition of translation by siRNAs and miRNAs is an important means of regulating gene expression, but it is far from the only such means. Other mechanisms exert control during and after synthesis of a protein.

One such mechanism is that of activation of previously **untranslated** mRNAs. Early in development, mRNAs in the cytoplasm of the fertilized egg are translated rapidly to provide the proteins needed for cellular proliferation and organogenesis. Mechanisms for this activation have been studied in flies and roundworms. They include elongation of the poly A tail and the prolongation of mRNA lifetime.

Two other systems function after translation has been completed. One involves addition of phosphate groups to the amino acid chain. In some instances **phosphorylation** activates a protein, in other instances it inactivates it. This mechanism is critically important in hormone action and other mechanisms of chemical signaling in the cells. The second system concerns **regulated protein destruction**. A particularly well-studied example involves **ubiquitin**, a small protein that functions as a marker. When ubiquitin is covalently bound to a protein (the process being called "**ubiquitination**"), that protein has been marked for destruction by a large multienzyme complex called the **proteasome**. A prime example of ubiquitination concerns separation of sister chromatids during cell division. The proteins that "glue" sister chromatids together must be removed so that the chromatids can separate. Such removal requires the ubiquitination and destruction of these proteins.

GENOMIC VARIATION

The human genome sequence, presented as the culmination of the HGP, was put together from pooling DNA from several individuals. Thus it is a "**reference genome**" as distinguished from any individual's genome. Subsequent work has revealed how similar we are to one another—and how different. When DNA sequences are compared between any two unrelated humans, about 99.9% of the aligned sequences are identical, but, given the large size of the diploid human genome (about 6 billion base pairs), this means that any two people (save identical twins) differ at about 6 million base pairs. This genomic individuality underlies the phenotypic individuality we all see around us. Superimposed on the DNA sequence and variations thereof, are chemical changes in the DNA which may affect genomic function without changing DNA sequence. Such chemical modification is called "**epigenetic**;" the total of all epigenetic changes in the genome is referred to as the **epigenome**. The nature and number of genomic and epigenomic variations is being actively explored, as are their biological and clinical significance.

Deleterious Mutations

The term **mutation** refers to any stable change in DNA sequence. Single-gene mutations are responsible for literally thousands of human disorders that are inherited as Mendelian traits. Such mutations have been the cornerstone of human genetic inquiry for more than a century and will be discussed in detail in Chapter 8. Most mutations, so-called **point mutations**, change one base pair for another (e.g., A to G). Other varieties of mutation exist as well. The point to emphasize here is that, as shown from recent studies of genomes, each human carries hundreds of these mutations in heterozygous form. Thus there is no "normal" or "wild-type" human. We are all "mutants."

Single-base Mutations

HapMap Project, a follow-on to the HGP, revealed that such single nucleotide substitutions occur, on average, every 800 base pairs. When present at a population frequency of more than 1%, such single nucleotide changes are called **single nucleotide polymorphisms (SNPs)**. Any two genomes differ by about 6 million SNPs. Some SNPs are "**synonymous**," meaning that the nucleotide change does not result in a change in the amino acid sequence of the protein involved. "**Non-synonymous**" SNPs do change amino acid sequence. Most SNPs occur in introns or in non-genic sequences in the genome, as would be expected from the small fraction of the total genome that encodes products. The importance of SNPs currently is their use in identifying susceptibility genes in common disorders, and in seeking genetic differences among different racial groups.

Chromosomal Variation

As shown in Table 7.1, SNPs are but one kind of common variation in the human genome. Other types effect larger structural changes in the DNA—ranging from a few base pairs to a Mb. **Short insertion/deletion polymorphisms** (called **indels**) may affect 0.5% of the genome. **Segmental duplications and inversions** have also been described. A recent, unexpected finding has to do with **sequence copy number**. We now know that many sequences—including genes—occur in more than a single copy. This has led to the appreciation that **copy number variation (CNV)** may affect as much as 10% of the genome and are likely to be involved in causing such important human disorders as schizophrenia and autism.

TABLE 7.1 Common Variation in the Human Genome

Type of Variation	Size Range (bp)
SNP	1
indel	1 to>1 million
CNV	10,000 to>1 million
Inversion	Few to>1 million
Segmental duplication	10,000 to>1 million

Abbreviations: SNP, single nucleotide polymorphism; indel, insertion/deletion polymorphisms; CNV, copy number variation.

Epigenomics

By analogy with genetics and genomics, the terms "**epigenetics**" and "**epigenomics**" refer to the mechanism by which gene and genome function are modified without an accompanying change in DNA structure. An individual's genome is determined at the time of zygote formation and is fixed thereafter. This genome is the same in every cell of every tissue. In contrast, the **epigenome** is plastic. It is different in oocytes than in spermatocytes. It varies from one cell and tissue type to another. It changes over an individual's life and in response to such environmental stimuli as nutrition and chemical exposure. It is involved in single-gene disorders and in cancer. In sum, it interacts with the genotype to produce the individual's phenotype. The chemical moieties and basic "rules" principally responsible for epigenomic activity are remarkably simple: methyl (CH_3) and acetyl (CH_3COO) groups covalently added to DNA and histones. The effects of methylation and acetylation, however, are complex and still being elucidated.

As stated earlier in this chapter, methylation of cytosine residues in CpG islands regulates transcription of some genes and some chromosomal regions. Those sites marked for methylation are determined in germ cells and are modified in early development, after which they are generally unchanged throughout life. DNA methylation is catalyzed by an enzyme called **DNA methyltransferase (DMT)** which uses 5-adenosylmethionine as substrate. Methylation of DNA inhibits initiation of transcription, thereby "silencing" adjacent gene(s). Other enzymes called **demethylases** remove these covalently attached methyl groups and permit transcription to occur (or recur).

The other principal chemical modification of chromatin affects **histones**, not DNA. The several histones that make up the nucleosome contain heavily charged tails. When these tails are acetylated or otherwise chemically modified, the super-coiled DNA becomes more accessible to transcription factors, thereby facilitating the initiation of transcription. Once again specific acetylase and deacetylase enzymes are responsible for the on–off addition and removal of acetyl groups from the histones.

DNA methylation and histone modifications take place in a cooperative, concerted way to modulate gene and genome function. The precise control of these concerted interactions is being investigated actively. Because these epigenomic modifications are reversible, may affect one allele differently than its partner, and change during life, there is great hope that understanding the genomic map of epigenomic marks will help elucidate disease mechanisms and offer new diagnostic and therapeutic opportunities.

REVIEW QUESTIONS AND EXERCISES

1. Choose the phrase in the right column that best matches the term in the left column.

a. initiation codon	1. process producing different mRNAs from same primary transcript
b. charged tRNA	2. gene expression depends on parent of origin
c. reading frame	3. codons which terminate translation
d. RNA splicing	4. a group of three bases coding for one amino acid
e. transcription	5. the linear sequence of base pairs in DNA corresponds to the linear sequence of amino acids in a protein
f. imprinting	6. removal of introns from the primary transcript
g. alternative splicing	7. a tRNA molecule to which an amino acid has been attached
h. co-linearity	8. AUG (for methionine) in humans
i. translation	9. mRNA→protein
j. miRNA	10. sequence of base pairs in primary transcript not found in mature RNA
k. nonsense codon	11. an RNA species that regulates gene expression
l. RNA-like DNA strand	12. the grouping of mRNA bases three at a time to generate codons
m. template strand	13. more than one codon specifies a single amino acid
n. degeneracy of the genetic code	14. DNA→RNA
o. codon	15. DNA strand having complementary base sequence to that in primary transcript
p. intron	16. DNA strand having the same base sequences as primary transcript

2. Order the following DNA elements in terms of size from smallest (1) to largest (8).
   a. Chromosome
   b. Nitrogenous base
   c. Exon
   d. Carbon atom
   e. Genome
   f. Codon
   g. Gene
   h. Nucleotide.
3. What is the "central dogma" of molecular biology?
4. List three structural differences between DNA and RNA.
5. List and describe briefly four RNA species found in human cells.
6. An organism's biological information is transmitted via replication, transcription, and translation. Which one of the three is involved in the following phenomena? Indicate briefly why you chose this particular step.
   a. Insertion of a nonsense codon into a gene
   b. Mitotic cell division
   c. A deleterious mutation in RNA polymerase

7. During transcription the polarity of the template strand runs opposite to that of the primary transcript.
   a. Explain this statement.
   b. Is the base sequence of the primary transcript identical to that in the template strand or complementary to it?
   c. Is the base sequence of the primary transcript identical to that in the non-template strand or complementary to it?
8. The human hormone, somatostatin, was among the first products of recombinant DNA technology. Bacteria were used to synthesize the hormone after its gene sequence had been spliced into the bacterial chromosome. The base sequence of a portion of the template strand was:

   CGA CCA ACA TTC TTG AAG...

   What is the amino acid sequence of this portion of somatostatin?
9. List the nine possible mRNA triplets and corresponding amino acids that would be formed by changing a single base in the UUU codon for phenylalanine. Such single base changes are the most common type of mutation, which will be discussed in Chapter 8.
10. Nobel Prizes were awarded to scientists who discovered DNA replication (Kornberg and Ochoa), DNA structure (Watson, Crick, and Wilkins), RNA splicing (Sharp and Roberts), and the genetic code (Holley, Khorana, and Nirenberg). No Nobel Prize was awarded for the discovery of mRNA, however. Why did this obvious oversight occur?
11. X chromosome inactivation and genomic imprinting represent the two most important examples of epigenetic regulation in mammalian cells. What are the two most important differences between X inactivation and imprinting?

CHAPTER 8

Mutation

CHAPTER OUTLINE

CORE CONCEPT

A **mutation** is a heritable change in the nucleotide sequence or arrangement of DNA. In the positive, evolutionary sense, mutations are responsible for the **selective advantage** that one species gains over another. In the negative sense, mutations cause or increase susceptibility to thousands of human disorders. Three general categories of mutations exist: **genome**, **chromosome**, and **gene**. Genome mutations result from **missegregration** (failure to properly segregate) during meiosis and produce changes in chromosome number. Chromosome mutations are caused by rearrangements in the structure of a chromosome, such as translocations or deletions. Gene mutations alter the base sequence of a gene. Mutations occur in **germ cells, somatic cells**, and **mitochondria**. Germ-line mutations affect DNA in all cells and are transmitted to offspring. Mitochondrial mutations affect mitochondrial DNA, all of which is inherited from the mother's egg; accordingly, they are inherited maternally. Somatic mutations affect cells in a single tissue and are not transmitted to the next generation. Mutations act, generally, by perturbing gene expression or the function of proteins, and they may do so from the earliest to the latest stages of life.

DNA is not static. Rather its sequence changes at a slow but measurable rate all the time. Mutations are of central importance to all life forms, including humans. Without mutations, our species would not have evolved over several billion years. Without mutations, humankind also would not be faced with a myriad of spontaneous abortions, inherited disorders, or cancers. Therefore, we must comprehend both the nature of mutations, and the effects they produce.

NATURE OF MUTATIONS

Categories and Frequencies

Three categories of mutations exist (Table 8.1): those that alter the number of chromosomes in the cell (**genome mutations**); those that alter the structure of single chromosomes (**chromosome mutations**); and those that alter individual genes (**gene mutations**). Genome mutations are the most common, occurring as often as 4×10^{-2} per meiotic cell division.

Human Genes and Genomes. DOI: 10.1016/B978-0-12-385212-0.00008-1

TABLE 8.1 Categories and Frequencies of Human Mutations

Category	Mechanism	Frequency
Genome	Chromosome missegregation	2 to 4 × 10^{-2} per cell division
Chromosome	Chromosome rearrangement	6 × 10^{-4} per cell division
Gene	Base pair change	10^{-10} per base pair per cell division 10^{-5} to 10^{-6} per locus per cell division

Chromosome mutations are about 100 times less common (6 × 10^{-4}/cell division). Gene mutations take place in about 10^{-10} per base pair per cell division or 10^{-5} to 10^{-6} per locus per generation (four orders of magnitude less frequent than chromosome mutations.) Genome and chromosome mutations will be discussed further in Chapter 11. Gene mutations will be detailed here and in Chapter 12.

Causes

Most mutations are spontaneous, that is, they occur in the absence of any known cause. We are aware of some causes. Water, for example, can cause a purine base (G or A) to separate from DNA's backbone (called **depurination**). The DNA repair system, to be discussed later in this chapter, may replace, for instance, G with A, thereby leading to a base change after the next round of replication. **Oxidation** (the addition of oxygen to a molecule), too, can lead to base changes. Other mutations are induced by **chemicals, ultraviolet light**, or **radiation**, all of which can break nucleotide chains or damage nucleotide bases. The vast majority of DNA damage is repaired by specific enzymes almost immediately and has no deleterious effects. Those new mutations that go unrepaired are copied during subsequent rounds of replication and may be transmitted to subsequent generations of cells.

Mutations that occur in cells destined to be germ cells are called **germ-line mutations**. Such mutations are transmitted during meiosis and mitosis to all cells of the body. Other mutations, called **somatic mutations**, arise in cells of only one tissue or organ. They affect, therefore, only that organ's structure and function and are not transmitted to the next generation. Finally, mutations can also occur in the mitochondrial chromosome. These mutations are transmitted, but only by the mother, because the egg is the source of the zygote's mitochondria.

A number of studies in flies, worms, mice, and humans have come to the same general conclusion: a given gene mutates in about 1 in 100,000 (1 × 10^{-5}) gametes. Because humans have about 21,000 genes, this rate computes to $2 \times 10^4 \times 10^{-5} = 0.2$ mutations per generation per haploid gamete. The mutation rate of some genes is decidedly higher than 10^{-5}; for other genes, decidedly lower. It seems likely that diploid organisms, like humans, can tolerate a relatively high mutation rate without ill effect because we have two alleles for each gene.

Molecular Types

Many kinds of mutations can change the usual (called the **wild-type**) DNA sequence to a mutant one (Figure 8.1). A **substitution** occurs when a base at a certain position on one strand of the DNA molecule is replaced by any of the other three bases (A→G, C, or T). During the next round of replication, the substituted base will be paired with its appropriate base-paired partner. Substitution of one purine for another purine (for example, A to G) or of one pyrimidine for another pyrimidine (e.g., C to T) is called a **transition**. Substitution of a purine for a pyrimidine (such as A to C) is called a **transversion**.

Other types of mutations are more complex (and rarer) than substitutions. A **deletion** occurs when one or more nucleotide pairs in a DNA molecule are lost. Most deletions remove one

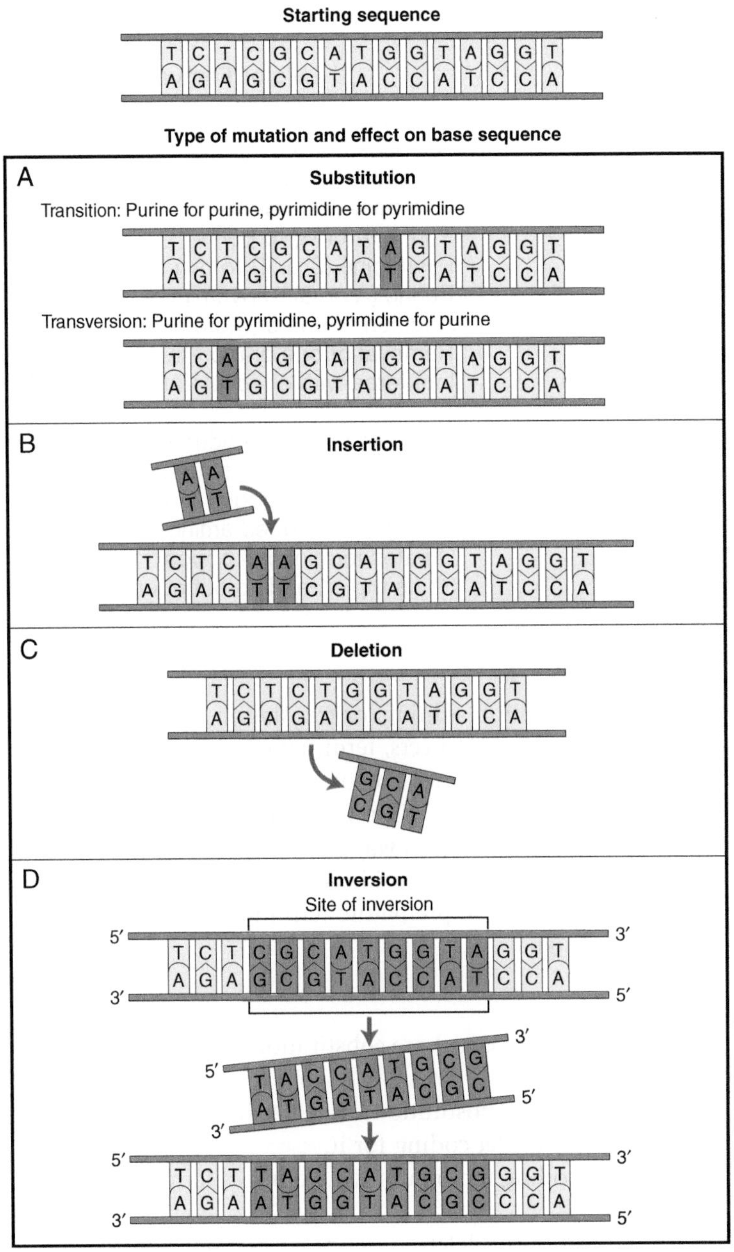

FIGURE 8.1
Classes of mutations affecting DNA structure. See text for further definition of substitution, insertion, deletion, and inversion.

to six nucleotides, but a few are much larger, removing whole genes or chromosomal regions. An **insertion** is just the opposite: addition of one or more nucleotide pairs to the DNA molecule. **Inversions** are more complex still: a 180° rotation of a segment of DNA, without either loss or gain in nucleotide number.

Effect of Mutations on Transcription, Editing, and Translation

Mutations can, and do, affect every step in the sequence of events required for gene expression (see Chapter 7). Most mutations eliminate or reduce gene function and are called **loss-of-function** mutations. Mutations that completely eliminate function are called **null** or "**knockout**" mutations. A mutation that reduces but doesn't eliminate expression of a gene is referred to as a **hypomorphic** one. In contrast, mutations that increase gene expression are referred to as **gain-of-function** mutations (Table 8.2).

TABLE 8.2 Classes and Distinguishing Features of Mutations

Parameter Affected	Types of Mutations	Major Features
Gene function	Loss-of-function (null)	Eliminates normal function
	Hypomorphic	Reduces normal function
	Gain of function	Increases normal function
Molecular change	Nucleotide substitution (point)	One base pair replaces another
	Transition	Purine (A or G) to purine Pyrimidine (C or T) to pyrimidine
	Transversion	Purine (A or G) to pyrimidine
	Insertion	Addition of one or more nucleotides
	Deletion	Removal of one or more nucleotides
Transcription	Promotor, enhancer, start site elongation, termination	Interferes with or accelerates process
Editing	Changes splice sites or exon– intron borders	Alternate splicing produces mRNA with changed exon structure
Translation	Silent (synonymous)	No change in encoded amino acid
	Missense (nonsynonymous)	Change in encoded amino acid
	Nonsense	Creates termination codon
	Frameshift	Shifts triplet reading out of correct phase

Loss-of-function mutations can affect transcription, editing, or translation. Many affect the transcriptional apparatus: start sites, promoters, enhancers, termination signals. Those affecting RNA splicing can reduce the rate of mRNA formation or inhibit splicing at a given site entirely. If splicing is qualitatively changed, the mRNA will not contain the information to encode precisely its polypeptide or protein. In analogous ways, translation may be altered. Mutations can interfere with the ability to initiate polypeptide chain formation, can result in incorporation of the "wrong" amino acid, can cause premature termination, or can lead to failure to add post-translational chemical signals.

Another set of terms is used here:

- **Missense mutations** (or "**non-synonymous**" ones) are substitutions that change one amino acid into a different one.
- **Silent mutations** (or **synonymous** ones) are substitutions in DNA that do not change the amino acid even though they change the triplet coding for it.
- **Nonsense mutations** convert a triplet coding for an amino acid into a terminator codon, thereby truncating the polypeptide.
- **Frameshift mutations** result from insertion or deletion of one or two nucleotides and throw off the triplet reading frame, thereby changing all subsequent amino acid residues in the growing polypeptide chain.

Special Types of Mutations

TRINUCLEOTIDE REPEATS

The human genome contains many trinucleotide repeats (including, but not limited to, CGG, CGG, CGG) as part of its normal sequence. In 1992, a novel kind of mutation was discovered by human geneticists studying an X-linked disorder called the Fragile X syndrome. They discovered that affected males had more than 200 CGG repeats at a site on the X chromosome, compared to less than 50 repeats in normal subjects. Mothers of the affected males had 50–100 of these repeats. Such trinucleotide expansion was not stable—it tended to increase in length from one generation to the next (Figure 8.2). Subsequent study revealed that such expansion of trinucleotide repeats is responsible for another 10 human disorders, and has been found in other animals as well.

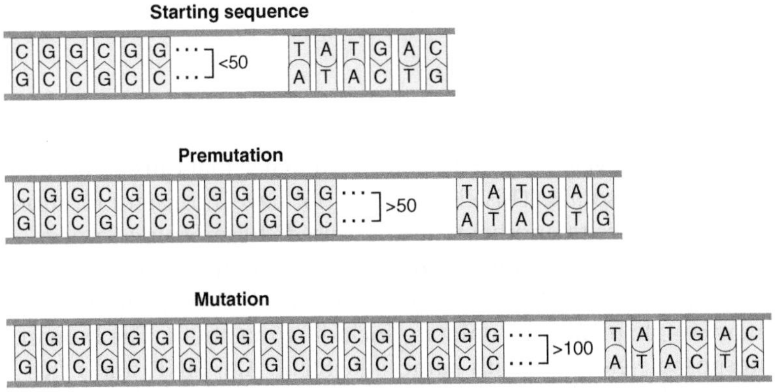

FIGURE 8.2
Trinucleotide repeat expansion. The starting sequence in healthy subjects has fewer than 50 CGG/GCC repeats, while those with triplet repeat premutations have > 50 and those with disease-causing mutations > 100 repeats, and the size of the expansion increases with each generation.

The rules governing the cause, expansion, and instability of trinucleotide repeats are not completely understood. We know the following: the larger the number of repeats at a site, the higher the probability that expansion will occur and function as a mutation; the longer the expansion, the earlier the onset of signs and symptoms; and large tracts of repeats affect DNA structure such that replication errors occur, leading to aberrant starting and stopping, called "**replication stuttering.**"

DNA REPAIR DEFECTS

Every minute the human genome is damaged at approximately three sites (1 per billion nucleotides). Such damage may be produced by physical or chemical agents in the environment or by endogenous errors in the machinery that conducts DNA replication. This amount of DNA damage would be intolerable to the survival of the organism, so all organisms are equipped with a number of different enzymatically controlled systems that scan DNA molecules for errors and repair them. Table 8.3 summarizes some of the types of DNA damage and the corresponding DNA repair system. For example, a base modified by a chemical is removed by the excision-repair enzyme system. A different system, called mismatch repair, identifies a mismatch in the double-stranded DNA (such as A→G, rather than A→T) and corrects it.

These DNA repair systems are highly effective. They deal with damage before it becomes a transmitted mutation. Every cell has its own complement of DNA repair enzymes. In rare instances, however, one DNA repair system or another is, itself, mutated, resulting in unrepaired changes—mutations throughout the genome that are deleterious. One such disorder, xeroderma pigmentosum (XP), is characterized by a large number of skin cancers. XP results from a specific defect in a repair enzyme that removes thymine dimers (T—T), abnormally bonded bases caused by exposure to ultraviolet light. Such individuals must avoid exposure to

TABLE 8.3 Types of DNA Damage and Mechanism of Repair

Type of Damage	Major Mechanism of Repair
Nicks in DNA strand	Repaired by DNA ligase
Uracil present in DNA	Uracil removed by DNA uracil glycosylase
Mismatched bases	Corrected by mismatch repair (excision and resynthesis)
Apurinic or apyrimidinic site	Repaired by AP endonuclease repair system
Pyrimidine dimers (from UV light)	Enzymatically reversed
Damaged region of DNA	Excision repair (excision and resynthesis across partner strand)

sunlight to compensate for strong predisposition to cancer. Another DNA repair defect leads to familial colon cancer.

TRANSPOSABLE ELEMENTS

In the 1940s, Barbara McClintock discovered a genetic element in maize that caused mottling in kernels as well as breaks in chromosomes. This element did not have a constant location; rather, it was transposed from one place to another where it caused chromosome damage. Since then, such **transposable elements** (or **transposons**) have been found in most organisms. Numerous families of transposable elements are now recognized. They are responsible for mutagenic events in many simple plants and animals, including mammals. The human genome contains many transposable elements with names like SINE, Alu, LINE, and LTR. Together they constitute nearly half of genomic DNA.

Whereas transposable elements are responsible for a significant fraction of mutations up to and including the mouse, they cause fewer than 0.5% of mutations in humans. They appear to be evolutionary remnants, but likely retain some functional role. SINE elements may promote translation under circumstances of stress; Alu sequences are found in abundance near coding sequences, suggesting that there is some reason for their being there. Importantly, there is little evidence that any of these elements are movable in humans—that they retain their capacity to transpose.

IMPLICATIONS: MOLECULAR BIOLOGY AND DISEASE

Pneumococcal DNA Morphs to Dodge Vaccines

For nearly 30 years researchers around the world have studied the evolution of a single strain of *Streptococcus pneumoniae*. More than 240 samples of this strain have been characterized since 1984, when it was first identified in Spain. Observations from laboratories in seven countries have given us insight as to why it is so difficult to create a vaccine or antibiotic that will continue to fight this bacterium. The answer to this question emerged with the development of new technology that allows speedy genomic sequencing.

It was found that the DNA of *Streptococcus pneumoniae* is constantly mutating and that the strain has turned over about three-quarters of its genome since it was first identified. The mutations in the DNA are a result of both recombination, in which pieces of DNA are moved around, and substitutions, where individual nucleotides in the DNA sequence are changed.

Vaccines and antibiotics, which target specific gene clusters in the DNA of the bacterium, are thrown off the track and cease being effective as a result of these mutations. Any mutation in the DNA, however, must be such that the bacteria can still be viable and infect its human host. It is, then, remarkable that viability and pathogenicity has been sustained after so much change in the DNA.

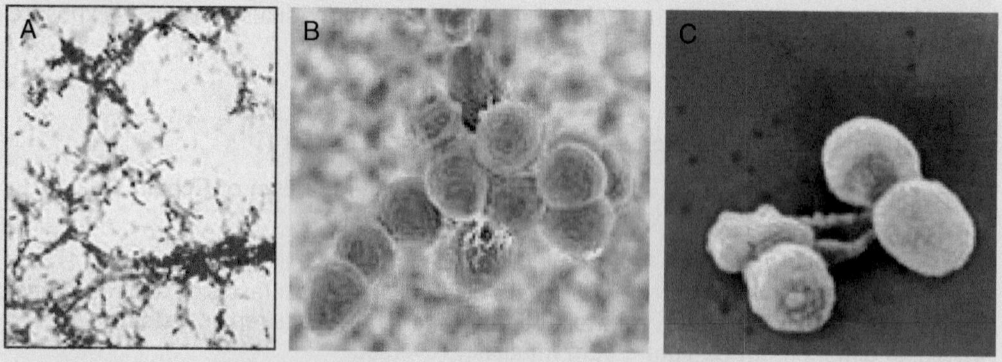

(A) Lung tissue showing streptococcal infection; (B) & (C) Enlargement of Streptococcus pneumoniae.

REVIEW QUESTIONS AND EXERCISES

1. Choose the phrase in the right column that best matches the term in the left column (each term in the left column refers to a particular type of mutation).

a. transition	1. occur in the absence of any known cause
b. mitochondrial	2. addition of one or more bases
c. silent mutation	3. occur in tissue cells, not gametes
d. deletion	4. missegregation of chromosome(s)
e. chain terminator	5. maternally inherited
f. spontaneous	6. substitution of purine for pyrimidine
g. genomic	7. caused by environmental exposure
h. frameshift	8. does not change amino acid
i. trinucleotide repeat	9. occur in cells destined to become gametes
j. chromosomal	10. substitution of one purine for another purine
k. induced	11. removal of one or more bases
l. gene	12. affects single gene
m. germ-line	13. results in formation of terminator codon (also called nonsense)
n. insertion	14. change in a single base
o. somatic	15. results in a structural rearrangement of whole chromosome
p. substitution (missense)	16. affects triplet reading frame thereby changing all subsequent triplets toward the 3' end of the mRNA
q. transversion	17. expansion of trinucleotide units in a gene

2. Suppose that the most common mutation in a gene responsible for a particular genetic disorder has changed an isoleucine residue to an asparagine. Assuming single base substitution mutations:
   a. How many different substitutions could change isoleucine to asparagine?
   b. List the codon changes that expand your answer to part (a).

3. Describe two general forms of RNA splicing mutations and their respective effects on mRNA formation.

4. Single base substitutions often cause no deleterious phenotype. Describe briefly three ways this could occur.

5. If a highly mutable human gene has a mutation rate of 5×10^{-4} per generation and if that gene is followed generation after generation, how many generations, on average, will it take before the gene undergoes a mutation? Show your work.

6. As far as the genetic code is concerned, the successive triplets beyond the point of a frameshift mutation become random.
   a. Of the major classes (not categories) of mutations discussed, are there any that may not change the reading frame?
   b. How often, on average, would you expect a stop codon to appear after a frameshift mutation?

7. In some disorders caused by trinucleotide repeat expansion mutations, the mutation occurs within an intron, in others within an exon. Speculate on the ways these different mutations could produce phenotypic abnormalities.

8. The most common mutation causing cystic fibrosis results in deletion of a single phenylalanine residue at position 508 of the protein involved. Describe the molecular mutation responsible.

9. The following normal amino acid sequence represents part of a protein:
   –lys–arg–his–his–tyr–leu–
   a. What is the one double-stranded DNA sequence of the gene (of many possible ones) corresponding to this amino acid sequence?
   b. What kind of mutation would produce the following amino acid sequence?
   –lys–glu–thr–ser–leu–ser–
   c. What kind of mutation would produce the following amino acid sequence?
   –lys–arg–ile–ile–ile–

10. Describe how mutation in each of the following might be expected to interfere with normal gene function.

a. Promoter

b. Initiator codon

c. DNA helicase.

CHAPTER 9

Biological Evolution

CHAPTER OUTLINE

CORE CONCEPT

Biological evolution is the central organizing principle of modern biology and has revolutionized our understanding of life on Earth. It has provided a scientific explanation for why there are so many different kinds of organisms on our planet and how all these organisms became part of a continuous lineage. DNA is central to understanding biological evolution because all traits in virtually all organisms originate from DNA and are transmitted from generation to generation through it. New mutations in DNA result in the variation of traits that enable the species to adapt more successfully to its environment. Through increased **reproductive fitness**, such traits will increase in frequency in a population from one generation to the next, as will the frequency of the DNA changes responsible for them. The process by which **genotypes** best suited to survive and reproduce in a given environment, and so gradually increase the overall ability of the population to thrive, is called **natural selection**.

HISTORICAL EVIDENCE SUPPORTING THE THEORY OF BIOLOGICAL EVOLUTION

Charles Darwin (Figure 9.1) electrified the world in 1859 with his theory of **biological evolution**. In 1858, Alfred Russel Wallace had published a paper co-authored by Darwin which expressed similar ideas, but it didn't gain much attention until after Darwin published his remarkable volume, 'The Origin of Species'. Each proposed that all life on our planet arose from a single, primordial ancestor, and that natural selection was the driving force behind what Darwin called "**descent with modification**." In the ensuing century and a half, the theory of evolution has penetrated every walk of human life—science, religion, business, law, education, and politics. Over time, paleontologists assembled an extensive fossil record supporting critical aspects of the theory: the geology of finds, the identification of extinct species, and the geographical distribution of organisms.

The evidence for evolution has come from virtually all scientific disciplines. Astrophysics and geology have shown that the Earth is old enough for biological evolution to have given rise to

Human Genes and Genomes. DOI: 10.1016/B978-0-12-385212-0.00009-3

FIGURE 9.1
Charles Darwin (left) and Alfred Russel Wallace (right) who proposed, independently, the theory of natural selection. They corresponded briefly in the late 1850s before each published his remarkable work.

today's species. Physics and chemistry have developed dating methods that have defined the timing of critical evolutionary events. Anthropology has provided insights into human origins and the shaping of human behaviors and social systems. And, most recently, the biological sciences have presented powerful data revealing the molecular biological and genetic forces responsible for **adaptation**, **selection**, and **speciation**. It is these forces that we shall now focus on.

EVOLUTIONARY PRINCIPLES

Origin of Life on Earth

Paleontologists have discovered layered rocks that appear to have resulted from the actions of bacteria at least 3.4—3.5 billion years ago. When life began is easier to answer than how. Despite many ideas and experiments, consensus about the "how" question is lacking. We know that the Earth's simplest chemical compounds—water, oxygen, nitrogen, hydrogen, carbon, and phosphorus—could have reacted to form some of the building blocks of life, including nucleic acids, proteins, and cell membranes. But how did they accomplish this phenomenal task?

Three conditions had to be met for life to begin.

- First, molecules that could reproduce themselves had to come together.
- Second, copies of these molecular assemblies had to show variation that some were better adapted to the environment than others.
- Third, the variations had to be heritable so that some species would increase and others decrease depending on environmental circumstances.

It appears likely (but not certain) that RNA was the first biochemical species that met these requirements. It is capable of self-replication, chemical variation, formation of cell precursors, and selective advantage. It is composed of a four-letter alphabet of bases (A, U, G, C) capable of providing the origin of evolutionary information.

Because of certain disadvantages (instability and mutability of the molecule) inherent in this notion of an "RNA world," pressures appeared so that information storage and functional machines were separated. **Evolutionary forces** led to the biosynthesis of DNA and protein. DNA became the master information molecule and protein, the **functional machine**. The duplex structure of DNA, the role of RNA as intermediate, and the triplet code for translation evolved. Selective advantage, too, led to the evolution of cell membranes, which resulted in the separation of intracellular from extracellular chambers.

Unity of Life

All living things share these characteristics: genetic information encoded in DNA, transcription into RNA, translation into protein, a common triplet code, and a set of functional enzymes and other proteins with common amino acid sequences and three-dimensional structures. This sharing has a straightforward explanation: all creatures share a common origin. From an evolutionary perspective, a primitive common ancestor, whose molecular mechanisms were in place, gave rise to successive populations of organisms (and species) with different genetic characteristics. Each species on Earth is the product of an evolutionary lineage, meaning that it is derived from a pre-existing species, and so on back through time.

This common ancestry is often depicted using **trees of relationships**. As noted in Figure 9.2, three major kingdoms of life forms are inferred from similarities in DNA sequence: Bacteria, Archaea, and Eukarya. Bacteria and Archaea lack a nucleus and are, therefore, called prokaryotes. In Archaea, machinery for DNA replication and transcription resembles that in Eukarya, but their metabolism is like that of Bacteria. Eukarya are

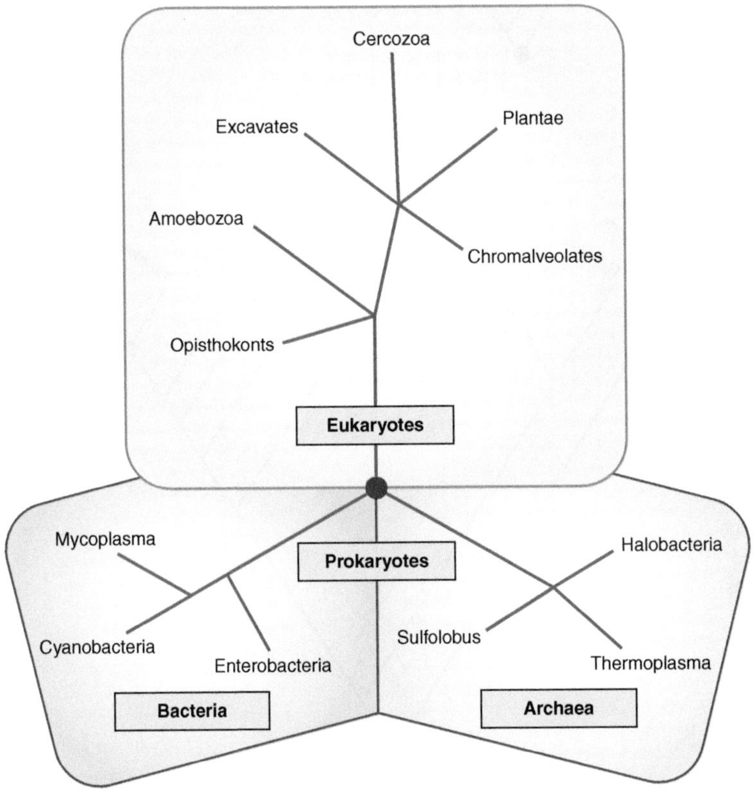

FIGURE 9.2
The three major kingdoms of life forms, each evolving from a primordial common ancestor (shown as the red dot). Their evolutionary relationships are inferred from similarities in DNA sequence. For illustrative purposes, the great diversity among eukaryotes is shown. Much more diversity among bacterial and archaeal organisms exists than is shown here.

characterized by a membrane-bound nucleus, a network of membranous cytoplasmic organelles, and the presence of mitochondria that themselves are derived from the simpler kingdom of bacteria. In turn, each major kingdom diversified into thousands—even millions—of other groups.

Animal Evolution

Opisthokonts are the groups of **eukaryotes** that include **amoebae**, **fungi**, and **animals**. Humans, of course, are among the animals. If we look at the animal group more carefully, we discern trees of life. For any two species living today (Figure 9.3), it is possible to trace back in time their evolutionary lineages until the two lineages intersect. At that intersection, the most recent common ancestor of the two modern species can be found. Figure 9.4 shows how such a tree can be used to identify the last common ancestor for all four-legged creatures, called

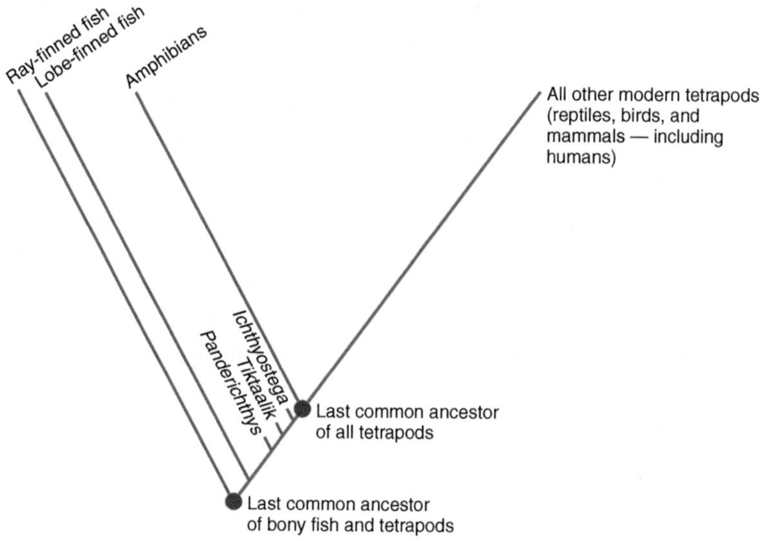

FIGURE 9.3
An evolutionary tree of life built on the concept of the last common ancestor (LCA). The LCA of bony fish and tetrapods (four-legged animals) lived many hundreds of millions ago. The recently discovered Tiktallik had features both of fishes and early tetrapods and is believed to be an intermediary form between them. Time is represented by the length of the lines. Modern groups of organisms are shown at the top.

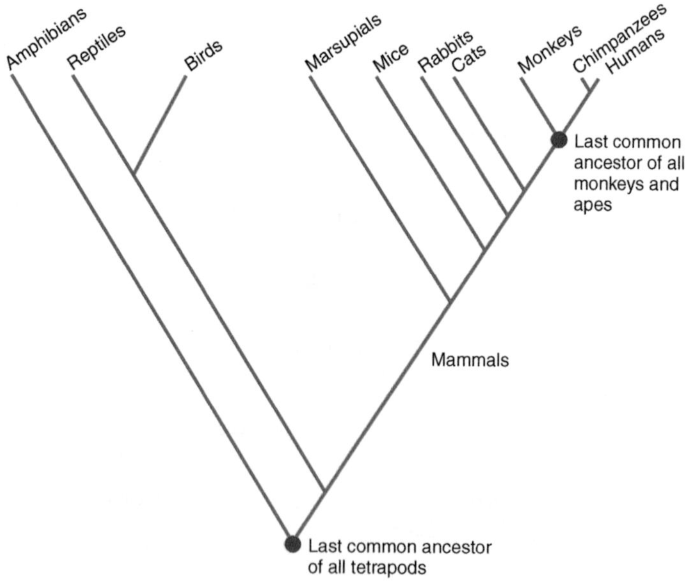

FIGURE 9.4
Last common ancestor (LCA) of the four-legged organisms (tetrapods) living today. This LCA emerged from the ocean more than 375 million years ago. It gave rise to amphibians and was the predecessor of reptiles. Birds and mammals evolved from different lineages of reptiles.

tetrapods. Amphibians were the earliest members of this group. Later, reptiles and birds appeared, followed by the many kinds of mammals shown. The last common ancestor of all monkeys and apes lived about 40 million years ago, while the ancestral species common to chimpanzees and humans lived 6–7 million years ago, as noted in Figure 9.5. If the common ancestor for any two species lived relatively recently, those two species will likely share more physical and behavioral characteristics, and genetic signatures, than two species with a more distant common ancestor. Thus, humans are more similar to chimps than to New World monkeys and far more similar to New World monkeys than to fish.

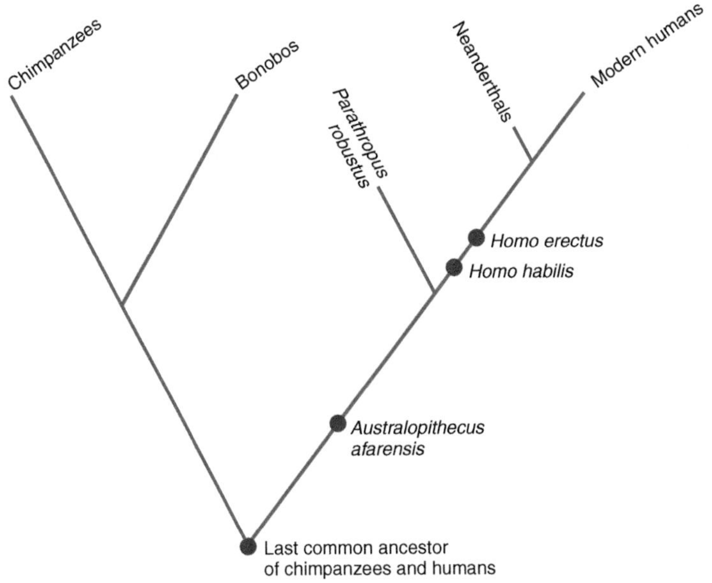

FIGURE 9.5
All monkeys and apes (including humans) evolved from a last common ancestor (LCA) living about 40 million years ago. The LCA of chimpanzees and humans lived about 6 million years ago. The monkeys and apes living today are shown at the top. A number of species (*Austrolopithecus atarensis*, *Homo habilis*, *Homo erectus*) are closely related on the human side of the family tree. They evolved about 4.2, 2.3, and 1.8 million years ago, respectively. These species and others are thought to represent evolutionary links between modern humans and the LCA common to humans and chimpanzees. The nearest evolutionary neighbor to humans was the Neanderthal, a species that appeared about 520 thousand years ago and became extinct less than 24,000 years ago. Recent evidence proves that Neanderthals and humans interbred.

Evolution of *Homo Sapiens*

The detailed phylogenetic tree leading to humans is shown in Figure 9.5. Based on paleontologic evidence, it posits that about 4.1 million years ago a species appeared in Africa, called **Australopithecus** (meaning "southern ape"). Its brain was comparable in size to that of modern apes, and its short legs and upper-arm features indicate that it spent time climbing in trees. But Australopithecus also walked upright, as indicated by well-preserved fossil footprints.

About 2.3 million years ago, the genus **Homo** evolved in Africa. Its first species was *Homo habilis* ("handy man"). Its brain size, determined from 2-million-year-old skulls, was about 50% larger than that of Australopithecus. Approximately 1.5 million years later (that is, 0.8 million years ago), *Homo erectus* ("upright man") evolved in Africa and spread to Eurasia. As best we know, the last common ancestor leading to *Homo sapiens* existed about 700,000 years ago. One of its evolutionary limbs yielded the Neanderthals, the other, *Homo sapiens* ("wise man"). Neanderthals and *Homo sapiens,* both of which evolved in Africa, split about 500,000 years ago. The Neanderthals migrated toward Europe where, by 150,000 years ago, they were populous and widespread. They moved to the Middle East and Asia as well. Modern humans evolved between 150,000 and 200,000 years ago in Africa. Their brains and bodies were much like ours, but different from Neanderthals in several ways. *Homo sapiens* dispersed through Africa and then—in successive migratory waves—to Asia, Australia, Europe, and the Americas. They co-existed with Neanderthals and bred with them, after which the Neanderthals became extinct (for reasons still not understood).

AMPLIFICATION: HUMAN PRE-HISTORY

How Neanderthals and* Homo Sapiens *are Related

Sequencing the genomes of Neanderthals and humans has begun to provide new answers to an old question: Were *Homo sapiens* descended from *Homo neanderthalensis*? Neanderthals appeared on Earth about 400,000 years ago and disappeared 30,000 years ago. Humans appeared some 250,000 years ago and thus could have overlapped with Neanderthals for over 200,000 years.

In 1856, a skull and a number of other bones were discovered in a limestone cave deposit in the Neander River Valley near Dusseldorf, Germany. At first, these bones were assumed to have been from a bear or an old Roman, or a Dutchman, or even a Central Asian soldier in the service of the Russian czar during the Napoleonic wars of the early 19th century. It was finally recognized that these bones were much, much older. The publication of Charles Darwin's *On the Origin of Species* in 1859 led to the idea that since species evolve over time, these bones were from an early ancestor of humans. In his publication, *Descent of Man*, Darwin himself made this suggestion.

For many decades, paleoanthropologists assumed that these Neander people, called Neanderthals (or Neandertals), were dull-witted, brutish, ape-like creatures that walked hunched over with a shuffling gait—the classic caveman.

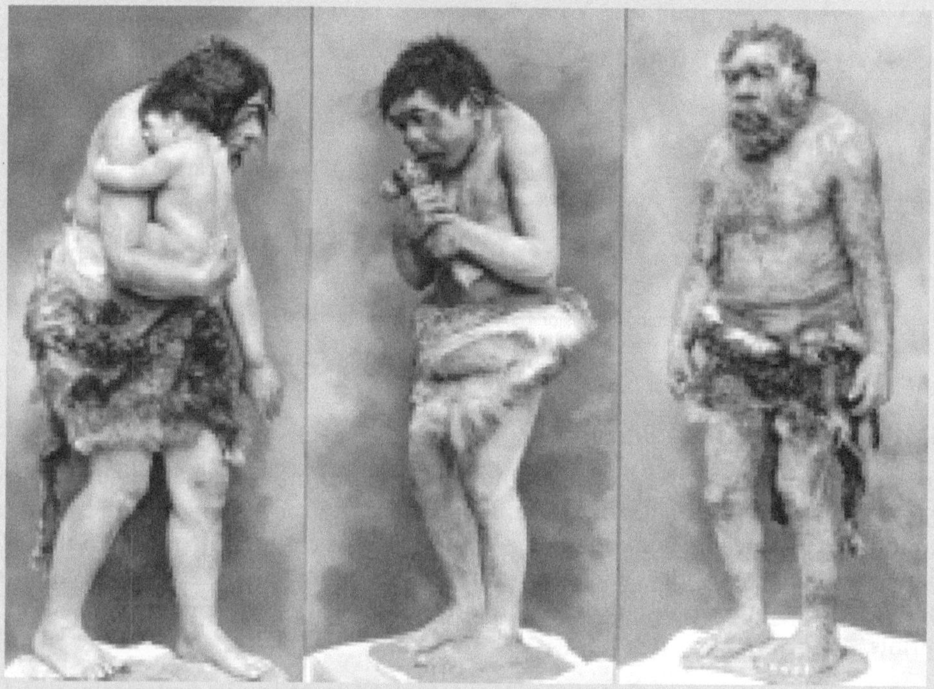

An earlier idea of what Neanderthals might have looked like was shown by a diorama of Neanderthals in an American museum during the 1930s. (permission needed from The Field Museum, Chicago, IL, negative numbers #SA66706, CSA gg834, and CSA 66700)

The consuming question about Neanderthals was whether they were members of the same species as humans, and thus our ancestors, or whether they had descended from a common ancestor but then developed independently. In 1997, Mark Stoneking's studies of mitochondrial DNA extracted from Neanderthal bone indicated that Neanderthals did not contribute mitochondrial DNA to modern humans.

The reconstruction of the Neanderthal genome in 2010, a monumental project led by Svante Pääbo of the Max Planck Institute for Evolutionary Anthropology in Leipzig, Germany, has confirmed the fact that humans and Neanderthals had a common ancestor some 600,000 years ago, but then split and

AMPLIFICATION: HUMAN PRE-HISTORY—Cont'd

developed separately. At first, there was no significant evidence that interbreeding between the two species occurred. However, after further study, Dr Paabo and his group have discovered that about 2.5% of the human genome (non-Africans exclusively) has derived from Neanderthals. After sequencing some 60% of the Neanderthal genome, only about 100 Neanderthal genes have contributed to the evolution of modern humans since the split. Nonetheless, the existence of these genes proves that Neanderthals mated with some modern humans and left a contribution to the human genome before they became extinct.

It now seems that Neanderthals and humans overlapped for only a brief period in Europe about 40,000 years ago. New evidence, resulting from an improvement in the radiocarbon dating of fossils by researchers Thomas F.G. Hingham at Oxford and Ron Pinhasi at the University College Cork in Ireland, has shown that most or all Neanderthal bones in Europe are found to be around 39,000 years old. Given that modern humans existed in considerable numbers in Europe at that time, and that Neanderthals became extinct about 30,000 years ago, the two species co-existed only briefly. It has been proposed that modern humans, because of increased numbers, greater intelligence and technological sophistication, may have caused the Neanderthals to become extinct almost immediately on contact. This extinction can be theorized to have resulted from less than friendly acts on the part of *Homo sapiens* against *Homo neanderthalensis.*

MOLECULAR EVOLUTION

Power of Findings

Within the past 30 years, and most impressively within the past decade, the study of genomes, genes, and gene products has provided unassailable evidence in support of the principles of evolution just described. These molecular studies have buttressed earlier work on life's origins, the first living things, common ancestors, and population migrations. Without doubt, they will help answer many questions that remain about evolutionary mechanisms and details.

Mutations

As discussed in Chapter 8, the sequence of nucleotides in DNA can change from one generation to the next because of mutations. If these changes produce **beneficial traits** (that is, if they improve survival and reproductive fitness), they are likely to spread within a population over multiple generations. If such mutations produce sufficient **selective advantage**, they will, over time, become fixed in and expressed throughout the population and constitute a new species. By comparing the DNA sequences of two different species, scientists are able to discern the genetic changes that have occurred since those organisms shared a common ancestor. The more recent the common ancestor, the more similar the DNA sequences. Thus mutations are the natural agent of **evolutionary change**.

Evolution of Genomes

As noted in Chapter 6, the complete genomes of many organisms have now been sequenced and compared. Numerous findings have emerged from this work. First, genome size has grown along with organismal evolution. The *E. coli* genome is 5 million base pairs (Mb) long, and the human genome 3,000 Mb long. The genomes of yeast, flies, and worms are intermediate in length. Second, eukaryotic genomes are far more complex than prokaryotic ones. Genes make up less than 5% of the total genome in eukaryotes, compared to 100% in prokaryotes. Much of this difference reflects the presence of intervening sequences (introns) in eukaryotic genes that don't exist in prokaryotes. However, as species have evolved, a large number of other changes have appeared: gene duplications, genomic transpositions, proliferation of repeating units, silencing of genes, and more. Third, the regulatory systems controlling gene expression have proliferated. Promoters, enhancers, locus control regions, and transcription factors are some of the means by which expression of genes in higher eukaryotes has been modulated during evolutionary time. How these regulatory changes have produced, for example, the remarkable evolution of the human brain is a question we are still far from answering.

Gene Conservation

Genomic expansion and **diversification** are prominent elements of molecular evolution, but not the only ones. **Conservation of gene structure** and **function** also tell us important things. Many genes are common across the genomes of all living creatures. Such conservation means that some genes are essential for life on our planet. These conserved genes have often changed in sequence over time, but their similarity—referred to as **homology**—has been notable and useful. For instance, the ability to recombine human genes with those of flies or fish has been a powerful tool in the study of animal models of human traits and disorders.

The gene ***CFTR***, which codes for a transport protein in the membranes of lung cells and whose mutations cause cystic fibrosis, underscores the information that can be obtained in this way (Figure 9.6). The data displayed here compare 10,000 nucleotide pairs composing the *CFTR* gene in 8 species, from chimpanzees to puffer fish, with that in humans. There is striking similarity between the human sequence and those of the chimp and orangutan. There is less similarity to humans in the sequences of baboon and marmoset, and even less in more dissimilar organisms. Study of other genes has revealed similar conservation of gene structure, thereby providing powerful evidence for biological evolution from common ancestors.

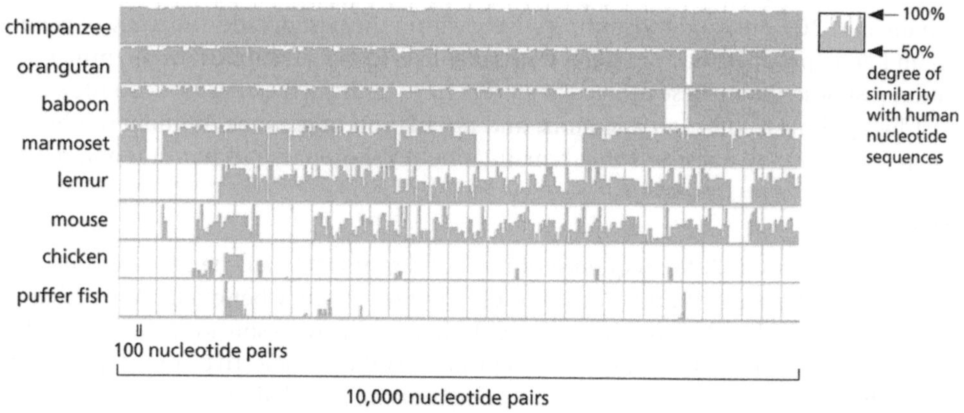

FIGURE 9.6
DNA sequence similarity in eight species for 10,000 nucleotides of the CFTR gene (whose mutations cause cystic fibrosis). Degree of similarity (inset on right) is shown by the height of the green bars, all compared to the sequence in humans. Note the remarkable similarity between chimps and humans, contrasted with some similarity with lemurs and very little with chickens. Sequence comparisons using other genes and proteins corroborate evidence shown here for conservation of genetic information through evolutionary time.

REVIEW QUESTIONS AND EXERCISES

1. Choose the phrase in the right column that best matches the term in the left column.

a. fitness	1. common ancestry of two or more species
b. RNA world	2. concept that RNA was first informational macromolecule
c. trees of relationships	3. organizing principle of modern biology
d. homology	4. property that increases survival and capacity to reproduce
e. mutation	5. the ultimate source of evolutionary change
f. biological evolution	6. similar genes in different species derived from common descent

2. Life on Earth is estimated to have appeared first about 4 billion years ago. The first species that evolved into humans appeared about 4 million years ago. What fraction of life on Earth has been occupied by humans and their progenitors? Show your work.
3. The eminent evolutionary biologist, Theodosius Dobzhansky, said "nothing in biology makes sense except in the light of evolution." Do you agree or disagree with this assertion? State briefly the reason(s) for your position.
4. Which of the following statements supports the hypothesis that RNA was the earliest replicator molecule during evolution?
   a. RNA is more chemically stable than DNA
   b. RNA can encode information
   c. RNA molecules can function as enzymes
   d. RNA is composed of the same nitrogenous bases as DNA.
5. Assume that all human genes stopped mutating completely. How might you know that this fanciful phenomenon had occurred?
6. List the three conditions that had to be met for life to begin on Earth.
7. Five branches of science have each produced information supporting Darwin's and Wallace's theory of evolution. What are the five branches, and what is the essence of their evidence?

CHAPTER 10

Human Individuality

CHAPTER OUTLINE

CORE CONCEPT

Each human being has his or her unique genome. This includes monozygotic twins whose DNA sequences are identical, but whose **epigenomes** differ. This genetic uniqueness expresses itself in the form of physical, chemical, and behavioral individuality. In the past, an understanding of this individuality was barely hinted at by the existence of rare, **single-gene disorders** and by estimating the frequency of **polymorphic protein variants** in a population of healthy people. As we enter the genomic era, genome-wide studies, including complete DNA sequences on a relatively small number of people, have already attested to considerable differences in single nucleotide polymorphisms (SNPs), small insertions and deletions, and copy number variations (CNVs), and in **heterozygosity** for rare disease genes. As such studies are conducted on more individuals and groups, they will lead to greater understanding of normal traits and common disorders. Ultimately, this will get us closer to finding out how each of us differs genetically from anyone else.

UNIQUENESS

All you need to do is look around you to know that you are **unique**. No one else looks and behaves exactly as you do. No one else has exactly the same susceptibility to disease or the same response to medicines and nutrients as you do. Even monozygotic twins differ in small—and sometimes not so small—ways. What accounts for this uniqueness, this individuality? A superficial answer would be "a combination of nature and nurture" —that is, some amalgam of genetic and environmental influences. But that is not scientifically satisfying. Which genes? Which genomic arrangements? Which environmental issues and interactions? We are still very far from the answers to these profound questions, but we are closer than we were last year, and closer than we were the year before and the year before that. We are scratching the surface and are inching forward.

What we know for sure is that each of us has a genome unlike anyone else's. This **genomic individuality**—even between identical twins—is the first thing that identifies us as a human

Human Genes and Genomes. DOI: 10.1016/B978-0-12-385212-0.00010-X

being and must be the blueprint on which our physical, chemical, and behavioral individuality is built. In turn, this genomic blueprint is acted upon by a host of environmental forces: UV radiation from the sun; chemicals in the air and water; nutrients we ingest; medicines we take; organisms we contact; exercises we perform; beliefs, facts and concepts we learn. If we are ever to comprehend human individuality, we must start where each of us starts— with our genes.

EARLY OBSERVATIONS REGARDING UNIQUENESS

Archibald Garrod

Early in his career, that is, between 1900 and 1910, Garrod articulated the concept of "**inborn errors of metabolism**" which, he proposed, were rare, recessive disorders caused by specific enzyme deficiencies (see Chapter 12). Such deficiency led, in turn, to chemical abnormalities and, in some instances, to clinical disturbances as well. By proposing that these affected individuals differed biochemically and genetically from those who were unaffected (and who vastly outnumbered the affected), Garrod took a step toward an understanding of chemical individuality, using disease as the phenotypic marker. Garrod, as we've said earlier, did not know that genes were made of DNA and that the disorders he studied resulted from specific mutations at single gene loci. But even without this information, he proposed in 1909 that these inborn errors reflected a general truth in the following words:

> the thought naturally presents itself that these [inborn errors] are merely extreme examples of variations of chemical behavior which are probably everywhere present in minor degrees, and that just as no two individuals of a species are absolutely identical in bodily structure, neither are their chemical processes carried out on exactly the same lines.

His prescience extended to the young field of evolution:

> Nor can it be supposed that the diversity of chemical structure and process stops at the boundary of the species ... Such a conception is at variance with any evolutionary concept of the nature and origin of species.

In 1931, shortly before his death, Garrod returned to his preoccupation with individuality in his second classic work, *Inborn Factors in Disease.* He emphasized that human biochemical uniqueness extended beyond particular disorders to variability in response to infectious agents, foreign antigens, and drugs. Humankind, he insisted, did not possess a single, evolutionarily optimal genotype, but one with an infinite array of variations.

Harry Harris

By the 1960s, the ability to assay enzyme activity, to purify proteins, and to separate proteins by electrophoretic methods permitted Harry Harris, a true disciple of Garrod, to add significantly to the scant information about human individuality. He and a colleague, D.A. Hopkinson, looked for **electrophoretic variants** of enzymes coded for at 104 human loci in a large number of healthy people. They found that 32% of these loci were **polymorphic**, meaning that the most common allele is present with a frequency of less than 99%. Further, they suggested that each individual would, on average, be heterozygous at about 6% of the loci studied. Their conclusions: genetic variation is common, not rare; such variation occurs in healthy people, not only in those with inborn errors of metabolism.

LESSONS FROM GENOMICS

Human genomics has added considerably to our understanding of our genetic uniqueness. This information has come from many kinds of studies, which we will summarize here.

Single Nucleotide Polymorphisms (SNPs)

Single nucleotide differences between any two people are scattered along the entire human genome, occurring in about 1 in 500 bases. When such changes occur at a population frequency greater than 1%, they are called **SNPs**. Population studies have shown that any two individuals differ in such SNPs at 0.1% of their sequences—that is, there are about 6 million differences between any two of us. Most of these differences occur in DNA not coding for products and, therefore, are not likely to be subject to evolutionary selection pressure.

Insertions and Deletions (indels)

Indels are short sequences of DNA containing insertions, deletions, and inversions. When indels were scored in populations, considerably more variation was noted than that found for SNPs. As many as 30 million differences in indels exist between any two people, or 0.5% of the total DNA.

Copy Number Variations (CNVs)

CNVs are sequence changes covering many thousands to millions of base pairs. CNVs occur remarkably frequently, involving at least 5–10% of the genome. These data and those from assessment of SNPs and indels, show that with each new tool used to measure genomic variation comes progressively more evidence for genomic uniqueness.

Complete DNA Sequences in Individuals

As of the time of writing, only a small number of people had had their complete genomic DNA sequenced. From this small sample, our view of human uniqueness has been expanded. The first two people to have their complete DNA sequence measured were James Watson (co-discoverer of the double helical structure of DNA) and Craig Venter (founder of Celera Genomics, the company that competed with the Public Consortium to be the first to sequence the human genome). Their sequences differed in many ways: Watson had 200 more non-synonymous SNPs and 30% more newly identified SNPs than Venter did and was heterozygous for more disease genes. Venter had nearly 100 more indels than did Watson. These differences were found in the small fraction of the genome coding for proteins (called the **exome**). Surely, much more variation exists in the much larger non-coding portion.

The most comprehensive analysis of genomic variation comes from complete sequencing of both haploid genomes in a single individual, designated **HuRef.** Over 4 million nucleotide variants were found, spanning some 12 Mb of DNA. Several hundred thousand indels and several hundred CNVs were discovered. Non-synonymous SNPs were found at 850 loci implicated in inherited disorders. Heterozygosity was found in at least 17% of loci coding for proteins (and perhaps as many as 40%).

Table 10.1 identifies the sources of our current information about human genetic uniqueness, reminds us that this kind of work has been ongoing for a century, and identifies the kinds of information obtained.

TABLE 10.1 Contributions Toward Understanding Human Genetic Uniqueness

Contributor	Date	Nature of Evidence
Garrod	1908	Presence of "inborn errors of metabolism"
Harris	1960	Frequency of polymorphic serum and enzymatic proteins
Watson/Venter	2008	Frequency of SNPs, indels, and heterozygosity for disease genes in the exome (DNA sequence coding for proteins)
HuRef	2006	Complete sequencing of both haploid genomes in a single individual

Implications

As more complete sequences are made available, our foggy glimpse of human individuality will clear and expand. This will occur soon, because the cost of complete sequencing is approaching a relatively affordable $1,000. Large numbers of complete sequences will do more than underscore human genomic uniqueness. They will give us an inventory of the changes that predispose to and cause disease at every age and of all severities, as well as those changes that promote and maintain health.

IMPLICATIONS: TECHNOLOGY OUTSTRIPS UTILITY

Complete Genome Sequencing for Everyone

The cost of DNA sequencing is falling so fast that within a decade, anyone will be able to have his or her complete genomic sequence determined for $100–500, that is, for the price of a versatile tablet computer. Will this technologic revolution be accompanied by a simultaneous revolution in understanding our genetic uniqueness?

The answer: an unequivocal "no." Some parts of our genome, such as that concerned with single-gene disorders, carrier states, and that relevant to cancer risk from oncogenes and mutant tumor suppressor genes, will be interpretable, provided the technology for interpreting genomic data has kept pace with producing the "raw" nucleotide order. However, such annotation currently does not exist except in a few highly specialized centers.

Even more daunting will be sorting out the meaning of all the SNPs, indels, CNVs, and other genomic variants that are ubiquitous in "everyman's" genome and predispose to a plethora of common traits and disorders. Here, we need even more capability to interpret the DNA sequence such that it is understandable and clinically useful.

Finally, the value of complete genome sequencing will be directly proportional to its intelligent, nondiscriminatory use by health providers, health insurers, employers, and life insurance companies, among others. The health of the individual and the public will surely benefit from genomic sequence information when, and only when, society catches up with the technology. The lag time will be considerable, at least as long as any other important medical advance has taken to be validated, used, and scrutinized—about 20–30 years.

Our individual genetic uniqueness, factored by our individual environment, will take a long time to decipher and decode, but we at least know how to begin to get this profound information.

REVIEW QUESTIONS AND EXERCISES

1. Each of us is genetically unique. From a biologic perspective, where is this uniqueness generated?
2. Any two humans, save monozygous twins, are 99.9% identical in genomic DNA sequence. Then why is each of us so different?
3. Even identical twins differ from one another. Why is this so?
4. In a few decades we will understand almost completely the human genome and its counterpart, the human proteome, yet we will still be far from understanding individual uniqueness. Why is this so?

PART 2

Genetic Disorders

CHAPTER 11

Chromosome Abnormalities

CHAPTER OUTLINE

CORE CONCEPTS

The 46 chromosomes found in the nucleus of humans cells—22 pairs of autosomes and one pair of sex chromosomes—are best examined during mitotic **metaphase**. Metaphase chromosomes are visualized by staining them with **Giemsa** or quinacrine mustard. Even more definitive identification employs **fluorescence *in situ* hybridization** (**FISH**). Chromosomes may be examined *in situ* using light microscopy on intact cells. More often, a **karyotype**, prepared by cutting out and mounting individual chromosomes from a metaphase spread, is studied. In a typical karyotype, chromosomes are displayed in homologous pairs according to size and to the position of the centromere.

Abnormalities in chromosome number (**aneuploidy**) or structure are found in 0.7% of newborns, but this figure is a gross underestimate of the significance of such abnormalities to human health: approximately 75% of all human conceptuses are aborted spontaneously due to such chromosomal defects. Cytogenetic testing (the study of chromosomes) is indicated under the following circumstances:

- advanced maternal age;
- previous infertility;
- still births;
- neonatal deaths;

Human Genes and Genomes. DOI: 10.1016/B978-0-12-385212-0.00011-1

- birth defects;
- chromosome abnormalities in first-degree relatives.

Most kinds of **aneuploidy**—**polyploidy** (extra sets of all chromosomes), **trisomy** (extra copy of one chromosome), or **monosomy** (loss of a single chromosome)—are found in aborted fetuses. Those aneuploid states compatible with full-term gestation include: trisomy for chromosomes 13, 18, and 21; different sex chromosome trisomies (XXX, XXY, XYY) and **quadrisomies** (XXXX; XXYY); and monosomy for the X chromosome. A variety of structural rearrangements are also observed cytogenetically. These include **insertions, deletions, inversions,** and **translocations.** The clinical consequences of such aneuploidies or structural rearrangements, most of which result from **non-disjunction** or chromosome instability occurring during meiosis, depend on several factors:

- which chromosome (or chromosomes) is involved;
- which genes have been perturbed;
- how much chromosomal material has been added or subtracted;
- what compensatory mechanisms exist (such as X chromosome inactivation);
- whether genomic imprinting is involved.

To amplify these general points, the following disorders are discussed in greater detail: Down syndrome; trisomies 13 and 18; sex chromosome aneuploidies; and deletion of part of the long arm of chromosome 15, leading to Prader-Willi or Angelman syndromes.

At a number of points in Part 1, we have mentioned in passing a few human inherited disorders (e.g., hemophilia, Huntington disease, sickle cell anemia) to help illustrate the 10 core concepts. In Part 2, we reverse the emphasis: we describe and examine a number of categories and selected examples of human genetic disorders, using the core concepts to explain and account for the clinical findings.

Genetic disorders are generally classified into three main types: **chromosomal abnormalities, single-gene defects,** and **multifactorial conditions.** Table 11.1 provides an overview of the frequency and the impact of these three categories in populations of live-born babies, hospitalized children, and childhood deaths. Multifactorial conditions are responsible for considerably more morbidity and mortality than the other two categories, but this "bottom line" assessment is incomplete, as will now become apparent in our discussion of chromosomal disorders.

TABLE 11.1 Impact of Major Categories of Genetic Disorders in Children*

Category	Neonatal Incidence (%)	Cause of Hospitalization (%)	Cause of Death (%)
Chromosome	0.7	0.5	2.5
Single Gene	0.4	5.4	8.5
Multifactorial	~4**	39	31

**Numbers derived from several studies in the United States and the United Kingdom.*
***Estimate only, based on overall frequency of birth defects rather than on measurement in 1 million live newborns.*

INTRODUCTION TO CYTOGENETICS

We have commented on chromosomes in several different contexts during Part 1: chromosomal behavior during meiosis and mitosis in Chapter 4; chromosome and gene segregation in Chapter 5; chromosome chemistry and gene content in Chapter 6; chromosomal mutations, as a class, in Chapter 8; and glimmers of genetic individuality from studies of chromosomal and genomic plasticity in Chapter 10. Now we will discuss the anatomy—microscopic and molecular—of human chromosomes and the remarkable variety of abnormal human phenotypes. We begin with discussion of **cytogenetics**, the science concerned with the study of chromosomes.

Historical Perspective

Chromosomes were first observed at the end of the 19th century by Theodor Boveri. He had no idea what their function was. In 1916, William Bateson, one of the scientists who helped rediscover Gregor Mendel's experiments and their significance, was remarkably off base in his comments about chromosomes. He wrote, "it is inconceivable that particles of chromatin ... can possess the powers that must be assigned to our factors (genes)." Several years later, Theophilus Painter reported that human cells had 48 chromosomes, an error that would go uncorrected for 30 years until Joe Hin Tjio and Albert Levan reported, somewhat apologetically, that they could see only 46 chromosomes in human cells in culture. This breakthrough depended on three methodologic modifications: use of the chemical colchicine to arrest cells in mitosis; use of phytohemagglutinin (a plant reagent) to stimulate cell growth; and hypotonic buffer solutions to lyse (break down) cells reproducibly.

As is so often the case, technological advance stimulated cytogenetics greatly (Table 11.2). By 1960, several human disorders caused by chromosomal abnormalities had been described, and others followed swiftly thereafter. In the 1970s, new staining techniques (referred to as banding) permitted unambiguous identification of different chromosome pairs and revealed a variety of not-before-seen structural modifications (**translocations**, **deletions**, and **inversions**). In the 1980s, use of **fluorescence *in situ* hybridization** (FISH) made chromosome "painting" possible and, with it, a variety of powerful detection methods. And in the first decade of the 21st century, **comparative genomic hybridization (CGH)**, a technique to detect genomic copy number changes (alterations of genomic DNA caused by deletion or addition of parts of chromosomes) in a subject's DNA, has made it nearly routine to identify variations in chromosome fine structure previously not identifiable by microscopy.

TABLE 11.2 Technologies that Advanced Cytogenetics

Technology	Date(s)	Advance
Colchicine, hypotonic cell lysis, phytohemagglutinin	1950s	Allowed accurate count of chromosome number
Banding with Giemsa and quinacrine	1970s	Reduced uncertainty in identifying chromosomes, homologues, and rearrangements
Fluorescence in situ hybridization	1980s	Permitted unambiguous chromosome identification by painting whole chromosomes and regions thereof
Comparative genomic hybridization (CGH)	2000s	Facilitated understanding of chromosome fine structure not detectable by microscopy

Chromosome Identification and Morphology

The condensed chromosomes of a dividing human cell are best seen during mitotic metaphase (or prometaphase). Such a chromosome spread, contrasted with the diffuse chromosomal staining seen during interphase, is shown in Figure 11.1A. In this common technique, the individual chromosomes in such a mitotic spread are cut out and mounted into a **karyotype**, shown in Figure 11.1B. By convention, the largest chromosome (1) is positioned at the upper left and the smallest (21) at the lower right, and the others arranged by decreasing size from 1 to 22. The sex chromosomes (here X and Y) are set off by themselves.

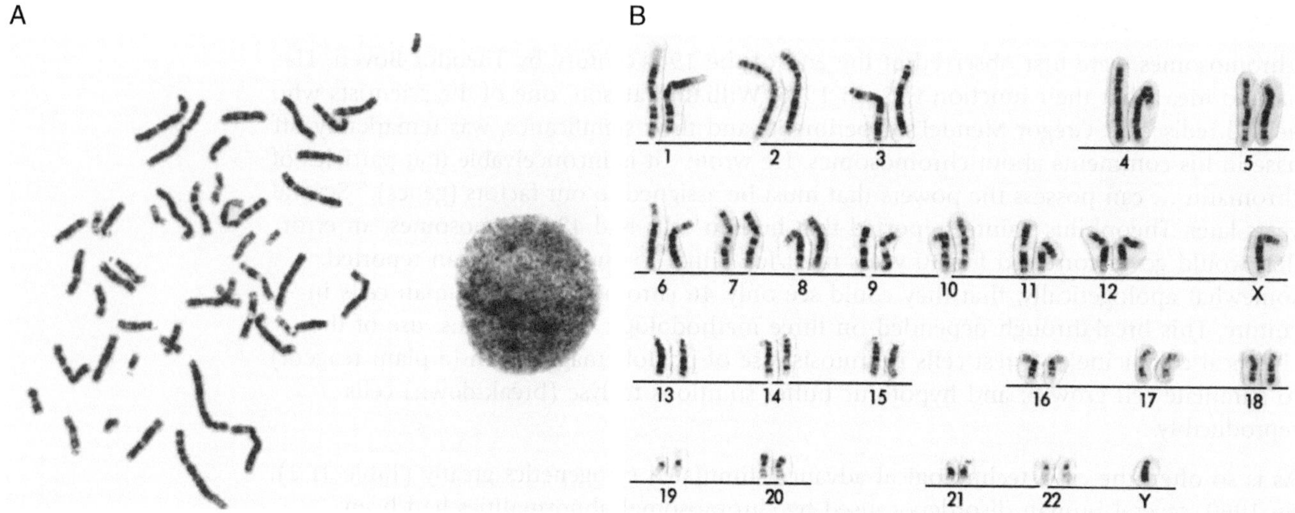

FIGURE 11.1
(A) Mitotic metaphase spread of human chromosomes in blood lymphocyte compared to appearance of a non-mitotic lymphocyte nucleus. (B) Karyotype prepared from spread shown in 11.1a. Chromosomes are arranged in groups according to size and position of centromere.

Most chromosomes can be distinguished from one another by the position of the **centromere**, as well as by differences in length. The centromere is a constriction or pinching in of the sister chromatids and is required for formation of the mitotic spindle (see Chapter 4). The centromere is a morphologic landmark that divides the chromosome into two arms, a **shorter one (designated p)** and a **longer one (designated q)**. Chromosomes are **metacentric** if the centromere lies near the middle of the chromosome, such that the p and q arms are of about equal length, **submetacentric** if the centromere is somewhat distant from the center, and **acrocentric** if the centromere lies near the end of the chromosome. Looking at Figure 11.1B, chromosomes 1, 2, 3, are metacentric; 4 through 12 are large submetacentric; 16 through 20 are small submetacentric; and 13 through 15, 21, and 22 are acrocentric. Finally, the X is large and submetacentric, and the Y is a small acrocentric.

A fourth type of chromosome, called **telocentric**, is characterized by the centromere being at the terminal end of the chromosome, thus leading it to have only one arm, not two. Telocentric chromosomes are not seen in healthy humans, since they are unstable and arise by misdivision or breakage near the centromere and are usually eliminated within a few cell divisions.

Staining Techniques

Unambiguous chromosome identification depends on the use of various staining techniques. Staining is responsible for the alternating dark and light bands on the chromosomes noted in Figure 11.1B. The most routinely used technique stains the metaphase chromosomes with **Giemsa** (after using the enzyme trypsin to digest proteins). Each chromosome pair stains with its own characteristic banding pattern. The bands **(G bands)** correlate approximately with the DNA sequence underlying it: AT-rich areas stain darkly, GC-rich areas lightly. An idealized picture of such a G-banded karyotype (called an **ideogram**) is shown in Figure 11.2. About 400 dark bands per haploid genome are seen in this way.

There are several other staining techniques used for more specialized purposes. One, called **Q-banding**, stains chromosomes with **quinacrine mustard** and views them fluorescently. The

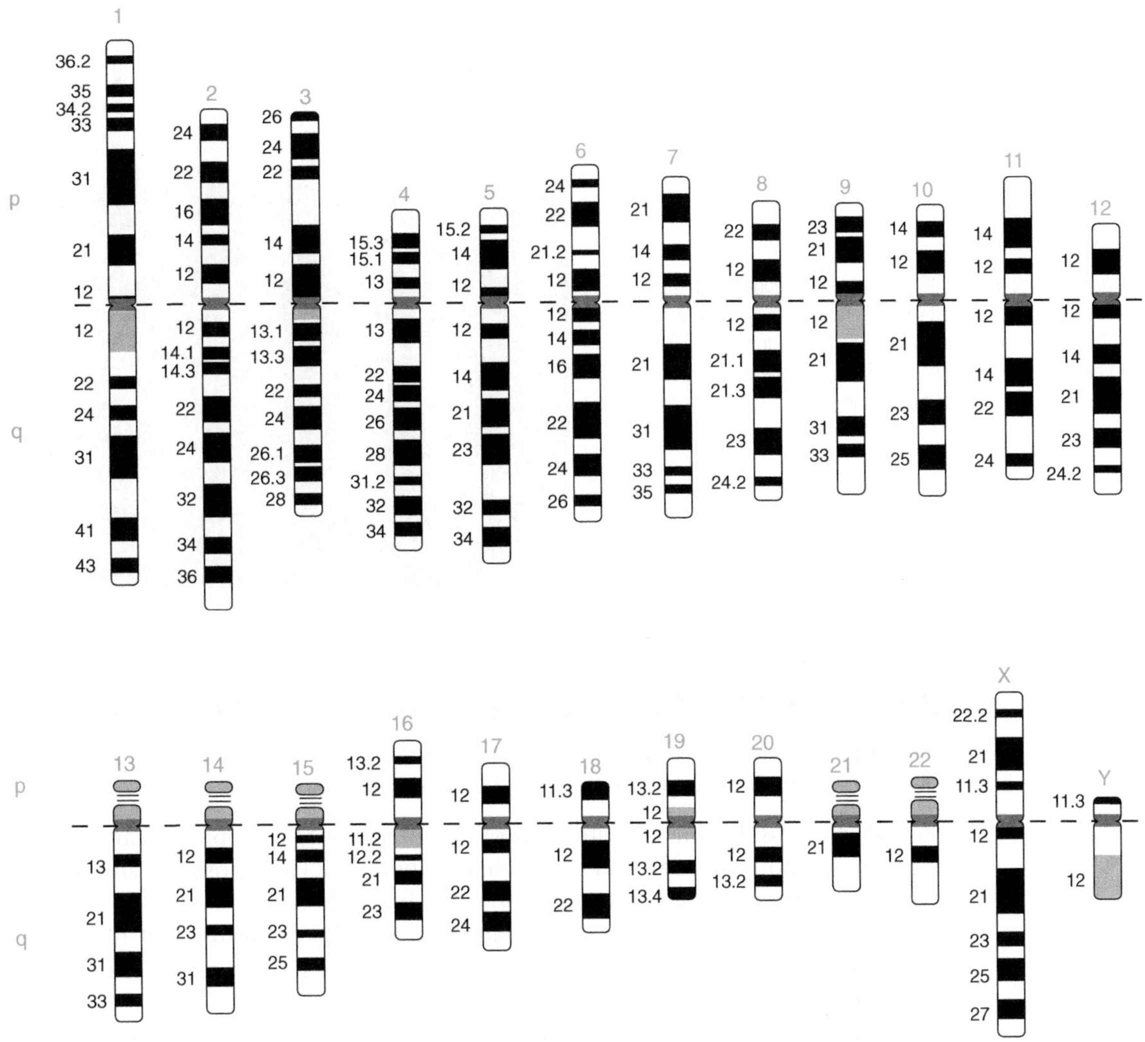

FIGURE 11.2
Ideogram showing G-banding patterns of human metaphase chromosomes. About 400 bands are observed per haploid set. Centromeres are indicated by the dark, gray regions separating the short (p) arms from the long (q) arms.

bright Q bands correspond almost exactly with the dark bands seen with Giemsa. A second method, called **R-banding**, treats chromosomes in such a way that the dark and light G bands are reversed.

Third, and most recent, as well as revolutionary, is the **FISH** (fluorescence *in situ* hybridization) technique. FISH deploys DNA probes specific for each chromosome (or subchromosomal region or single locus). These probes are fragments of DNA or RNA, usually 100–1,000 bases long, used to detect the presence of nucleotide sequences that are complementary to the sequence in the probe. The probes are labeled with modified nucleotides that fluoresce under particular conditions. By using different fluorochromes, a karyotype can be "painted" as desired (Figure 11.3). As shown, 24 different probes and fluorochromes produce a visually striking karyotype that has powerful uses in cytogenetics. It allows, for instance, direct karyotyping of a metaphase spread without the laborious cutting and arranging process usually employed.

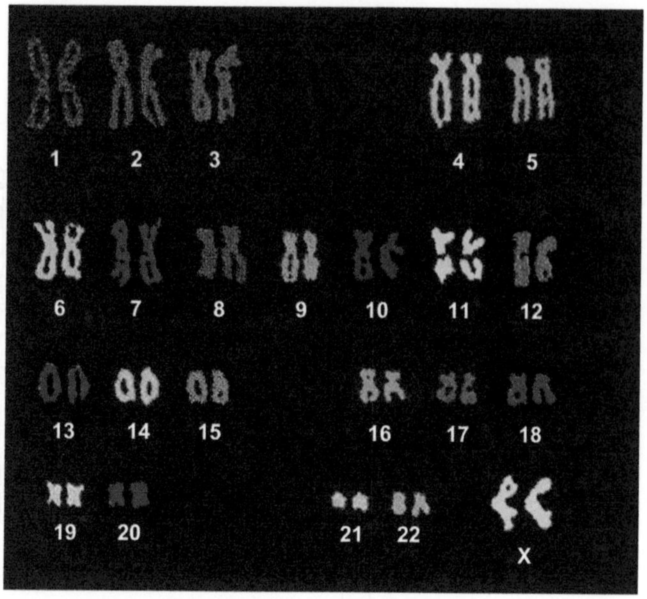

FIGURE 11.3
Fluorescence *in situ* hybridization (FISH) of human metaphase chromosomes. Twenty-four different probes labeled with different fluorochromes are used to "paint" each chromosome unambiguously. Karyotype emphasizes distinct appearance of each pair of chromosomes. This uniqueness greatly facilitates chromosome identification in the clinical setting.

CLINICAL INDICATIONS FOR CHROMOSOME ANALYSIS

Obtaining karyotypes is not part of routine clinical care. However, there are numerous situations that call for chromosome analysis. The major indications are as follows:

- *Fertility problems.* In couples with a history of infertility or multiple miscarriages, karyotyping both the man and the woman is indicated. Chromosomal abnormalities are seen in a significant fraction (up to 6%) of such couples.
- *Stillbirth and neonatal death.* Chromosome abnormalities occur in about 10% of stillbirths or neonatal deaths, compared to 0.7% of live births. A karyotype on the affected child may aid genetic counseling.
- *Problems of early growth and development.* A small fraction of infants show one or more of the following phenotypes: multiple congenital malformations, failure to thrive, short stature, ambiguous genitalia, and others. Karyotyping is indicated for children with one or a combination of such manifestations.
- *Family history.* Under certain circumstances, the presence of a chromosome abnormality in a first-degree relative calls for karyotyping other family members.
- *Advanced maternal age.* Women older than 35 years of age who become pregnant have an increased risk of chromosome abnormalities in their fetuses. Fetal karyotyping has become a routine part of appropriate prenatal care in this setting.

IMPACT OF CHROMOSOME ABNORMALITIES ON HEALTH

As stated earlier, chromosome abnormalities occur in about 0.7% of live births—but this number does not begin to define the magnitude of such abnormalities on pregnancy loss, birth defects, and mental retardation. Major chromosome defects are found in fully one-half of spontaneous abortuses. Because chromosomal abnormalities must also be present in miscarriages occuring even before a woman knows she is pregnant, it is estimated that fully 25–40% of all conceptuses have chromosome abnormalities incompatible with full term gestation and postnatal life. This profound wastage bears witness to the fact that meiosis and mitosis are error-prone, and that full gestation is the exception rather than the rule. It also shows that the body has a most effective way of ridding itself of severely damaged embryos and fetuses.

TYPES OF CHROMOSOME ABNORMALITIES

General Features

Chromosome abnormalities may be numerical or structural. They may affect autosomes or sex chromosomes, or both simultaneously. Numerical abnormalities far outnumber structural ones and most often lead to having an extra or missing chromosome. Translocations—the exchange of chromosome segments between **non-homologous** chromosomes—are not rare, but their clinical consequences vary depending on location and whether the translocation is "balanced." With a microscopically **balanced translocation**, no chromosome material has been lost in the rearrangement.

A series of terms is used to describe chromosome abnormalities. An exact multiple of the haploid number (N) is called **euploid**. The normal human chromosome complement of 46 is then both diploid (2N) and euploid. Other abnormal euploid states are **triploidy** (3N), characterized by having three copies of each chromosome, and **tetraploidy** (4N), with four copies of each chromosome. **Aneuploidy** refers to any situation in which there is an extra or missing chromosome, such as 2N + 1 or 2N − 1. Table 11.3 presents a list of abbreviations commonly used to describe normal and abnormal cytogenetic findings.

TABLE 11.3 Abbreviations Used to Describe Chromosomes and Their Abnormalities

Abbreviation	Meaning	Representative Example
1–22	autosomal numbers	
X, Y	sex chromosomes	46XX or 46XY
p	short arm (petite)	
q	long arm	
del	deletion	46XX, del(5p)
der	derivative, a structurally modified chromosome	der(1)
dup	duplication	
i	isochromosome	46X, i(X)(q10)
ins	insertion	
inv	inversion	Inv(3)(p25q21)
r	ring chromosome	46X, r(X)
rob	Robertsonian translocation	Rob(13;21)(q10:q10)
t	translocation	46XX, t(2;8)
+	gain of	47XX+21
−	loss of	45XX-22
/	mosaicism	46XX/47XX+8

The phenotypic consequences of these abnormalities depend on many things: the nature of the abnormality, the particular chromosome involved, the genes affected by the structural change, and the likelihood of its transmission to subsequent generations. Table 11.4 offers a snapshot of the nature and frequency of human chromosome abnormalities.

TABLE 11.4 Outcome of 100,000 Recognized Pregnancies

Outcome	Pregnancies	Spontaneous Abortions	Live Births
TOTAL	100,000	15,000	85,000
Normal chromosomes	92,000	7,500	84,400
Abnormal chromosomes	8,000	7,500	500
Trisomies:			
1–3	1,920	1,920	0
4–5	95	95	0

Continued

TABLE 11.4 Outcome of 100,000 Recognized Pregnancies—continued

Outcome	Pregnancies	Spontaneous Abortions	Live Births
6–12	561	561	0
13	145	128	17
14–17	1,832	1,832	0
18	236	223	13
19–20	52	52	0
21	463	350	113
22	424	424	0
Translocations:			
Balanced	178	14	164
Unbalanced	277	225	52
Polyploid:			
Triploid	1,275	1,275	0
Tetraploid	450	450	0
Sex chromosomes:			
XXY	48	4	44
XYY	50	4	46
XXX	65	21	44
X monosomy	1,358	1,350	8
Other	339	280	49

Inspection of this table reveals the following:

- 15% of recognized pregnancies abort spontaneously in this particular population.
- 50% of the spontaneous abortions result from chromosomal abnormalities (7,500/15,000).
- **Autosomal trisomies** for all autosomes except numbers 1 and 5 are found in abortuses, but only three autosomal trisomies are found in live births (numbers 13, 18, 21).
- Trisomy for 13, 18, and 21 occur much more frequently in abortuses than in live births (75% of the total for trisomy 21; 88% for trisomy 13; 95% for trisomy 18).
- Extra copies of X or Y chromosomes are more compatible with being live-born than are extra autosomes.
- **Monosomy** for an autosome is not observed in live births.
- Having a single copy of the X chromosome is compatible with live birth; even then, most X monosomy results in abortion.

Abnormalities of Chromosome Number

POLYPLOIDY

Triploidy (3N) and tetraploidy (4N) occur in human embryos not infrequently. Triploidy, for instance, may be seen in 1–3% of conceptions. But triploidy and tetraploidy are not compatible with extra-uterine life (see Table 11.4). Triploidy usually is caused by two sperm fertilizing a single egg, but meiotic non-disjunction has also been implicated.

ANEUPLOIDY

At least 5% of recognized pregnancies are characterized by aneuploidy. Most aneuploid humans have either three copies of a single chromosome (trisomy) or one copy of a single chromosome (monosomy). About 1 in 260 live births have some form of trisomy. Viable trisomies are identified for only three autosomes (13, 18, and 21) and for the X and the Y. Why only these five? We don't know, but some observations are worth mentioning. For the X, inactivation of two of the three X chromosomes seems the likely explanation for the viability of some XXX fetuses. Chromosomes 13, 18, 21, and Y are the most gene-poor chromosomes (that is, among human chromosomes, they contain the fewest genes). Perhaps this explains why an extra copy of the chromosome is not always lethal, as is trisomy for all the remaining autosomes. Monosomy, on the other hand, is uniformly lethal, with the single exception of the X.

Aneuploidy usually results from **meiotic non-disjunction**, meaning failure of a pair of chromosomes to separate properly during one of the two meiotic divisions, usually meiosis I. As shown in Figure 11.4, the consequences of non-disjunction occurring in meiosis I or meiosis II are different. When non-disjunction occurs in meiosis I, the gamete with 24 chromosomes contains both the paternal and the maternal homologues. If the non-disjunction takes place in meiosis II, however, the gamete with 24 chromosomes contains two copies of either the paternal or maternal homologue. Figure 11.4 makes clear how non-disjunction can produce monosomy as well.

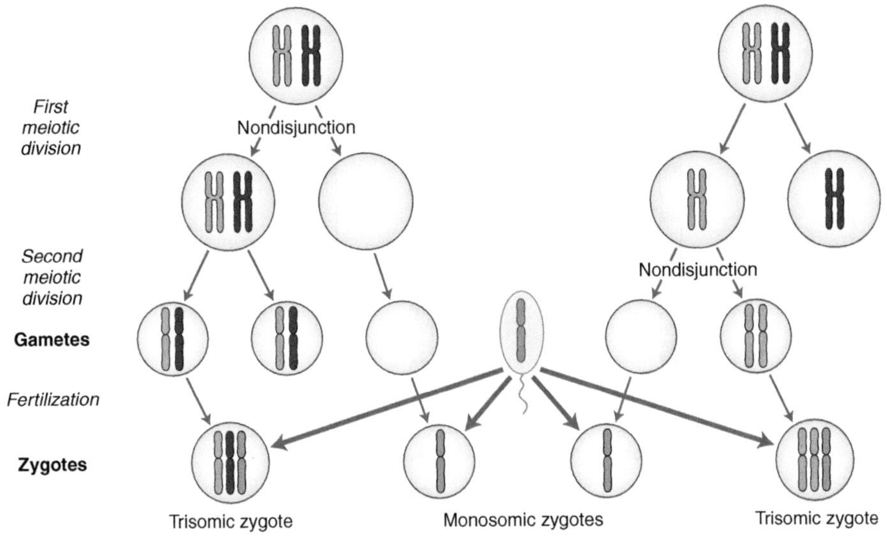

FIGURE 11.4
Consequences of non-disjunction occurring in either meiosis I or meiosis II. A single pair of chromosomes is depicted—the paternal homolog is in pink, the maternal homologue in red. As discussed in the text, the homologues present in the two trisomic zygotes are not identical.

Occasionally, the non-disjunction occurs in an early mitotic, rather than meiotic, division. In this circumstance, the embryo will have two populations of cells, 2N and 2N + 1. This occurrence is known as **mosaicism**, a condition where cells in one individual have a different genetic makeup. We will have more to say about the mechanisms underlying aneuploidy when we discuss Down syndrome and Turner syndrome later in this chapter.

Abnormalities of Chromosome Structure

Structural abnormalities of chromosomes occur in about 1 in 400 live births. They can involve one chromosome, two chromosomes, and, rarely, more than two. They are caused by aberrant chromosome breakage and reunion. When a single break in only one chromosome is involved, the result may be a deletion of some chromosomal material. The deletion may involve the end of the chromosome (**terminal deletion**) or an internal portion (**interstitial deletion**). Many other kinds of changes affecting one chromosome are noted in Figure 11.5. A portion of the chromosome may be **duplicated**, with the duplicated segment retaining its original direction (tandem) or being rotated by 180° (mirror). **Inversions** may involve the centromere (pericentric inversion) or only one of the arms (paracentric inversion). **Isochromosomes** arise from abnormal centromere division and result in duplication of either the short arm or the long arm.

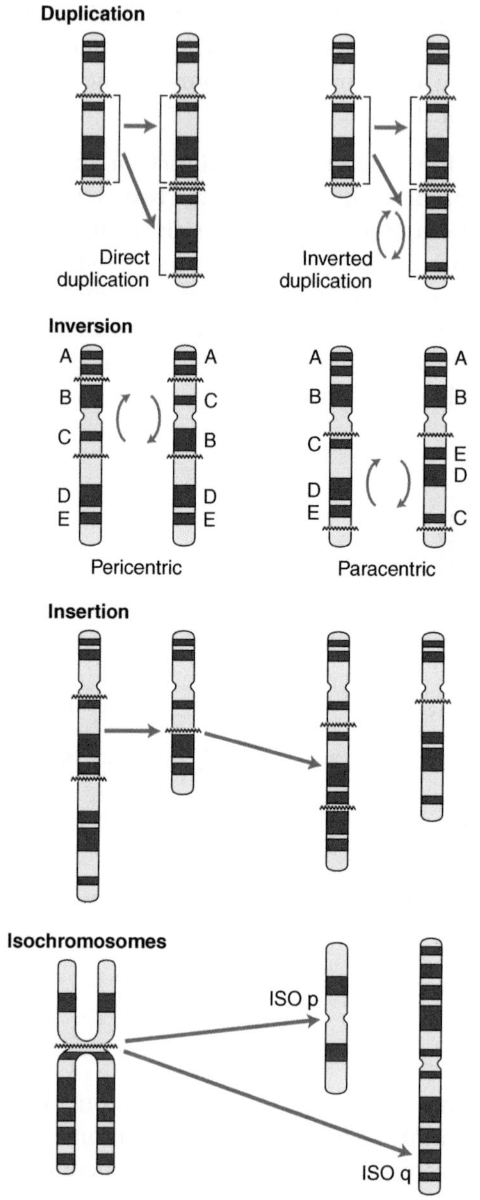

FIGURE 11.5
Structural abnormalities of chromosomes. Deletions (not shown) are caused by a single chromosome break in one chromosome. Duplications require two breaks in one chromosome. Insertions result from at least three breaks in two chromosomes. Isochromosomes reflect abnormal centromere division. See text for more detail.

When two or more chromosomes are involved, different changes occur as shown in Figure 11.6, and a portion of one chromosome may be inserted into a different one. More commonly, there is exchange of chromosomal material between two non-homologous chromosomes. This is called a translocation. Two general types of translocations occur. One, called a **reciprocal translocation**, is depicted in Figure 11.6. In the example shown, a portion of chromosome 21 is translocated to the end of the long arm of chromosome 3. Reciprocally, a significant fraction of the long arm of chromosome 3 is moved to the end of chromosome 21.

The other type, a **Robertsonian translocation**, involves centric fusion between two non-homologous, acrocentric chromosomes as shown in Figure 11.7. When this occurs, the short arms of each chromosome are lost, and the long arms of each are joined together.

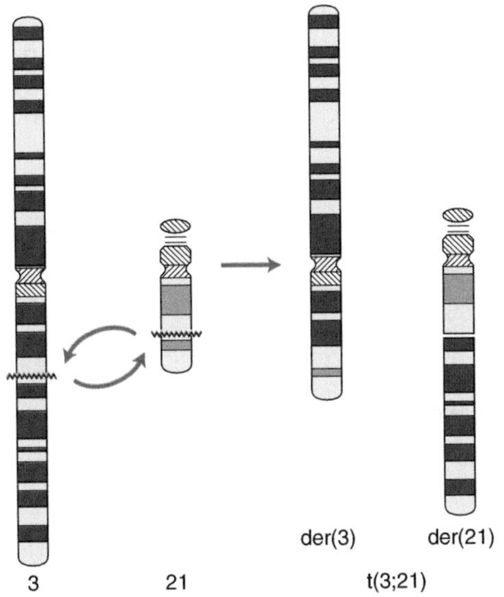

FIGURE 11.6
Reciprocal translocation between chromosomes 3 and 21. Chromosome 3 is depicted in red, chromosome 21 in green. The diagram illustrates that the translocation has resulted in no loss of chromosomal substance, that is, it is balanced.

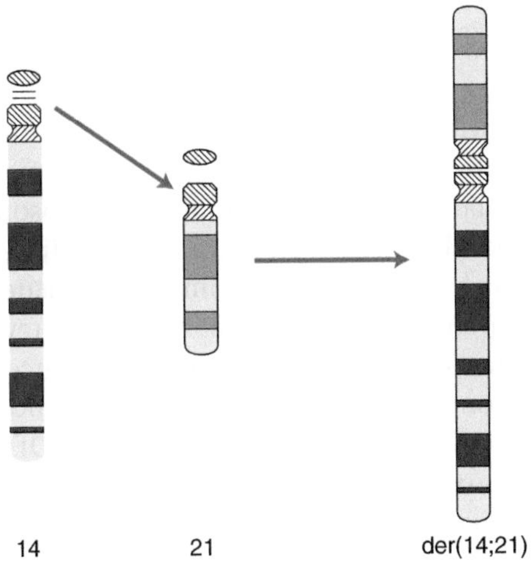

FIGURE 11.7
Robertsonian translocation (or centric fusion) involving the long arms of chromosomes 14 and 21. Chromosome 14 is depicted in red, chromosome 21 in green. The derivative (der) chromosome has lost the short arms of both acrocentric chromosomes.

The phenotypic consequences of reciprocal and Robertsonian translocations are important to understand. In both instances, if no essential chromosomal material is lost and if no genes are damaged by the rejoining, those who carry the translocation will be clinically normal. But these individuals are at increased risk of having chromosomally unbalanced offspring. The reason for this is detailed in Figure 11.8 and Figure 11.9. In the case of a reciprocal translocation between chromosomes 3 and 21 (Figure 11.8), a cross-shaped (or **quadrivalent**) structure is formed during meiosis. Under this circumstance, the homologous chromosomes can separate in several different ways. In **alternate segregation**, the two normal chromosomes go to one daughter cell and the two translocated chromosomes go to the other. In this instance, two types of gametes are formed: normal, and **translocation carrier**.

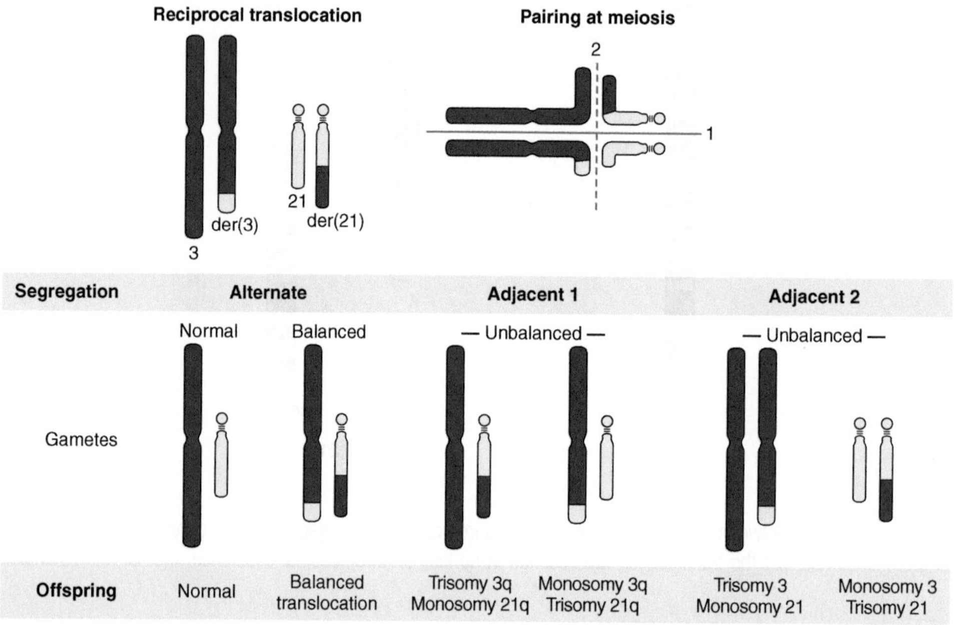

FIGURE 11.8
Meiotic segregation, gamete formation, and outcome of offspring in a reciprocal translocation between chromosomes 3 and 21. Of the six possible outcomes, only two are compatible with being live-born. See text for details.

Upon fertilization, offspring will be phenotypically normal, being either genotypically normal or a translocation carrier like the parent. However, if meiotic segregation leads to gametes containing one normal and one translocated chromosome, then an unbalanced gamete will be formed. Two kinds of this separation (called **adjacent**) are known: adjacent 1 and adjacent 2. In adjacent 1, homologous centromeres segregate. At fertilization, offspring will either be trisomic for 3q and monosomic for 21q or the reverse. In adjacent 2 segregation, homologous centromeres do not separate. This occurrence leads to different kinds of unbalanced gametes that, upon fertilization, yield offspring who are trisomic for chromosome 3 and monosomic for chromosome 21, or the reverse. Of the six possible outcomes shown, two are viable—the products of alternate segregation. Neither adjacent 1 nor adjacent 2 segregation produces live offspring, rather yielding early or later spontaneous abortions.

Figure 11.9 presents a similar picture for a carrier of a Robertsonian translocation. In this instance, the two normal chromosomes and the translocated one (centric fusion between the long arms of 14 and 21) form a triradial configuration at meiosis. Again, there are three possible kinds of segregation. In alternate segregation (line A on the diagram), one gamete will receive each of the normal chromosomes and the other gamete the translocated one. Upon fertilization, the zygote will either be normal or be a phenotypically normal carrier of the translocation. On the other hand, adjacent segregation following line B will yield one gamete with a normal chromosome 14 but no 21; the other gamete will have a normal 21 and the translocation chromosome. Upon fertilization, two possible outcomes are partial trisomy 21 (also called translocation Down syndrome) and monosomy for 21. Finally, following adjacent segregation shown by dotted line C, the offspring will be either trisomic for chromosome 14 or monosomic for 14. Accordingly, of the six possible outcomes, two are phenotypically normal, one has Down syndrome, and three are non-viable because of monosomy or trisomy.

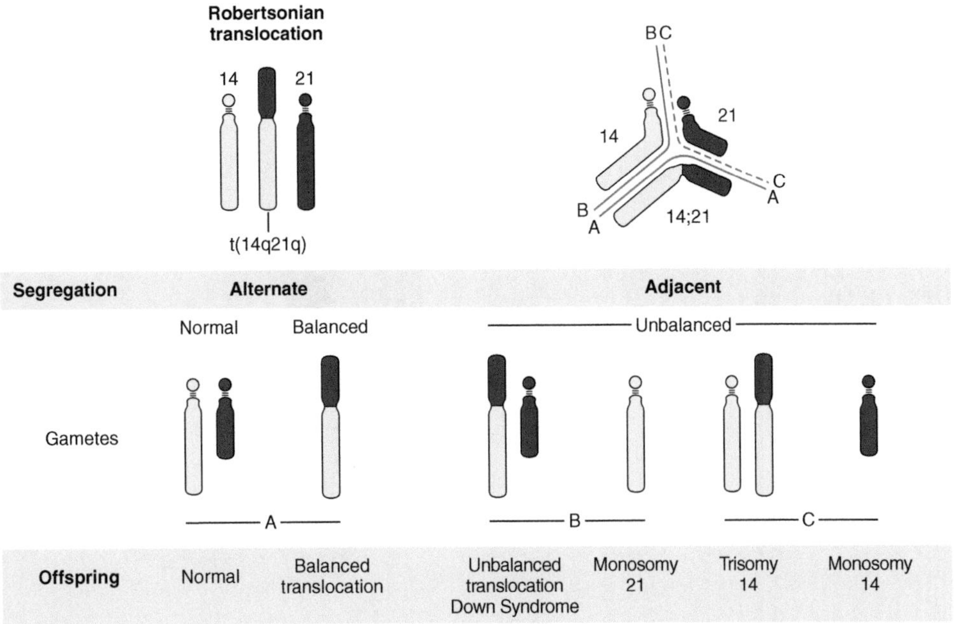

FIGURE 11.9
Meiotic segregation, gamete formation, and offspring outcome in Robertsonian translocation between chromosomes 14 and 21. Of the six possible outcomes, three are compatible with being live-born: normal, a balanced translocation, and translocation Down syndrome. See text for details.

ILLUSTRATIVE EXAMPLES

As stated earlier, chromosome abnormalities may produce a large number of clinical problems: unexplained miscarriages, birth defects, developmental delay, mental retardation, and more. From a very long list of such disorders, we have selected several because they illustrate scientific principles, emphasize clinical consequences, and underscore societal dilemmas.

Down Syndrome

If we can use the expression "classic" chromosome disorder for only one condition, **Down syndrome** is surely it. The syndrome was described clinically by Langdon Down in 1866 (the year Mendel published his work on peas). It is, by far, the most common and widely recognized chromosome disorder. It occurs in about 1 in 700 live births and is much more common than that in pregnant women over age 35. It was the very first condition shown, in 1959, to be caused by an extra copy of a chromosome. It was also the first human disorder detected prenatally, by obtaining fetal cells from amniotic fluid. And it has engendered major societal debates concerning abortion rights and education of the retarded.

CLINICAL FINDINGS

The characteristic features of Down syndrome are usually recognized at birth (Figure 11.10). They include hypotonia (poor muscle tone), short stature, a flattened occipital surface of the skull, a single transverse palmar crease, and a collection of facial abnormalities, including a flat face, lowset ears, upward slanting eyes, an extra skin fold at the medial aspect of the eyes (epicanthal folds), a small mouth, and a protruding tongue.

Developmental and cognitive delay are the major clinical consequences of Down syndrome. Delay is usually obvious during the first year. Most often, the IQ ranges from 35 to 70. In addition, 40% of patients have congenital heart defects, 5% have gastrointestinal anomalies, and there is an overall incidence of leukemia of 1%. In those with Down syndrome who have

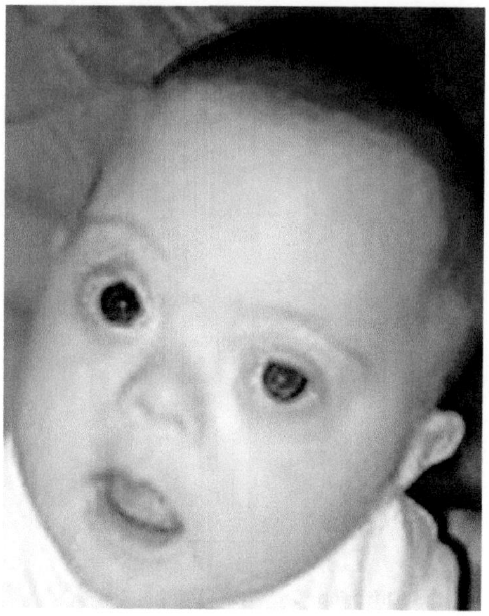

FIGURE 11.10
A child with Down syndrome. Note the epicanthal folds, small ears, and protruding tongue.

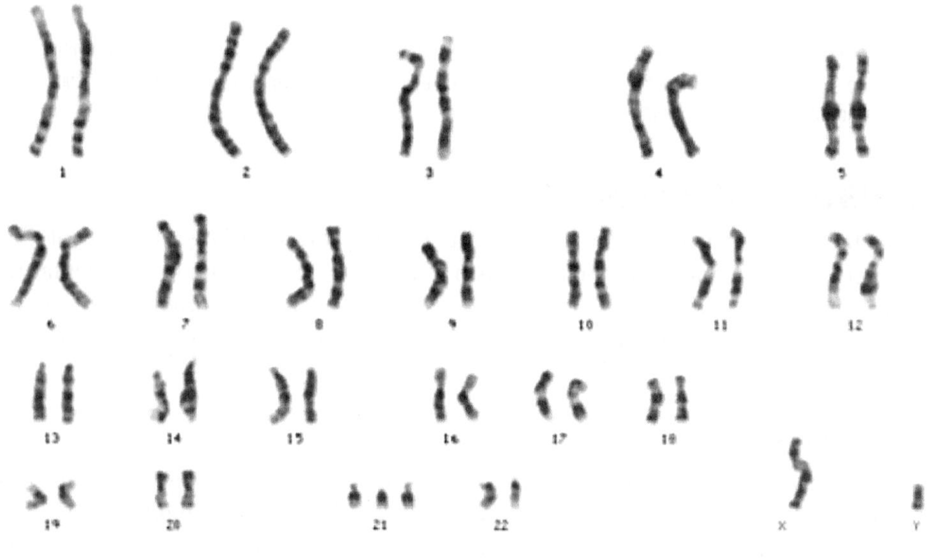

FIGURE 11.11
A karyotype, stained with Giemsa, of a male with Down syndrome due to trisomy 21 (47,XY,+21).

come to autopsy, virtually all have the histopathologic disease (plaques and tangles) seen in patients with Alzheimer disease.

Fifty years ago, most patients with Down syndrome were institutionalized. In those settings, an increased susceptibility to infections and a pattern of substandard care resulted in an average life span of less than 25 years. Today, the average life span of people with Down syndrome is nearly 60 years, reflecting the determination of parents to care for these individuals at home or in increasingly independent surroundings. Today, most children with Down syndrome go to regular public and private schools, where they receive special attention from their teachers. There is also a growing trend for adults with Down syndrome to seek employment where their generally positive temperament makes them an asset to the workforce.

CHROMOSOME ABNORMALITIES

Trisomy 21

Karyotyping is performed to confirm the clinical impression of Down syndrome and to counsel families regarding risk of recurrence risk. Ninety-five percent of patients have trisomy

21 (47,XX, + 21 or 47,XY, + 21) caused by meiotic non-disjunction (Figure 11.11). In the vast majority (90%), the extra chromosome is maternal in origin and reflects non-disjunction occurring during meiosis I.

Robertsonian Translocation

About 4% of patients with Down syndrome have 46 (rather than 47) chromosomes, one of which is a Robertsonian translocation between the long arm of chromosome 21 (21q) and the long arm of another acrocentric chromosome, usually 14 or 22. The translocation chromosome replaces one of the normal acrocentric chromosomes (Figure 11.12). Thus, the karyotype of a girl with Down syndrome due to a Robertsonian translocation would be denoted XX,rob(14; 21)(q10; q10),+ 21. Such patients are trisomic for only those genes on 21q, but their phenotype is indistinguishable from those patients with Down syndrome who have 47 chromosomes and trisomy for the entire chromosome 21.

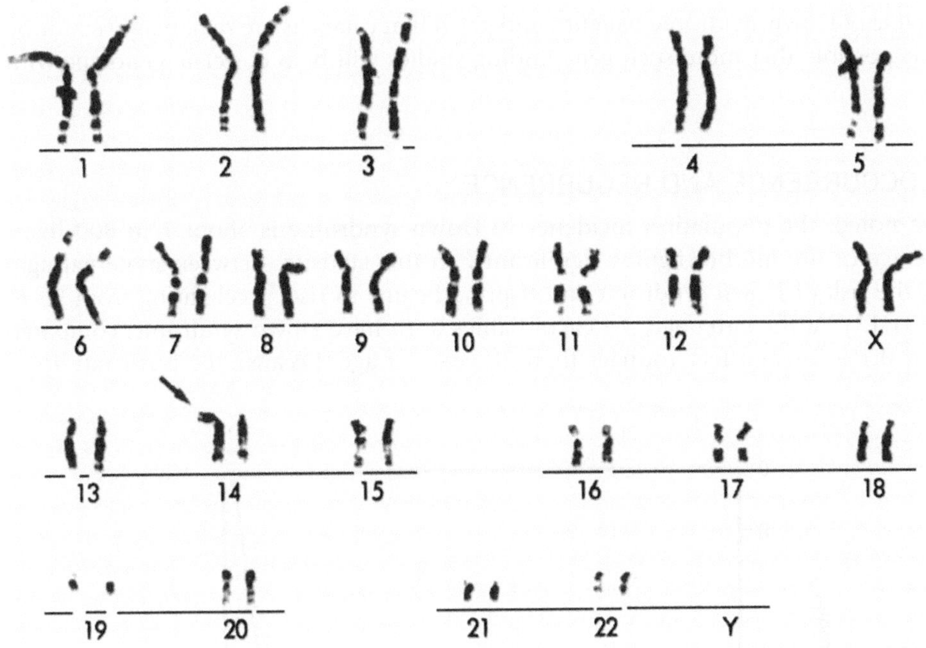

FIGURE 11.12
A karyotype, stained with Giemsa, of a girl with translocation Down syndrome resulting from a Robertsonian translocation between two acrocentric chromosomes, 14 and 21. The translocation chromosome is identified by the arrow.

21q21q Translocation

A very small fraction of people with Down syndrome have a translocation composed of two chromosome 21 long arms and one normal 21. Most such events occur post-zygotically—that is, they represent a new event in the zygote. Rarely, however, a parent may be a carrier for a 21q21q translocation chromosome. Under this circumstance, each gamete will contain either the translocated chromosome or no number 21, which would yield only two kinds of offspring: 21 monosomy, which is lethal, and 21q21q + 21 Down syndrome.

Mosaic Down Syndrome

About 1% of patients with Down syndrome have two populations of cells, 47 + 21 and 46. The trisomic cell line presumably arises in an early mitotic division rather than a meiotic one. The phenotype in patients with **mosaic** Down syndrome is often less severe than in the more frequent trisomy 21, but this outcome depends on the fraction of cells that have 46 chromosomes relative to those with 47.

MOLECULAR UNDERSTANDING OF PHENOTYPIC ABNORMALITIES

We have known for more than 50 years that Down syndrome results from having three copies of chromosome 21, but what is it about being trisomic for all (or a fraction) of the genes on 21 that causes the many facial, cerebral, cardiac, and other features of this syndrome? Despite considerable study, the answer is, we don't know. We are getting closer to the answer, however. In those people who have three copies for only a portion of 21, either because of a reciprocal or a Robertsonian translocation, it is possible to ask, for instance, what is the minimal part of 21 needed to cause the phenotype? But this line of inquiry still leaves a major task. Even though chromosome 21 is among the smallest human chromosomes and it contains only about 250 genes, that still leaves open many possibilities for single-gene or multiple-gene effects. A small group of genes, in particular, is being examined. In people with Down syndrome who have an extra copy of a small part of the long arm (q21.2 to q21.3) of chromosome 21, dosage effects for several genes that reside in this region have been identified. Intriguingly, one of these genes codes for β-amyloid, implicated in Alzheimer disease. Another codes for α-crystallin (found in the optic lens). Given that pathologic abnormalities characteristic of Alzheimer disease are always found in the post-mortem brains of Down syndrome patients, and given that cataracts are remarkably common as well, it is possible that more such gene-finding studies will help correlate genotype with phenotype.

RISK OF OCCURRENCE AND RECURRENCE

As we have noted, the population incidence of Down syndrome is about 1 in 800 live births. The age of the mother figures significantly in this statistic. Between maternal ages 30 and 35 the risk of Down syndrome in offspring begins to rise, accelerating from 35 to 45 (Figure 11.13). At the latter age, 3–5% of babies born have Down syndrome, compared to less than 0.1% for mothers younger than 20 years of age. Because the birth rate for

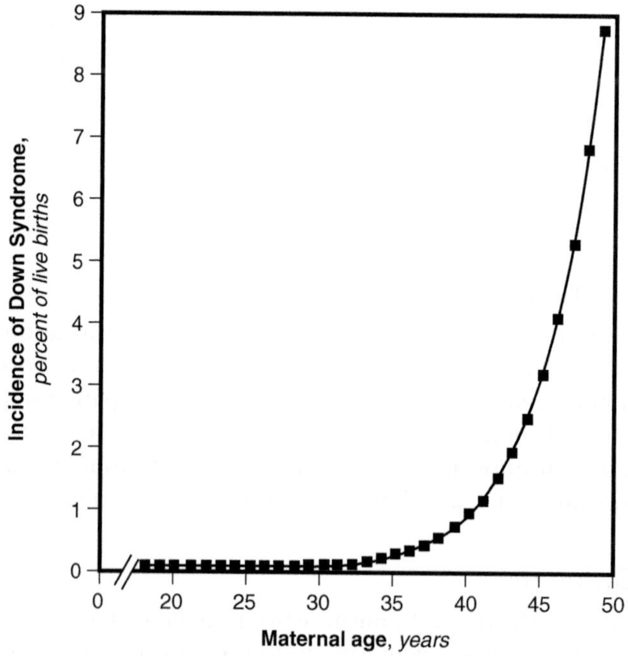

FIGURE 11.13
Increasing risk of Down syndrome in pregnancies of mothers over age 35 years. The risk accelerates and reaches a value of about 9% for mothers between 45 and 50 years of age.

younger mothers is much greater than for older ones, however, more than half of all people with Down syndrome have mothers under age 30. There is no comparable paternal age effect in Down syndrome. What is it, then, about female meiosis that results in increasing meiotic disjunction as a function of maternal age? The likely, though not incontrovertible, explanation concerns major differences in timing of meiotic divisions in the two sexes. As discussed in Chapter 4, male meiosis begins at puberty, and the two meiotic divisions occur sequentially over a brief time span to yield sperm. In females, meiosis I begins early in fetal life but is interrupted at important times thereafter: primary oocytes are halted at prophase I until after menarche, when meiosis I is completed; meiosis II is initiated at ovulation and completed only after fertilization. The net effect of this sequence is that oocytes may stay in meiosis I or II for many years, thereby increasing the likelihood of effects on the spindle or other accessory meiotic structures that predispose to non-disjunction.

The risk of recurrence of Down syndrome in a family that has had one affected child is about 1%. In mothers under 30, the risk of recurrence is about 1.4%, whereas in older mothers, recurrence risk approximates the already high occurrence risk. Why recurrence risk increases in younger women is not understood.

PRENATAL DIAGNOSIS

Prenatal detection of genetic disorders, which began in the 1960s, is becoming an increasingly routine part of good obstetrical practice. A broad perspective on this matter will be provided in Chapter 17. Here, we will discuss prenatal detection of Down syndrome and that for other chromosome abnormalities.

Ultrasound and Maternal Serum Screening

Currently, prenatal studies are carried out in the first and second trimesters. The earliest methodology uses **ultrasound** visualization of the fetus and **maternal serum screening** during the first trimester. Ultrasound focuses on looking for abnormal nuchal (back of the neck) folds characteristic of Down syndrome. Of course, other anatomical features of the fetus are looked at as well. The pregnant woman's serum is assayed for several proteins whose concentrations—elevated or decreased—have been found to correlate with particular pathologic states. It is now estimated that these methods can detect 85% of cases of Down syndrome with a false-positive rate of 5%. Such testing is also useful in detecting other autosomal trisomies. Thus, first-trimester screening can identify high-risk fetuses, leading to more definitive procedures thereafter.

Chorion Villus Sampling

At 12 to 13 weeks gestation, it is possible to pass a small catheter through the vagina and, under ultrasound visualization, obtain a small sample of cells from the fetal portion of the placenta, called the chorion (Figure 11.14). These cells are prepared for karyotyping, and other chemical or genetic tests. In about 1 in 300 cases, **chorionic villus sampling** (**CVS**) may produce a miscarriage. Nonetheless, some women prefer CVS because it can be done before their pregnancy is apparent to others.

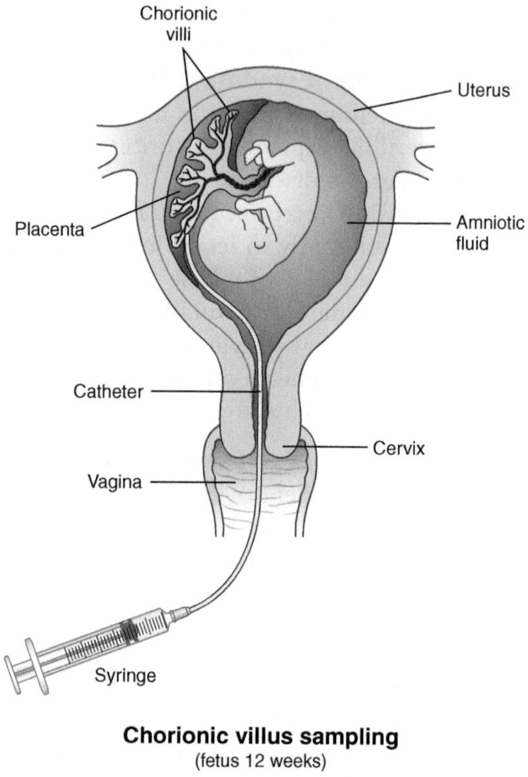

FIGURE 11.14
Prenatal testing using chorionic villus sampling (CVS). This procedure is usually performed at about 12 weeks of gestation.

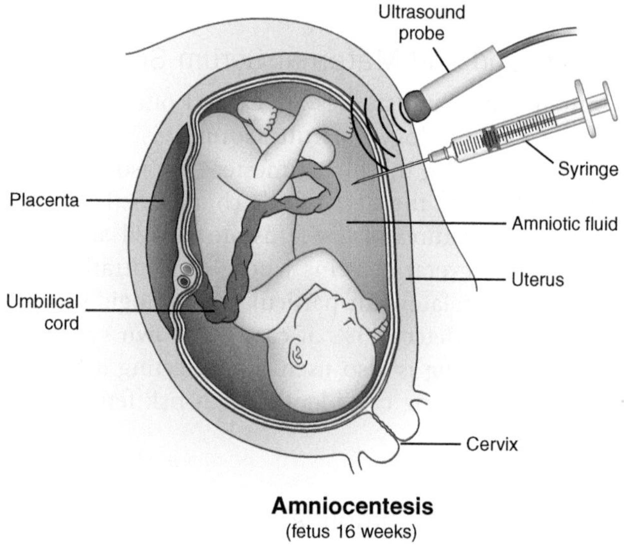

FIGURE 11.15
Prenatal testing by amniocentesis. This procedure is usually performed at or after 16 weeks' gestation.

Amniocentesis

Amniocentesis, the first and still most widely used and proven prenatal diagnostic test for Down syndrome, involves removal of a small amount of amniotic fluid (Figure 11.15). It is not carried out until about 16 weeks' gestation, so that there is sufficient amniotic fluid to sample without danger to the fetus. This procedure involves inserting, under ultrasound visualization, a needle through the abdominal wall and into the amniotic space surrounding the fetus. Then, a sample of amniotic fluid containing fetal cells is withdrawn and prepared for culture and analysis. The risk of inducing an abortion is about 1 in 600, a small risk when compared with the risk of spontaneous abortion at this stage of pregnancy (between 2 and 3%). Since its first use in 1966, amniocentesis has been used for the diagnosis of many genetic disorders, both chromosomal and single gene. Today, in North America, more than 50% of pregnant women over age 35 undergo amniocentesis.

INTERSECTIONS: GENETICS AND BIOETHICS

Inside the Senate Hearing Room: A Bill Regarding When Human Life Begins

On April 23, 1981, hearings began on United States Senate Bill S.158 entitled "A Bill to Provide that Human Life Shall Be Deemed to Exist from Conception." The bill's author, Senator John P. East of North Carolina, made clear at the outset that the purpose of the bill was to give the human conceptus a legal right to protection. In practical terms, he said, enactment of this bill would overturn the decision made by the United States Supreme Court in 1973 in *Roe versus Wade*. That decision held that the right to an abortion belonged exclusively to the pregnant woman and her physician, up to 20 weeks' gestation. Clearly, if life began at conception under the law, then voluntary abortion at any stage thereafter would constitute murder.

Senator East called for testimony from seven physicians and scientists, seeking incontrovertible scientific evidence that actual human life begins at conception. Six witnesses, handpicked by the Senator, testified that the zygote was obviously alive, and they all supported the bill. One of this book's authors, Dr Rosenberg (then chair of Yale's Department of Human Genetics) then testified that there was no agreement among scientists as to when "actual" human life—as distinguished from "potential" human life—begins. Life is continuous, he said. The sperm is living, as is the egg. What then determines that human life begins at conception, not before or after? He argued that the bill would prohibit the use of contraceptives preventing uterine implantation. Further, it would stop all amniocentesis for prenatal diagnosis because obstetricians would not be willing to take the risk of being sued for manslaughter or murder in the rare event that a miscarriage was caused by the amniocentesis. Accordingly, a family with one child with a serious genetic disorder would not be able to determine if a subsequent fetus also had the disorder. He concluded his testimony as follows:

> *... we all know that this bill is about abortion and nothing but abortion. If this matter is so compelling that our society cannot continue to accept a pluralistic view which makes women and couples responsible for their own reproductive decisions, then I say pass a constitutional amendment that bans abortion and overturns the Supreme court decision in* Roe v. Wade. *But don't ask science and medicine to help justify that course because they cannot. Ask your conscience, your minister, your priest, your rabbi, or even your God, because it is in their domain that this matter resides.*

Dr Rosenberg's testimony was met with resounding applause by the gallery of virtually all women—much to the dismay of Senator East, who, as a result, then pledged he would keep the hearings open and invite other scientists to testify. Subsequently, many other scientists testified that science didn't concern itself with such metaphysical matters as the beginning of life. Ultimately, this led the bill's supporters to accept defeat, and the bill never came to a vote.

This vignette cannot be concluded without an ironic coda. On the day that Dr Rosenberg testified, his wife Diane—this volume's co-author—found out she was pregnant for the first time. Because Diane was 36 years old, the Rosenbergs elected to have the pregnancy monitored by amniocentesis. The karyotype of the fetus was normal (46,XX); this result was confirmed at birth. Today their daughter is 30 years old and the mother of a healthy, 1-year-old boy.

SOCIETAL ISSUES

Because Down syndrome is common and because its clinical manifestations, however significant, are compatible with long and meaningful life, it has been the subject of considerable societal debate over the past 25 years. We'll consider two such issues: abortion and education. From its inception to the present, prenatal diagnosis of Down syndrome has been caught up in our society's unending argument about abortion. Those who believe that abortion is morally wrong abhor prenatal diagnosis because it may lead to results that can encourage parents to abort an affected fetus. People with this view see no difference between terminating a pregnancy involving an impaired fetus and terminating one for any of a number of reasons reflecting a woman's right to choose whether to be pregnant or not. Those who view Roe *v.* Wade as the law of the land hold that parents who will have the responsibility to care for a child with a serious problem, like Down syndrome, have the right to make an informed decision using CVS or amniocentesis. Although most prospective parents will choose to abort a fetus with Down syndrome, a not insignificant fraction will decide to continue the pregnancy and use the diagnostic information to prepare for the baby's birth.

A second, more recent, issue concerns education of those with special needs. A generation or two ago there were few educational options for children with Down syndrome because most were institutionalized. Today the reverse is true, and a progressively louder question is being asked: Where should children with Down syndrome be educated, and who should bear the costs, financial and otherwise? When children with Down syndrome go to public schools, they will need special attention, and that attention may come at the expense of children without such disability; it may increase the cost of public education. On the other hand, if children with Down syndrome go to special schools, should their parents bear the entire financial brunt of this education?

We have written about the societal issues posed by Down syndrome because they are important, but also because they serve to highlight issues that pertain to children with many other genetic disorders.

Trisomy 13 and Trisomy 18

These two trisomies found in live births are both more rare than trisomy 21 and clinically more deleterious. **Trisomy 13** is found in about 1 in 20,000 births; **trisomy 18** in about 1 in 7,500. Clinically, as shown in Figure 11.16, those with trisomy 13 have the following abnormalities: failure of forebrain development; cleft lip and palate; extra fingers and toes; congenital defects of the heart and urogenital system; and failure to thrive. Half of those

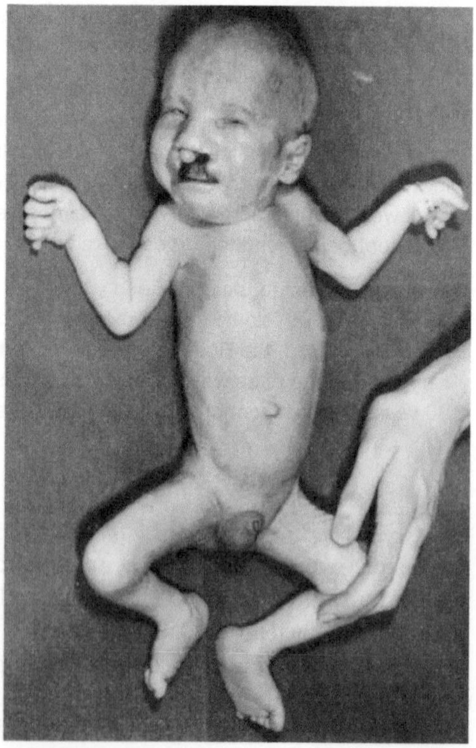

FIGURE 11.16
An infant with trisomy 13. The bilateral cleft lip and polydactyly are obvious.

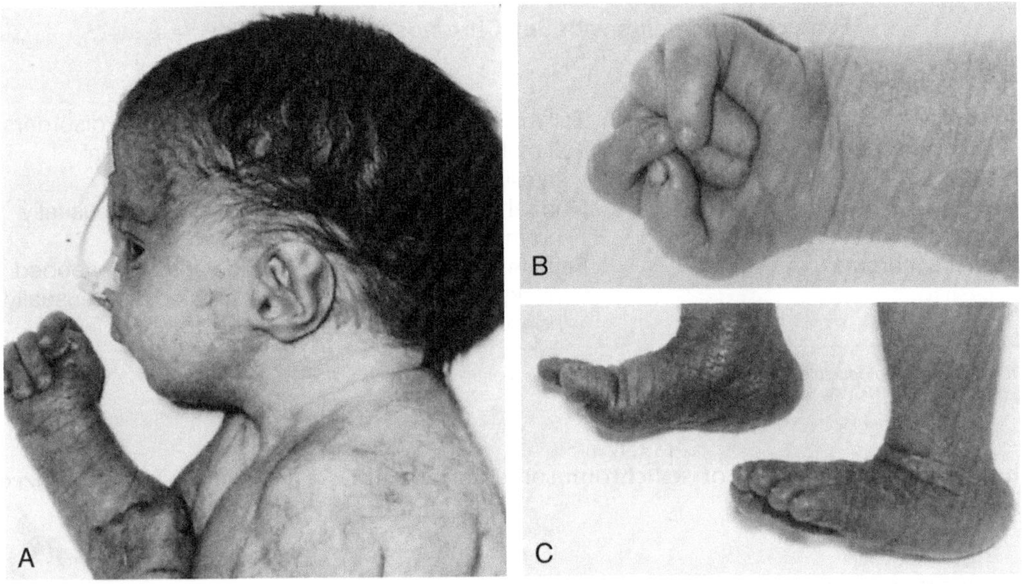

FIGURE 11.17
An infant with trisomy 18. Note the receding jaw and malformed ears in (A), the clenched fist in (B), and the rocker-bottom feet in (C).

severely malformed newborns die by age 1 month, 95% fail to live past 1 year. Newborns with trisomy 18 fare no better. They are recognized at birth (Figure 11.17) because of a prominent cranial occiput, receding jaw, a characteristic clenched fist, "rocker bottom" feet, and congenital heart defects. Again, most die during the first month, and few survive to 1 year. As with trisomy 21, advanced maternal age is often seen in mothers of those with trisomy 13 or 18. Also, as with trisomy 21, unbalanced translocations are seen in addition to typical trisomy.

Sex Chromosome Abnormalities

Abnormalities of the X and Y chromosomes, like those of autosomes, may be either structural or numerical. Aneuploidy for the sex chromosomes is common, occurring in about 1 in 400 males and 1 in 600 females. This makes this kind of genetic disorder among the most common of any category. As noted in Table 11.4, having an extra copy of either the X or the Y is found more frequently in live births than in abortuses, testifying to their lesser impact on fetal viability. By far the most common sex chromosome abnormalities in live births are the trisomies XXY, XYY, and XXX. Monosomy X is the single most common form of aneuploidy in abortuses. Some are live-born, but fully 95% of documented conceptuses that have a single X chromosome abort. In addition to the just mentioned trisomies and monosomy, more extreme sex chromosome aneuploidies exist, including XXYY, XXXY, XXXX, and XXXXX.

Phenotypically, those with sex chromosome aneuploidy tend to have mild abnormalities compared to those with autosomal abnormalities. How to account for this? The best guesses we have relate to the biology of the X and Y chromosomes. In cells with more than one X chromosome, inactivation of all but one X occurs to provide dosage compensation with males. Thus in females with three, four, or five Xs, only one is active. Likewise, in those with the XXY genotype, one of the two Xs would be silenced. As regards extra copies of the Y, we surmise that the paucity of its genes results in less phenotypic imbalance. Abbreviated comments about the clinical findings in people with

TABLE 11.5 Features of Patients with Sex Chromosome Aneuploidy

Disorder	Karyotype	Phenotype*
Klinefelter syndrome	47,XXY	Tall male, hypogonadism, infertility, learning disorders
XYY syndrome	47,XYY	Tall male, normal sexual development, normal intelligence, frequent behavioral problems
Trisomy X	47,XXX	Normal female, normal sexual development, usually normal intelligence
Turner syndrome	45,X	Female, neonatal edema of hands and feet, webbed neck, short stature, infertility, streak ovaries, usually normal intelligence

**Depicts most reproducible findings. No phenotypic feature is found in all patients with each syndrome.*

the more common forms of sex chromosome abnormalities follow and are summarized in Table 11.5.

KLINEFELTER SYNDROME (XXY)

About 1 in 1,000 males are born with an extra X chromosome. They generally appear normal at birth and until puberty. Then, the characteristic phenotype is seen. Affected males tend to be tall and thin. Their testes are small and secondary sex characteristics are underdeveloped. Affected males are almost always infertile because of failure of germ cell development. Although there is wide phenotypic variability, Klinefelter syndrome is associated with mild-to-moderate degrees of intellectual impairment: lower IQ scores, learning difficulties, and below-average verbal skills. In about half of the cases, errors in paternal meiosis I occur; in the other half, non-disjunction occurs in maternal meiosis I.

If one of the two X chromosomes in these patients is inactivated, what accounts for these phenotypic findings? We don't know, but it almost surely relates to some function of the extra X, because patients with Klinefelter syndrome with the rare aneuploid karyotypes (such as XXXY or XXXXY) tend to have more defective sexual development and greater mental impairment.

XYY SYNDROME

About 1 in 1,000 males are born with an extra Y chromosome, resulting from non-disjunction during paternal meiosis II. Males with XYY cannot be distinguished from XY males physically or behaviorally. Males with XYY tend to be tall, have normal fertility, and are not dysmorphic. Males with XYY may have an increased incidence of behavioral problems.

XXX SYNDROME

In about 1 in 1,000 females, three X chromosomes are noted. These girls have no obvious phenotype. Menarche and fertility are unaffected. A minority has mild-to-moderate learning difficulties. Not in keeping with simple ideas of X chromosome inactivation, females with four or five X chromosomes usually have significant mental retardation and physical deformities. The observation that individuals with XXXY or XXXX chromosome constitution tend to have more phenotypic abnormalities than their counterparts with fewer X chromosomes suggests that the small portion of the X chromosome which is not normally inactivated contains genes that may cause phenotypic abnormalities when they are present in increasing dosage.

TURNER SYNDROME (MONOSOMY X)

About 1 in 4,000 live-born females has a single X chromosome (45X). As mentioned earlier, this is the only human monosomy seen in live births; however, this is misleading because more than 95% of 45X conceptuses abort. In all, 1–2% of all conceptuses are monosomic for the X chromosome.

As shown in Figure 11.18A, those with Turner syndrome are often diagnosed at birth because of pronounced webbing of the neck and swelling of the hands and feet. With time, more important phenotypic abnormalities are noted (Figure 11.18B). These findings include short stature (usually less than 5 feet), an exaggerated carrying angle of the arms, amenorrhea, infertility, and vestigial (or "streak") ovaries. These patients also have a high incidence of congenital heart defects, other internal anomalies, and, infrequently, modest learning difficulties.

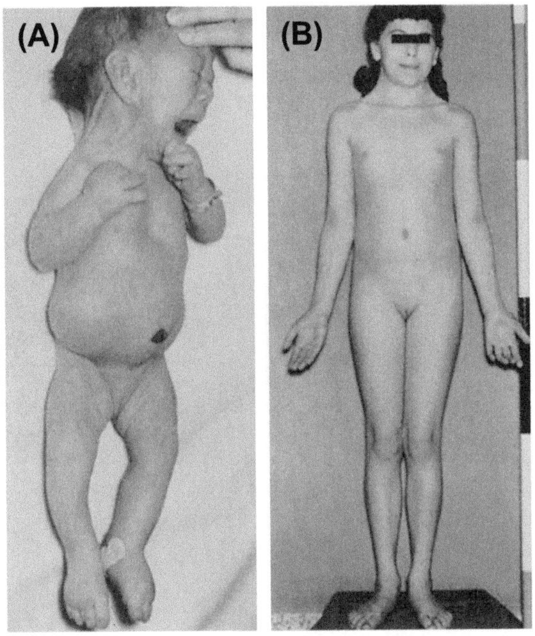

FIGURE 11.18
Clinical features in 45,X Turner syndrome. (A) Newborn infant with prominent webbing of the neck and swelling of the hands and feet. (B) An 18-year-old girl with Turner syndrome showing several typical features: webbed neck, widely-spaced nipples, short stature, and delayed sexual maturation.

Complete absence of one X chromosome due to meiotic non-disjunction is found in only about half of patients with Turner syndrome. The other half has a variety of different chromosome abnormalities involving the X: ring chromosome X (a chromosome whose arms have fused together), mosaicism, and deletion. Why do the vast majority of X monosomies abort, and what explains the phenotype in those reaching term? Firm answers to either of these questions are lacking. The small region of the short arm of the X (which is homologous to a portion of the Y) has been implicated, because only a single copy of this region is present in monosomy X, while it is present in two doses in both XX and XY fetuses.

Chromosome Deletions and Genomic Imprinting

In Chapters 5, 6, and 7, we have mentioned genomic imprinting, which involves epigenetic silencing of either the maternal or paternal copy of a particular gene. This means that expression (normal or abnormal) of that gene will depend on its parent of origin. As more has been learned about the number and nature of imprinted genes, a striking confluence of the fields of imprinting, chromosome deletions, and clinical phenotypes has taken place.

We will use discussion of two syndromes, Prader-Willi and Angelman, to illustrate this joining.

PHENOTYPES

Prader-Willi syndrome (Figure 11.19A) is relatively common and has a distinctive set of phenotypic abnormalities: short stature, developmental delay, small hands and feet, obesity, and hypogonadism. The phenotype in Angelman syndrome is very different (Figure 11.19B):

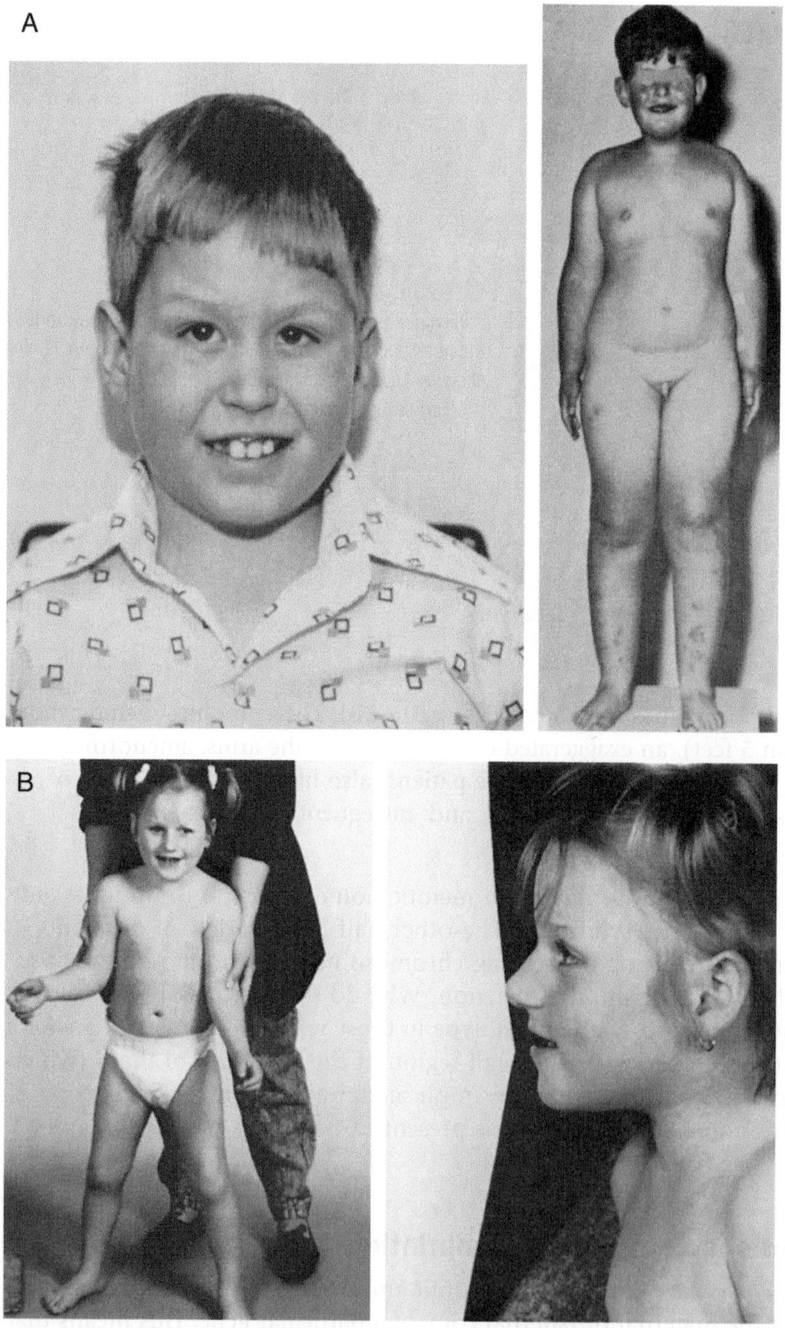

FIGURE 11.19
(A) A 9-year-old boy with Prader-Willi syndrome. Note the obesity, small hands, and hypogonadism. He also demonstrated short stature and developmental delay. (B) A 4-year-old girl with Angelman syndrome. Note the wide stance. Most patients have seizures and severe developmental delay.

early failure to thrive, an autistic-like personality, seizures, spasticity, and, often, severe developmental retardation.

CHROMOSOME ABNORMALITIES

The reason to discuss these phenotypically different syndromes together has to do with the chromosome abnormalities that cause them. In about 70% of people with Prader-Willi syndrome, there is a deletion involving the proximal long arm of chromosome 15 (15q11-q13). Patients with Angelman syndrome also have a deletion of this same region of chromosome 15. A critical distinction between these seemingly similar cytogenetic abnormalities has to do with the deleted chromosome's parent of origin. In Prader-Willi syndrome, the deletion affects chromosome 15 inherited from the father; in Angelman, the chromosome 15 inherited from the mother. Stated the other way, 15q11-13 genetic information in Prader-Willi is exclusively from the mother; 15q11-13 information in Angelman syndrome exclusively from the father. Other findings support the above. A small fraction of patients with Prader-Willi syndrome have **uniparental disomy** for chromosome 15, having inherited both homologues from their mother. Uniparental disomy is also encountered in Angelman syndrome, only here both homologues are inherited from the father.

GENOMIC IMPRINTING

The explanation for these puzzling findings involving parent of origin comes from understanding genomic imprinting of 15q11-13 (Figure 11.20). We now know that at least three

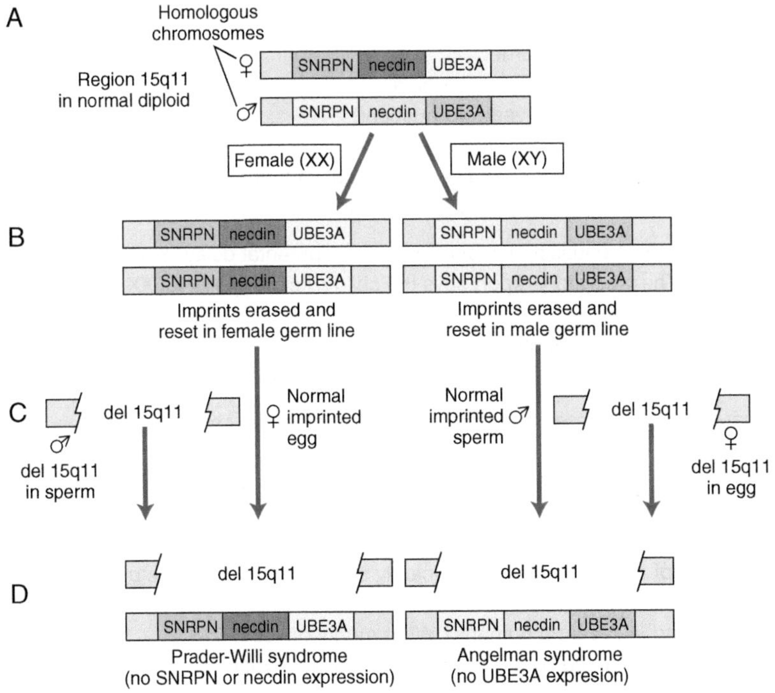

FIGURE 11.20
Imprinting of genes in 15q11-13. (A) In a normal person, SNRPN and nectin are imprinted in the female homologue; UBE3A is imprinted in the male homologue. (B) Imprints are erased and reset in germ line of female and male. (C) and (D) A person who inherits a maternally imprinted chromosome along with a paternal number 15 with a q11-13 deletion will have Prader-Willi syndrome. Angelman syndrome occurs if the deletion is on the maternal homologue and the normal imprint involves the paternal one. See text for additional details.

genes in this region are imprinted: one, called UBE3A, on the paternal homologue, two (*SNRPN* and *necdin*) on the maternal one. For these genes, then, each normal person has only one active copy. If a normal female gamete is fertilized by a sperm with a 15q11-13 deletion, the embryo will have no active copy of *SNRPN* or *necdin* and will have Prader-Willi syndrome. Conversely, if a normal male gamete fertilizes an egg with a 15q11-13 deletion, the embryo will have no active copy of *UBE3A* and will have Angelman syndrome. In support of this schema, point mutations of *UBE3A* have been reported to produce Angelman syndrome in the absence of a15q11-13 chromosome deletion.

REVIEW QUESTIONS AND EXERCISES

1. Choose the phrase in the right column that best matches the term in the left column.

a. aneuploid	1. missing one copy of a chromosome
b. Robertsonian translocation	2. having an extra copy of one chromosome
c. non-disjunction	3. loss of DNA sequence from a chromosome
d. duplication	4. exact multiple of haploid number of chromosomes
e. acrocentric	5. failure of two homologous chromosomes to separate during meiosis
f. deletion	6. chromosome in which centromere lies near the end
g. reciprocal translocation	7. having three copies of each chromosome
h. polyploidy	8. a translocation between two acrocentric chromosomes
i. triploid	9. chromosome in which centromere lies near the middle
j. monosomy	10. having an extra segment of a chromosome
k. uniparental disomy	11. any chromosome number not an exact multiple of haploid number
l. metacentric	12. transfer of segment of one chromosome to another non-homologous one
m. euploid	13. inheriting both copies of a chromosome from one parent
n. trisomy	14. any multiple of haploid number of chromosomes other than diploid
o. mosaicism	15. Having two or more genetically different cell lines

2. You see couples as a genetic counselor. In which of the following situations would you recommend a karyotype? Who would you suggest performing karyotypes on? In each instance, what kind of chromosomal abnormality would you be looking for?
   a. A couple whose first child has Down syndrome
   b. A couple who have had three miscarriages in succession
   c. A pregnant 43-year-old woman and her 40-year-old husband
   d. A couple with two children with albinism
   e. A couple whose son is very tall, somewhat cognitively impaired, with abnormally small testes
   f. A couple with two children who are severely cognitively impaired.

3. You are evaluating a child with a most unusual face and mild developmental delay. You send off a blood sample for karyotyping. The laboratory report says the child's karyotype is 46, XX, del(13)(q9).
   a. How do you interpret this karyotype?
   b. How might this karyotype account for the child's findings?
   c. Would karyotyping the child's parents assist you in answering (b) above?

4. Down syndrome (DS) is caused by the presence of three copies of the entire chromosome 21 or a considerable portion thereof.
   a. What evidence is there that DS is more often produced by a chromosomal error in oogenesis than spermatogenesis?
   b. What biological error is most often responsible for DS and when does that error occur?
   c. Why is the average life span of people with DS significantly shorter than that in the general population?
   d. What other trisomies are found in live born humans? What monosomies?
   e. What kinds of sperm could be produced by a man with DS whose karyotype is 47XY + 21? What kinds of offspring can this man have?
   f. Why should one obtain a karyotype in a child with obvious manifestations of Down syndrome?

5. Chromosomal mosaicism is rarely found in Down syndrome and often found in Turner syndrome.
   a. Using standard symbols, depict a karyotype found in a boy with mosaic Down syndrome and in a girl with mosaic Turner syndrome.
   b. Regarding reproduction, when does the error yielding mosaicism usually occur?
   c. Does mosaicism confer any clinical advantage or disadvantage?

6. You are asked whether a male with trisomy 18 and a female with monosomy 18 could have normal offspring. What is your reply?

7. Robertsonian translocations often result in aneuploidy.
   a. Denote the karyotype of a woman with a Robertsonian translocation involving chromosomes 14 and 21.
   b. What kinds of gametes can this woman produce? Diagram them (for chromosomes 14 and 21 only).
   c. What outcomes of pregnancy would you expect this woman to have, assuming that her husband has a normal karyotype?

8. What is the difference between genomic imprinting and uniparental disomy? How does each come about?

9. Prader-Willi syndrome (PWS) and Angelman syndrome (AS) each result from a similar chromosomal error, namely del(15)(q11), yet their phenotypic features are very different.
   a. What are the key phenotypic features in these two syndromes?
   b. What accounts for the dissimilar phenotypes in PWS and AS?

10. A couple with a child with DS is seeking a divorce—each parent claiming that the other is the cause of the trisomy 21 in their child. You are an expert witness who has been asked to testify on this matter. Accordingly, you have studied restriction fragment length polymorphisms (RFLPs) on chromosome 21. One such RFLP in normal subjects has alleles at 8, 6, 4, and 2 kb, as shown on the left in the diagram of the Southern blot. The pattern for the father, mother, and child are shown along with size markers (M).

kb	M	father	mother	child
8	–		–	–
6	–		–	
4	–	–		–
2	–	–		–

   a. In which parent did the non-disjunction occur?
   b. What would you say to the woman who wanted to attribute the reason for DS to her husband's drinking?

CHAPTER 12

Single-Gene Defects

CHAPTER OUTLINE

CORE CONCEPTS

Mutations of single genes have been documented at more than 10% of the 21,000 genetic loci in the human genome that code for protein products. These single-gene mutations have a considerable effect on child health: they occur in 0.4% of newborns; they are responsible for 5% of hospitalizations; and they cause 8% of deaths. The study of these disorders—once called "inborn errors of metabolism" and now generally referred to as "inherited metabolic diseases"—has been of value in several ways: by elucidating normal biochemical pathways of anabolism and catabolism; by defining the biochemical mechanisms of myriad disorders and the nature of the gene mutations that cause them; and by using this information to develop diagnostic tests and therapeutic strategies.

This value has been extracted by studying disorders caused by single mutations at four logically constructed, ascending levels: the clinical phenotype; the metabolic pathway or specific reaction; the

Human Genes and Genomes. DOI: 10.1016/B978-0-12-385212-0.00012-3

protein affected; and the genetic locus perturbed. Although each disorder is unique, reflecting as it does the locus and its product, some generalities deserve mentioning:

- Most disorders are genetically heterogeneous, that is, each results from a wide variety of different mutations (missense, nonsense, frameshift, and splicing) which interfere with the function of a single protein. A few are caused by trinucleotide repeats.
- They reflect modification of the structure and function of one or more of the kinds of proteins found in the human body: enzymatic, structural, regulatory, circulating, and membrane. Organ dysfunction follows from differential gene expression.
- Some are inherited as dominants, others as recessives; some are autosomal, others sex-linked.
- Some, like red hair, are benign traits; others are uniformly fatal during childhood. Still others are compatible with extended life but impair organ function in serious ways.
- Whereas some conditions are found with near equal prevalence in all ethnic groups, most show ethnic clustering.

To illustrate these general features, the following traits or disorders are discussed in greater depth: sickle cell anemia, Duchenne muscular dystrophy, familial hypercholesterolemia, Huntington disease, and red hair.

Single-gene disorders represent one of the three major categories of genetic diseases. It is estimated that 1 in 250 newborns has some kind of disorder due to mutations of a single gene, and that such conditions are responsible for more than 5% of children's hospitalizations and more than 8% of childhood deaths.

This snapshot raises several important questions. How many single gene disorders are there? How do they perturb the body sufficiently to be called a defect or disorder? How are they detected, treated, and prevented? How do they differ, if at all, from the large number of single-gene mutations of no clinical consequence? Complete answers to these questions are not at hand, but we can say this much. The most recent tabulations in **Online Mendelian Inheritance in Man (OMIM)** listed 2,145 phenotypes (autosomal, X-linked, and Y-linked) with Mendelian patterns of inheritance (presumably reflecting mutations at a single locus). This tabulation emphasizes that single-gene variants and disorders are already known to affect more than 10% of the 21,000 genes in our genome. Additional ones will surely be discovered.

As for how single-gene variants perturb normal function, we can safely say that the vast majority act by perturbing the function of a single polypeptide. Regarding means of detection, treatment, and prevention, it becomes increasingly clear that as we understand the biochemical and molecular bases of these disorders, we fashion new ways to intervene clinically. This subject will be dealt with extensively in Chapter 17. As for harmless single-gene mutations, we shall illustrate this class of variants at the end of this chapter, using red hair as an example.

In this chapter, as in previous ones, we shall combine some informative history, some general features, and a few specific examples to give the reader a perspective on this class of disorders which has just passed its centenary and continues to accrue information at an impressive rate.

ARCHIBALD GARROD

It is much easier to know how to begin this presentation than how to end it. It begins with a British physician named Archibald Garrod (Figure 12.1).

Garrod's Life

Archibald Garrod was the son of a prominent British physician, Alfred Baring Garrod, who discovered that patients with gout had elevated concentrations of uric acid in their blood. The younger Garrod, born in 1857, benefiting from his family's wealth and prominence, completed his formal education at the University of Cambridge, where he majored in chemistry. After some years of uncertainty, he received his medical degree at St Bartholomew's Hospital, where he was to carry out his scientific work. By his own admission, he was more

FIGURE 12.1
Sir Archibald Garrod, the British physician who created the field of human biochemical genetics in 1908 by his study of alkaptonuria and other "inborn errors of metabolism."

interested in his patients' chemical abnormalities than in their signs and symptoms. This predilection led to his studies of alkaptonuria, the contribution that demonstrated his prescience and brilliance.

Alkaptonuria

When Garrod began his work on this rare disorder in 1898, he was unaware of Mendel's work, which, by that point, had been neglected for more than three decades. (Recall that Mendel's laws established that a single gene (Mendel's word "factor") can control a visible phenotype, but did not explain the mechanism by which genotype affects phenotype.) **Alkaptonuria** means black urine, the hallmark of the condition. Garrod observed that affected patients were generally healthy except for one prominent sign: their urine turned black when exposed to oxygen in the air (Figure 12.2). Using his background in chemistry, Garrod isolated **homogentisic acid**, the substance in the urine responsible for the urinary discoloration (Figure 12.3). He also demonstrated that upon exposure to the air, homogentisic acid is oxidized to its black product.

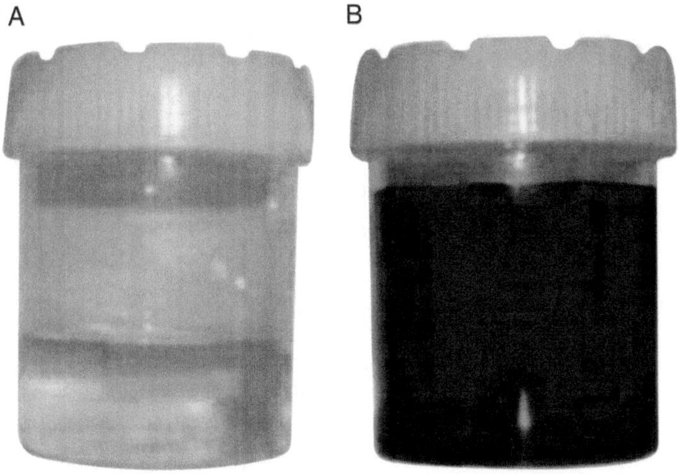

FIGURE 12.2
The chemical hallmark of alkaptonuria, black urine (B), compared to normal colored urine (A). This chemical phenotype led directly to Garrod's hypothesis that the substance (homogentisic acid) responsible for the discoloration of urine reflected a block in the metabolism of phenylalanine and tyrosine.

Examination of the structure of homogentisic acid, with its six-carbon phenyl ring, suggested to Garrod that it might be related to the amino acids phenylalanine and tyrosine, each with a phenyl ring as a major substituent. Therefore, Garrod embarked on a series of feeding experiments in alkaptonuric patients and in healthy controls.

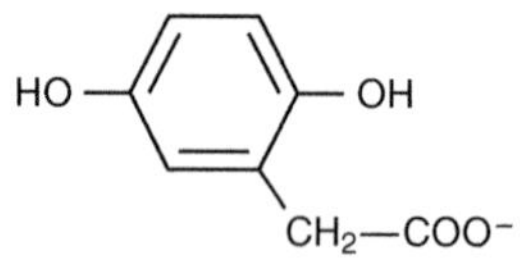

FIGURE 12.3
Structure of homogentisic acid. The six-membered phenyl ring is identical to that found in phenylalanine and tyrosine.

- He fed homogentisic acid to patients and controls. Patients excreted all they ingested in their urine, controls excreted none.
- He fed phenylalanine and tyrosine to both groups and found that homogentisic acid excretion increased markedly in patients, but none was found in controls.

From these results, Garrod proposed that homogentisic acid was an intermediate in the pathway through which phenylalanine and tyrosine are broken down, and that alkaptonuric patients had a block in this pathway which resulted in accumulation of homogentisic acid (Figure 12.4). He went on to propose that there must be an enzyme responsible for this

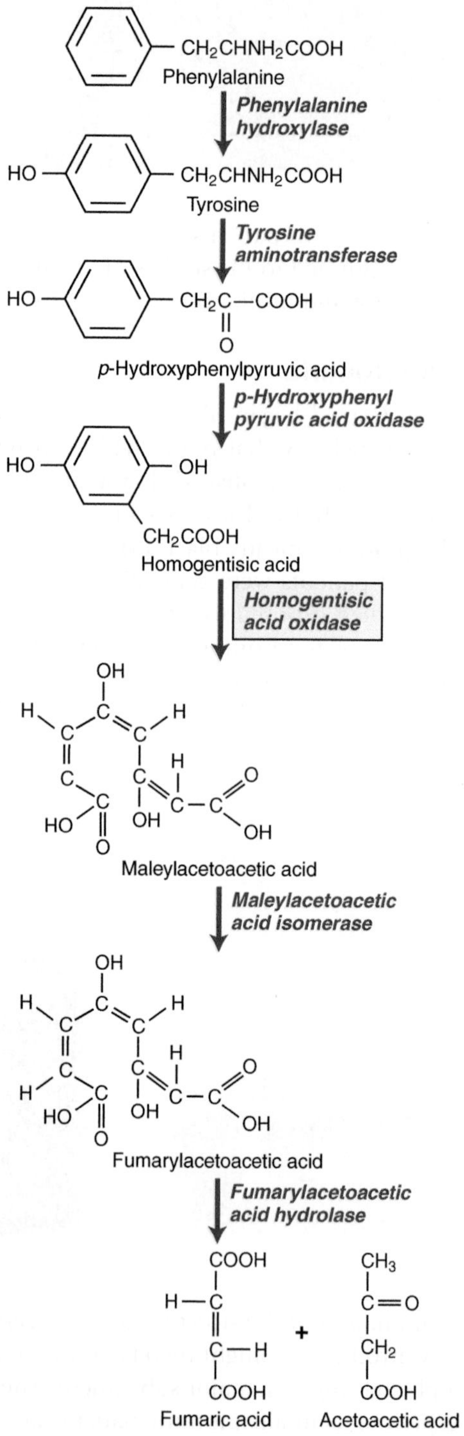

FIGURE 12.4
The metabolic pathway by which the essential amino acids, phenylalanine and tyrosine, are broken down to acetoacetic acid. Each step is catalyzed by a specific enzyme. Homogentisic acid oxidase, the enzyme deficient in alkaptonuria, is highlighted in the box. Additional details are found in the text.

breakdown of homogentisic acid and that the enzyme was deficient in people with alkaptonuria.

Garrod's studies of 17 families, each with at least one member with alkaptonuria, were equally pathfinding.

- He observed that brothers and sisters of patients were "apt" to have it as well, but that parents "do not exhibit the anomaly."
- He noted that affected patients were not infrequently the offspring of first cousins or other **consanguinous** matings.

Fortuitously, he consulted his colleague, William Bateson, a biologist who had been instrumental in rediscovering Mendel's work and who coined the word "gene," which means "giving birth to." The two men realized that alkaptonuria was behaving like a rare Mendelian recessive.

Between 1902 and 1908, Garrod studied three other disorders characterized by increased excretion of one or more substances in the urine: albinism, cystinuria, and pentosuria. In his now famous Croonian lectures delivered in 1908, Garrod called the four conditions he had studied "**inborn errors of metabolism**," a term used synonymously with "**inherited metabolic diseases**" to this day. In these lectures, Garrod uttered some timeless truths:

- Any inherited disease in which cellular metabolism is abnormal results from an inherited defect in an enzyme.
- Each successive step in the building up or breaking down of proteins, carbohydrates, and fats is the work of special enzymes. If any one step in the process fails, the intermediate product will escape further change and is likely to be excreted as such.
- The existence of chemical individuality follows necessarily from that of chemical specificity. Even those idiosyncrasies with regard to drugs and articles of food which are summed up in the proverbial saying that "what is one man's meat is another man's poison" presumably have a chemical (and genetic) basis.

Garrod referred to alkaptonuria as a "sport"—a term meant to define a harmless trait. He was incorrect about this. Alkaptonuric patients deposit homogentisic acid in cartilages, where it is responsible for black ears and arthritis, and in the kidney, where it forms kidney stones. He was also incorrect in proposing that all metabolic disorders resulted from enzymatic defects. Cystinuria, for example, is caused by mutations of a transporter protein. Now, of course, we know that defects in a wide array of proteins other than enzymes can cause inherited metabolic disorders. But this does not diminish Garrod's prescience.

Garrod's Legacies

Like Mendel, Garrod was unappreciated at the time he did his work. Toward the end of his life he was knighted, not for his science but for his work as the Regis Professor of Medicine at Oxford. In fact, his scientific work went almost unnoticed for 40 years until Beadle and Tatum's Nobel Prize-winning work referred to as the "**one gene, one enzyme**" hypothesis (discussed in Chapter 7). In his Nobel lecture, Beadle credited Garrod for his formidable contributions which long antedated theirs. A decade later, Garrod's hypothesis about alkaptonuria was confirmed when assays of liver from alkaptonuric patients revealed complete absence of activity of the enzyme homogentisic acid oxidase, which catalyzes the breakdown of homogentisic acid. Since the 1950s, Garrod has been recognized as the "father of human biochemical genetics," a field that has contributed enormously to our understanding of the molecular and metabolic basis of human disease. Finally, Garrod can rightly be called the father of pharmacogenetics, which concerns itself with genetic factors affecting the use of pharmaceuticals (to be discussed in Chapter 19).

TERMS AND PRINCIPLES

Garrod's expression, "inborn errors of metabolism," is now understood to mean any genetically determined biochemical disorder, usually in the form of an enzyme defect that produces a metabolic block. This definition leads directly to the need for several others:

- **metabolism**—the series of physical and chemical processes involved in the maintenance of life;
- **metabolic pathway**—a set of chemical reactions that take place in a definite order to convert a particular starting molecule into one or more specific products;
- **anabolism**—the metabolic process by which simple substances are synthesized into the complex materials of living tissue;
- **catabolism**—the metabolic process by which complex molecules are broken down;
- **disease gene**—a gene whose mutation(s) produce pathologic consequences.

Over the 20th century, the notion of inborn errors of metabolism was broadened and deepened. Today we talk about the field of biochemical genetics, defined as the inherited variation of gene products, their metabolic consequences and their resulting phenotypes. From this construct has come the term "disease gene," defined as any gene whose mutations produce pathologic consequences. It is important to remember, however, that the normal genes themselves do not cause disease: it is mutations in those genes that disrupt normal function and result in pathology. Single-gene defects, then, can be studied at four principal levels: the phenotype, the metabolic abnormality, the altered gene product, and the mutated gene.

For the first half of the 20th century, investigations were confined to the phenotype, such as black urine, mental retardation, or sickle-shaped red blood cells. Then tools were discovered to assay enzyme activity and identify qualitative changes in proteins. This development led to the ability to move one rung up the ladder of discovery and show, using the same examples, that black urine resulted from the deficient activity of homogentisic acid oxidase; that the cognitive disability seen in phenylketonuria was produced by phenyalanine hydroxylase deficiency; and that sickled red cells were a result of a physical change in the structure of hemoglobin. Subsequently, the availability of cultured cells and methods to purify and isolate proteins gave investigators tools to study a wide variety of proteins and polypeptides directly and deduce from their amino acid sequences the likely gene mutations responsible for their structural variation. Finally, the discovery of recombinant DNA, restriction enzymes, and the polymerase chain reaction —all occurring between 1972 and 1985—made it possible to study disorders at the level of the gene (refer to Chapters 6 and 7). Now, just a bit more than 100 years after Garrod's Croonian lectures, it is possible to sequence the entire genome or exome of a patient and find the causative mutation without any prior knowledge or hypothesis about the particular gene or pathway responsible for the disorder.

PANORAMA OF SINGLE-GENE DEFECTS

So what can we see from this vantage point? The answer: a great deal, but not everything.

- We see that single-gene disorders may be caused by all the kinds of mutation mentioned in Chapter 8: substitutions; insertions; deletions, duplications, frameshifts, splicing; missense, nonsense, trinucleotide expansion. Most of these mutations result in loss of function, but some are gain-of-function alterations.
- We see that mutations at more than 2,500 loci have been documented, accounting for more than 10% of loci in the human genome. Any given disorder may be produced by many different mutations at the involved locus, a situation called **genetic heterogeneity.**
- We see that, collectively, these mutations cause dysfunction of every class of protein present in human cells—structural, enzymatic, circulatory, membrane, transcriptional regulators, etc.—and that these dysfunctions have been observed in every tissue and every cell type.

- We see that some of these disorders are inherited as dominant traits, others as recessives, and still others are X-linked or mitochondrial (Figure 12.5).

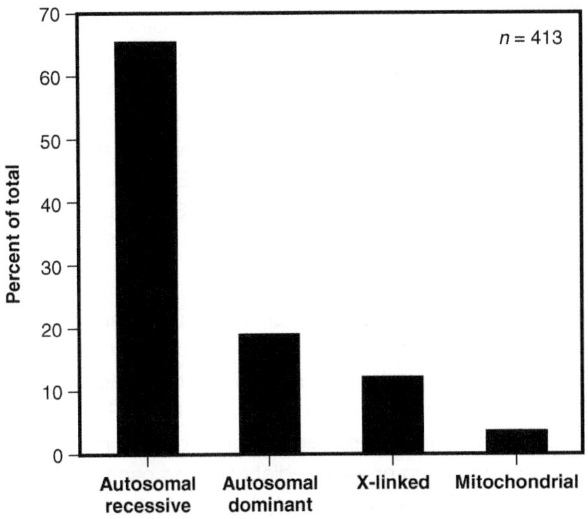

FIGURE 12.5
Mode of inheritance of 413 well characterized inborn errors of metabolism. The preponderance of recessives means that having half normal activity of an enzyme (or other protein) is sufficient to prevent chemical or clinical disturbance.

- We see that many of these disorders show ethnic clustering; for example, Tay-Sachs disease in Ashkenazi Jews, and cystic fibrosis in northern Europeans.
- We see that the frequency of these disorders varies widely, with some as common as 1 in 700 people, others as rare as 1 in 1,000,000.
- We see that some disorders have clinical consequences as early as fetal life or the neonatal period while others have no clinical consequences until adolescence or adulthood.
- We see that about one-third of these inborn errors are clinically harmless (Figure 12.6), another third result in severe dysfunction, and the remainder cause mild or moderate abnormalities.

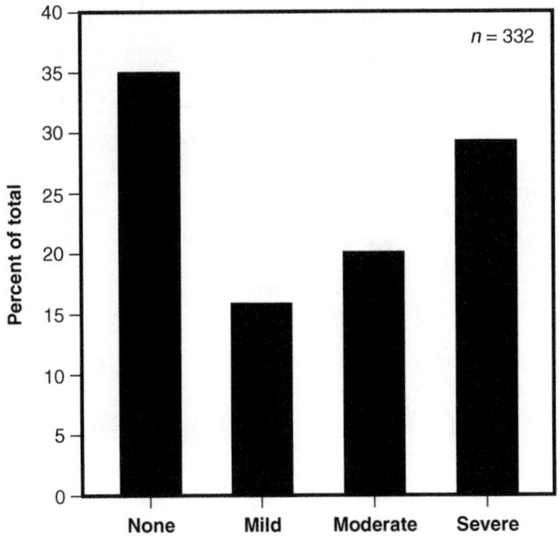

FIGURE 12.6
Clinical consequences of 332 well characterized inborn errors of metabolism. Note that one-third are benign variants while two-thirds have consequences of varying severity.

- We see that molecular and biochemical characterization of single-gene disorders has led to major advances in their detection and treatment.
- We see that we still have a great deal to learn about the relationship between genotype and phenotype and about more successful treatments and cures.

AMPLIFICATION: MUTATION AND PHENOTYPE

Positive Effects of a Deleterious Gene

How would it be to know that you would almost certainly not get cancer or diabetes in your lifetime? What if this benefit came from having a mutation that made you less than 4 feet tall? Since 1987, Dr Jaime-Guevara Aguirre, an Ecuadoran diabetes specialist, has been studying 99 people with Laron syndrome who have a complex phenotype exactly like that posed in the questions just asked. In the nearly 25 years since being discovered, these patients—whose height rarely reaches 4 feet—have no recorded instances of cancer or diabetes. Their 1,600 relatives of normal height, however, have prevalence rates of cancer (17%) and diabetes (5%) similar to those in the general Ecuadoran population. This inbred population probably descended from Sephardic Jews from Spain and Portugal who, after being forced to convert to Catholicism by the Inquisition in the 1490s, fled to Ecuador.

Laron syndrome, an autosomal recessive condition, is caused by a mutation in the gene encoding the cell surface receptor for growth hormone (GH). The mutant receptor doesn't bind GH, thus it doesn't carry out its normal function, which is to stimulate formation of insulin-like growth factor one (IGF-1)—the protein directly responsible for the growth-promoting effect of GH. But the IGF-1 deficiency caused by the mutation in the GH receptor does more than interfere with growth: it protects against two prominent, deleterious consequences of the aging process—diabetes and cancer. Laron syndrome patients are said to look "youthful", but increased longevity has not been documented thus far.

Support for these clinical studies comes from laboratory investigation with model organisms. Roundworms and mice with mutations that lead to low IGF-1 concentrations live much longer than is normal for their respective species, and they are protected from cancer and other manifestations of aging.

This story raises another profound question. If it was possible to interfere with IGF-1 formation or activity in healthy human children or adults, might that prevent diabetes and cancer? This question won't be answered any time soon, but we owe the fact that it can even be asked to a small group of people in the mountains of Ecuador.

A 32-year-old community leader and artist who has Laron-type dwarfism with his bride, 17.

A 67-year-old man who has Laron-type dwarfism with his daughter, 5, and sons, 7 and 10.

We'll now discuss, in some detail, five different single-gene defects, with the aim of illustrating the principles and generalizations just articulated. Four are disorders: the **hemoglobinopathies**, **Duchenne muscular dystrophy**, **familial hypercholesterolemia**, and **Huntington disease**. One is a benign trait: **red hair**.

ILLUSTRATIVE EXAMPLES

For many years, disorders of **human hemoglobin** structure and function were the only ones amenable to study at the biochemical and molecular levels. This was the case for several reasons. First, red blood cells (RBCs, or erythrocytes) are easy to obtain, merely by withdrawing a sample of blood. Second, hemoglobin is so abundant that it can be studied without extensive purification. Third, the globin (the protein constituents of hemoglobin) mRNAs are easily obtainable, again reflecting the abundance of the product. Fourth, a large number of genetic disorders of hemoglobin have been well characterized. And, fifth, the genes responsible for synthesizing hemoglobin turn out to be small and of relatively simple structure.

Hemoglobinopathies

HEMOGLOBIN

The major function of circulating red blood cells (RBCs) is delivering oxygen to the tissues. This function is carried out by the protein hemoglobin, which makes up 70% of the protein of RBCs. RBCs are produced in the bone marrow and are among the very few non-nucleated cells in the human body. Their nucleus is lost shortly before these cells (sometimes caricatured as "sacs of hemoglobin") enter the bloodstream. Hemoglobin is a tetramer (four subunits) consisting of two alpha (α) globin chains and two beta (β) globin chains (Figure 12.7). Each globin chain contains a heme molecule, which, when complexed with iron, is the oxygen-carrying moiety.

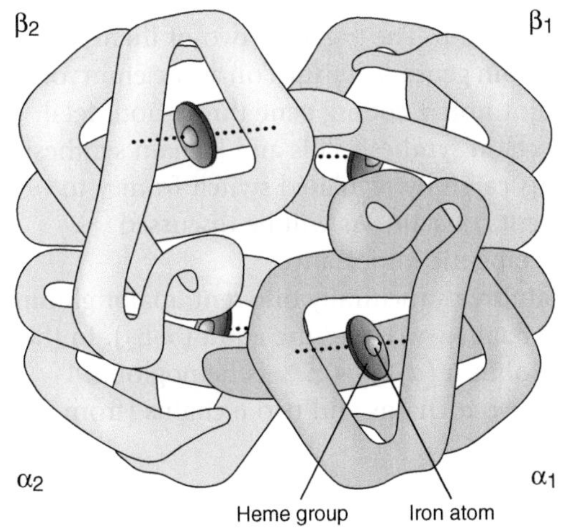

FIGURE 12.7
Three-dimensional structure of human hemoglobin, a tetramer composed of two α chains (green) and two β chains (blue). Each chain is complexed with an iron-containing heme group (red).

GLOBIN GENES

Organization and Diversity

The human genome contains 13 globin genes, four of which are **pseudogenes** (remnants of once-functioning genes) and one a locus of unknown function. The remaining eight genes are expressed sequentially. The α genes (those coding for α globin chains) are found in a cluster on chromosome 16 (Figure 12.8). The β cluster is found on chromosome 11. These clusters contain, in addition to the α and β genes active in adults, many other genes used at different points during embryonic and fetal development. In Figure 12.8 diagramming these gene clusters, the genes are arrayed in a 5′ to 3′ direction (left to right) both regarding the direction of transcription and the order in which they are used during development (that is, the genes used earliest in life are found at the 5′ end of the cluster and those used latest at the 3′ end).

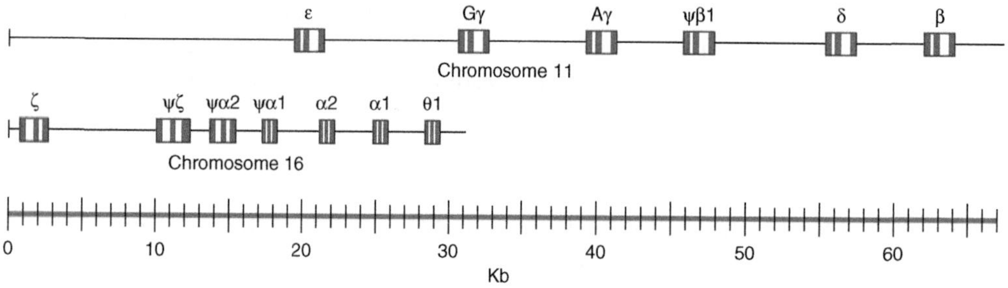

FIGURE 12.8
Map of the β-globin gene cluster on chromosome 11 and of the α-globin cluster on chromosome 16. For both clusters the genes are arranged from 5′ to 3′ in order of their developmental participation. The four genes designated by Ψ are pseudogenes (inactive gene remnants). The eight globin genes active during embryonic, fetal, and adult life are discussed in the text.

Development and Diversity

Figure 12.9 depicts the pattern of expression of the active globin genes in both clusters as a function of developmental age. The essence of this information follows:

- In the first weeks of embryonic life, hemoglobin is made in the yolk sac. The major hemoglobin formed at this time is a tetramer of two zeta (ζ) chains (encoded in the α cluster) and two epsilon (ε) chains (encoded in the β cluster).
- Soon thereafter, production of the ζ and ε chains diminishes and stops, and α chains begin to be produced encoded by α genes. (There are two α genes, called α1 and α2, but their coding sequences are identical, as is the protein they encode.) The α genes then stay turned on for the remainder of the individual's life.
- The β-globin cluster contains two sets of genes: γ and β. The γ genes (two of them) are active during fetal life and are called fetal hemoglobin genes. They turn on as the embryonic globin genes are turned off and become the major non-α coding gene throughout fetal development. Beginning shortly before birth, γ chain synthesis falls and β chain synthesis (relatively minor up to this point) increases. This carefully regulated switch from γ to β chains keeps the circulating hemoglobin content constant. As will be discussed subsequently, this switch is reversible and of therapeutic importance.
- Overall, this developmental schedule means that three structurally different, major globin tetramers are formed: the embryonic ($\zeta_2\varepsilon_2$), the fetal ($\alpha_2\gamma_2$), and the adult ($\alpha_2\beta_2$). In the adult, 97% of circulating hemoglobin is hemoglobin A ($\alpha_2\beta_2$), 0.5% is hemoglobin F ($\alpha_2\gamma_2$), and 2% is hemoglobin A_2, composed of two α chains and two δ chains (from the modestly expressed δ locus.)

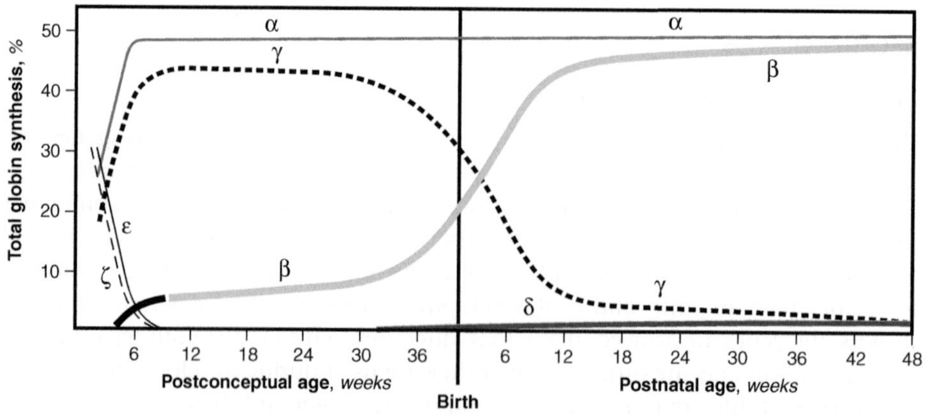

FIGURE 12.9
Developmental schedule of human hemoglobin expression. The time-course of production of the several globin chains is shown. Note that two hemoglobins ($\delta_2\varepsilon_2$ and $\alpha_2\zeta_2$) are made during embryonic life, two hemoglobins ($\alpha_2\gamma_2$ and $\alpha_2\beta_2$) during fetal life and three ($\alpha_2\beta_2$, $\alpha_2\gamma_2$, and $\alpha_2\delta_2$) during childhood and adult life. Additional details are discussed in the text.

Structure and Expression

Because each of the globin genes has a similar size and organization, we will use the β globin gene to illustrate features discussed generally in Chapters 6 and 7. Figure 12.10 depicts the β globin DNA and the RNA it codes for. The gene is small, only 1.6 kb in length. It is composed of three exons and two introns. Gene expression is controlled by a number of sequence motifs at or near the 5′ end of the gene. These elements include the promoter region, and sites for transcription factor interaction; they determine where and when transcription will start. Moreover, they are responsible for the tissue expression of the gene being limited to RBC precursor cells in the bone marrow. The primary transcript of the β gene is edited by the usual addition of a cap site and a signal for the poly A tail. Then, the edited transcript is spliced: the two introns are removed and the three exons are joined together to make β-globin mRNA. The mRNA exits the nucleus and is translated on cytoplasmic ribosomes to form the β-globin protein, which contains 146 amino acid residues.

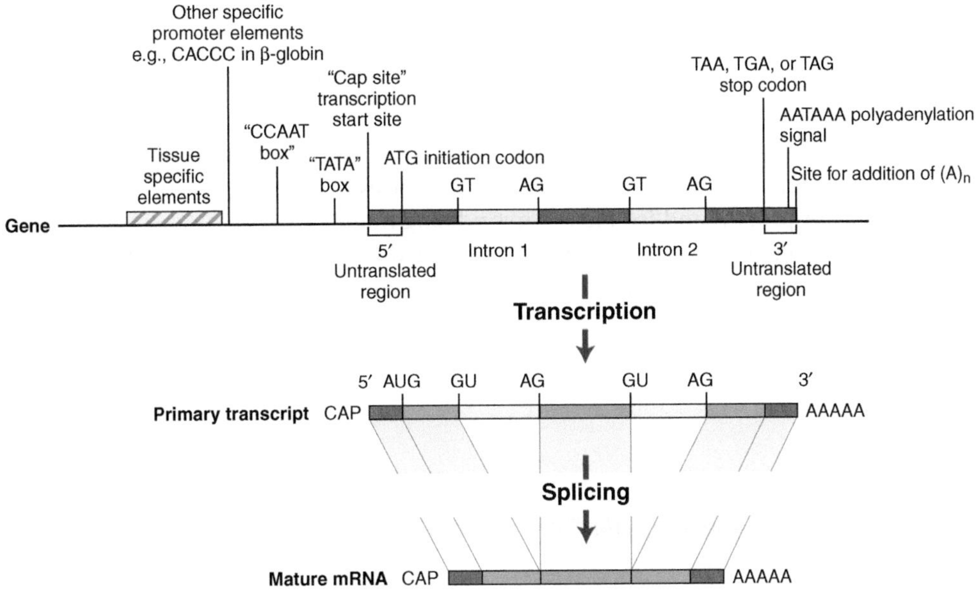

FIGURE 12.10
Schematic representation of the structure and expression of the human β globin gene. At the top, the genomic DNA is pictured with its many 5′ flanking sequences. Below are shown the various RNA products of transcription, editing, and splicing. The three exons and two introns (IVS or intervening sequences) are emphasized, as are the cap site, the polyadenylation signal, and the poly A tail. Further details are provided in the text.

Isolation

Because hemoglobin is so abundant, determining its protein structure was relatively straightforward. Not so for the genes. For example, the β globin locus makes up only about 6.7×10^{-5} percent of the human genome. How was it to be located and dissected to provide the information shown in Figure 12.10? The answer: by a technique called **functional cloning** (Figure 12.11). This name derives from the fact that the process begins with the disease phenotype, moves to the protein responsible for the function, and then uses information about the protein (and its mRNA) to finally isolate the gene. The steps required to move from the mRNA to the gene are as follows (Figure 12.12):

1. Isolate cells making the protein whose gene you are searching for.
2. Obtain the mRNA for that protein.
3. Reverse transcribe that mRNA using a viral enzyme called **reverse transcriptase.**
4. Purify the single-stranded **complementary DNA(cDNA)** product of that reaction, and convert it to a double-stranded form using DNA polymerase.

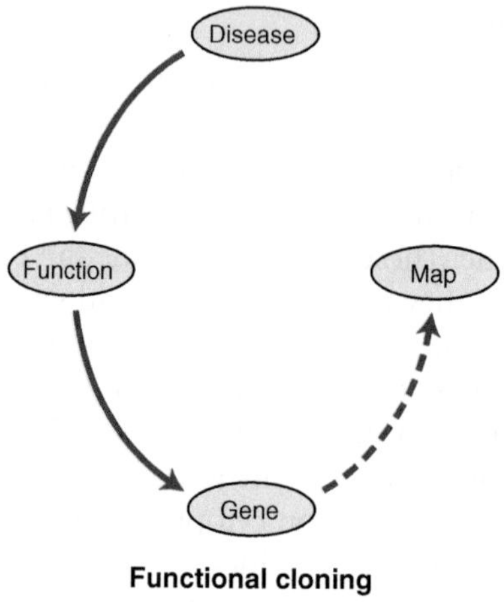

FIGURE 12.11
Functional cloning of a gene. The isolation process starts with the disease phenotype, proceeds to the functional protein and then to the gene. Once the gene has been cloned, it may be mapped. More detail is provided in the text.

5. **Hybridize** the cDNA to a library of DNA fragments representing the entire human genome.
6. Select those colonies that contain nucleotide sequences of interest (because they hybridize with the cDNA probe), and isolate the hybridizing fragments.
7. Sequence the fragments, piece their sequences together, and align them to finally produce the complete structure of the gene—nucleotide by nucleotide.

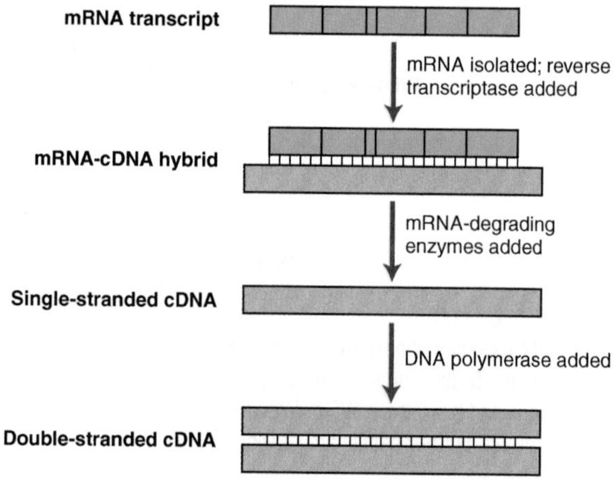

FIGURE 12.12
Steps in isolating complementary DNA (cDNA) from mRNA. A critical step involves conversion of mRNA to cDNA in a reaction catalyzed by a viral enzyme, reverse transcriptase. Once the double-stranded cDNA is made, it can be used as a probe to identify its genomic DNA by hybridization.

However arduous this sounds, cloning the globin genes was relatively straightforward because globin mRNA is abundant in young RBCs (called reticulocytes) and amenable to isolation.

Because no other protein is found in anything like the abundance of hemoglobin, functional cloning depended on such techniques as "fishing out" the mRNA from its polypeptide product using antibodies, or deducing mRNA sequence from the amino sequence of the polypeptide it encodes. These methods made functional gene cloning a time- and labor-consuming, rare event. Fewer than 100 genes were cloned using any and all of these techniques between the early 1980s and the late 1990s.

Mutations

For nearly a decade between 1978 and 1988, globin genes were the only genes capable of being subjected to intense mutational analysis in humans. A very large number of mutations were found, encompassing all the kinds of changes described in Chapter 8 (substitutions, deletions, frameshift, splicing, etc.). These changes in DNA resulted in a myriad of effects on the formation and structure of α- and β-globin chains. One can summarize this mutational analysis as follows:

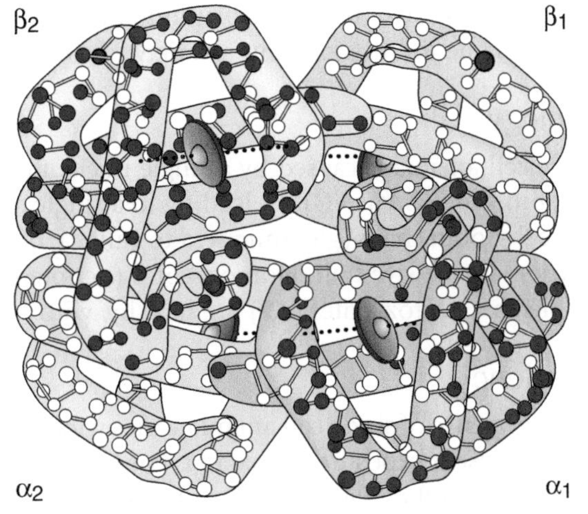

FIGURE 12.13
Missense mutations of the α- and β-globin subunits. The figure is based on substitutions identified by 1982. Many more have been found since. The red circles represent amino acid residues at which functionally significant missense mutations have been observed. The open circles show missense mutations of little or no functional significance.

- Two general classes of mutations have been identified—qualitative ones that change the tetrameric structure, and quantitative ones that affect the amount of hemoglobin produced.
- Missense mutations constitute the largest class of qualitative mutations. Some of these cause phenotypic changes, others do not. These missense mutations "pock mark" the α and β globin chains (Figure 12.13), meaning that most of the amino acids (more than 75%) in each chain have undergone substitutions.
- Missense mutations often affect one or more of hemoglobin's crucial properties, including its ability to form tetramers, carry oxygen, or be chemically stable.
- Frameshift mutations constitute the second largest class of mutations in the globin genes. Here, insertion or deletion of one or two nucleotides will produce mRNA that is not translated properly using the triplet genetic code. Such frameshifts often lead to major changes in amino acid sequence, depending on where the frameshift occurs in the mRNA. Thus there may be little, if any, normal chain synthesized, or the chain may be truncated by a premature stop codon, or the ability to fold will be impaired.

We will now present in greater detail a classic hemoglobinopathy, **sickle cell anemia**, which epitomizes a mutation resulting in a qualitative change in the hemoglobin molecule.

SICKLE CELL ANEMIA

This condition well deserves to be called "classic." It was described a century ago in people of African descent living on Caribbean islands. It was named for the elongated, sickle-shaped RBCs seen in blood smears (Figure 12.14). In 1949 it became the first human disease to be understood at the molecular level, by a Nobel Prize-winning team led by Linus Pauling. It was the first disease in which the mutant β-globin chain was sequenced and a single amino acid

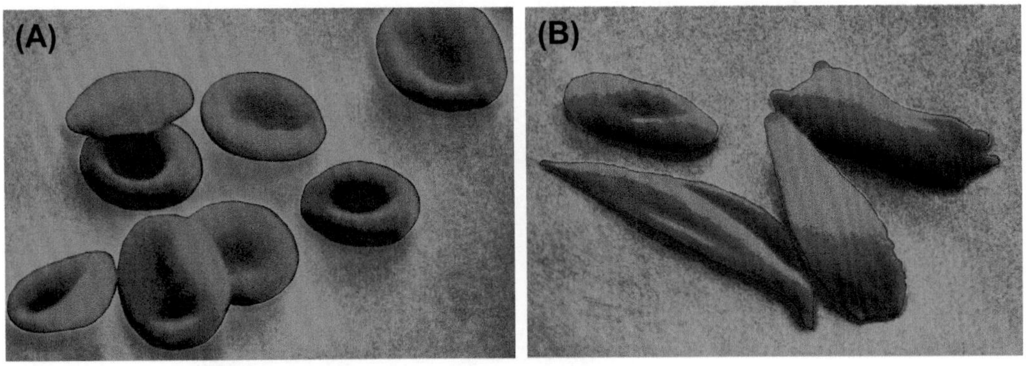

FIGURE 12.14
Scanning electron micrographs of normal, biconcave disc-shaped RBCs (A) and RBCs from a patient with sickle-cell anemia (B). The distorted architecture of the latter gave the disease its name.

substitution shown. It was the first human genetic system proving that the triplet genetic code was the same in humans as in a variety of simpler model organisms.

Etiology

Sickle cell anemia is caused by a single nucleotide subsitution in the sixth codon of the β-globin gene (GAG → GTG), leading to a single amino acid substitution in the sixth position of the β-globin chain (glutamic acid to valine), often abbreviated Glu6Val. It is inherited as an autosomal recessive trait. Homozygotes (denoted SS) almost always have serious clinical problems. Heterozygotes (AS) are healthy, with only a few rare exceptions.

Frequency

The prevalence of sickle cell anemia varies widely among populations, in direct relationship to past or present exposure to malaria. For example, in Nigeria 1 in 50 people has the SS genotype. Contrast that with the following statistics: sickle cell anemia occurs in 1 in 700 African Americans; 1 in 150,000 Europeans; and in fewer than 1 in 200,000 Asians. Accordingly, the frequency of those who are carriers for the S allele ranges from 40% in some African countries, to 10% in African Americans, to 0.4% in Asians.

Pathogenesis

The molecular basis of the problems seen in homozygotes for the hemoglobin S gene is depicted in Figure 12.15. The Glu to Val replacement changes hemoglobin tetramers such that,

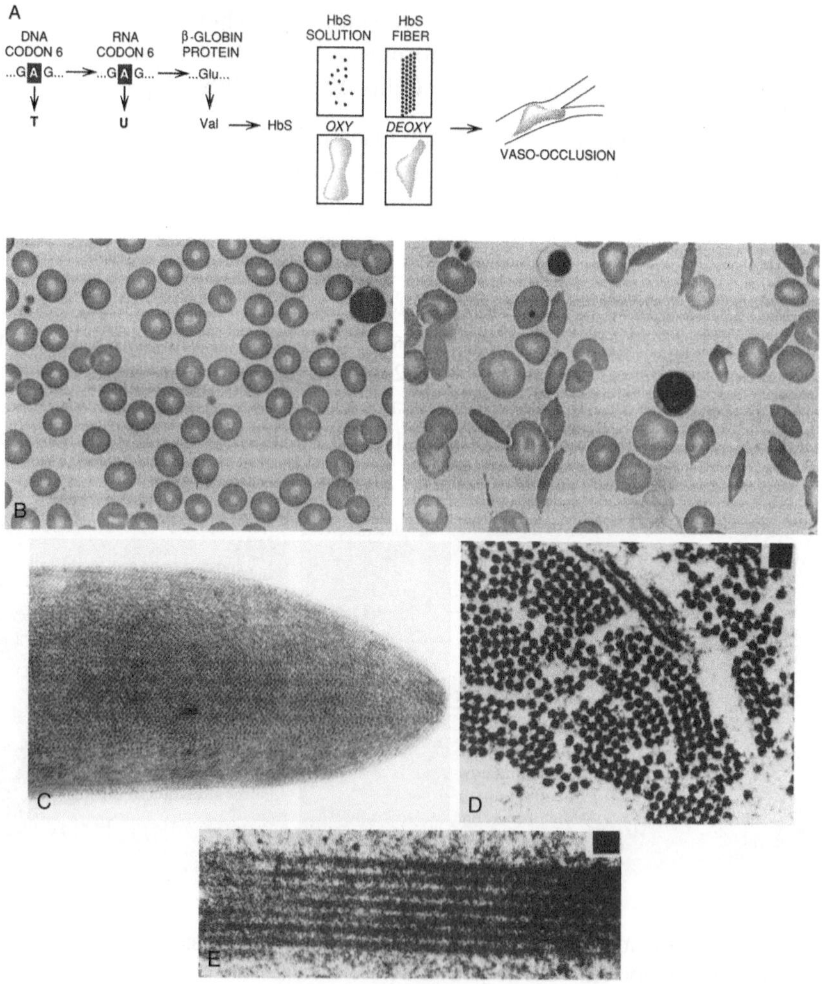

FIGURE 12.15
(A) Schematic diagram of the pathogenesis of sickle cell anemia. (B) Micrograph of normal RBCs (left) and sickled RBCs (right). Deoxygenated sickle hemoglobin precipitates within RBCs, deforming their structure. (C) Electron micrograph of deoxygenated sickle cells shows fiber of precipitated hemoglobin. (D) Higher magnification of hemoglobin fibers seen transversely. (E) High magnification of fibers seen longitudinally.

when deoxygenated in the tissues, they become insoluble and aggregate. This, in turn, deforms the shape of the usually biconcave RBC, producing distorted cells, some of which conform to a sickle shape. After repeated cycles of solubility (when oxygenated) and insolubility (when deoxygenated), the precipitated hemoglobin molecules assume a long fiber formation and the RBC becomes irreversibly sickled. This process results in plugging of capillaries and removal of the damaged cells by the spleen.

Clinical Features

The clinical abnormalities follow logically from the molecular pathology and may affect virtually any tissue. Accumulation of sickled RBC impairs the tissue's blood supply (and that of vital oxygen). This, in turn, leads to strokes; heart attacks; infarction of the spleen, kidney, bone, or skin; and increased susceptibility to bacterial infections. Episodes of severe pain (called sickle cell crisis) ensue, usually resulting in repeated hospitalizations. In the clinical setting, use of narcotics to relieve pain and transfusions to treat anemia are the mainstays of supportive care. Sickle cell anemia affects homozygotes from early childhood through adult life, leading to a shortened mean life span, and a high incidence of narcotic addiction resulting from drug use to relieve the pain of a crisis. It is ironic that, despite all we know about the molecular sequence leading to these clinical consequences, no truly satisfactory treatment for sickle cell anemia has been discovered (a topic that will be discussed in Chapter 17).

Heterozygous carriers for sickle cell anemia are at no clinical risk, except in such unusual circumstances as flying in under-pressurized airplanes or becoming dehydrated on the battlefield or in strenuous athletic competition. In each of these situations, hypoxia is the likely cause of pathologic sickling usually seen only in those who are homozygous for the mutation.

Diagnosis

Sickle cell anemia can usually be diagnosed simply by looking at a blood smear (Figure 12.14). In those suspected of being heterozygous carriers (usually called sickle trait), deoxygenating blood will often lead to the ability to see some distorted RBCs. In recent years, electrophoretic and molecular hybridization techniques have made diagnosis of homozygotes and heterozygotes routine. This is critical for accurate genetic counseling, as discussed in Chapter 5 and to be discussed again in Chapter 17.

Societal Issues

Because sickle cell anemia and sickle cell trait are so much more common in African Americans than in other racial and ethnic groups in our population, racism and racial discrimination have been part of the American experience with this genetic disorder. In the late 1950s, when it became easy to detect sickle cell trait by simple hemoglobin electrophoresis, many African Americans refused to be tested out of fear that the information would be used against them. There were, in fact, some examples where employment discrimination followed identification of the trait. This shouldn't have surprised anyone, given the long history of slavery, Jim Crow laws, heinous medical experiments like that of the Tuskegee syphilis experiment, and job discrimination. Even more recently, African Americans were hesitant about neonatal screening for sickle cell anemia unless all neonates—not just African Americans—were tested. Even the knowledge that 10% of patients with sickle cell anemia die in the first 6 months of life from treatable bacterial infections did not overcome the mistrust within the African American community, which found it hard to believe that those delivering healthcare would do something unambiguously in their interest.

Duchenne Muscular Dystrophy

Duchenne muscular dystrophy (DMD) is one of several human disorders characterized by abnormalities in skeletal muscle. We have chosen DMD for special discussion because of the way the gene responsible for it was identified and isolated. Such understanding will be crucial in testing new forms of treatment.

ETIOLOGY

DMD is caused by mutations of the gene coding for **dystrophin**, a protein expressed in skeletal, smooth, and cardiac muscle. DMD is inherited as an X-linked recessive trait. About 1 in 3,500 males are affected, meaning that there are about 1,000 new cases in the US yearly. The disorder is found in all ethnic groups.

GENE ISOLATION

In sharp contrast to sickle cell anemia, in which a plethora of information permitted gene isolation, mutational analysis, and genotype/phenotype correlation, none of these features were known for DMD as late as the mid-1980s. Thus, the formidable scientific problem was how to identify a gene when all that was known about it was the tissue in which it was expressed and the chromosome on which it was encoded. As noted in Figure 12.11, the globin genes were isolated by functional cloning, which moved the inquiry from the disease to the function and, finally, to the gene. In the absence of information about function, however, a new approach called **positional cloning** had to be devised. Here, the idea was to go from the disease to its map position on the genome, then to the gene, and, finally, to that gene's function (Figure 12.16).

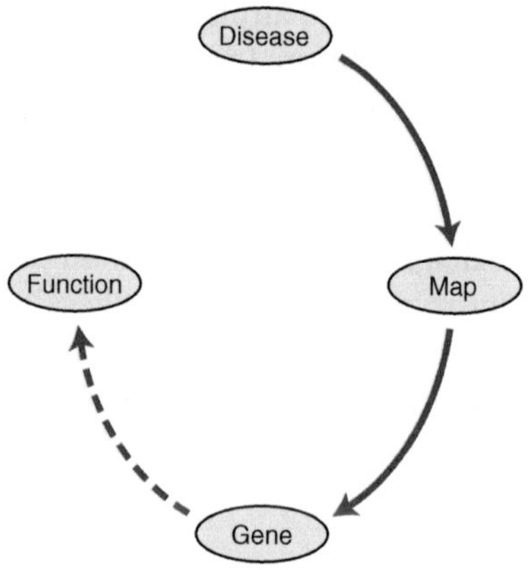

FIGURE 12.16
Positional cloning of a gene. In contrast to functional cloning (Figure 12.11), this type of cloning is used to find genes whose function is unknown. It proceeds from disease to map position on a chromosome to identification of the gene and then to defining its function. See text for details.

When the prodigious effort to clone the DMD gene using positional cloning was begun, scientists knew that the causal gene was on the X chromosome because the disease was inherited as an X-linked trait. Linkage studies had placed it somewhere on the short arm (p)—but the short arm of the X is about 60 million bases long and contains about 400 genes. How to find the needle in this haystack? The answer: by carrying out "saturation" linkage analysis, which identified markers ever closer to the DMD disease gene (Figure 12.17), thereby narrowing the region of the short arm of the X chromosome in which the DMD locus could reside. A second key tool came from identifying people with chromosomal translocations and deletions that interrupted the gene for DMD. Very rare females with DMD proved to be of particular value. These women had X/autosome translocations where the breakpoint on the X chromosome interrupted the gene responsible for DMD. This translocation was expressed because X inactivation preferentially silences the normal X rather than the translocated X, for reasons not understood. This kind of study placed the DMD gene at p21. Another patient of value was a boy who had DMD and several other disorders mapped to the p21 region of the X. From these informative patients, it was possible (with great difficulty) to clone pieces of the DMD gene and, ultimately, all of it. In this way the **dystrophin** gene was isolated.

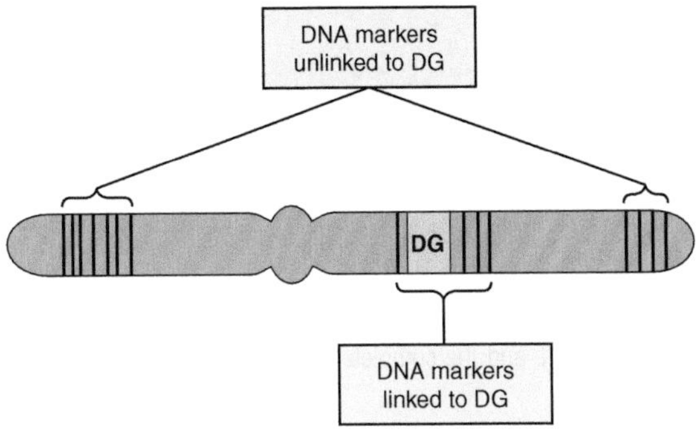

FIGURE 12.17
Scheme showing how linkage analysis locates a disease gene (DG). The method: DNA markers close enough to DG will be inherited together (i.e., linked); the closer a marker is to DG, the more likely they will be inherited together; the larger the number of useful markers one has, the closer one can get to DG.

GENE AND PROTEIN CHARACTERIZATION

Dystrophin is the largest human gene isolated to date, about 2.3 million bases (Mb) long. It contains 79 exons and occupies more than 1% of the X chromosome. The dystrophin protein is extremely large as well—some 420 kiloDaltons (kDa). Dystrophin is involved in the contractile apparatus of muscle proteins, where it interacts with the cytoskeleton (Figure 12.18). Several other proteins make up this complex of contractile proteins, mutations of which have been shown to be responsible for other variants of muscular dystrophy.

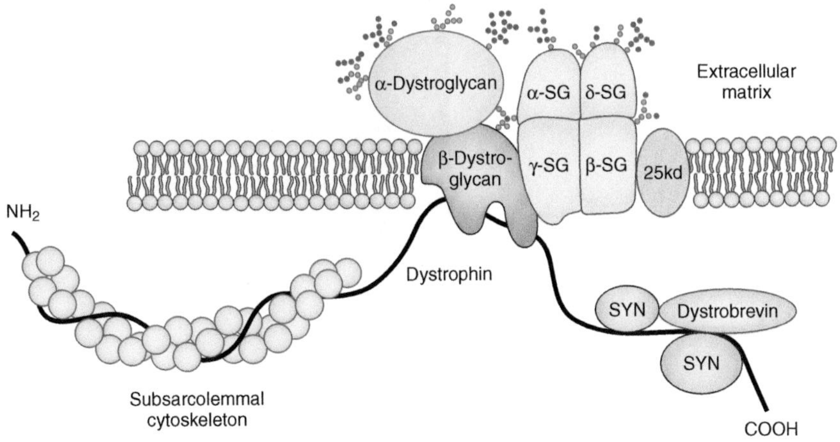

FIGURE 12.18
Model of dystrophin and dystrophin-associated proteins in and around the sarcolemmal cytoskeleton of striated muscle cells. Dystrophin (thick black line) interacts with several other proteins. Dystrophin deficiency causes DMD; inherited deficiency of some of the other proteins results in other forms of muscular dystrophy.

MUTATIONS

The majority of mutations in DMD patients are large or small deletions in the dystrophin gene. They can occur throughout the gene but tend to cluster in the 5′ half of this gene. As a rule, these deletions affect the reading frame of dystrophin's mRNA such that truncated proteins are made that are non-functional. Thus, most DMD patients have no detectable dystrophin in their muscle fibers (Figure 12.19). A variety of different gene duplications, insertions, and point mutations constitute the remainder of structural DNA alterations in people with DMD.

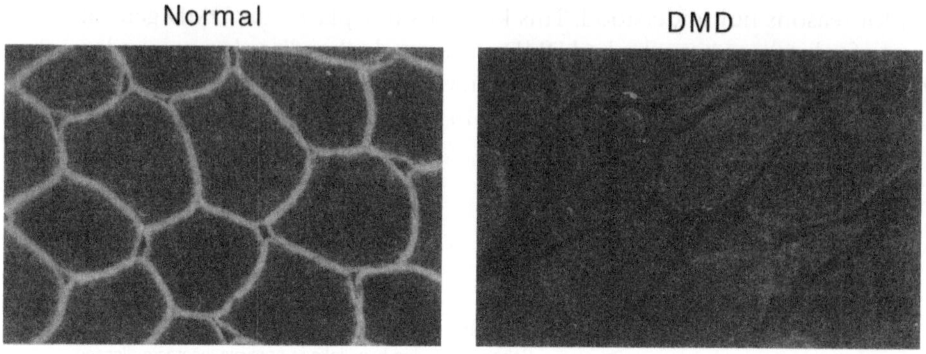

FIGURE 12.19
Detection of dystrophin by immunofluorescence. Histologic sections of normal muscle and muscle from a DMD patient were labeled with antidystrophin antibody and identified using fluorescence. Note the intense staining around the periphery of normal muscle fibers and the complete absence of staining in DMD.

CLINICAL FEATURES

As would be expected for this X-linked recessive disorder, the great majority of people with DMD are males. Typically, they are well until age 3–5, when they begin to have difficulty rising from a sitting position or climbing stairs. DMD patients exhibit one characteristic physical finding, called calf pseudohypertrophy (Figure 12.20), which is enlargement of the calf of the legs due to deposition of fat and connective tissue in the muscle. By age 10, affected patients are often limited to a wheelchair because of muscular weakness. Most patients die of impaired pulmonary function, pneumonia, or cardiac disease (dystrophin is also expressed in heart muscle). The average life span is only 18 years.

DMD is seen rarely in females heterozygous for dystrophin mutations. As discussed earlier, some such females have disadvantageous X chromosome inactivation so that their cells are skewed toward expressing the X having the mutation. In other affected females, X/autosome

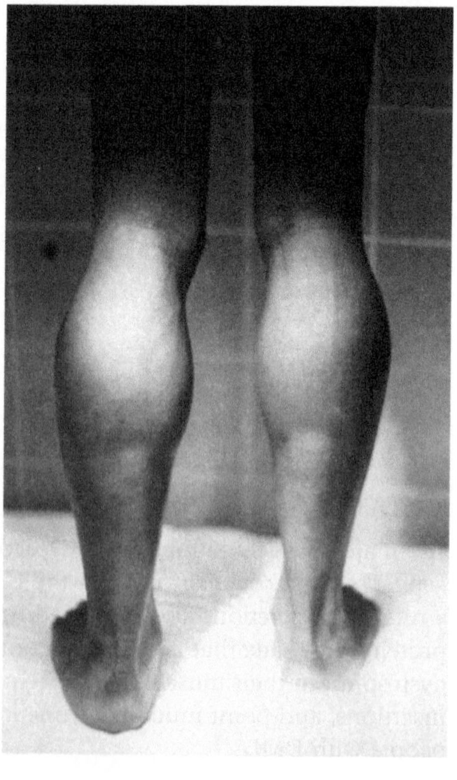

FIGURE 12.20
Pseudohypertrophy of calf muscles in a boy with DMD. Normal muscle tissue has been replaced by fat and connective tissue.

translocations result in impaired dystrophin activity. Clinically, females tend to be more mildly affected than hemizygous males.

DETECTION AND RISK ASSESSMENT

In families in which the mother is a carrier, 50% of her sons will be affected and 50% of her daughters will be carriers. Such risk assessment is straightforward provided the mother has two or more affected sons (thereby proving that she is a carrier). But what of the woman who has only one affected boy? She may be a carrier or she may not. It has been shown that in X-linked disorders in which affected males do not reproduce, about one-third of mothers are not carriers; rather, they reflect a new mutation occurring in their germ cells. For this reason, genetic counseling in DMD is often not simple, underscoring the need for accurate detection of affected males prenatally and of carrier females, using sophisticated molecular analyses.

Familial Hypercholesterolemia

Coronary artery disease (CAD) leading to myocardial infarcts (heart attacks) is the leading cause of death in the United States and other developed countries. In 20% of such patients, moderately or markedly increased concentrations of low-density lipoprotein (LDL) cholesterol are seen. Such elevation can be produced by a variety of genetic and environmental factors. We will discuss one such disorder (familial hypercholesterolemia) because it illustrates how the study of a rare disease led to major advances in our understanding of cholesterol metabolism and in prevention of heart attacks.

ETIOLOGY

Familial hypercholesterolemia (FH) is caused by mutations in the gene that encodes the cell surface receptor protein that transports LDL cholesterol. FH is inherited as an autosomal, incompletely dominant trait. This is an exception to Mendel's simple concept of dominance and recessiveness (see Chapter 5) and is based on the following: each of the three genotypes (homozygous normal, heterozygous, and homozygous affected) can be distinguished from one another clinically and by measuring serum cholesterol. Whereas affected homozygotes are found in only about 1 in a million people, heterozygotes are found in 1 in 500 people worldwide. Heterozygotes account for 5% of people who develop heart attacks due to elevated serum cholesterol concentration.

PATHOGENESIS

In elegantly planned and executed experiments, which lead to the Nobel Prize, two young scientists named Michael Brown and Joseph Goldstein unraveled the metabolic basis for FH. They began by studying cultured cells from young children whose serum cholesterol concentrations were as much as 10 times normal and who developed heart attacks and subcutaneous cholesterol deposits before age 10. These children rarely lived to the end of their second decade. They turned out to have a complete defect in the ability of their cells to transport cholesterol into cells via a cell surface receptor (the LDL receptor, or LDLR). This defect not only impaired their ability to move cholesterol out of the blood and into tissue cells; it also led to excess synthesis of cholesterol by cells because the regulatory pathway by which internalized cholesterol suppresses synthesis of endogenous cholesterol was deficient (Figure 12.21). Cells from parents of such severely affected children had LDLR activity greater than in their children, but less than in controls.

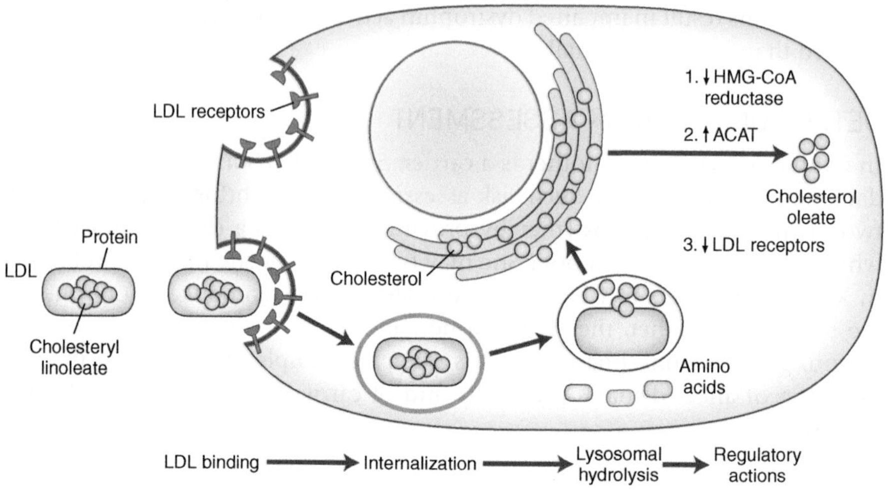

FIGURE 12.21
Sequential steps in the LDL (low density lipoprotein) pathway in human cells. LDL binds to its receptor on the cell surface; it is then internalized and delivered to lysosomes where it is degraded, freeing its cholesterol cargo. The internalized cholesterol then regulates intracellular synthesis of cholesterol by inhibiting the first enzyme in that pathway, HMG-CoA reductase. See text for additional details.

GENE AND PROTEIN CHARACTERIZATION

Subsequent to the discovery of the LDLR, it was possible to clone its gene and characterize the product it encoded. The gene is located on chromosome 19; it is of moderate size (45 kb) but of great complexity (Figure 12.22), having specific domains for binding its ligand, LDL cholesterol, for internalizing it, for attaching sugar molecules, for spanning the cell membrane, and for producing its cytoplasmic tail. Each of these domains is coded for by a number of the 18 exons in the *LDLR* gene.

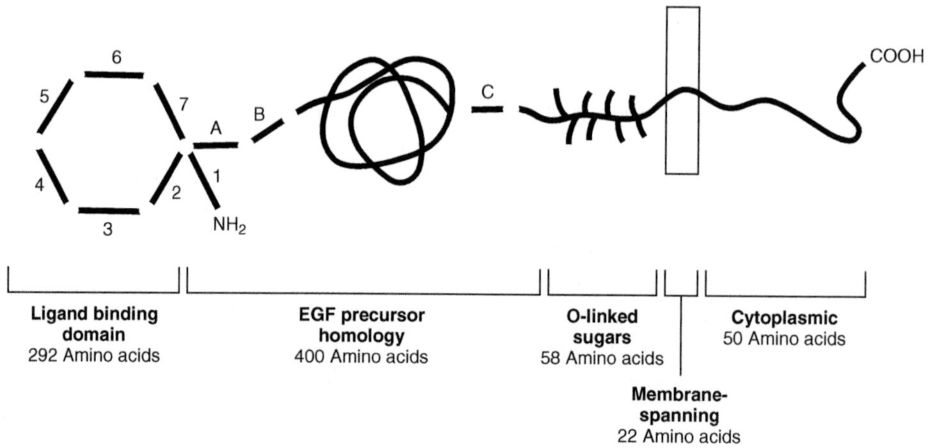

FIGURE 12.22
Structure of LDL receptor. Note the several domains: ligand binding; epidermal growth factor homology; sugar attachment; membrane spanning; and cytoplasmic component. LDLR mutations have been identified in every domain.

MUTATIONS

By now, more than 700 different mutations in the *LDLR* gene have been identified and characterized. Point mutations are found throughout its length; a smaller number of insertions, deletions, and more complex structural rearrangements occur as well. These mutations fall into six functional classes affecting each step in the pathway by which the LDLR is synthesized, transported intracellularly, packaged, recycled, and bound to specific regions of the cell membrane.

CLINICAL FEATURES

The medical importance of FH derives directly from its name, hypercholesterolemia. Normal subjects have serum cholesterol concentrations between 120 and 180 milligrams per deciliter (mg/dL), FH heterozygotes display serum cholesterol concentrations between 250 and 500 mg/dL, and affected homozygotes show concentrations from 600 to 1,000 mg/dL

(Figure 12.23). These serum values correlate directly with clinical abnormalities. FH heterozygotes typically develop CAD in their thirties to fifties. Further, they often deposit cholesterol in their Achilles tendons and their eyelids. FH homozygotes exhibit a variety of cholesterol deposits in the first decade of life and often die of heart attacks before age 15 (Figure 12.23).

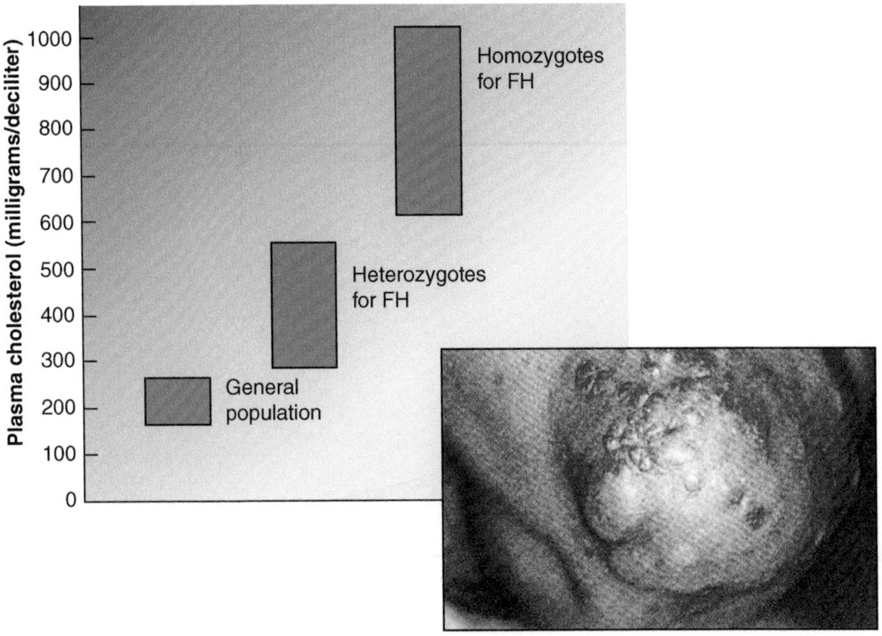

FIGURE 12.23
Hallmarks of familial hypercholesterolemia (FH). Serum cholesterol is moderately elevated in heterozygotes and dramatically elevated in FH homozygotes. As noted in the lower right, a picture of a knee with calcium deposits, affected homozygotes deposit cholesterol in their skin, over their joints, in their eyelids, and, most dangerously, in their coronary arteries.

The work with FH and the complex metabolic system that underlies it led to the discovery of the statins, a most effective form of treatment of many forms of elevated serum cholesterol, including that resulting from being heterozygous for LDLR mutations. The statins are the most widely used pharmaceutical in the world today and have been shown to prevent first, as well as subsequent, heart attacks. Unfortunately, FH homozygotes do not respond to statins and can be saved only by heart or liver transplantation, as we will discuss further in Chapter 17.

Huntington Disease

Huntington disease (HD) is a single-gene defect of special interest for several reasons. First, it is caused by unstable trinucleotide repeat mutations. Second, its age of onset correlates with the size of the triplet expansion in the mutant gene. Third, the clinical onset of the disorder is usually in midlife, after one's children have been born, thereby complicating genetic counseling and family decision-making.

ETIOLOGY

Huntington disease is caused by mutations of the HD gene, found at the tip of the short arm of chromosome 4. The disease occurs in about 1 in 20,000 Europeans; it occurs in other ethnic groups as well, but at much lower frequencies (in Japan, for instance, the incidence is 1 in 250,000). Huntington disease is inherited as an autosomal dominant trait with almost complete penetrance.

PATHOGENESIS

The gene for HD is one of a small number of human genes characterized by multiple triplet repeats (see Chapter 8). Specifically, the normal HD gene has up to 35 CAG repeats in exon 1. CAG codes for glutamine, meaning that the protein product of the gene, called huntingtin, normally has up to 35 glutamine residues near its amino terminal (5′) end. The functional role of this long stretch of glutamines is not understood. Equally murky is the function of huntingtin, which is expressed in most tissues, including the brain.

Patients with Huntington disease have more than 40 CAG repeats in exon 1 of the *HD* gene, and the number of repeats ranges as high as 120. There is a general correlation between the number of CAG repeats and the age of onset of clinical manifestations (Figure 12.24): the

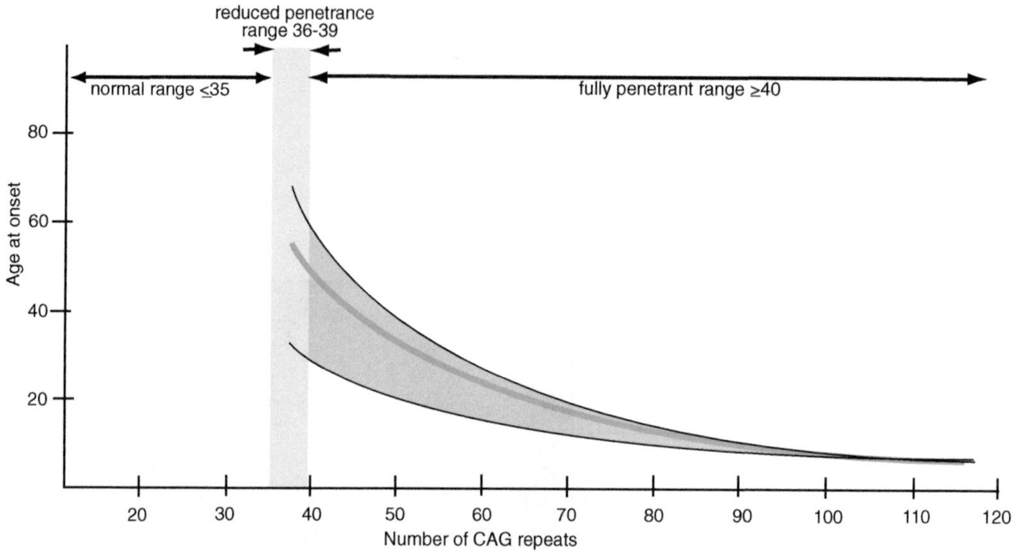

FIGURE 12.24
Graph correlating approximate age of onset of clinical abnormalities in Huntington disease (ordinate) with number of CAG repeats in the HD gene (abscissa). The heavy, solid line is the average age at onset, the shaded area shows the range of age of onset for any given number of repeats. Note that clinical consequences are late (or not seen at all) in those with 36 to 39 CAG repeats.

larger the number of repeats, the earlier the onset. This relationship between repeat number and clinical consequences has two other ramifications. First, as shown in Figure 12.25, people with 36–39 repeats sometimes are clinically affected (see individual I1) but usually are not. Second, the number of CAG repeats in the mutant gene tends to increase as the mutation is passed from one generation (individual I1) in an affected pedigree to another (individuals II2, II4, and II5). As the number of repeats increases, the age of onset of symptoms decreases (a phenomenon called anticipation). Interestingly, such anticipation is seen only when the mutation is transmitted by an affected father. The mechanism behind this gender-specific phenomenon is unknown.

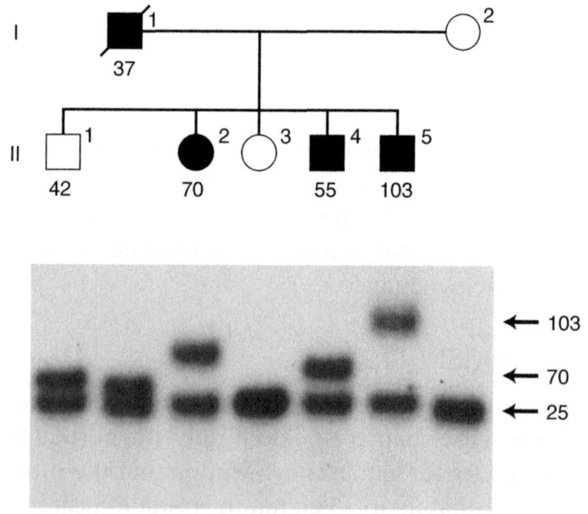

FIGURE 12.25
Pedigree of family with Huntington disease (top) and number of CAG repeats in individual family members (bottom). Each person has a band (developed using Southern blotting) at about 25 repeats. Those with HD (solid symbols) have a second band (compatible with heterozygosity) at 37 (II1), 70 (II2), 55 (II4), and 103 (II5) repeats, respectively. Individual II1 with 42 repeats is likely to develop clinical HD. This kind of test is now available to individuals at risk.

CLINICAL FEATURES

Huntington disease is among the most dreaded human afflictions. Its onset is usually between 40 and 50 years of age and is generally heralded by involuntary movements of the hands and feet (called chorea). These neurologic manifestations are followed over time by personality changes, a gradual loss of memory and other cognitive skills, and, finally, progressive muscle weakness and death, usually within 10–15 years. In affected individuals who have 70 or more CAG repeats, the onset of clinical manifestations can be as early as 20 years, with demise also earlier in life. At present, there is no effective treatment. At autopsy, pathologic findings are limited to the brain. Widespread degenerative changes are found in the cerebral cortex and in the basal ganglia—areas that coordinate fine movements.

DETECTION AND GENETIC COUNSELING

Elucidation of the triplet expansion that is characteristic of Huntington disease has greatly facilitated diagnosis. Southern blot analysis (Figure 12.25) can identify affected individuals (II2, II4, and II5) and exclude the disease in others (II3). However, availability of this diagnostic capability has created a major dilemma for family members. For example, assume that you have a parent with Huntington disease and know that you have a 50% chance of developing it too. Given its lethal outcome and lack of effective treatment, would you want to have a diagnostic test while you were feeling well, or would you prefer to wait until you developed symptoms or, more positively, moved beyond the age where risk decreases to near zero (about 55 years)? It is estimated that 80% of those in this situation choose not to be tested. This dilemma becomes even more excruciating if the grandson of an affected person chooses to be tested and his at-risk parent refuses. If the grandson is found to be affected, a diagnosis of Huntington disease will have been confirmed in his parent as well.

IMPLICATIONS: SERENDIPITY IN SCIENCE

Inside the Hospital and Laboratory: The Case of Lorraine

In 1968, the study of two families by Dr Rosenberg with seemingly disparate clinical phenotypes converged. In the first, three boys in succession, each born healthy, lapsed into coma during the first week of life and died. In the second, a 20-month-old girl named Lorraine was admitted to Yale–New Haven Hospital in coma. She was found to have an extraordinary increase in blood ammonia (NH_3), a ubiquitous end product of protein metabolism known to be toxic to the brain when present in high amounts. When the blood ammonia level was corrected by dialyzing it from her intestines and peritoneal cavity, Lorraine woke up, only to relapse partially when small amounts of protein were returned to her diet.

Because the normal pathway of ammonia detoxification in the human was known to occur in the liver, each of the pathway's five enzymes was assayed in a tiny piece of Lorraine's liver. Activity of one enzyme, ornithine transcarbamylase (OTC), was distinctly reduced; the other enzymes were normal. Subsequently, Lorraine had a brother who, like the boys in the first family, died before he was 1 week old. His liver had no OTC activity. Lorraine's family history was also remarkable in that two of her mother's brothers had died in the neonatal period of unknown causes.

These findings fit perfectly with the idea that OTC deficiency is an X-linked trait: hemizygous affected males succumb to ammonia intoxication early in life; heterozygous females have variable phenotypes depending on how X inactivation affects OTC expression in liver cells. Biochemical and genetic studies confirmed this hypothesis, paving the way for several "spin offs" of this work:

- Today, all neonates with coma are tested for hyperammonemia.
- A few males with complete OTC deficiency have been rescued by liver transplantation.
- Heterozygous females are treated with a low protein diet and medications to siphon off ammonia, and many of them do well.

(*Continued*)

- OTC deficiency can be detected prenatally using DNA analysis on cells obtained by CVS or amniocentesis.
- Molecular analysis of OTC deficiency has revealed more than 100 different mutations.

... and, best of all, Lorraine is alive and well in her fifth decade of life.

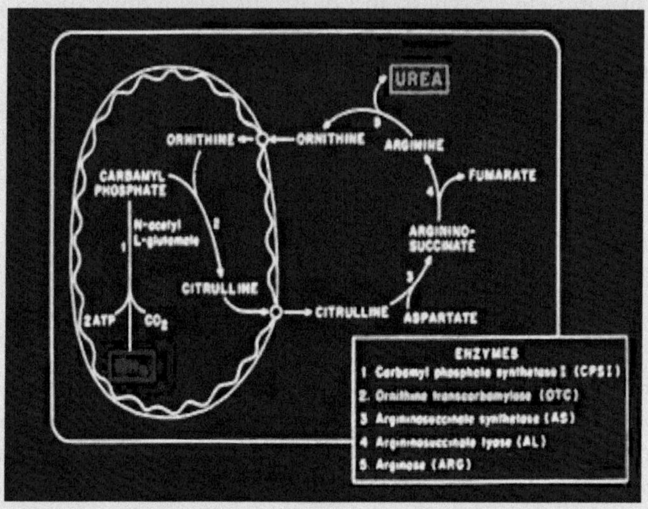

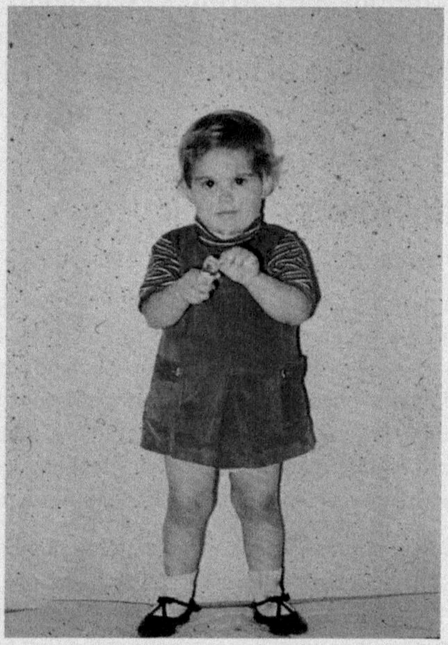

Red Hair

As stated on the first page of this book and again at the beginning of this chapter, many single-gene mutations cause phenotypic variation not associated with clinical disturbance. Red hair is such a phenotype about which we know a considerable amount.

ETIOLOGY AND FREQUENCY

Red hair is generally caused by reduced function of alleles at the locus coding for the melanocortin receptor-1 (MCR-1) protein. The gene for MCR-1 has been cloned and is located on chromosome 16. Red hair, an autosomal recessive trait with typical Mendelian inheritance, is much rarer than black, brown, or blond hair. Between 2% and 6% of people in the United States have red hair, compared to 13% in Scotland, 10% in Ireland, similar frequencies in Scandinavia, and virtual absence in sub-Saharan Africa, most of Asia, Brazil, Argentina, and Chile. More than 25% of Scottish and Irish adults are carriers for alleles that produce red hair.

BIOCHEMICAL MECHANISM

Melanin is the pigment that produces the wide variation seen in skin and hair color in humans. Melanin is synthesized by cells in the skin and hair follicles called melanocytes. Two major classes of melanin are known: eumelanin, a brown-black pigment; and pheomelanin, an orange-to-red pigment. Melanin is stored in cytoplasmic organelles within melanocytes, called melanosomes. As shown in Figure 12.26, melanin synthesis is under hormonal control. The pituitary hormones adrenocorticotropic hormone (ACTH) and melanocyte stimulating hormone (MSH) signal melanocytes to make melanin. The hormones bind to a group of transmembrane receptors, the melanocortin receptors. A major receptor of this class is MCR-1. When normal MCR-1 binds its hormone ligand, a series of intracellular signaling reactions

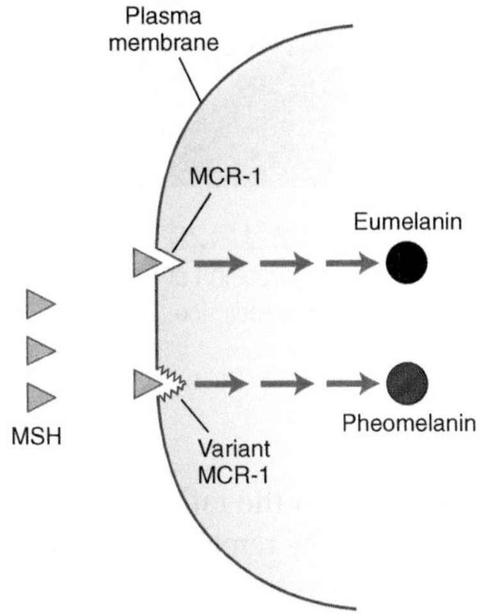

FIGURE 12.26
Model of a melanocyte showing the relationship between the melanocortin receptor-1 (MCR-1) and the synthesis of eumelanin (black) or pheomelanin (red) in melanosomes. When the melanocyte-stimulating hormone (MSH), made in the pituitary, binds to normal MCR-1, it initiates a signaling pathway leading to the synthesis of brown/black eumelanin. If the MCR-1 is mutated, MSH binding results in synthesis of pheomelanin. Pheomelanin is the pigment found in red hair.

takes place, ultimately leading to formation of eumelanin in melanosomes. If the structure of MCR-1 has been changed by one of several allelic variants, however, the signaling pathway will lead instead to the synthesis of pheomelanin, the pigment found in red hair.

The structure of these variant forms of MCR-1 has been elucidated. They are different point mutations leading to single amino acid substitution in the protein. People with red hair are either homozygous for the same mutant allele or are compound heterozygotes (meaning that they have two different mutant alleles.)

PHENOTYPIC FEATURES

Red-haired people (Figure 12.27) generally have light-colored eyes (blue, green, hazel), pale skin, and (sometimes) freckles. Characteristically, they tan poorly and burn easily when exposed to ultraviolet (UV) light from the sun and are at an increased risk for melanoma. Their hair color may range from auburn, to bright red, to copper. The range of hair colors found in

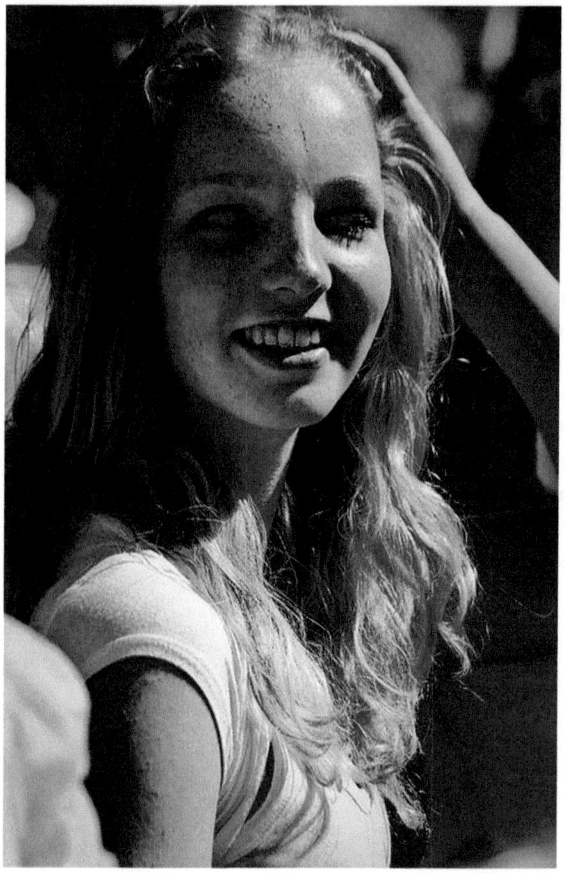

FIGURE 12.27
A young woman with red hair, light skin, and freckles.

FIGURE 12.28
Queen Elizabeth I of England. Her parents, King Henry VIII and Ann Boleyn, were redheads, too.

those with red hair must reflect changes in the ratio of pigments made in their melanocytes, but the details of this interesting variability remain to be elucidated.

EVOLUTION

Somewhere between 20,000 and 100,000 years ago, the mutations responsible for red hair appeared. They almost surely enhanced formation of vitamin D in the skin following UV exposure, with the consequent advantage of preventing rickets. This capability may have been the selective advantage that led to the high frequency of red hair in northern Europeans, who encounter little sun during much of the year. Another theory posits that Africans with red hair were at a reproductive disadvantage because they couldn't tolerate the sun, thereby leading to virtual disappearance of the trait there and in other parts of the world.

HISTORY AND SOCIOLOGY

Red hair has been written about from the Greeks on. Homer mentioned that Achilles and Menelaus were redheads. The monstrous Roman emperor, Caligula, had red hair. Artists, particularly Titian, favored red-haired females in their paintings. Queen Elizabeth I was a redhead as well (Figure 12.28). Much has been said about the relationship between red hair and temperament. Red-haired women have been described as "fiery;" red-haired men as "sexually beastly." Vampires have been pictured with red hair. These characterizations have no scientific basis and likely reflect particular temperaments in individuals who are noticeable because of their hair color, rather than any statistical association between hair color and behavior.

REVIEW QUESTIONS AND EXERCISES

1. Choose the phrase in the right column that best matches the term in the left column.

a. homogentisic acid	1. matings that affect frequency of recessive disorders
b. cDNA	2. protein composed of α and β chains
c. huntingtin	3. product of reverse transcriptase
d. allelic heterogeneity	4. catalog of human Mendelian traits
e. IGF-1	5. process by which complex materials are synthesized
f. functional cloning	6. protein with expanding triplet repeats
g. genotype AS	7. isolating gene through its genomic map location
h. LDL receptor	8. protein that stimulates growth
i. consanguineous	9. substance discovered by Garrod
j. catabolism	10. confers heterozygote advantage
k. disease gene	11. deficiency leads to red hair
l. OMIM	12. single phenotype produced by several mutations at a locus
m. dystrophin	13. genes whose mutations cause disorders
n. positional cloning	14. isolating gene through its protein product
o. hemoglobin	15. regulates cholesterol transport
p. MCR-1 receptor	16. process by which complex materials are broken down
q. anabolism	17. largest protein in human cells

2. Genetic heterogeneity is commonly found in inherited metabolic disorders. In some instances the heterogeneity is allelic, in other instances non-allelic.
   a. What is the difference between allelic and non-allelic heterogeneity?
   b. Suppose two individuals who have been deaf since birth married and had several children successively with normal hearing. What kind of genetic heterogeneity does this suggest? Why?
   c. Suppose the individuals in (b) had several children in a row with hearing difficulties ranging in severity from complete to moderate. What kind of genetic heterogeneity does this suggest? Why?

3. A man has a maternal grandfather with sickle cell anemia. His wife's paternal uncle has sickle cell anemia, too. Neither the man nor the wife is affected. What is the risk that their first child will have sickle cell anemia?

4. Ornithine transcarbamylase (OTC) deficiency is inherited as an X-linked recessive trait. The OTC enzyme is a trimer of identical subunits.
   a. Why is ammonia intoxication due to OTC deficiency so much more common in males than females?
   b. Describe briefly four mechanisms that would explain clinically severe OTC deficiency in females.
   c. How would you determine which of the mechanisms described in (b) accounted for clinically significant OTC deficiency in a young girl?
   d. Hemizygous affected males have been treated successfully by liver transplantation. Why has this been accomplished in only a handful of infant boys?
   e. What does it mean to say that OTC deficiency in males is a "genetic lethal?"

5.

$$\begin{array}{c} N \rightarrow I \rightarrow H \\ E_{NI} \;\; E_{IH} \end{array}$$

Suppose N is an essential nutrient that is converted to hormone H via intermediate I. E_{NI} is the enzyme catalyzing the conversion of N to I. Likewise, the enzyme E_{IH} catalyzes the formation of H from I. Suppose, too, that you have discovered an inherited metabolic disorder caused by deficiency of E_{NI}, and have shown that the condition is inherited as a mitochondrial trait.
   a. Draw a three-generation pedigree of a family with this disorder, distinguishing affected individuals from unaffected ones.
   b. What metabolic disturbances referable to substances N, I, and H would you expect to find in affected patients.
   c. Describe briefly two therapeutic approaches you would consider.
   d. Would you be surprised if clinical severity in affected individuals differed considerably? Why or why not?

6. A mutant allele at the LDL receptor gene locus (responsible for familial cholesterolemia) encodes a receptor protein 180 amino acids longer than normal.
   a. Describe briefly two possible mechanisms to explain this abnormality.

b. How many additional nucleotides were added to the LDL receptor mRNA to account for the observed change in protein length?

c. Familial hypercholesterolemia (FH) is described as an "incompletely" dominant trait. What does this mean?

7. More than 80% of people with red hair have a deficiency of the melanocortin-1 receptor (MCR-1) which is inherited as an autosomal recessive trait. Yet hair color in redheads can vary from auburn to copper.

a. List three mechanisms that might account for the variable hair color in red heads.

b. Why is red hair so often associated with fair skin and freckles?

8. Would you expect a nonsense mutation in the first exon of a gene to have a greater effect on that gene's function than a missense mutation of the same position in exon one? Why?

CHAPTER 13

Multifactorial Traits

CHAPTER OUTLINE

CORE CONCEPTS

Most common human phenotypes reflect interactions between genes and the environment and are termed "**multifactorial**." Some multifactorial phenotypes are ubiquitous physiologic traits such as height, weight, and mathematical aptitude. Others are a multitude of common disorders encountered at birth (cleft lip, neural tube defects), in children (asthma, juvenile diabetes mellitus), or in adults (high blood pressure, coronary artery disease, schizophrenia). A small number, like eye color and fingertip ridge count, are called **polygenic traits** because they result from the action of two or more genes with little or no environmental contribution.

Multifactorial traits run in families, but not according to Mendelian modes of inheritance. Some, like height, are **quantitative traits** in that they vary continuously over a range of measurement and display a normal (Gaussian) distribution curve. Others, such as cleft lip or schizophrenia, are **qualitative traits** with two

Human Genes and Genomes. DOI: 10.1016/B978-0-12-385212-0.00013-5

classes of people—affected or unaffected. A variety of mathematical tools are employed to determine that genes play a part in quantitative or qualitative multifactorial conditions, and to estimate the magnitude of that genetic contribution. These tools include **empiric study**, **intrafamilial correlation**, and **twin concordance**.

Identifying the particular genes involved and determining how they interact with one another and with the environment has proven to be much more difficult. Although there is not a single instance in which complete understanding of a multifactorial trait's biological basis has been produced, progress is being made. Such progress uses **candidate gene studies** (seeking mutations in genes whose function is relevant to the trait), **linkage analysis** (examining pedigrees for linkage between the trait and one or more genetic markers), and **genome-wide association studies** (in which the frequency of single nucleotide polymorphisms is compared in controls and those with a particular trait or disorder). As complete genome sequencing becomes more widespread and financially affordable, more comprehensive understanding is expected.

The following disorders are used to illustrate the key points made here: **bipolar disorder, Alzheimer disease, diabetes mellitus**, and **age-related macular degeneration.**

More than 2,000 different human disorders are caused by mutations of single genes or by chromosome abnormalities. Collectively, however, these two categories together have an incidence of 1.5% at birth and a population prevalence of about 2.5%. In sharp contrast, conditions that cannot be ascribed to mutations of single genes or discernable chromosome abnormalities but are caused at least in part by genetic factors have an incidence in newborns of more than 5% and a population prevalence of 60%. The remarkable latter statistic reflects the fact that most physiologic traits (height, weight, mathematical aptitude, artistic talent) and most common disorders of children (birth defects, diabetes mellitus, asthma, growth, and cognitive disability), and adults (coronary artery disease, Alzheimer disease, emphysema, high blood pressure, and cancer) are complex traits determined by multiple genes that interact with each other and with the environment. We call these complex traits **multifactorial** because more than one factor—genetic or environmental—contributes to the phenotype.

Multifactorial conditions tend to "run in families," meaning that close relatives are apt to have the condition more often than it is found in the general population. For example, consider the pedigree shown in Figure 13.1. Multiple individuals of both genders in multiple generations are affected with this common condition. Further inspection reveals the following: not every affected person has an affected parent; affected males have affected sons; affected females have unaffected children. Thus, we can exclude autosomal dominant, autosomal recessive, X-linked, or mitochondrial transmission. Cytogenetic studies, too, are unrevealing. In fact, this is a pedigree of bipolar disorder (also known as manic depression) in an Amish family. Bipolar disorder is a common multifactorial trait occurring in about 1% of the general population. It is distinctly more prevalent in the Amish sect in which individuals generally marry others within their small population subgroup.

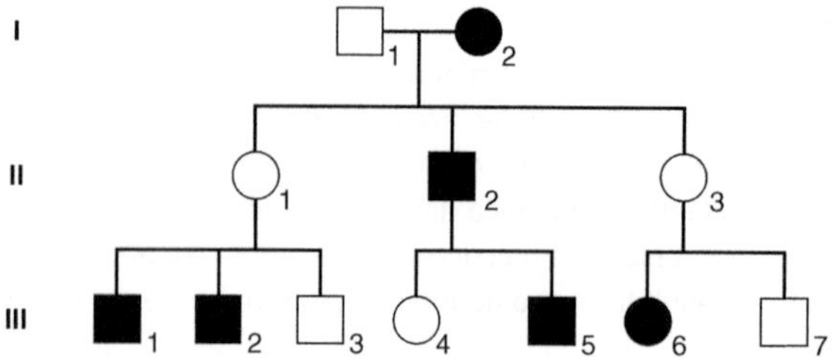

FIGURE 13.1
Three-generation pedigree of Amish family with bipolar disorder (manic depression). The distinguishing features of this pedigree compared to those for single-gene disorders are discussed in the text.

Multifactorial traits are not only much more prevalent than those due to single-gene mutations or chromosome problems, they are also much less well understood. It is no exaggeration to say that there is not a single multifactorial trait or disorder about which we understand fully the genes that contribute or the environments with which those genes interact. This ignorance makes multifactorial conditions arguably the most important area of research in human genetics. It is in this area that we expect the greatest contributions of human genomics. Soon you'll understand why we say this.

DEFINITIONS AND FORMULATIONS

Terms (Table 13.1)

Multifactorial (or, synonymously, complex) traits are those determined by two or more factors, often multiple genes interacting with each other or with the environment. Multifactorial traits determined solely by two or more genes (with no environmental component) are termed **polygenic** traits. Eye color and fingerprints are among the best examples of "pure" polygenic traits in humans. Multifactorial traits are generally grouped under the terms quantitative and qualitative. **Quantitative traits** vary continuously over a range of measurement from one extreme to the other, with no discernible breaks in between. Height, weight, and blood pressure exemplify this kind of trait. In contrast, **qualitative traits** are either present or absent; one has them or one doesn't. Examples include such physiologic traits as right- or left-handedness, birth defects such as cleft lip and adult disorders such as schizophrenia.

TABLE 13.1 Terms Used to Describe Multifactorial Traits and Disorders

Term	Definition
Multifactorial trait	One determined by two or several factors; often multiple genes interacting with each other or with the environment
Complex trait	Synonymous with multifactorial trait.
Polygenic trait	One determined by two or more genes with no environmental component
Quantitative trait	One that varies continuously over a range of measurements from one extreme to the other with no discernible breaks in between
Qualitative trait	One that is either present or absent

Models

THE NORMAL DISTRIBUTION

By definition, quantitative traits vary continuously over a range of measurement. Take height, for example. If one graphs the number of individuals in a population (y-axis) having a particular height (x-axis), the familiar bell-shaped (or Gaussian) curve is produced (Figure 13.2). This is called the normal distribution. The location of the peak of the graph and the shape of the curve are determined by two numerical quantities, the mean and the variance. The mean is the arithmetic average of all the values. Because more people have heights near the mean, the curve has its peak at the mean height for the population. The variance (or its square root, the standard deviation) is a measure of the degree of spread (of heights) on either side of the mean. The smaller the variance, the steeper is the curve. In many situations, the normal range is determined from the bell-shaped distribution. Thus, for height, basic statistical theory states that if height is normally distributed as in Figure 13.2, only 5% of the population will have heights more than 2 standard deviations above or below the population mean. This small percentage is said to have values outside the normal range.

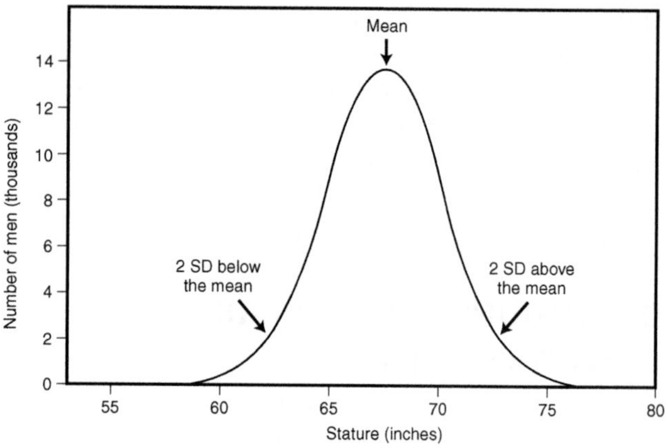

FIGURE 13.2
Normal distribution of height. Stature (in inches) on the x-axis is plotted against number of men (in thousands) on the y-axis. The mean is the arithmetic average from this sample of 91,163 English men. The variance and standard deviation are measurements of the degree of spread from the mean. The normal range is conventionally defined as the mean +/− 2 standard deviations.

EXPANSION OF MENDELIAN INHERITANCE

Can one approximate a bell-shaped curve for any phenotype using Mendelian terminology? The answer is yes. Assume that a small number of genetic loci (A, B, C) determine plant height. Each gene has two alleles (A_1, A_2; B_1, B_2, etc.). Assume first that height is determined by alleles at a single locus and that the allele with the subscript 1 contributes nothing to height and that with subscript 2 contributes one unit. If one carries out a simple Mendelian cross with pure-breeding A_1 and A_2 plants and self-crosses the F_1, three discrete phenotypes would be produced (Figure 13.3A): short, medium, and tall with a phenotypic ratio of 25% short, 50% intermediate, and 25% tall. If we assume instead that plant height is determined by two genes, each with two alleles (Figure 13.3B), the same types of crosses would yield five phenotypic classes ranging from short to tall with a fractional distribution of 6% at the low end, 6% at the high end and intermediate height classes making up 25%, 38%, and 25%, respectively. Further expansion of this example to three loci (Figure 13.3C), each with two alleles, yields seven phenotypic classes and a population distribution that resembles a bell-shaped curve. The more loci, the more alleles; the more alleles, the greater the resemblance to a normal distribution curve characteristic of a quantitative trait. This does not prove, of course, that the bell-shaped curve observed for human height, weight, and blood pressure is explained by the action of only a few genes. Rather, it says that relatively modest extension of Mendelian genetics can produce the kind of variation observed for many multifactorial traits.

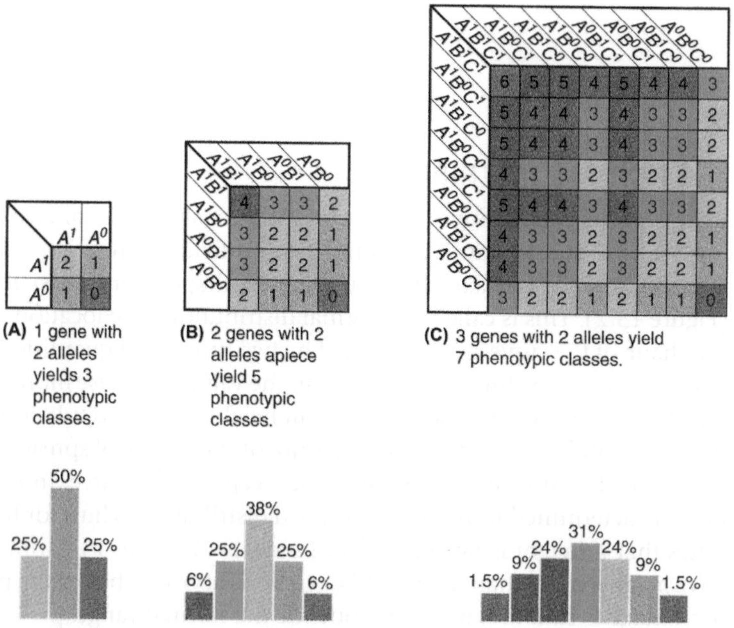

FIGURE 13.3
A Mendelian explanation of continuous variation. (A) Frequency distribution of height produced by a single gene with two incompletely dominant alleles. (B) Distribution produced by two genes, each with two incompletely dominant alleles.
(C) Distribution produced by three genes, each with two incompletely dominant alleles. See text for details.

THRESHOLD DISTRIBUTION

Qualitative traits have only two or, at most, a few phenotypic classes, yet their inheritance is presumed to be determined by the same kind of gene–environment interaction seen with quantitative traits. How can this situation, which is encountered for many common traits like cleft lip, congenital heart defects, and Alzheimer disease, be explained? A widely held theory proposes that each person has an underlying risk (or liability) for the qualitative trait in question and that this liability is distributed in a bell-shaped way (Figure 13.4). Although the underlying liability is not observed directly, it is assumed that if it exceeds a certain threshold, the trait will be expressed. This theory goes on to propose that the liability distribution reflects an increasing number of deleterious alleles, which ultimately produce the trait in question.

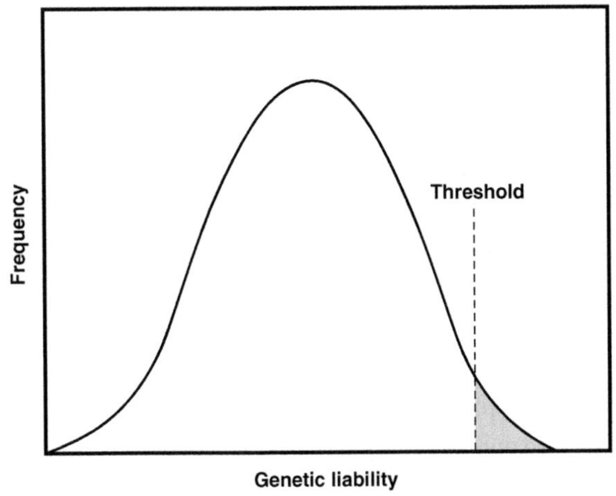

FIGURE 13.4
Threshold model of multifactorial inheritance. The model proposes that genetic liability is continuously distributed and that the qualitative trait will be expressed when liability exceeds a threshold. Additional details can be found in the text.

Critical Questions

To understand multifactorial traits and disorders, we must answer the following four questions (Table 13.2). First, how can we determine whether genes play a part in these common traits and disorders? Second, how can we estimate the magnitude of the role genes play for any given trait? Third, how do we identify the specific genes, the specific environments, and the interactions between them? Fourth, how can this information be used to advise individuals and families regarding prevention, diagnosis, and treatment? We will weave answers to these questions into the remainder of this chapter.

TABLE 13.2 Critical Questions Concerning Multifactorial Conditions

- How can we determine that genes play a part?
- How can we estimate the magnitude of the genetic contribution?
- How do we identify the specific genes, the specific environments, and the interactions among them?
- How can this information be used to advise individuals and families?

DETERMINING THE ROLE OF GENES

Quantitative Traits

The demonstration that genes play a role in complex traits, and the magnitude of that role, comes from family and population studies. These studies use intrafamilial correlations, empiric observations, studies of twins, and derived functions such as heritability. We'll demonstrate using particular examples and particular concepts.

AMPLIFICATION: POLYGENIC TRAITS

Keeping an Eye on Eye-Color Genes

Among the very first things we notice about people is the color of their eyes. Perhaps this is because eye color is more interesting than nose size or mouth shape or chin prominence, or perhaps it is because we usually look people in their eyes to detect clues to their character. Eye color ranges from light blue to dark brown (see figure below), with intermediate shades of amber, hazel, green, and grey. Brown eyes predominate among the world's people, being virtually the only eye color seen in Asia, Oceania, sub-Saharan Africa, and much of Central and South America. Blue eyes are most common in Scandinavia and in Northern, Eastern, Central, and Southern Europe. It is estimated that about one-sixth of the United States population has blue eyes.

What gives eyes their color? The simple answer is, melanin pigment in the iris. Soon after birth, melanocytes in epithelial cells of the iris begin to synthesize a black pigment called eumelanin and transport it to the stromal layer of the iris (consisting of fibrous, supporting cells). Eye color is determined by the amount of eumelanin present in the epithelial cells and by the thickness of the stromal layer. If the epithelial layers contain an abundance of melanin and the stromal layer is thick, most incoming UV light will be absorbed, and eye color will be perceived as brown. If the converse is true—that is, little pigment and a thin stromal layer—little UV light will be absorbed, and the eye will appear blue. All the intermediate shades reflect variations on this physiologic theme.

The genetics of eye color was thought formerly to be simple: it involved a single locus with two alleles—one dominant, the other recessive. Brown eyes, this idea held, were dominant, blue eyes, recessive. This is incorrect and oversimplified, as exemplified by the not infrequent circumstance in which two blue-eyed parents have a brown-eyed child. In fact, almost any combination of parental eye color can give rise to offspring with different eye-color shades. Eye color is polygenic. It is determined by a major locus on chromosome 15, abbreviated *OCA2*. People with two fully functional *OCA2* alleles have brown eyes. When both *OCA2* alleles are knocked out, very little melanin is produced, and the eye appears red (as in albinism). When the *OCA2* gene's function is reduced—either by point mutations within it or by a mutation in a nearby gene called *HERC2*, which controls expression of *OCA2*—various eye colors from blue to hazel may result. Further, genome-wide association studies have identified another six loci that influence eye color. Hence, the interplay between *OCA2* and a small orchestra of other genes ultimately is responsible for different eye colors.

One other point is of interest. People with blue eyes have a single common ancestor. About 6,000–10,000 years ago, a mutation in the *HERC2* gene arose in a person living near the Black Sea. This mutation regulated *OCA2* activity and impaired its ability to make brown eyes. Why did this mutation spread? What was its selective advantage, if any? Some have suggested, not facetiously, that blue-eyed people were considered more sexually attractive partners and therefore out-reproduced those with brown eyes. Other, more prosaic explanations, such as gene migration and inbreeding, have been advanced as well.

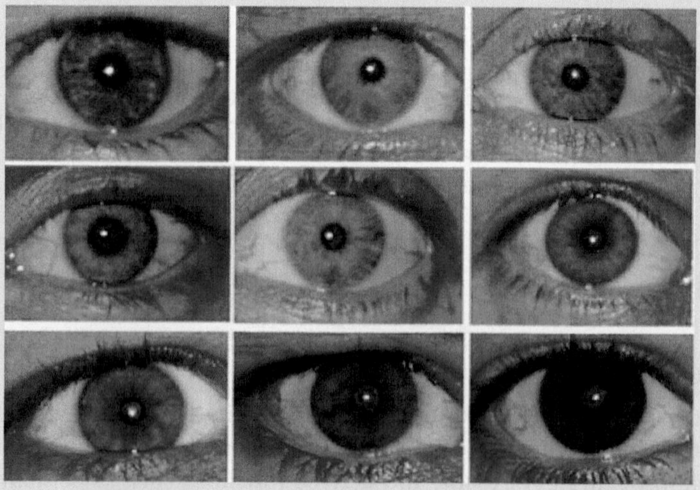

From Discover Magazine, *March 2007 issue.*

CORRELATION: FINGERTIP RIDGE COUNTS

Early in prenatal development, ridges appear on each of our fingertips. The number of ridges and the patterns they produce (fingerprints) change little thereafter. The total number of ridges on the 10 fingertips, counted under specific rules, constitutes an individual's fingertip ridge count. In a population, the individuals' total fingertip ridge counts conform to a normal, Gaussian distribution. The overwhelming role that genes play in this phenotype was determined by the observation that the degree of **correlation** of fingertip ridge count among paired relatives corresponded almost perfectly to the number of genes the pair shares in common. (A correlation is a statistical tool applied to a set of paired measurements.) Table 13.3 depicts the kind of data obtained. Monozygotic (MZ) twins, who share 100% of their genes, had a correlation of 0.95 for their total fingertip ridge counts. Dizygotic twins, siblings, and parent–child pairs, who share 50% of their genes, had ridge-count correlations of 0.49, 0.50, and 0.48 respectively. In contrast, spouses, who share none of their genes, had a correlation in total ridge count of 0.05. The almost perfect relationship between the correlation of ridge count with that defined by the number of shared genes, plus the normal distribution of total ridge count in the population, means that the variance in total finger ridge count is purely polygenic; that is, it is determined by multiple genes with no environmental contribution.

TABLE 13.3 Familial Correlation of Fingertip Ridge Counts

Relationship	Observed Correlation*	Expected Correlation**
Monozygotic twins	0.95	1.0
Dizygotic twins	0.49	0.5
Siblings	0.50	0.5
Parent/child	0.48	0.5
Spouses	0.05	0.0

**A correlation is a statistical tool applied to a set of paired measurements. A correlation of 1.0 means that there is complete agreement between the measurements in the members of the pair; a correlation of zero means that there is no relationship between the paired measurements.*

***The expected correlation is based on the proportion of shared genes.*

HERITABILITY: HEIGHT

Total fingertip ridge count is the only quantitative trait in humans that has been shown to be purely polygenic. All the others have intrafamilial correlations less than that defined by the fraction of shared genes, meaning that their variances are produced by a combination of genetic and environmental factors. Estimating the relative contributions of genes and the environment is difficult and imprecise. Take height as an example. As noted in Figure 13.2, in any given population height is a quantitative trait displaying a normal distribution. There may be 20 or more genes that affect height. Some are known, such as those for growth hormone, the growth hormone receptor, thyroid hormone, and insulin and its receptor. Many are not. Further, it is intuitively obvious that environmental influences also exist, including total caloric intake, protein and vitamin intake, and more.

To estimate numerically the influence of genes, we turn to a theoretical construct called heritability. **Heritability** is defined as the fraction of the total variance of a quantitative trait in a population that is caused by genes. It is a measure of the extent to which alleles at different loci contribute to phenotypic variation. Heritability is symbolized as h^2. The higher the h^2, the greater is the genetic contribution to the trait's variability. An h^2 value of 1 means that genes account for all of the phenotypic variance (and that the environmental contribution to the observed variance is negligible); an h^2 of 0 means that genes contribute nothing to the variance (implying that the variance reflects environmental factors exclusively). Mathematically, heritability may be estimated in two ways. In the first, h^2 is determined from intrafamilial correlation between measurements among pairs of relatives of known relatedness, such as parents and children, or sibs. In the second, twin pairs are employed, using the assumption that MZ twins share all their genes, while DZ twins share only 50%. The following formula is employed:

$$h^2 = \frac{(\text{Vp in DZ twins}) - (\text{Vp in MZ twins})}{(\text{Vp in DZ twins})}$$

where Vp is the variance in measurement observed between different kinds of twin pairs. If Vp chiefly reflects environmental influences, Vp in DZ twins will approximate Vp in MZ twins; the numerator (h^2) will approach 0. If Vp chiefly reflects genetic influences, Vp in MZ twins will approach 0 and h^2 will approach 1.

Using one or both of these methods, h^2 has been estimated for a number of quantitative traits arranged in decreasing h^2 values (Table 13.4). By inspection, variance in fingertip ridge count, height, maximum heart rate, and numerical ability largely reflect genetic influences while variance in weight, verbal ability, and total serum lipids is produced almost equally by genetic and environmental factors. Heritability has also been estimated in MZ and DZ twins separated at birth and reared apart. This should remove shared environment as a complicating factor in estimating h^2. When such a study was done estimating the variance in weight, h^2 was 0.7—to 0.8, values somewhat higher than those shown in Table 13.4 for twins reared together.

TABLE 13.4 Heritability of Human Quantitative Traits

Trait	Heritability (h^2)*
Total fingertip ridge count	0.95
Height	0.85
Maximum heart rate	0.84
Numerical ability	0.76
Weight	0.63
Verbal ability	0.63
Total serum lipids	0.44

Adapted from Hartl, D.L. and Jones, E.W. (2009). *Genetics*, 7th edn, Jones and Bartlett, San Francisco, CA.
**h^2 values are obtained from twin studies.*

We must recognize the limitations of heritability estimates. First, as just mentioned, family members share environments as well as genes, meaning that intrafamilial correlations may not reflect only genetic relatedness. This is particularly true for twin studies which assume that MZ and same-sex DZ twins reared together differ only in the fraction of genes they share; that is, that their environments are identical. This assumption may approximate the truth in young twins, but does not hold as twins grow and become increasingly independent. Second, h^2 estimates made in one population group do not necessarily agree with those in other groups with different diets, living conditions, and other environmental exposures. Third, h^2 values, even higher ones, give no information about the number of genes involved, their nature, or the interaction among them. High h^2 values suggest that the search for specific genes may be fruitful. Low values for h^2 suggest that environmental manipulation may affect the particular phenotype in a beneficial way.

Qualitative Traits

Estimating the contribution of genes to qualitative (or discontinuous) traits, such as cleft lip or neural tube defects, makes use of familial aggregation, twin studies, and empiric observations, just as studies of quantitative traits do. Yet some terms and approaches differ.

APPROACHES USED

Familial Aggregation

Disorders displaying qualitative phenotypes (that is, those in which the condition is either present or absent) tend to cluster in families. Such clustering suggests but does not prove that the condition has a genetic component. As mentioned previously, family members share environments as well as genes. Nonetheless, a variety of approaches can be used to point toward or away from a genetic contribution.

Concordance and Discordance

When two members of a family have the same trait or disorder, they are said to be **concordant**. If one member of the pair is affected and the other is not, they are termed **discordant**. Because twin pairs share their environments more than do other family pairs, concordance rates are generally estimated in MZ and DZ twins. The higher the concordance rate in MZ twins, and the greater the difference between MZ and DZ twins, the greater the genetic contribution is presumed to be. Concordance estimates cannot be equated absolutely with a genetic contribution because even such overwhelmingly environmental conditions as a bacterial infection can affect two members of a family (that is, can be concordant). Conversely, discordance in no way excludes a genetic contribution because the condition depends on the relative contributions of genes and the environment. If two members of a family have a comparable disease-producing genotype but a different constellation of environmental influences, they may exhibit discordance. Nonetheless, twin concordance of the type shown in Table 13.5 is a valuable indicator of genetic contribution. The values indicate, for example, that autism is overwhelmingly genetic in origin and that genes play little if any role in acne.

TABLE 13.5 Concordance Rates for Multifactorial Conditions Displaying Qualitative Phenotypes

Disorder	Concordance (%)* MZ	Concordance (%)* DZ
Autism	90	5
Alzheimer disease	78	39
Epilepsy (non-traumatic)	70	6
Bipolar disorder	62	8
Anorexia	55	7
Type 1 diabetes	40	5
Acne	14	14

**Concordance rates are arranged in decreasing order for MZ twins.*

Relative Risk (λ)

The familial aggregation of a disorder can be measured by comparing the frequency of the condition in the relatives of an affected proband (the first member of a family affected with a condition) with the frequency (or prevalence) of the condition in the general population. This leads to the notion of relative risk. **The relative risk ratio (λ_r)** is defined as follows:

$$\lambda_r = \frac{\text{prevalence of the disease in the relatives of an affected person}}{\text{prevalence of the disease in the general population}}$$

(The subscript "r" refers to relatives. Most estimates of λ employ sibs and use the notation λ_s.) The larger the value for λ_r or λ_s is, the greater the familial aggregation. A value of $\lambda_r = 1$ signifies that a relative of an affected individual is no more likely to have the disorder than is any individual in the population. In contrast, a value of 100 means that a relative is 100 times as likely to have the disorder as someone in the population. We'll now turn to some examples demonstrating how these analytical tools are used.

ILLUSTRATIVE EXAMPLE: CLEFT LIP

About 1 in 1,000 newborns in the United States and Europe is born with a cleft lip (Figure 13.5). A fraction of these children have a cleft palate as well as a cleft lip. A still smaller fraction has bilateral clefts of the lip and/or palate. Over the past 40 years, a number of studies have assessed the genetic contribution to this common malformation.

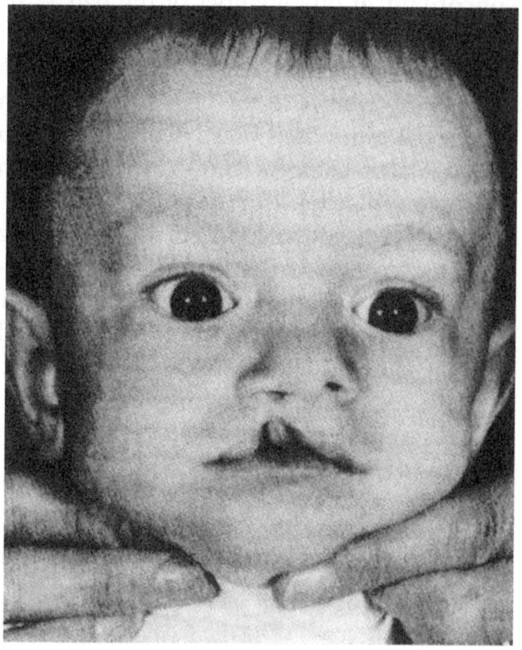

FIGURE 13.5
A child with unilateral cleft lip. In some children the cleft is bilateral and may affect the palate as well as the lip.

One such study was purely observational. It enrolled a large number of probands with cleft lip and counted the number of their family members who had this birth defect. The risk of occurrence or recurrence in family members is called the **empiric risk**. The kind of data obtained is shown in Table 13.6. Clearly, the frequency of cleft palate in relatives of affected children is much higher than in the general population (for example, 4.1% in sibs compared to 0.1% in the population). Noteworthy, too, is the sharp decline in the frequency of second- and third-degree relatives compared to first-degree relatives. These data do not conform to any Mendelian mode of inheritance, but they do indicate that genes are involved in the etiology of the conditions.

The relative risk (λ_r) was calculated from this kind of study and is shown in Table 13.6. Values far greater than 1.0 were seen consistently, again indicating a genetic component to this trait. Twin studies supported this conclusion as well. Concordance for cleft lip was 40% in MZ twins and 4% in DZ twins. If cleft lip was caused totally by genetic factors,

TABLE 13.6 Famly Studies of the Prevalence of Cleft Lip

Relatives	Affected Relatives (%)	Prevalence Relative to General Population*
First degree:		
Siblings	4.1	×40
Children	3.5	×35
Second degree:		
Aunts/uncles	0.7	×7
Nephews/nieces	0.8	×8
Third degree:		
First cousins	0.3	×3

Adapted from Carter, C.O. (1969). Genetics of common disorders, *Br. Med. Bull*. 25: 52, 1969.
**Prevalence of cleft lip in the general population is 0.1%. ×40 means 40-fold increase in sibs compared to that in the general population.*

one would have expected 100% concordance in MZ twins and 50% concordance in DZ twins. Hence, all these results lend support to the idea that cleft lip results from a still unknown set of genetic and environmental factors acting during early embryonic development.

Other observational studies of cleft lip support the multifactorial threshold model of liability discussed earlier in this chapter and depicted in Figure 13.6. This model makes a number of predictions about recurrence risks for conditions that are qualitative in their character.

- Recurrence risks such as those depicted in Table 13.6 represent average risks and will vary among different families. The model holds that affected individuals fall above a certain threshold in the population distribution of genetic liability, although their exact position is unknown. As shown in Figure 13.6, the model also predicts that recurrence risk will fall sharply as the degree of relatedness from the proband decreases. First-degree relatives, who share half their genes with the affected proband, will have a distribution of genetic liability that is shifted considerably to the

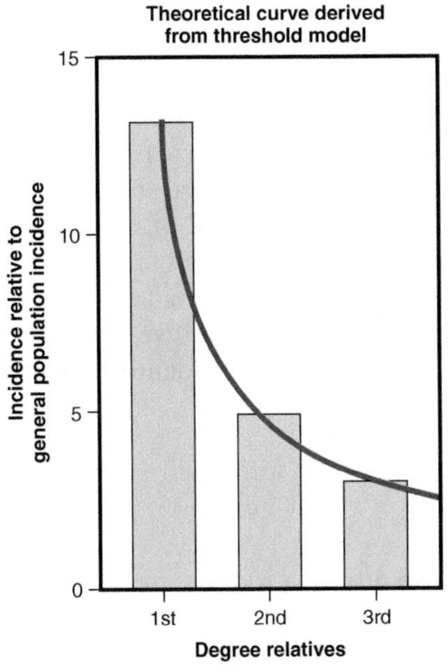

FIGURE 13.6
Theoretical curve of recurrence risk for a qualitative trait using the threshold model. Solid pink line approximates incidence in first-, second-, and third-degree relatives of proband (abscissa), relative to general population incidence (ordinate). See text for details.

right of that for the general population. Second-degree relatives, who share a quarter of the proband's genes, will have a distribution curve for genetic liability shifted back closer to the population mean. Third-degree relatives, sharing one-eighth of their genes with the proband, have a distribution even closer to that of the population. The result: the proportion of first-, second-, and third-degree relatives falling above the threshold will be progressively smaller. Because the normal distribution has geometric properties, one would expect a markedly increased risk to first-degree relatives, a slightly increased risk to third-degree relatives, and a risk to second-degree relatives that is moderately increased. This prediction is in keeping with the data in Table 13.6.

- The risk increases with the number of affected relatives. The empiric recurrence risk to a sibling of a child with cleft lip is about 4%. If there are two affected individuals in the family, however, the risk to another sib rises to more than 10%, suggesting that this family's liability curve is shifted further to the right of that in the population.
- The risk increases with the severity of the condition. The recurrence risk for a sibling of a child with unilateral cleft lip without cleft palate is about 2.5%. That risk rises to about 6% if the proband has bilateral cleft lip and cleft palate. Again, this is explained by proposing that the affected individual is shifted further to the right on the distribution curve.

These predictions and others do not prove the validity of the threshold model, but the empiric data that led to it offer an important quantitative tool when providing genetic counseling to families who have a child with a cleft lip or other relatively common multifactorial conditions, such as congenital heart defects or neural tube defects, or who have a common multifactorial disorder with adult onset, such as type 2 diabetes, bipolar disorder, or Alzheimer disease.

IDENTIFICATION OF SPECIFIC SUSCEPTIBILITY ALLELES

We have gone some distance in answering three of the four questions posed earlier. That genes play a part in common traits and disorders is supported by several kinds of studies: familial aggregation, correlations between phenotypes and shared genes, and twin concordances. That we can estimate how large a role genes play for any such trait is supported by the above kinds of studies and by estimates of heritability and relative risk. That we can use this information to advise and counsel families is made clear by the utility of the empiric risk estimates. But nothing said up to now addresses the fourth question: How much do we understand about the specific genes, the specific environments, and the gene–gene or gene–environment interactions. The simple answer: very little, but more than we did a decade ago. Some experts have said that progress has been slow, even halting, because the problem is inherently very difficult. If many genes increase risk or susceptibility to a given condition and if each gene exerts only a small effect, then finding the genes will be daunting, particularly if their action occurs in a network of interacting genes and environments. Others have said that our experimental tools have too little power to locate the genes involved. Nonetheless, some progress has been made and more will come, probably at an accelerating rate. Before presenting some illustrative "case" studies, we need to understand the general types of studies employed to identify susceptibility alleles.

Types of Studies

CANDIDATE GENE ANALYSIS

This approach carries out detailed analysis of genes whose protein products are known or thought to be involved in the disorder or trait in question. Finding that a particular allele (or mutation) occurs in a significantly greater frequency in patients than in controls suggests that the allele plays a role in the etiology of the trait. In a number of the multifactorial traits to be discussed subsequently, such **candidate gene** approaches have identified a subset of patients, small in each case, whose disorder follows Mendelian laws, thereby proving that its cause is mutation at a single gene locus. Obviously, each condition has its own list of logical candidates, such as the insulin gene in diabetes or the serotonin transporter in depression.

LINKAGE STUDIES

Linkage studies search for the co-transmission, in families, of a marker locus with the multifactorial trait. At its best, linkage analysis shows directly that a particular locus is involved with the condition or is very close to such a locus. As the Human Genome Project proceeded in the 1990s, a remarkably increased number of linkage markers were developed for each region of each chromosome. This armamentarium, now composed largely of millions of SNPs scattered throughout the genome, greatly increased the power of linkage studies, which have proven their value in many qualitative traits, including type 1 diabetes and macular degeneration.

ASSOCIATION STUDIES

While candidate gene approaches examine individual patients and linkage studies employ families, association studies seek to correlate, in populations, the occurrence of specific alleles with the trait or disorder being studied. They determine whether a given allele is found statistically either more frequently or less frequently in a group of affected individuals than in the general population. It is crucial to understand the nature of such **association studies**. They say that one event tends to occur when another event does. They do not say that one event causes the other.

Subsequent to the HGP's feat of sequencing the entire human genome, it was found that any two individuals differed at about 0.4% of their haploid genomic structure (that is, at about 12 million of their nucleotides). Most of this normal variation is in the form of changes in single nucleotides (see Chapter 6). A single nucleotide change is any substitution of one base pair for another; for example, some individuals will have an adenine (A) at a site, others a thymine (T). If such a single nucleotide change occurs in at least 1% of the general population, it is referred to as **single nucleotide polymorphism** (or **SNP**). Any SNP may be found on both homologues or just one. Thus, a SNP can be said to have two "alleles" and three genotypes, (A/A, A/T, and T/T, for example). SNPs occur throughout the genome at a frequency of about 1 in 800 base pairs. This means that each person has about 3 million SNPs in his or her haploid genome (or put another way, that any two individuals differ at 0.1% of their DNA sequence regarding this kind of common variation).

Most SNPs are neutral, meaning that they mostly occur in the non-coding region of the genome (for example, in introns or in sequences between genes) and are therefore not likely to be of functional significance. Others occur within exons where they may be **synonymous** (they result in no change in the polypeptide product) or **non-synonymous** (they *do* produce a single amino acid change). Because SNPs are so numerous, they suggested themselves as useful signposts for **genome-wide association studies (GWAS)**. To achieve such utility, **microarray chips** (also called "SNP chips") have been developed that can score up to 500,000 SNPs on a single microchip. The process is carried out as shown in Figure 13.7: a short oligonucleotide containing each SNP "allele" is synthesized and tagged with a fluorescent or radioisotopically-labeled probe; a short stretch of the genomic DNA to be tested is amplified by PCR; samples of the genomic DNA are fixed to a solid support; the tagged, allele-specific oligonucleotides are hybridized to the DNA; and the genotype is read by the fluorescence pattern or by autoradiography.

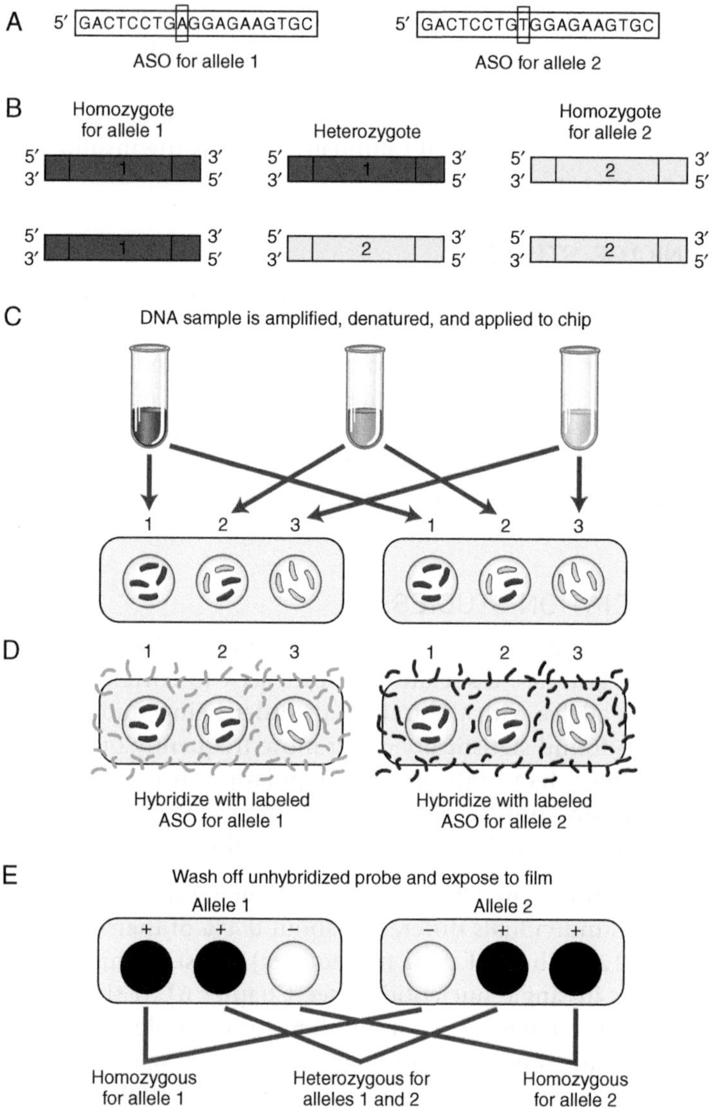

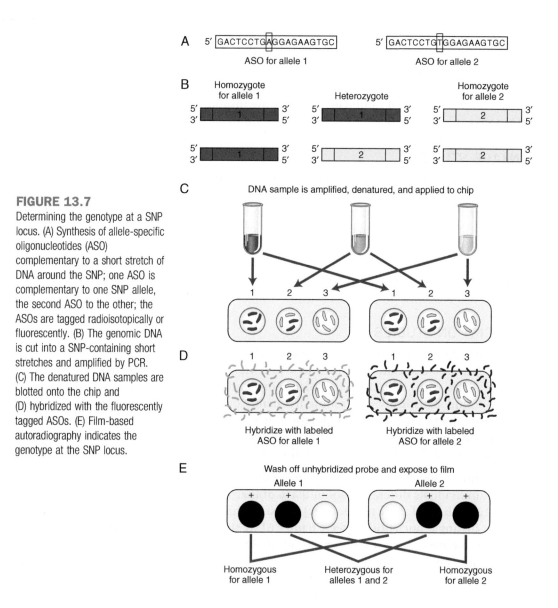

FIGURE 13.7
Determining the genotype at a SNP locus. (A) Synthesis of allele-specific oligonucleotides (ASO) complementary to a short stretch of DNA around the SNP; one ASO is complementary to one SNP allele, the second ASO to the other; the ASOs are tagged radioisotopically or fluorescently. (B) The genomic DNA is cut into a SNP-containing short stretches and amplified by PCR. (C) The denatured DNA samples are blotted onto the chip and (D) hybridized with the fluorescently tagged ASOs. (E) Film-based autoradiography indicates the genotype at the SNP locus.

Within the past 6 years, thousands of genome-wide association studies have been conducted using SNPs; these studies have been powered by the chip technology just described. Some of these studies have used hundreds of patients and controls matched for age, gender, and ethnicity; others have employed as many as 100,000 subjects. As you will see, genome-wide association studies have successfully identified major susceptibility alleles for a few disorders and minor susceptibility alleles for many other conditions. That is, in a few situations SNP studies have found associations that account for a large fraction of the genetic risk, while in the majority of diseases associated SNPs explain only a small fraction of the genetic risk. This raises the critical question of how to find and explain the remainder of that risk, called the "**missing heritability.**"

We will now present several case studies that amplify these general comments. Each disorder is manifestly multifactorial and common. Each has major health consequences. They differ, however, in how much we know about the genes that cause or predispose to them. We will start with a condition about which we've learned little and end with one where we've learned a great deal. In this way, the future scientific and clinical paths will become evident.

ILLUSTRATIVE EXAMPLES

Bipolar Disorder

Not surprisingly, a considerable number of common, serious mental illnesses are multifactorial. Included on the list are addiction to drugs or alcohol, anorexia, bipolar disorder, depression, obsessive-compulsive disorder, and schizophrenia. In each of these conditions (and others not mentioned) abnormalities in thinking, mood, or behavior are prominent. None of them has chemical, physical, or pathologic hallmarks, in contrast to other disorders of the brain such as Alzheimer disease, Parkinson disease, multiple sclerosis, or amyotrophic lateral sclerosis (ALS). We will focus on **bipolar disorder (BPD)**, but it should be understood that the important conclusions drawn would hold for many conditions mentioned above that affect thinking, mood, or behavior.

CLINICAL FINDINGS

As suggested by its name, BPD is characterized by abnormally wide mood swings, ranging from mania to depression. **Mania**, defined as an abnormal and persistently elevated, expansive mood lasting for 1 week or more, is accompanied by prodigious energy for work, feelings of grandiosity, decreased need for sleep, hypersexuality, buying sprees, and evidences of poor judgment. At the other pole of this mood disorder lies **depression**. Major depression (not to be confused with the occasional "blues") is characterized by pervasive sadness and loss of interest and pleasure in life lasting for at least 2 weeks. Other manifestations include difficulty in concentrating or making decisions, disturbed sleep, decreased energy, feelings of worthlessness, and recurrent thoughts of suicide. These extremes of mood may follow one another closely in time, but there also may be long periods of normal mood between the episodes of mania and depression. Manic episodes are more common in the summer; depressive episodes in spring and fall.

The onset of symptoms in BPD occurs most often between 18 and 25 years of age. The frequency and severity of mood swings tend to increase over time. Onset during childhood occurs as well, but distinguishing the mood swings of BPD from those of normal childhood development is difficult. It is widely held that fewer than half of BPD patients get diagnosed and treated. Of those who go undetected, 15–25% attempt suicide, unable to endure the excruciating psychic pain of depression. About half of those attempting suicide complete it.

ASSOCIATION WITH CREATIVITY

There exists a scientifically validated association between BPD and unusual creativity that bears mention. Notable writers (Poe, Melville, Ruskin), composers (Beethoven, Schumann, Morrison, Cobain), artists (van Gogh, O'Keefe, Pollock), statesmen (Alexander the Great, Napoleon, Churchill), and scientists (Newton, Freud, Luria) are believed to have been affected. The biological basis for this association is enigmatic, but it likely involves the expansive mood of mania and the decreased need for sleep that accompanies it, along with the gift we call talent or brilliance.

TREATMENT

Psychopharmacologic medications and psychotherapy offer patients with BPD unusually effective treatment. For more than 40 years, lithium carbonate has been the drug of choice. In addition, medications developed for other conditions of the central nervous system have proved beneficial (valproic acid and lamotrigine particularly). For completely unexplained reasons, a small fraction of BPD patients have a severe adverse reaction to selective serotonin reuptake inhibitors (SSRIs) characterized by rapidly cycling mania and depression, restlessness, and suicidal tendency. This reaction may be lethal if not recognized. Combining medications with psychotherapy has been shown to be more beneficial than either modality alone.

STIGMA

Given that the clinical signs of BPD are generally obvious and that effective treatment is at hand, why does such a large fraction of affected patients remain undiagnosed? The answer: stigma. Rather than seeing BPD and other mental illnesses as brain disorders in the same way that we accept disorders of the heart, lung, liver, and other organs, too many in our society still see mental illness as a sign of weakness, as imaginary rather than real. As long as this view persists, it will affect the job opportunities, careers, and personal lives of BPD patients, as well as their willingness to be identified. Genuine parity in acceptance and treatment of mental illness is not yet at hand, although real progress is being made to reduce—and ultimately erase—stigma.

GENETICS AND GENES

There is ample evidence from a number of directions that genes play a role in BPD (Table 13.7). The condition clusters in families, but not in recognizable dominant, recessive, or X-linked modes (see Figure 13.1). Empiric risk studies have shown that first-degree relatives of affected probands have about a 6–8% chance of being affected, compared to 1% in the general population. The relative risk measured in sibs of affected patients (λ_s) is about 7, in contrast with a λ_s of 1 for a purely environmental condition. Concordance in MZ twins is 62%, compared to 8% in DZ twins. Importantly, this marked difference is observed in twins reared apart as well as in twins reared together.

TABLE 13.7 Nature of Evidence Indicating Role of Genes in Bipolar Disease

- Increased risk to first-degree relatives (6–8%) compared with risk in general population (1%)
- Higher relative risk for sibs (λ_s) of ~7, compared to value of 1 for environmental condition
- Greater concordance in MZ twins (62%) than in DZ twins (8%)

Despite this plethora of evidence for genetic involvement, attempts to identify genes that cause or increase susceptibility to BPD have yielded disappointing results. For two decades, a number of linkage studies have been performed in large pedigrees, but no consistent localization(s) has been found. In the past few years, a large genome-wide association study has been conducted using hundreds of thousands of SNPs in 3,000 controls and 2,000 patients with BPD. A single SNP on chromosome 16 was found more often in patients than in controls. It explains only a small fraction of the heritability, however, and the locus in question has not yet been defined.

Why has it been so difficult to find "**mood genes**"? Two possible answers deserve mention. First, BPD may be a group of etiologically different conditions—a collection of syndromes rather than a specific condition. Given the absence of any specific diagnostic test or biomarker, and given the known presence of patients with unilateral depression and other affective disorders in families with BPD, this is likely. This answer would make data collection and interpretation of genome-wide association studies most difficult to carry out. Second, BPD may be caused by a large number of genes interacting among themselves and with the environment. This could mean that any SNP associated with the condition would require huge populations to find, and that each associated SNP would explain a very small amount of the genetic risk. Until we have more precise information, patients and families can rely on the empiric risk and concordance results to avail themselves of useful genetic counseling regarding occurrence and recurrence.

Alzheimer Disease

FREQUENCY AND EPIDEMIOLOGY

An estimated 5.4 million Americans of all ages have **Alzheimer disease (AD)** in 2011. This figure includes 5.2 million people aged 65 and older, and 200,000 individuals under age 65 who have early-onset AD. One in eight people over age 65 has AD. Nearly half of people over age 85 have AD. This means that AD is the most common cause of **dementia** in the elderly, responsible for at least half of all patients with dementia. The disorder is panethnic. In about 5% of cases, AD is familial, has an early onset (third to fifth decade) and is inherited as an autosomal dominant trait. In the remainder, onset is typically later than 60 years of age and is usually sporadic, although familial cases (without Mendelian inheritance) are seen not infrequently.

CLINICAL AND PATHOLOGIC HALLMARKS

AD is a slowly progressive, fatal neurodegenerative disorder. It is characterized by progressive loss of memory, accompanied by other cognitive impairments (in abstract reasoning, concentration, and visual perception). Some patients perceive their cognitive decline and become frustrated and frightened; others are unaware. Eventually, patients can no longer work and require supervision. Ultimately, most patients develop muscle rigidity, social withdrawal, loss of speech, and other neurologic manifestations. AD usually runs its fatal course in about 10 years, but this may vary from 2 to 20 years.

At present, definitive diagnosis can be made only at autopsy. The pathologic hallmarks of AD are extracellular, neuritic plaques (Figure 13.8, left) and intracellular neurofibrillary tangles (Figure 13.8, right) that are widely distributed in the cerebral cortex. The major constituent of these plaques is a small (39- to 42-amino acid) peptide, called amyloid beta (Aβ), derived from cleavage of a normal protein, the beta amyloid protein precursor (βAPP). A protein called tau is the major constituent of the tangles. Recently, new means of diagnosing AD have been reported, coming from studies employing magnetic resonance imaging (MRI) and cerebrospinal fluid. When confirmed, such studies will permit antemortem diagnosis, with its attendant benefits (and potential drawbacks).

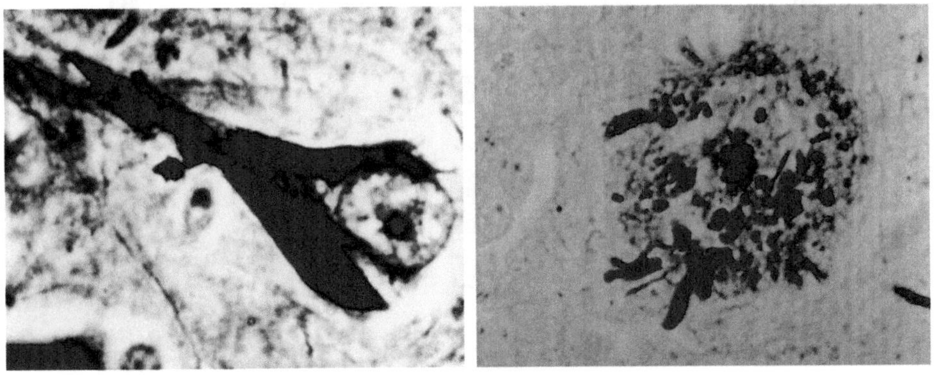

FIGURE 13.8
A neuritic plaque (left) and a neurofibrillary tangle (right) found on histopathologic examination of the brain of a patient with Alzheimer disease. The amyloid peptide is the major consitituent of plaques; the tau protein is the major constituent of neurofibrillary tangles.

TREATMENT

Although there are several medications approved for the treatment of AD, none of them has been shown to modify the course of the disease. Supportive and institutional care are the only modalities available.

GENES AND GENETICS

There is abundant evidence that genes play a role in the etiology of AD. This is most striking in early-onset AD (that is in the third or fourth decade of life), in which autosomal dominant inheritance is observed. In late-onset AD, the hallmarks of multifactorial inheritance are apparent: familial clustering; λ_s of 4—5; three- to four-fold increased risk in first-degree relatives of affected patients; and concordance of about 50% in MZ twins compared to about 18% in DZ pairs.

EARLY-ONSET AD

Mutations at one of three different loci cause early-onset AD. Each locus codes for a protein concerned with the expression of the beta amyloid precursor protein (Figure 13.9). The first locus encodes βAPP itself; the other two encode enzymes that produce cleavages at specific sites in βAPP. βAPP is a transmembrane protein that is subject to three distinct proteolytic events, catalyzed by the activity of three different proteases (α secretase, β secretase, and γ secretase). Normally βAPP is cleaved by α secretase to form an intramembrane product of 40 amino acids (AB_{40}). In one form of dominantly inherited AD, mutations within βAPP result in its aberrant cleavage to a 42-amino acid product (AB_{42}). In the other two dominantly inherited forms of AD, the mutations affect the β and γ secretase respectively, leading (again) to production of AB_{42}, a major constituent of the extracellular plaques that are toxic to neurons and which constitute the pathologic hallmark of AD. Most investigators believe that the common, late-onset form of AD also results primarily from some disturbance in cleavage of βAPP, but this remains unproven, even contentious.

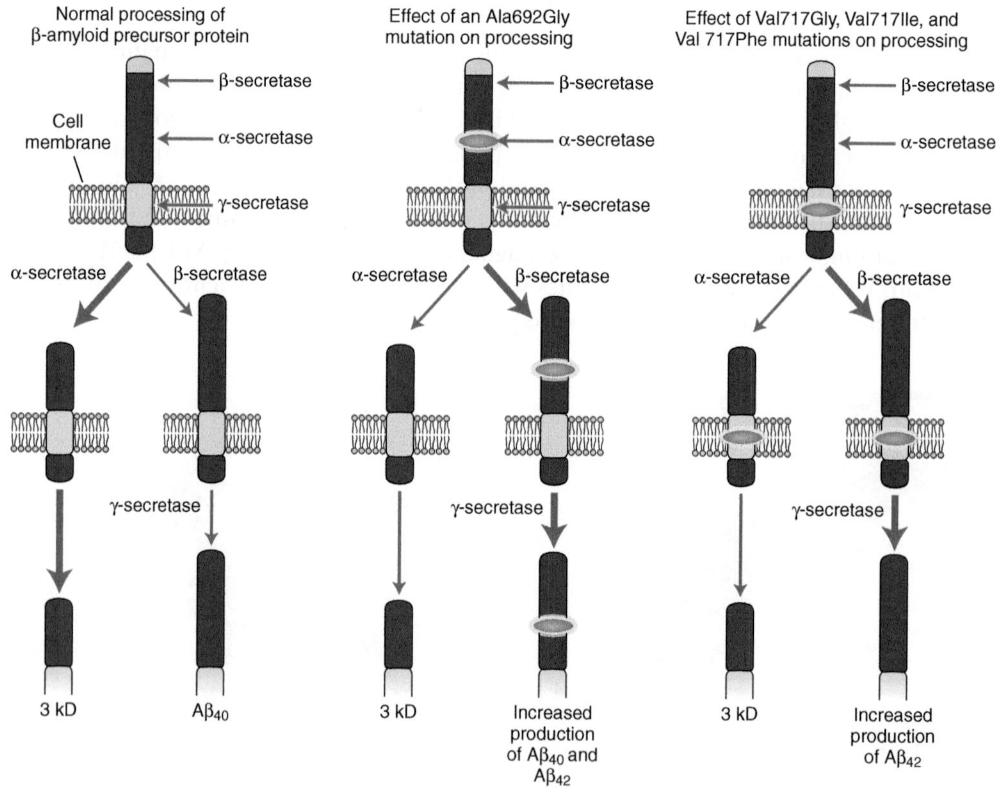

FIGURE 13.9
Normal processing of the β-amyloid precursor protein (βAPP) on the left and the effect of missense mutations in βAPP in the center and on the right. The mutations are denoted by the pink ovals in βAPP. Each mutation results in the formation of a toxic 42-amino acid peptide ($A\beta_{42}$), rather than the normal, 40-amino acid peptide $A\beta_{40}$. Other point mutations in β or γ secretase affect processing similarly, all yielding formation of $A\beta_{42}$.

LATE-ONSET AD

The most significant susceptibility locus for common, late-onset AD is that coding for **apolipoprotein E (APOE)**. APOE is a protein component of the low-density lipoprotein (LDL) particle that carries cholesterol into hepatocytes by receptor-mediated endocytosis. A variety of evidence revealed its relationship to AD: linkage analyses in late-onset, familial cases; the discovery that APOE is a component of the amyloid plaques in AD; and the finding

TABLE 13.8 Association of Apolipoprotein E Genotypes with Alzheimer Disease

Genotype	Frequency*			
	United States		Japan	
	AD	*Control*	*AD*	*Control*
ε4/ε4; ε4/ε3; ε4/ε2	0.64	0.31	0.47	0.17
ε3/ε3; ε3/ε2; ε2/ε2	0.36	0.69	0.53	0.83

From Nussbaum, R.L., McInnes, R.R. and Willard, H.F. (2007). *Thompson and Thompson, Genetics in Medicine*, 7th edn. W.B. Saunders, Philadelphia, PA.

**Frequency of genotypes with and without the ε4 allele among Alzheimer disease (AD) patients and controls from the United States and Japan.*

that APOE binds to the AB peptide. The APOE locus has three common alleles, ε2, ε3, and ε4. When the genotypes at the APOE locus were analyzed in AD patients and controls (Table 13.8), a genotype with at least one ε4 allele was found more than twice as frequently in AD patients as in controls, in populations from the United States and from Japan. More impressive, as shown in Figure 13.10, is the effect of genotype on age of onset in AD. Patients homozygous for ε4 had an earlier onset than those with any other genotype. By age 80, only 10% of patients with two ε4 alleles remained disease-free. While it is clear that ε4 is an important susceptibility allele, it must be emphasized that many ε4/ε4 homozygotes don't develop AD and that 50 to 75% of all heterozygotes with one ε4 allele remain disease-free. It has been estimated that the ε4 allele may constitute 30–50% of the genetic risk for AD. The remaining susceptibility loci are being sought actively using genome-wide association studies.

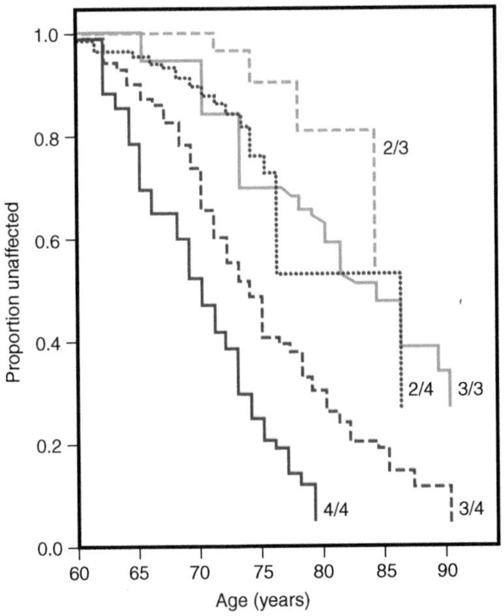

FIGURE 13.10
Effect of ApoE genotypes on chance of not developing Alzheimer disease as a function of age. At one extreme, ε4/ε4 homozygotes (solid red line) have less than a 10% chance of not being affected by age 80 years. At the other extreme, those heterozygous for ε2/ε3 (dashed green line) have more than 80% chance of not being affected by age 80 years.

The APOE locus is the first major susceptibility gene for late-onset AD, but a large genome-wide association study employing 100,000 patients and meticulously matched controls from all over the world has identified five additional loci conferring susceptibility for AD. Interestingly, some of these loci involve genes regulating cholesterol metabolism, others inflammatory responses. It is too early to know whether each of the susceptibility loci identified acts alone or in concert with one another—and with APOE. It is premature as well to discern whether the new findings will lead to new diagnostic tests or therapies.

GENETIC COUNSELING

In the three early-onset forms of AD demonstrating autosomal dominant inheritance, each child of an affected patient has a 50% chance of developing AD. Molecular testing for these mutations is available.

Regarding late-onset AD, counseling relies on empiric risk estimates. Individuals with an affected first-degree relative have a three- to four-fold increased risk. In the presence of two affected first-degree relatives, the risk is seven to eight times higher than in the general population. APOE testing could be done to identify those with ε4 genotypes, but in the absence of any effective treatment, and knowing that the testing would produce uncertainty, it has not been deemed advisable to carry out such genotyping.

Diabetes Mellitus

Diabetes mellitus (DM) is a disorder of carbohydrate metabolism characterized by impairment of insulin formation, secretion, or physiologic action. DM is among the most studied and observed of all human disorders; its earliest descriptions date back to ancient Egyptian and Greek tracts. Its name means "sweet flow" in Latin, almost surely reflecting the presence of sugar in the urine. The chemical hallmarks of DM have been written about for centuries: elevated blood glucose (hyperglycemia), glucose in urine (glycosuria), excess urine output (polyuria), and dehydration leading to coma and death.

More than a century ago, it was shown that insulin was produced in the pancreas, an organ with both digestive and endocrine functions (Figure 13.11). Within the pancreas, there are small collections of cells called the islets of Langerhans. Insulin is produced in one cell type (β cells) within these islets, and these β cells are the sole source of the body's insulin. A succession of brilliant experiments established that insulin could be extracted from the pancreas, that patients with DM lacked insulin, and that insulin administration corrected the abnormalities in glucose metabolism mentioned above.

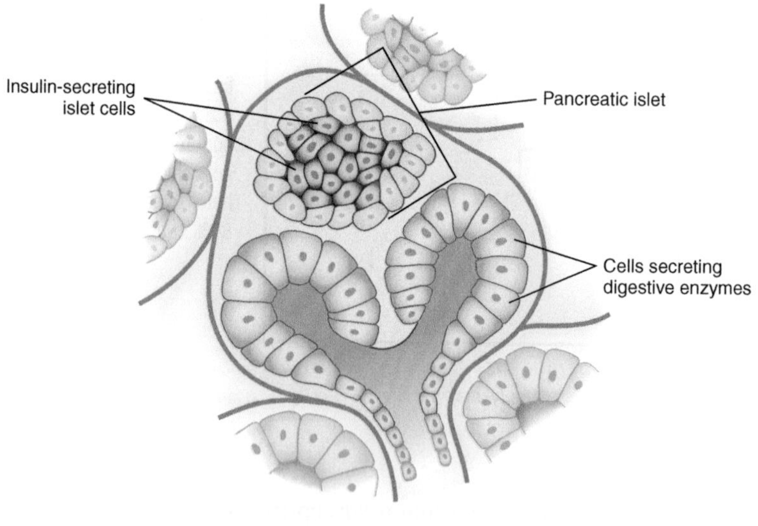

FIGURE 13.11
Anatomic drawing of the pancreas focusing on the islet cells that secrete insulin. Adjacent to the islet is a portion of the pancreas that contain cells which secrete digestive enzymes needed for breaking down proteins, fats, and carbohydrates.

Despite knowledge of insulin's role in DM and insulin's ability to correct the homeostatic dysfunction in DM, the delayed consequences of DM continue to plague patients and frustrate physicians. Three organs particularly are the targets of these late effects. Damage to the retina (retinopathy) is common, and its progression can be used as a gauge of DM's progression (Figure 13.12). Progressive destruction of renal glomeruli (the kidney's filtering system) is the most common indication for renal dialysis and transplantation. Damage to peripheral nerves is a prominent cause of sensory and motor dysfunction.

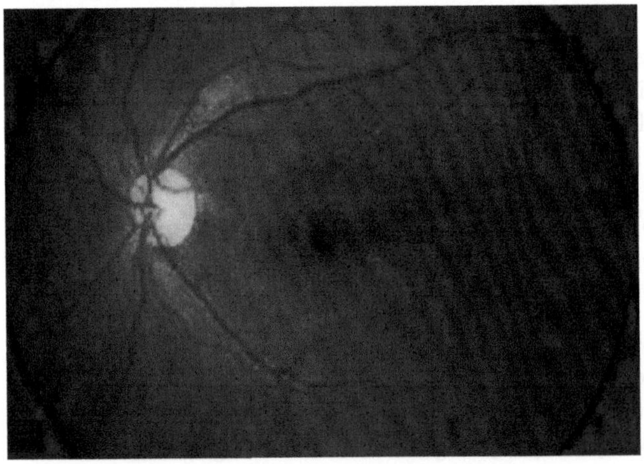

FIGURE 13.12
Pathologic consequences of diabetes mellitus on the retina. The pale yellow spots scattered across the retina are called diabetic exudates. This kind of retinal damage may proceed to the development of hemorrhages, with accompanying severe visual impairment. Also shown are the optic nerve head in bright yellow, and blood vessels (red) radiating from it.

CLINICAL CLASSIFICATION: TWO TYPES

DM is subdivided into two forms: type 1 (insulin-dependent diabetes mellitus or IDDM); and type 2 (non-insulin-dependent diabetes mellitus or NIDDM). These two types are abbreviated T1D and T2D. We will discuss them individually because they differ so much in etiology, epidemiology, genetics, and treatment (Table 13.9).

TABLE 13.9 Comparison of Type 1 Diabetes (T1D) and Type 2 Diabetes (T2D)

Parameter	T1D	T2D
Frequency	0.4%	8%
Age of onset	Childhood/adolescence	Adult
Affected first-degree relatives	Uncommon	Common
Concordance:		
MZ twins	33–50%	69–90%
DZ twins	1–14%	24–40%
Plasma insulin	Reduced to absent	Normal to elevated
Mode of treatment	Insulin	Oral hypoglycemic agents

T1D

This form of DM is caused by autoimmune destruction of insulin-secreting β cells in pancreatic islets. For as-yet poorly understood reasons, the body's immune system recognizes these cells as foreign (or "non-self") and acts to destroy them as it would a bacterium or virus. Typically, T1D has its onset during the first decade of life (which is why it is sometimes called juvenile-onset DM) at which time the insulin-secreting cells are damaged and ultimately destroyed. From then on, patients are dependent on daily injections of insulin. Good control of blood glucose concentrations corrects the disturbed glucose metabolism and mitigates damage to the eyes, kidneys, and peripheral nerves.

Frequency and Epidemiology. T1D accounts for about 5% of patients with DM. It has a prevalence of about 1 in 2,500 Caucasians at 5 years of age, increasing to 1 in 300 by age 18. Its frequency is lower in African Americans and Asians.

Genetic Susceptibility. Multiple observations indicate genetic predisposition in T1D. Ten percent of patients have at least one affected sibling. Concordance among MZ twins is about 40%, far exceeding concordance in DZ twins (5%). The risk for T1D in sibs of an affected proband is about 7%, resulting in an estimated λ_s of about 35. The prevalence in some population groups differs considerably from that in other groups. One major and several minor susceptibility genes have been established. The first, and most important, involves the major histocompatibility complex (MHC). The MHC consists of a large cluster of genes on the short arm of chromosome 6. MHC genes are remarkably polymorphic, meaning that multiple alleles at a given locus are the rule. Within the MHC are two classes of genes, class I and class II, which correspond to the human leukocyte antigen (HLA) genes important in tissue transplantation. The HLA genes encode cell-surface proteins that play a critical role in the initiation of an immune response by complexing with antigens and "presenting" them to lymphocytes capable of destroying such antigens or stimulating an antibody response to them. Hundreds of different alleles of HLA genes have been identified, and their association with a number of autoimmune disorders has been shown.

In T1D, 95% of all patients were shown to be heterozygous for HLA DR3 and DR4 alleles, compared to 50% of those in the general population. Subsequent work showed that DR3 and DR4 are not susceptibility alleles; rather, they are closely linked to other alleles called DQ. Two DQ alleles increase susceptibility to T1D, perhaps by single amino acid changes in their encoded proteins that may bind to islet cells inadvertently and mark them for destruction. Population studies suggest that these HLA alleles may account for 30–to 60% of the genetic risk in T1D.

Other susceptibility genes have been claimed for T1D, including changes in the insulin gene itself, but these claims remain unconfirmed. Based on the finding that T1D concordance in MZ twins is only about 40%, environmental factors must be involved in this disorder, but their nature remains unclear. Some have suggested that "mimicry" between viral antigens and pancreatic islet cell-surface proteins may be responsible for initiating the autoimmune response.

Genetic Counseling. Use of HLA typing, coupled with empiric risks, offer families with T1D some information but not nearly enough (Table 13.10). For example, the risk to siblings (7%, empirically) can be sharpened considerably using HLA typing: if a sibling has no HLA DR haplotype in common with the proband, the risk for T1D is 1%; if the sib has two DR haplotypes in common, the risk is 17%. In the future, the ability to prevent autoimmune destruction of β cells would make information about individual risks to first-degree relatives more useful.

TABLE 13.10 Genetic Counseling in Type 1 Diabetes (T1D)

Relationship to affected individual	Risk for T1D (%)
MZ twin	40
Sibling	7
Sibling with no shared DR allele	1
Sibling with one shared DR allele	5
Sibling with two shared DR alleles	17
Child	4

Adapted from Nussbaum, R.L., McInnes, R.R. and Willard, H.F. (2007). *Thompson and Thompson, Genetics in Medicine*, 7th edn. W.B. Saunders, Philadelphia, PA.

T2D

This form of DM, sometimes called adult-onset DM or non insulin-dependent DM, differs from T1D in several important ways (Table 13.9): it is much more prevalent; it usually does not appear until adolescence and often much later; it is more common in African Americans and Native Americans than in whites; it is characterized by resistance to insulin's action, not lack of insulin production; it is commonly associated with obesity; it usually responds to dietary treatment and oral hypoglycemic agents; and genetic factors are of greater significance in its

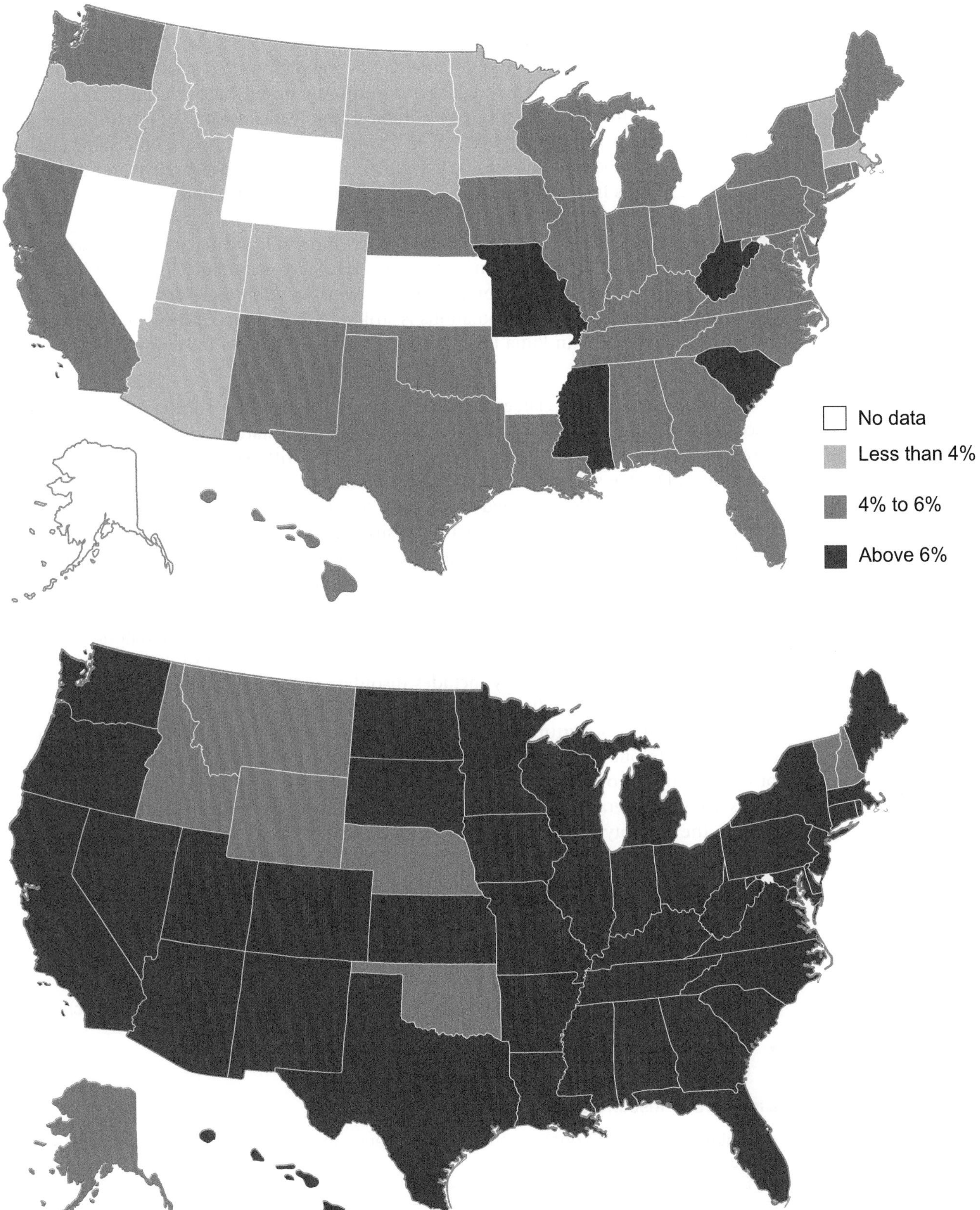

FIGURE 13.13
The dramatic increase in the prevalence of type 2 diabetes (T2D) in the United States between the year 1990 (top) and the year 2000 (bottom). Note the marked reduction in the number of states with a prevalence less than 4% and the striking increase in states with prevalence greater than 6%.

etiology. As with TID, good control of glucose metabolism can delay the onset of complications.

Frequency and Epidemiology. About 8% of the US population (that is, more than 25 million people) have T2D. Its frequency is higher in African Americans, Mexican Americans, and Native Americans than in whites. In one extreme example, nearly 50% of Pima Indians have T2D. T2D occurs throughout the world, with an estimated 400 million people being affected. Startlingly, the number of adults in the US with T2D doubled between 1990 and 2000 (Figure 13.13). This increase mirrors the rapid increase in the prevalence of obesity in this country (and in others).

Genetic Susceptibility. A variety of pieces of information indicate that a genetic susceptibility is prominent in T2D—more prominent than in T1D. MZ concordance for T2D is 70–90%, far greater than the 25–40% concordance in DZ twins. In addition to familial clustering, prevalence of T2D in different populations is striking: from 1% in Chinese and Mapuche Indians in Chile to 50% in Pima Indians in the southwest US and in Nauru tribes in the South Pacific.

In T2D, we recognize two classes of genes: those that cause T2D and those that increase susceptibility. Table 13.11 lists a number of single-gene mutations that cause T2D. They were identified, as expected, from examining pedigrees with Mendelian patterns of inheritance for T2D. One subset is called **MODY** (maturity-onset diabetes of the young). This group of disorders is distinguished by its early onset, autosomal dominant mode of inheritance, and lack of association with obesity. Candidate gene and linkage studies have revealed at least six different genes and gene products with demonstrable mutations. The first was glucokinase, an enzyme that is rate-limiting in glucose breakdown in cells. Others are transcription factors expressed in pancreatic β cells. Their localization implies that they influence insulin production, secretion, and regulation. Precise information about the genes these transcription factors regulate is lacking. Two other categories of genes (beyond MODY) are known. One category includes disorders of insulin action in muscle and elsewhere. The second is a disorder of mitochondrial DNA in which T2D is part of a broader spectrum of perturbed energy metabolism. These monogenic causes of T2D have added greatly to our information about insulin sensitivity and regulation, but they account for only about 2% of patients with T2D. This says clearly that the vast majority of T2D results from the interaction of genes and the environment, with no single factor sufficient to produce the phenotype.

TABLE 13.11 Single-Gene Mutations Causing T2D

Maturity Onset Diabetes of the Young (MODY)
• Mutations of the glucokinase gene • Mutations of five to six different regulatory transcription factors expressed in pancreatic β cells • Autosomal dominant inheritance • Early age of onset
Disorders of Insulin Action
• Mutations of several genes resulting in resistance to insulin's actions in liver and muscle • Autosomal recessive inheritance • Early age of onset
Defect in Mitochondrial DNA
• Mutation perturbs cellular energy metabolism • T2D is part of a broader spectrum of phenotypes • Maternal inheritance

One of the hopes of the HGP was to identify genes in common disorders like T2D that increase susceptibility in a large way. This hope has not yet been realized for DM. Linkage studies using large pedigrees have been disappointing because loci linked to T2D couldn't be replicated by subsequent studies. Genome-wide association studies have been undertaken in recent years that have identified more than 10 susceptibility genes. One of them, *TCF7L2*, is a transcription factor that regulates secretion of glucagon, a β-cell hormone that opposes the action of insulin. Another, *PPARγ*, is a member of a family of nuclear receptor genes that regulate fat cells. In each case, the genome-wide association study indicated that the contribution of the susceptibility gene identified was very small, meaning that it explained a very small fraction of the genetic risk. Thus, most of the heritability for T2D is missing. Will other such studies find it, or will complete sequencing of patients' genomes be required? Some geneticists have called DM "a geneticist's nightmare." While it may no longer be a nightmare, it remains far from a good dream.

Counseling and Management. In the absence of more information about susceptibility genes, management of T2D depends on several kinds of medications and, perhaps more important, on controlling body weight with the help of diet and exercise. This is of particular importance to family members of affected patients, whose risk for T2D is very high: 25% in a sibling or child of a patient with T2D and more than 70% in a MZ twin of an affected individual.

Age-related Macular Degeneration

The fourth common, serious multifactorial disorder is of particular importance to those doing genetics and genomics. **Age-related macular degeneration** (**AMD**) is the best example to date in which genome-wide association studies have identified particular variants responsible for a large fraction of the genetic risk (in contrast to the other three conditions we've discussed). Thus, its path of discovery is worth noting in some detail.

FREQUENCY AND EPIDEMIOLOGY

AMD is among the most common causes of severe visual impairment and blindness in older adults. While it is very rare in people under age 55, more than 10% of people over age 75 have early or later stages of AMD, meaning that it afflicts approximately 10 million Americans. The disorder usually occurs sporadically, but familial cases occur not infrequently.

CLINICAL AND PATHOLOGIC FEATURES

Patients with AMD usually complain first of bilateral visual impairment. Typically, this impairment affects central vision, making reading and driving difficult. Over time, central blind spots appear which may progress to near total blindness. Diagnosis is made by ophthalmoscopy, which shows degenerative changes in the macular region of the retina, that portion responsible for the greatest visual acuity. AMD is subdivided pathologically into two forms, "dry" and "wet." The dry form, seen most often in mildly affected people, is characterized by the presence of drusen, which are large, soft, yellow deposits occurring beneath the retina in and around the macular area (Figure 13.14). As AMD progresses, retinal degeneration proceeds and new blood vessels form beneath the retina and extend into it, where they may bleed and fracture, thereby producing wet AMD and, with it, much greater visual loss.

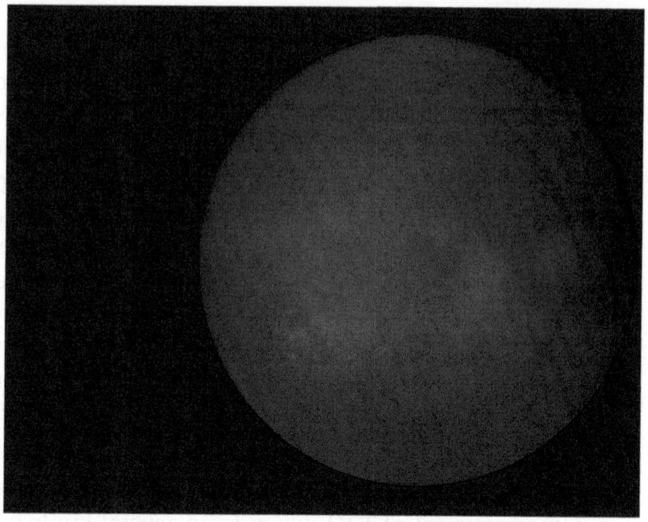

FIGURE 13.14
Appearance of the retina in the "dry" form of age-related macular degeneration. The soft yellow deposits are called drusen, which collect around the macular area—that portion of the retina responsible for the greatest visual acuity.

There is no specific treatment for AMD. The dry form doesn't require intervention. The wet form may respond temporarily to laser coagulation of the new blood vessels or to intra-ocular administration of an antibody to vascular endothelial growth factor (VEGF), a protein that promotes the growth of small blood vessels. It is too early to know whether the latter treatment can forestall visual loss in the long run.

MAJOR SUSCEPTIBILITY GENES

It has been known for some time that genes play a significant role in the pathogenesis of AMD. The relative risk in first-degree relatives of affected patients is about 4; concordance in MZ twins is about 40%, at least double that in DZ twins. Further, between 5% and 10% of patients have an early onset of the disease caused by monogenic defects in the *ABCR* gene, which encodes a protein required for the transport of the vitamin A derivative, retinol, into the retina. This knowledge prompts a question akin to those we've discussed throughout this chapter: What about the other 90% of patients?

In 2006, answers began to emerge. Candidate gene screening in search of variants in the *ABCR* gene was disappointing, but linkage studies in several large pedigrees implicated a region on the long arm of chromosome 1(lq31). Genome-wide association studies were then carried out independently by three groups. The first used 100,000 SNPs in 96 AMD patients and 50 controls (a remarkably small sample in retrospect). All of their patients and controls were white, none were Latinos; all patients had dry AMD; controls were carefully matched for age and gender. A strong signal was detected in chromosome lq31; that is, a SNP was found with much greater frequency in patients than in controls (Figure 13.15). This association was highly significant statistically ($P < 0.001$). These researchers then saturated lq31 with many more SNPs. Two other groups of investigators employed a similar experimental strategy, with similar findings.

A second SNP associated with AMD was subsequently found. Using data from the HapMap (a set of haplotypes distributed throughout the genome, usually inherited as a unit), these two SNPs were found to be linked in a region of lq31. Ultimately, both SNPs were found to lie in an intron of the gene coding for complement factor H (CFH). Further searching detected a SNP in the coding region of CFH that was non-synonymous, substituting a histidine for a tyrosine residue (Tyr402His). The significance of this observation became clear soon thereafter: those people heterozygous for Tyr402His had a two- to four-fold higher risk of developing AMD;

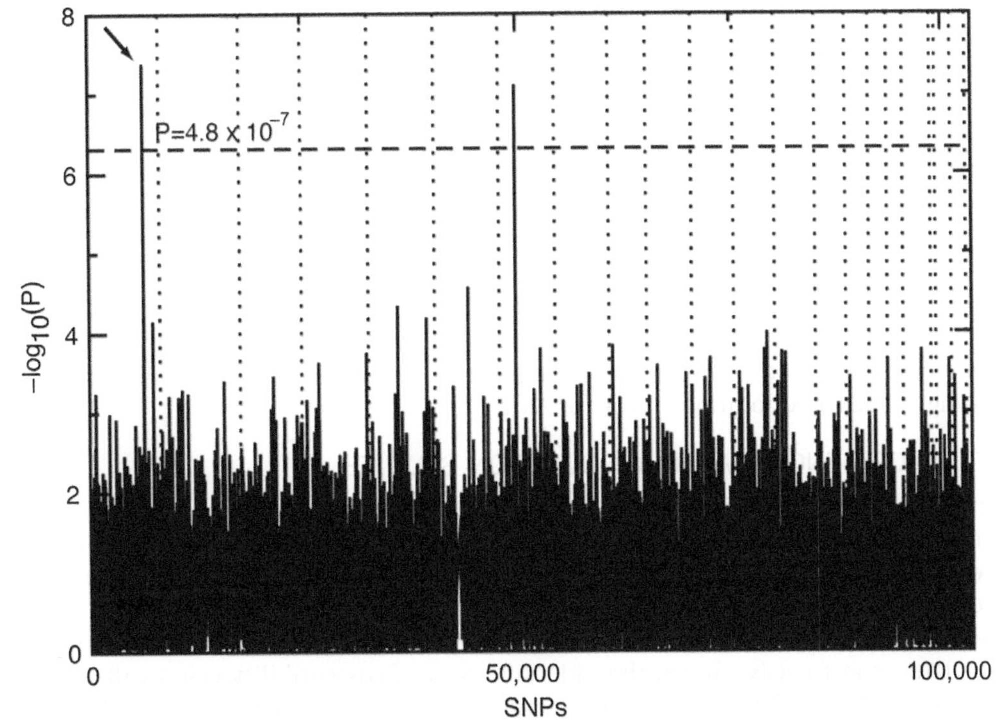

FIGURE 13.15
Genome-wide association study (GWAS) in age-related macular degeneration (AMD). 100,000 SNPs (abscissa) were evaluated in patients and controls. The dotted lines represent boundaries between chromosomes, which you can imagine as lined end-to-end along the abscissa. The SNP marked with the arrow was highly statistically more frequent in patients than in controls. This SNP led to other studies culminating in the discovery that the complement factor H gene had a mutation producing an amino acid substitution at residue 402 (tyr402his), which explains much of the genetic risk for AMD. See text for details.

those homozygous for Tyr402His had a five- to seven-fold increased risk. This variant accounts for 40–45% of the total genetic risk for AMD.

The biochemical basis for the increased risk of Tyr402His has been probed. CFH is a negative regulator of the complement cascade, which is involved with inflammation. CFH is found in drusen and around drusen. It seems likely, then, that CFH with the Tyr402His substitution is less effective at controlling inflammation than is normal CFH.

Subsequent work has identified other SNPs associated with AMD. One study identified a polymorphism in the promoter region of a serine protease called HTRA1, which is expressed in the retina. As shown in Table 13.12, the two variants just discussed increased risk synergistically—homozygosity for the polymorphisms in CFH and HTRA1 increased the odds ratio for AMD more than the product of the effects of each variant alone.

TABLE 13.12 Synergistic Risk for Age-Related Macular Degeneration Produced by Susceptibility Alleles at Two Loci*

	Complement Factor H	HTRA1 Promoter		
		GG	AG	AA
Susceptibility alleles	TT	1.0	1.8	3.4
	CT	1.1	2.3	7.3
	CC	3.1	4.0	31

**Homozygosity for the AA allele of HtRA1 increased the odds ratio (OR) for AMD to 3.4; homozygosity for the CC allele of CFH increased the OR to 3.1; homozygosity for both alleles increased the OR to 31.*

Equally interesting, recent studies have identified variants in other components of the complement pathway that are associated with decreased (not increased) risk for AMD. Much remains to be learned about this important system: How do CFH and HTRA1 interact? What is the normal function of HTRA1? What accounts for the protective effects of some complement factor variants, and the negative effect of others? These questions won't be answered easily, but they are precisely the kind of questions scientists need and want to address as we seek a greater understanding of this common condition and others like it.

ENVIRONMENTAL RISK

There is ample evidence that cigarette smoking increases the risk of developing AMD. The mechanism underlying this risk factor is unknown. Might smoking effect inflammation in the eye, for example by interacting with the complement cascade? Whatever the mechanism, smoking is ill advised for those at risk for AMD.

FUTURE OF GENOME-WIDE ASSOCIATION STUDIES

As mentioned elsewhere in this chapter, genome-wide association studies have achieved notable success in identifying major genetic risk factors in AMD and in a few other diseases, but not in other common, multifactorial conditions. Some believe that other disorders will be added to the list. Others posit that most multifactorial conditions result from the action of many rare variants rather than a few common ones. If the latter construct is correct, then identification of susceptibility genes will require either huge GWAS of the kind already mentioned for Alzheimer Disease or full genome sequencing.

REVIEW QUESTIONS AND EXERCISES

1. Choose the phrase in the right column which best matches the term in the left column.

a. empiric risk	1. phenotype reflecting gene–environment interaction
b. OCA2 and HERC2	2. disorder where GWAS defined large fraction of genetic risk
c. candidate gene	3. trait determined by two or more genes
d. autism spectrum	4. varies continuously over range of measurements
e. polygenic	5. compares frequency in patients with that in controls
f. qualitative trait	6. loci controlling eye color
g. correlation	7. value >1 indicates involvement of genes
h. single nucleotide polymorphism (SNP)	8. allows estimation of mean and variance
i. normal distribution	9. compares frequency in family members with that in general population
j. heritability	10. risk above which abnormal phenotype is observed
k. multifactorial	11. statistical tool applied to set of paired measurements
l. association study	12. major risk factor in Alzheimer disease
m. VanGogh, O'Keefe, Pollock	13. fraction of total variance attributable to genes
n. threshold	14. discontinuous: present or absent
o. apolipoprotein E	15. found in about 1/800 base pairs
p. relative risk (λ)	16. neurodevelopmental disorders
q. linkage analysis	17. had bipolar disorder
r. quantitative trait	18. present in two members of a family
s. age-related macular degeneration	19. analysis of genes whose products are involved with disorder
t. concordance	20. search for co-inheritance of marker locus with trait

2. Heritability (h^2) estimates the fraction of variance for a phenotype that can be attributed to genes.
   a. Describe two formulae used to estimate h^2.
   b. Why are values for h^2 estimates only, rather than firm values?
   c. Among whites in the US, h^2 for a particular congenital anomaly is 0.7. In whites in Australia, h^2 for the same condition is 0.4. How do you explain this?

3. Eighty years ago, Newman reported IQ scores on 19 pairs of monozygotic twins reared apart. The table below shows the data obtained: the twin pairs are displayed horizontally; the IQ values vertically.

1	2	3	4	5	6	7	8	9	10	11	12	13	14	15	16	17	18	19
106	116	95	96	91	106	85	92	101	78	90	97	93	115	127	88	102	102	116
105	92	94	77	90	89	84	77	99	66	88	85	89	105	122	79	96	94	109

a. Does being an identical twin affect average IQ score?
b. What is the average IQ difference within pairs?
c. What conclusions do you draw from these data?

4. In the table shown below are concordance values for three different disorders in monozygotic twins (MZ), dizygotic twins (DZ), and siblings (S).

	Concordance (%)		
Disorder	**MZ**	**DZ**	**S**
1	10	9	0.1
2	50	12	5
3	100	25	25

a. For which disorder do genes play the largest causative roles, and for which the smallest?
b. Can you estimate the number of loci involved in any of these disorders?
c. Describe two circumstances in which a pair of MZ twins with disorder (3) would have distinctly different phenotypes.
d. How do you explain the fact that DZ twins for disorder (2) are more concordant than sibs are?

5. Many siblings have taken this course over the years. It seems that earning an “A” on the final exam tends to run in families. In 50 cases where one sib got an “A”, 15 of their sibs also got an “A” (of course, this story is fictitious).
a. What is the heritability of this desirable trait?
b. Why might this estimate of h^2 not be an accurate reflection of the actual genetic contribution to course performance?

6. Imagine that 2% of the great grandchildren of people who like the taste of Brussels sprouts also like that taste. What is the estimate of h^2 for this sensory phenotype?

7. What fraction of your genome, on average, do you share with your first cousin once removed (e.g. your mother’s first cousin)?

8. During the past decade, genome-wide association studies (GWAS) have emerged as a major tool in identifying genes contributing to the etiology of multifactorial traits.
a. What is the difference between GWAS and linkage analyses?
b. Why have GWAS been conducted since about 2003 but not before?
c. What kind of gene is identified in GWAS compared to the kind found in studying single gene disorders?
d. In age-related macular degeneration (AMD), GWAS have revealed a few genes that account for more than half of the genetic risk. In diabetes mellitus, GWAS have identified many genes each of which accounts for a very small fraction of the genetic risk—all together accounting for less than 10% of that risk. Why is this so?
e. Describe briefly three ways to improve the productivity of GWAS.

	SNP						
Individual	**1**	**2**	**3**	**4**	**5**	**6**	**Phenotype**
1	A	C	A	T	G	A	Bald
2	A	G	A	G	A	T	Bald
3	G	C	A	T	T	T	Bald
4	G	G	T	T	T	A	Not bald
5	A	C	T	G	A	A	Not bald
6	G	G	T	G	G	T	Not bald

9. Imagine that you are trying to find genes associated with male-pattern baldness. In the table above the SNP pattern across a particular region of a chromosome linked to baldness is shown for six unrelated individuals.

a. Which SNPs are most likely to be associated with a locus predisposing to male-pattern baldness?

b. How would you proceed to identify the specific locus, and demonstrate its role as a susceptibility locus?

10. A series of children affected with a particular congenital malformation includes about equal numbers of males and females. What studies would you carry out to determine if this malformation is more likely to be multifactorial than autosomal recessive?

11. In recent years pundits of many types have written things paraphrased as follows: Genes explain obesity, criminality, intelligence, political leanings, and preferred dress styles. There are selfish genes, celebrity genes, gay genes, sinning genes, and couch potato genes.

a. What do we call this kind of pronouncement?

b. Comment briefly on your response to the above words.

c. Would the lay public at large likely respond the way you have responded in (b)? Why or why not?

CHAPTER 14

Disorders of Variable Genomic Architecture

CHAPTER OUTLINE

CORE CONCEPTS

The human genome has, until very recently, been thought of in classical terms:

- Chromosomes and genes occur in pairs—one from each parent.
- A pair of alleles at each locus produces dominant, recessive, and sex-linked traits.
- Mutations produce heritable changes in DNA that often have deleterious results.

The Human Genome Project has taught us, however, that these classic understandings are over-simplified and incomplete: the function of chromosomal DNA is sometimes affected by epigenetic chemical modification; some genes are duplicated and others have multiple alleles; some Mendelian disorders result from mutations at two different loci, or even three. From these observations has come the current view of the genome:

- that only 2% of it codes for proteins;
- that about half is present in single copy sequences;
- that more than half of it consists of a variety of repeating sequences;
- that variation in copy number affects 10–20% of the genome.

Human Genes and Genomes. DOI: 10.1016/B978-0-12-385212-0.00014-7

It is already safe to conclude that the human genome is dynamic—not fixed; that it is capable of great variability which is occurring continually. Although any two humans are 99.6% identical in genomic DNA sequence, the other 0.4% of non-identity is already beginning to provide a glimpse of what makes each human genetically unique.

These key points are expanded by more detailed discussion of α-thalassemia, Charcot-Marie-Tooth disease, and schizophrenia.

From Mendel and Garrod on, the field of genetics has viewed genes as occurring in pairs, one inherited from each parent. Alternate forms of a gene are called alleles. This view has led to some classic terms: homozygote (two identical alleles), heterozygote (two different alleles), dominant (expressed in heterozygotes), recessive (expressed in homozygotes), autosomal (involving chromosomes 1 through 22), and sex-linked (involving the X and Y).

Mutations were thought of in relationship to the paradigm just stated: that they were heritable changes in DNA sequence, that they affected a single gene, that they usually affected a single base pair, that unrepaired ones were rare, and that deleterious ones have been detected at about 10% of human genetic loci.

Chromosomes, too, were viewed through this prism, reflecting the history of discovery. Chromosomes occurred in pairs—one of each pair inherited from the father and the other from the mother. Germ cells were haploid, containing 23 chromosomes; all other body cells were diploid, containing 46 chromosomes. Any deviation from this configuration generally led to spontaneous abortion or major phenotypic abnormalities.

Today we know that these concepts are fundamentally correct as far as they go, but they do not go nearly far enough. Here is some evidence for this statement. Many of our genes have multiple alleles and may be present in two duplicated copies. Some Mendelian disorders result from mutations at two different loci, or even three. Classes of mutations range from single base changes to trinucleotide expansions, with base insertions and deletions in between. Chromosomes are more than linear molecules of DNA; they also contain a complement of associated proteins. Genes constitute only a small fraction of chromosomal DNA and are irregularly dispersed through the genome. Chromosomal abnormalities produce a wide variety of phenotypes, identifiable microscopically or using submicroscopic tools. These are but a few of the pieces of information attesting to the complexity and perplexity of our genome. Our genome can be likened to a blueprint that can be read in multiple ways or to a map that may have multiple routes going from point A to point B. It is a testament to the Human Genome Project that we have this richer (yet still woefully incomplete) view of the human genome.

ARCHITECTURE OF THE GENOME

Single-copy Sequences

Only about half of the 6 billion base pairs in the human genome are present as a single copy. About 4% of this fraction (or 2% of total DNA) is "genic"—constituting DNA that codes for a specific polypeptide. We know a great deal about the DNA sequence of this coding portion, although there are still major gaps in defining the function for each of the estimated 21,000 protein-coding genes in the human genome. Each gene has its unique structure and function, but there are many examples of genes occurring in families or super families—genes of related structure evolving from a single ancestral one. Some prominent examples (discussed previously) are the β-globin gene family, the HOX gene family (to be discussed in Chapter 15), and the variable region family of immunoglobulin genes.

What about the rest of the **single-copy DNA** that doesn't code for polypeptides? Some of it is composed of genes that synthesize functional RNAs. These RNAs are of several types: tRNAs, rRNAs, and miRNAs. As discussed in Chapter 7, these RNAs play critical roles in transcription, translation, and expression of genes. But this still leaves a large part of single-copy DNA unaccounted for. A fraction of it represents pseudogenes. Pseudogenes are DNA sequences closely related to active genes that do not produce any RNA or protein product. In the main, pseudogenes are thought to be byproducts of evolution, or "dead" genes that were once functional. Many thousands of these gene remnants are scattered across the human genome. The remainder of single-copy DNA, large in quantity, has resisted definition to date.

Repetitive Sequences

About half of the human genome consists of **repetitive sequences**. These sequences vary greatly in length, location, function, copy number, and origin. Brief descriptions of these various classes follow.

CLUSTERED REPEATS

Estimates are that approximately 10—14% of the human genome is composed of **clustered repeats**—stretches of repeating nucleotide units localized to a particular region of a chromosome. The repeating unit may be as short as 2 nucleotides (AGAGAG) and as long as 300 nucleotides. Clustered repeats are arranged in a tandem head-to-tail fashion. They may be as long as several Mb. Interestingly, more than half of the Y chromosome is composed of such repeats.

Two classes of clustered repeats deserve special mention. One is called **α-satellite DNA**. This class of repeats has units that are 171 nucleotides long, arranged in tandem in the centromere of each human chromosome. These repeats are normally required for formation of the mitotic spindle and for successful cell division. The second class is the **telomere,** which consists of repeating units of six nucleotides (TTAGGG) found at the ends of each chromosome. These repeats shorten in length with each cell division, thereby functioning as a "clock" for the reproductive capacity of a given cell (see Chapter 16).

DISPERSED REPEATS

An even larger component of total genomic DNA—as much as 30% to 35%—consists of repetitive units that are dispersed throughout the genome, rather than being localized into clusters. Two general types of **dispersed repeats** are recognized. The first type is called **SINE** (for "**short, interspersed nuclear element**"). The most prominent member of this family is called ALU. The **ALU family** is composed of 300-nucleotide units that are very similar, but not identical, in base sequence. There are more than 1 million ALU repeats scattered throughout the genome. This single class of repeats makes up about 10% of genomic DNA. A second major dispersed repetitive sequence family is called **LINE (long, interspersed nuclear element)**. LINE family elements range up to 6 kb in length and make up as much as 20% of the total genome.

SINE and LINE elements are ancient from an evolutionary perspective, and they have generally lost whatever function they once had; they may date as far back as the earliest **retroviruses**. Retroviruses are composed of RNA; their name derives from their ability to be "reverse" transcribed from RNA to DNA. This they accomplish by an enzyme called **reverse transcriptase**. The DNAs of SINES and LINES were likely formed initially by being reverse transcribed and then transposed into the genome of organisms more complex than viruses. They are rarely transcribed, but this does not mean that they are totally inactive. SINE elements are likely transcribed during stress. LINE elements, when transcribed and inserted into the genome at sites other than where they originally exist, have been shown to produce human disease by interrupting such genes as that for Factor VIII (the gene mutated in hemophilia).

SEGMENTAL DUPLICATIONS

Another class of repeats has been identified recently. It is composed of blocks of different sequences present in multiple copies in many locations in the genome. The blocks are called "**segmental duplications.**" They may contain genes, and they likely account for 5% of genomic DNA. The full medical significance of segmented duplications remains unclear, but they have already been associated with some developmental disorders. Such genomic rearrangements of gene-containing segmental duplications could result in deletions (or other modifications) of clinical significance. Table 14.1 presents an overview of the architecture of the human genome, emphasizing what we know and what we still need to find out.

TABLE 14.1 Human Genomic Architecture

Structural Class	Percent of Total DNA
Single copy:	
Coding genes	2
Non-coding RNAs, pseudogenes, and unidentified	48
Repetitive sequences:	
Clustered	~10–15
Dispersed	~30
Segmental duplications	~5

HUMAN GENOMIC VARIATION

It is widely held that no two humans are genetically identical— that each of us is unique (see Chapter 10). This means that variation in our genomes must exist, sufficient to account for this genetic uniqueness. We are beginning to identify the types of variation and are starting to probe the relationship between this variation and health.

Types of Variation

It is now estimated that any two randomly selected individuals have DNA sequences that are 99.6% identical. What about the 0.4% that is not identical—that is, the 12 million base pairs per haploid genome that differs? Some of this variation reflects changes in single base pairs. Some reflects more complex rearrangements. The rate of progress in cataloging the nature of this genomic variation has been accelerating for the past 5 years and will accelerate more in the months and years ahead. The types of variation observed are shown in Table 14.2.

TABLE 14.2 Common Variation in the Human Genome

Class of Variation	Size Range
Single nucleotide polymorphism (SNP)	1 bp
Insertion/deletion polymorphism (indel)	1 bp to 1 Mb
Inversions	Few bp to 100 kb
Copy number variation (CNV)	1 kb to several Mb

SINGLE NUCLEOTIDE POLYMORPHISMS (SNPs)

As discussed in Chapters 7 and 13, SNPs constitute the first and still most extensively studied form of DNA sequence variation. SNPs occur at about 1 in 800 base pairs, meaning that any 2 genomes would be expected to differ by about 6 million nucleotides.

INDELS AND INVERSIONS

Insertions and deletions (**indels**) and inversions of lengths varying from a few bp to 1 Mb occur throughout the genome and may involve as much as 1% of genomic DNA.

COPY NUMBER VARIANTS (CNVs)

The most recently recognized sources of human genomic variation are referred to as **copy number variants**. **CNVs** are segments of DNA, ranging in length from 10 kilobases to several megabases, that differ in the number of copies between any two individuals. CNVs affect two to three times more genomic DNA than do SNPs, and their medical implications are likely to equal or exceed those of SNPs as well. Thus, they require greater explanation.

Copy Number Variation

The time-honored bi-allelic view of the human genome is no longer tenable. The presence of so many human disorders obeying Mendel's first law of gene segregation means, of course, that many coding sequences are present in single copy number, one copy on each homologous chromosome. But we now know that somewhere between 10% and 20% of the human genome varies in the number of copies present. We know this because new technologies have uncovered the extent of this variation. The most widely used method is called **array-based comparative genomic hybridization,** or **CGH**.

CGH

Microarray or chip technology used in CGH has been employed widely since 2006. The basic idea is shown in Figure 14.1. As many as 1 million oligonucleotides (short nucleic acid

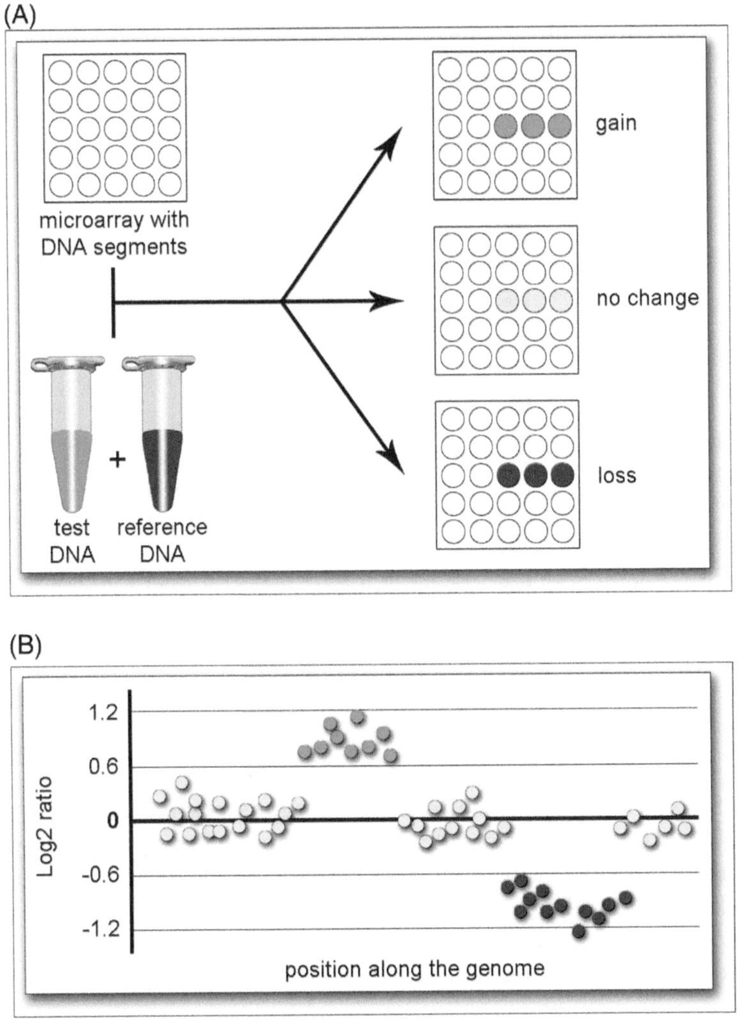

FIGURE 14.1
Array-based comparative genomic hybridization (CGH). (A) The test genome (green-labeled) and the reference genome (red-labeled) are mixed and hybridized to an array containing DNA sequences. Fluorescence detection yields green "spots" if test DNA contains sequences with copy number greater than reference, red "spots" if test DNA contains sequences with copy number fewer than reference, and yellow "spots" where test and reference DNAs contain sequences of equal copy number. (B) The identified variants in copy number are displayed as a function of their position on a chromosome or a series of chromosomes.

polymers, typically with 50 or fewer bases), representative of the entire human genome, are spotted on a microarray. The "test" genome is isolated and labeled with one type of fluorescent molecule (depicted as green). The reference or control genome is labeled with a different fluorescent substance (shown as red). The two genomic DNAs are then mixed, denatured (treated so their strands come apart), and hybridized (matched with correspondingly appropriate pieces) to the oligonucleotide array. Any spot with green fluorescence means that the test genome contains more copies of the sequence examined than does the reference one. If the test genome contains fewer copies, the spot involved will fluoresce red. If the two samples contain equal copy numbers, the spot corresponding to it will be yellow. Many refinements and expansions to CGH have been made in recent years, but the basic idea has not changed.

TYPES OF CNVs (FIGURE 14.2)

By their nature, CNVs cannot be seen with the microscope. They vary greatly in size and kind. CNVs ranging in size from 1 kb to several Mb have been identified throughout the genome. Many different rearrangements perturb the normal biological balance of the diploid state. Deletions, insertions, duplications, triplications, and translocations have been found. Deletions are more common than the other types of CNVs. Any two genomes differ by more than 1,000 CNVs, or about 0.8% of a person's genome. Among the more interesting and provocative aspects of the largest survey of CNVs conducted are these: 75 genomic regions have "jumped around" in different genomes; some CNVs are inherited, others occur *de novo* in germ cells; CNVs are biased away from coding genes, although clear examples of gene deletions and duplications have been observed as well. Strikingly, more than 250 genes in our genome can lose one of their two copies without obvious consequences. Genes concerned with early embryonic development and of control of the mitotic cell cycle seem to have been spared CNVs.

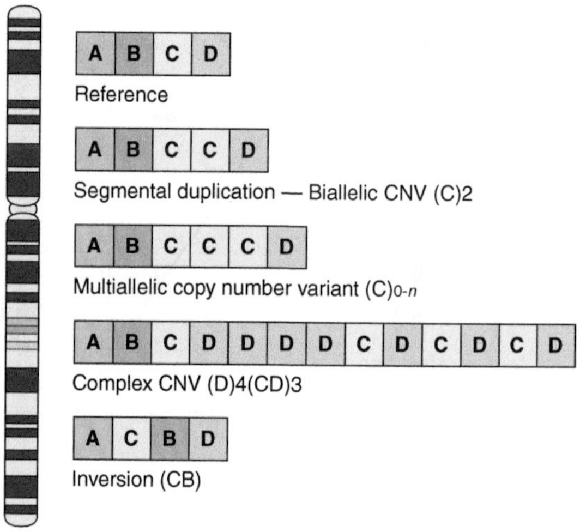

FIGURE 14.2
Schematic representation of types of structural variants in the human genome. Depicted are a segmental duplication, a multiallelic copy number variant, a more complex CNV, and an inversion.

MECHANISM RESPONSIBLE FOR CNVs

The most common cause of CNVs involves **non-allelic homologous recombination (NAHR),** depicted in Figure 14.3. In **NAHR,** also called unequal crossing over, two highly homologous, repetitive sequences of DNA, which are misaligned, can lead to duplications or deletion products with different numbers of copies of the genes contained between the repeats (producing CNVs). The size and content of the duplications and deletions will depend on the distance between the long sequence repeats whose misalignment produces them.

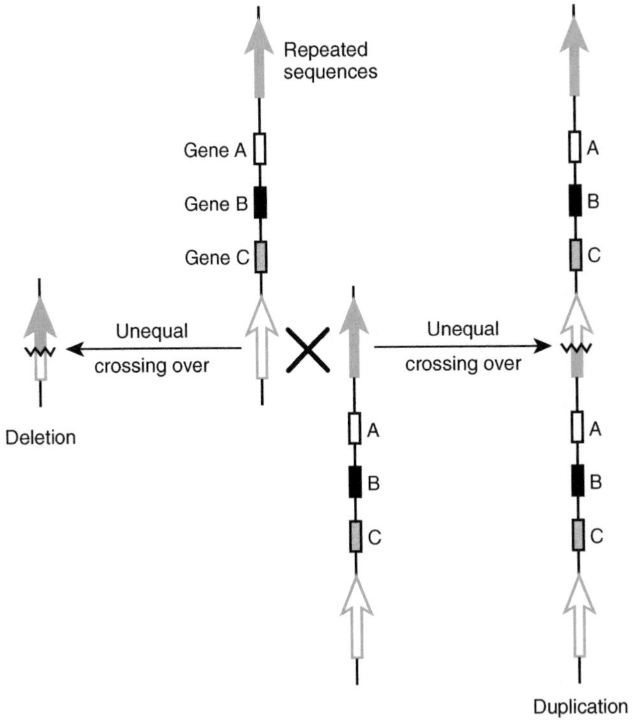

FIGURE 14.3
Schematic representation of non-allelic homologous recombination (NAHR) (also referred to as unequal crossing over) resulting in either deletion or duplication of the DNA segment containing three genes (A, B, C). Closed and open arrow heads represent highly homologous sequences near the three-gene segment.

CLINICAL SIGNIFICANCE OF CNVs

At least 1 in 17 children has a new CNV. Most of this genomic variation appears to have no clinical significance, but many scientists question the strength of this conclusion. Just as SNPs have been associated with many multifactorial traits, CNVs are being described in such conditions as well. The list of CNV-associated disorders is already long, and growing. It now includes autism, epilepsy, Parkinson disease, Alzheimer disease, schizophrenia, and various causes of intellectual disability. It is too early to tell what these associations mean or how they affect genomic function. Do CNVs cause disease, or do they identify regions increasing susceptibility to disease? It seems likely that they do both under different circumstances. It is particularly intriguing that associations of CNVs with disease have been biased to date toward disorders of the brain.

ILLUSTRATIVE EXAMPLES

Changes in gene dosage are a hallmark of CNVs, but such dosage effects are not new. For example, the α-globin locus is duplicated and has been studied extensively because of its relationship to α-thalassemia. The *PNP22* gene, normally a single-copy gene, is duplicated in Charcot-Marie-Tooth disease type 1; the duplication causes this neurologic disorder. Thus, in these and other situations, CNVs in the form of duplication have clear medical implications. In contrast, many CNVs have recently been found in schizophrenia, but they have not yet been shown to produce the mental illness characteristic of schizophrenia. Here, it must be remembered, association is not synonymous with causation. These three conditions will now be discussed in greater detail.

Alpha-thalassemia

The **thalassemias** represent two families of disease produced by mutations of the globin genes discussed extensively in Chapter 12. **Alpha thalassemia** affects the formation of α-globin chains; **beta thalassemia** of β-globin chains. In both instances, the disorders are characterized by quantitative reduction in globin chain synthesis, rather than by qualitative changes in the hemoglobin molecule (as typified by sickle cell anemia). Under normal circumstances, equal amounts of α-globin and β-globin chains are made by RBCs. This balance is interrupted in the thalassemias. We will concentrate on α-thalassemia because it is directly related to gene dosage.

FREQUENCY AND EPIDEMIOLOGY

Alpha-thalassemia is most common in Asia, India, and parts of the Middle East and Africa. The severe form of the disease is largely restricted to Southeast Asia where it may affect 30–40 people in 100.

ETIOLOGY AND CLINICAL FINDINGS

As discussed earlier, the α-globin gene cluster is found on chromosome 16 (see Chapter 12). In contrast to the single β-globin gene, the α-globin locus underwent duplication sometime in the evolutionary past and is present in the human genome in two copies, α1 and α2. Each locus codes for α-globin chain synthesis, meaning that, under normal circumstances, each human has four functioning alleles, not two. The α1 and α2 genes are virtually identical, as are their surrounding regions.

The α-thalassemias are caused most often by deletions. As shown in Figure 14.4, the molecular pathology closely mirrors the clinical consequences. In the normal situation, both α1 and both α2 alleles are functioning. If one α2 allele is deleted, the so-called "silent carrier" is produced, suffering no clinical disturbances. Being heterozygous for deletions of each α gene, or being homozygous for deletions of either α1 or α2, is next in genetic severity. In these instances, there are still two functioning α-globin gene copies, and anemia is mild or absent. When only one of the four copies remains, clinical consequences are severe—anemia from birth on. If all four

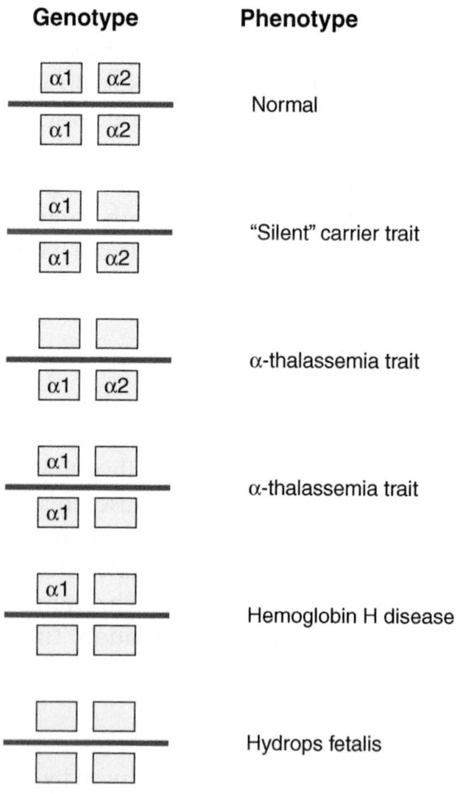

FIGURE 14.4

Genotypes and phenotypes in α-thalassemia. Top: normal people have four copies of α-globin genes (2α1, 2α2). Bottom: in the absence of all four copies (shown as empty boxes), infants succumb *in utero* or at birth with hydrops fetalis (meaning fluid accumulation in the fetus). Four intermediate genotypes reflect progressive loss of α1 or α2 copies. See text for details.

copies of the α-globin genes are deleted, stillbirth or neonatal death results. This is because α-globin chain synthesis occurs virtually throughout prenatal development of the embryo and fetus. Hence, in the absence of α-chains, the fetus is deprived of oxygen-carrying hemoglobin.

MANAGEMENT

As anticipated from the foregoing, people missing one or two copies of α-globin genes require no treatment. Those missing three require regular transfusions, and those with all four copies deleted cannot be saved. This means that couples who have had a baby die from fetal or neonatal α-thalassemia have a one in four chance of the same outcome with each pregnancy because each parent has a genotype of α1α2/deldel.

Charcot-Marie-Tooth Disease (CMT)

FREQUENCY AND EPIDEMIOLOGY

Charcot-Marie-Tooth (CMT) disease is a family of genetically heterogeneous disorders that cause neurodegenerative changes in the legs and arms. CMT affects about 1.5 in 10,000 people. It occurs in all ethnic groups worldwide.

ETIOLOGY

The best studied of the CMT diseases is termed **CMTIA**. It is caused, in most instances, by an inherited or a *de novo* duplication of *PNP22*, a gene whose product is required for integrity of myelin. *PNP22* is ordinarily found as a single-copy gene on chromosome 17 (Figure 14.5). When the copy number of *PNP22* is increased from two copies to three copies by unequal crossing over during meiosis, overproduction of the PNP22 protein leads to myelin degeneration and impaired nerve function. A closely related neuropathy is called **hereditary neuropathy with pressure palsies (HNPP)**. HNPP is caused by deletion of one copy of *PNP22*. Clearly, then, proper myelin structure and function requires careful control of the PNP22 protein: either too much or two little is deleterious. CMTIA was among the first Mendelian disorders shown to be due to a CNV affecting dosage of a single-copy gene.

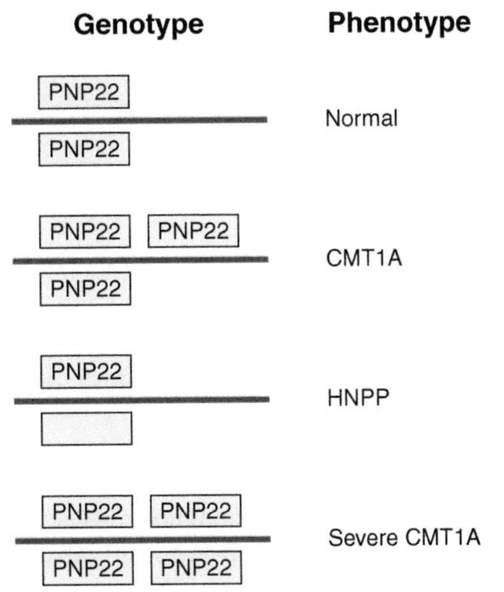

FIGURE 14.5
Genotype and phenotype in Charcot-Marie-Tooth disease type 1A (CMT1A) and hereditary neuropathy with pressure palsies (HNPP). Copies of *PNP22* gene are shown in boxes. Empty box denotes deletion of *PNP22* gene. See text for details.

CMTIA is inherited as an autosomal dominant trait and can be categorized as a gain-of-function mutation. It is fully penetrant, but about 20–30% of patients have *de novo* germline mutations rather than an inherited one.

PATHOGENESIS

Overexpression of *PNP22* results in an inability to form and maintain myelin, the complex sheath required by peripheral nerves to carry impulses properly. In the absence of myelin or if myelin is distributed aberrantly, the peripheral nerves to the arms, hands, legs, and feet degenerate.

CLINICAL FINDINGS

Though fully penetrant, CMTIA shows markedly variable expressivity. Some patients show neurologic effects during infancy or early childhood. Others may be so mildly affected that they require no medical intervention until the third decade. Typically, symptoms begin with slowly progressive weakness of the leg muscles, associated with wasting of the affected muscles (Figure 14.6). With time, walking is impaired, leading to a "flapping," dropped-foot gait. Atrophy of hand muscles usually occurs late in the disease. Treatment is limited to symptomatic management.

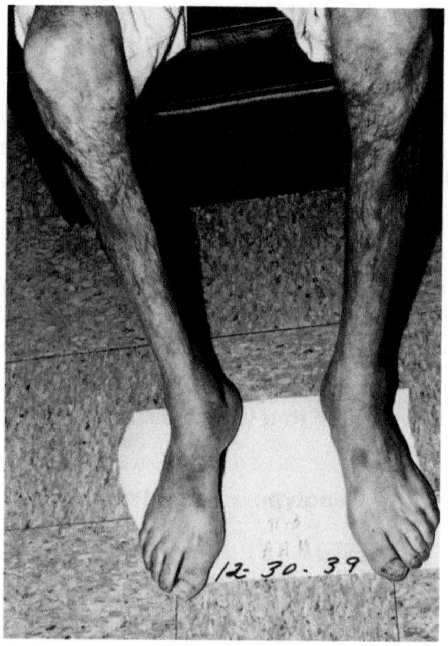

FIGURE 14.6
Distal leg muscle atrophy in man with Charcot-Marie-Tooth disease type 1A (CMT1A).

INTERSECTIONS: GENOME SEQUENCING AND DISEASE FINDING

Charcot-Marie-Tooth Disease: A Needle in a Haystack

Soon after the Human Genome Project produced the first complete sequence of the human genome, scientists assumed that the mutations that caused common diseases would themselves be common. Therefore, to find the genetic cause of a disease such as autism, which may affect 1 in 200 people, one should look at common mutations in the population. Once these common mutations were found, this thesis continued, it would be possible to compare patients' genomes with those of healthy people and identify new disease or susceptibility genes.

Using a shortcut called single nucleotide polymorphism (SNP) that only scanned a tiny portion of the genome and thus saved money and time, scientists found that about 2,000 sites on the human genome could be statistically linked with various diseases, but many of these sites were not found within working genes and were shown to account for only a small portion of the cases of any one disease. Thus scientists came to the conclusion that many common diseases are caused by rare mutations

INTERSECTIONS: GENOME SEQUENCING AND DISEASE FINDING—Cont'd

instead of common ones, and that one must look at the whole genome of people with a genetic disease whose cause was not revealed by any other method.

One of the first applications of this approach employing whole-genome sequencing was to investigate the genetic cause of a degenerative neurologic disorder known as Charcot-Marie-Tooth disease (CMT), which affects nerve function in the limbs, hands, and feet. A physician-scientist, James Lupski, at Baylor College of Medicine, who has the disease, investigated himself. First, he determined that his condition was not caused by duplication of the *PNP22* gene—the usual cause of CMT1A (described above). Then he proceeded to do complete genome sequencing on himself and to compare the observed variants in his genome with those of his parents and seven siblings. The results showed that Dr Lupski and his three siblings who had the disease carried two altered versions of the gene *SU3TC2*. Each of his parents, who did not have the disease, was heterozygous for one of the two different mutations of *SU3TC2*—but not the same one. His unaffected siblings inherited a single copy of one or the other mutant allele. Thus Dr Lupski and his affected siblings were demonstrated by genome sequencing to be doubly heterozygous for alleles at a previously undescribed locus. This finding opened the door to inquiry about the product of *SH3TC2* and about why its deficiency caused this form of CMT. Doubtless, complete genome sequencing will be used to find the molecular basis for other disorders as the cost of such sequencing continues to fall.

Schizophrenia

As mentioned in Chapter 13, a number of important and common mental illnesses display multifactorial characteristics. We discussed one such condition, bipolar disorder, in Chapter 13. We return to another serious mental illness here because recent use of array-based comparative genomic hybridization (CGH) has revealed important clues to understanding its genetic basis. That disorder is schizophrenia.

FREQUENCY AND EPIDEMIOLOGY

Schizophrenia occurs in about 1% of men and women in the United States. It is panethnic and is found throughout the world. Schizophrenia shows the hallmarks of a multifactorial trait. It clusters in families, but not according to any Mendelian pattern. Concordance in MZ twins is about 46%, while concordance in DZ twins is about 17%. As noted in Table 14.3, first-degree relatives of patients with schizophrenia have a much greater lifetime risk of developing schizophrenia than does the general population. Increased risk is also apparent in second- and third-degree relatives.

TABLE 14.3 Risk of Developing Schizophrenia

Relation to Person with Schizophrenia	Risk (%)*
Identical twin	46
Non-identical twin	17
Child	13
Sibling	10
Parent	6
Uncles/aunts	2
Nephews/nieces	2
Grandchild	2
Unrelated (general population)	1

**Lifetime risk of developing schizophrenia. All numbers are approximations based on several studies.*

CLINICAL FINDINGS

Schizophrenia is a devastating condition for patients and their families. Onset is usually in adolescence or young adulthood and is characterized by abnormalities in thought and behavior. The most characteristic of these abnormal behaviors are false beliefs (delusions) and false perceptions (hallucinations). This inability to distinguish what is real from what is not is a hallmark of psychosis, a term used generally to describe the behavior of patients with schizophrenia. The diagnosis of schizophrenia rests on several findings: hallucinations, delusions, or disorganized speech occurring for at least 1 month; social or occupational dysfunction occurring concomitant with those psychotic manifestations; and continuous signs of the disorder lasting for 6 months.

In most instances, psychotic behavior persists for the remainder of a patient's life, making it extraordinarily difficult to conduct a productive existence. During the past 30 years a variety of antipsychotic drugs, affecting one of several neurotransmitter systems, have provided relief for many patients, but not all by any means.

GENETIC STUDIES

The cause or causes of schizophrenia remain unknown, and genome-wide association studies using SNPs have been disappointing. Recently, however, CNVs of several types have been found. Each involves a submicroscopic deletion or translocation of a different chromosome.

- 22q11.2 deletion. A deletion of about 3 Mb results in three different conditions all affecting the face, the heart, and the brain. About 30 genes exist within the deleted region. Approximately 25% of patients with one of these syndromes develop schizophrenia, implying that one or more genes in this region may be involved. The vast majority of schizophrenic patients, however, have no detectable deletion of chromosome 22.
- (1:11)q42:q14.3 translocation. In a single family in which schizophrenia segregates, each affected member had a translocation involving chromosomes 1 and 11. Subsequently, a gene called "**disrupted in schizophrenia 1**" was located to this region. A second family (from Scotland) has also been identified with a closely related but not identical translocation involving chromosomes 1 and 11. Attractive candidate genes expressed in the brain have been identified in each of these families.
- 14q13.3 deletion. Recently, two large-scale association studies employing array-based CGH have been conducted. One group studied 3,391 patients with schizophrenia and 3,181 controls. The other group examined 4,728 patients and 4,120 controls. Both groups confirmed the association of the 22q deletion (discussed above) with schizophrenia. In addition, duplications in 14q11.2, not previously found, were associated with the disease.

This work is promising but leaves many unanswered questions. Which genes are involved? How many different genes must be affected to result in the phenotype? How do these putative genes interact with the environment? Are these genes responsible for the cortical atrophy that has been described in patients with schizophrenia using MRI?

SOCIETAL CONSIDERATIONS

Schizophrenia has a huge impact on our society. For patients, it almost always affects education, employment, and interpersonal relationships in a major way. For families, great suffering is caused by watching a loved one's irrational, sometimes frightening, and regularly debilitating signs and symptoms. For society, the costs are twofold: loss of 1% of the potential workforce at a young age, often permanently; and a financial burden estimated at 50—100 billion dollars per year resulting from hospitalization, medication, and outpatient psychiatric care.

Finally, it is worth mentioning that increasingly precise identification of etiologic or susceptibility loci for schizophrenia would, as for bipolar disorder, make it clear that these are brain diseases, akin to diseases of the heart or other organs. This would unquestionably reduce the stigma associated with common mental illnesses.

REVIEW QUESTIONS AND EXERCISES

1. Choose the phrase in the right column that best matches the term in the left column.

a. non-allelic homologous recombination	1. used to detect CNVs
b. dispersed repeats	2. makes up ~50% of genome
c. copy number variation (CNV)	3. disease mechanism found in Charcot-Marie-Tooth disease
d. SINE and LINE	4. repeating nucleotide subunits localized to a particular chromosome region
e. pseudogenes	5. mechanism of CNV formation
f. three copies of PNP22 gene	6. centromeric repeats
g. clustered repeats	7. most common dispersed repeats
h. gene deletion	8. non-functional sequences closely related structurally to active genes
i. α-satellite DNA	9. dispersed repeats of six nucleotide units (TTAGGG)
j. insertion/deletion polymorphism	10. repeats of varying length scattered throughout the genome
k. telomeric repeats	11. disease mechanism in α-thalassemia
l. comparative genomic hybridization	12. presence or absence of a short genomic segment
m. single copy DNA	13. segments of DNA ranging from 10Kb to several Mb that vary between any two people

2. Until recently, scientists called the vast majority of genomic DNA that doesn't code for proteins "junk DNA." List at least 10 DNA elements demonstrating the importance of this anything-but-junk DNA.

3. Chimpanzees, the closest primate relative to humans, have 48 chromosomes in contrast to our 46. Human chromosome 2 is formed by end-to-end fusion of two chimpanzee chromosomes, thereby accounting for the different chromosome number in the two species. This fusion event occurs at the telomeres. What is the nucleotide sequence at the site of the fusion on both DNA strands?

4. Name two examples of repetitive non-coding DNA whose function is understood. Name two examples of repetitive non-coding DNA whose function is not understood.

5. Copy number variations (CNVs) occur throughout the genome but their physiologic significance is, as yet, unclear in most instances.
   - **a.** What is the best tool currently available to identify CNVs? Describe briefly how this tool works.
   - **b.** What is the "ultimate" tool for detecting CNVs?
   - **c.** What kind of CNVs have been found in α-thalassemia, Charcot-Marie-Tooth disease type 1A, and schizophrenia? Discuss each briefly.
   - **d.** How might finding CNVs in two common multifactorial disorders increase understanding of that condition?

6. What would be the societal significance of finding specific, reproducible CNVs in schizophrenia and bipolar disorder?

CHAPTER 15

Birth Defects

CHAPTER OUTLINE

CORE CONCEPT

Birth defects occur in 5–7% of newborns. About half of these defects affect major organs, including the brain, heart, and limbs; the remainder are minor (crooked fifth finger, missing fingernails, etc). These birth defects reflect but a small minority of the lethal embryonic errors that occur during human gestation and which end in spontaneous abortion. Five critical cellular events must occur during successful embryonic and fetal development: proliferation; differentiation; migration; communication; and apoptosis. These events—some sequential, some concurrent, all critically timed—take place during the embryonic (weeks 1–8) and fetal (weeks 9–38) periods of gestation.

This program is controlled by a series of genes acting at precise intervals in precise fashion. Some of these genes (maternal-effect) are encoded by the mother's genome and synthesize products transferred first to the oocyte and then to the zygote and early embryo. The remainder of the genes controlling development are the embryo's own and control such critical processes as:

- segmentation (dividing the embryo into parts from head to tail);
- pattern formation (fate of cells in each segment);

Human Genes and Genomes. DOI: 10.1016/B978-0-12-385212-0.00015-9

- cell signaling (chemical communication between and within cells);
- apoptosis (programmed cell death essential for tissue remodeling).

When this developmental program works perfectly, a single-cell zygote ultimately becomes a 40-billion-cell neonate. But a myriad of accidents leads to the birth of children with defects. Each of the three major classes of genetic disorders (chromosomal, single gene, multifactorial) underlies a fraction of birth defects. The nature, severity, and outcome of any particular defect depends on what part of the genome is affected and to what degree.

These key points are illustrated by discussing three different kinds of congenital abnormalities: neural tube defects, achondroplasia, and Accutane embryopathy.

INTRODUCTION

A **birth defect** is an abnormality present at birth. Birth defects occur in 5–7% of newborns. About half of human birth defects (also called anomalies) are major, meaning that they reflect abnormalities in the development of the brain, heart, limbs, or other organs and organ systems that impair function. The remaining birth defects, minor in nature, include such things as a white forelock, crooked fifth fingers, and missing fingernails. It would be wrong to suggest from these birth defect statistics that 93–95% of zygotes complete gestation successfully. As pointed out in Chapters 4 and 11, as many as 75% of conceptions have chromosome abnormalities resulting in embryonic lethality, and this figure doesn't include an additional fraction of embryos that abort spontaneously because of single-gene or multifactorial conditions. Nonetheless, we should marvel at the idea that, of the fraction of human zygotes that come to term, 93–95% are born with all their organs, tissues, and cells intact.

Stop for a moment to think about this marvel, which, in many ways, is also a mystery. A single-cell zygote must, in 9 months, become a 40-billion-cell newborn. The undifferentiated fertilized egg must differentiate into 250 cell types, in organs as dissimilar as the brain, heart, and liver. During much of the 20th century, embryologists (those who study development) and dysmorphologists (those who study anatomic abnormalities) studied children with morphologic (anatomic) abnormalities and became expert at describing and classifying birth defects. Understanding such defects, however, seemed beyond the reach of science. How, it was asked, could cells with an identical set of genes and chromosomes become so many different types of cells, each with their own unique structure and function? After answering that difficult question, the next logical one seemed almost impossible to answer: What causes the myriad array of human birth defects?

Today, the answer to the former question is straightforward: not all genes are "turned on" in all tissues. Gene expression is regulated such that each gene's product is formed only when and where it is required. The latter question is much less well answered. We will not know the cause and mechanism of birth defects until we know which genes are active in which tissue, at which time, and in what amount.

As summarized in Table 15.1, five critical events occur during development: cell proliferation, cell differentiation, cell migration, cell communication, and apoptosis (programmed cell death). Mitotic cellular division occurs throughout development, resulting in a mass of some 40 billion cells by the end of gestation. Cell differentiation describes the process of specialization leading to the formation of brain, heart, and all other tissues from the single-cell zygote and the primitive, three-layered embryo. Cell migration is the movement of cells from one location to another. Cell communication involves the means by which one cell affects the behavior of a neighboring one, and apoptosis is required for tissue remodeling. Each of these processes must work perfectly if the end result is to be a single, healthy newborn. Each of these processes is under genetic control, a subject we will turn to now.

TABLE 15.1 Cellular Events Required for Human Development

Event	Nature and Importance
Proliferation	Mitotic divisions responsible for the formation of a 40-billion cell neonate from a single-cell zygote
Differentiation	Process of specialization whereby a cell acquires novel characteristics specific for a particular organ or tissue
Migration	Movement of cells from one location to another
Communication	Process by which cells exchange information regarding their relative position to one another
Apoptosis	Programmed cell death that occurs as part of normal development

Further on in this chapter we will support these general statements with specific clinical examples, but first we must discuss some salient features of human embryonic development and of developmental genetics.

EMBRYONIC DEVELOPMENT

Weeks 1 and 2

In Chapter 4, we discussed in some detail the critical events of the first 2 weeks of human development. These include: fertilization of an oocyte by a sperm in the Fallopian tube; migration of the zygote down the tube, accompanied by cell divisions producing 2, 4, 8, 16 cells, and so on; formation of postzygotic balls of cells forming a hollow blastocyst with an inner cell mass destined to become the embryo; and implantation of the blastocyst into the lining of the uterus.

Weeks 3 to 8

The most crucial period of development occurs during weeks 3–8. During the third week, the inner cell mass has become a flat, disc-shaped embryo (Figure 15.1). By a process called **gastrulation**, a series of orderly cellular migrations and juxtapositions has resulted in the formation of three embryonic layers: **ectoderm**, **mesoderm**, and **endoderm**. All the body's organs will ultimately form from these layers. The first event in organogenesis consists of a portion of mesoderm forming the **notochord**, a solid rod from the front end to the back end of the embryo. Above it, ectodermal cells form neural folds, then an enclosed **neural tube**. The neural tube is the antecedent of the brain and spinal cord. Soon thereafter, small mesodermal blocks, called **somites**, appear on either side of the developing neural tube. Somites will differentiate into muscle, connective tissue, blood vessels, reproductive organs, and excretory organs. The endodermal layer will later give rise to cells lining the digestive tract and lungs and to a variety of glandular cells in different organs.

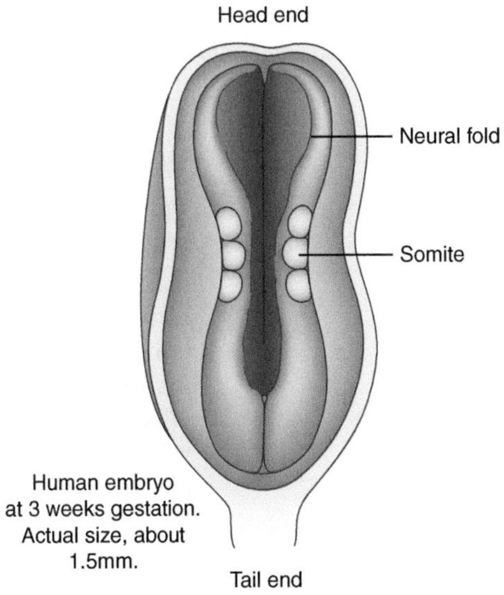

FIGURE 15.1
A 3-week-old human embryo undergoing gastrulation. The anterior–posterior axis has been developed, as have the neural fold and somites.

During week 4 (Figure 15.2), the flat, disc-shaped embryo bends to form a C-shaped structure revealing regions that will give rise to the eye, head structures, the heart, the arms and legs, and the tail bud (which is subsequently lost in the second month of development). By the

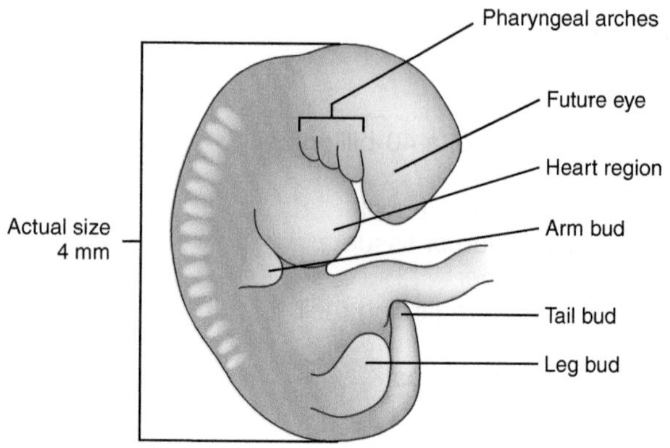

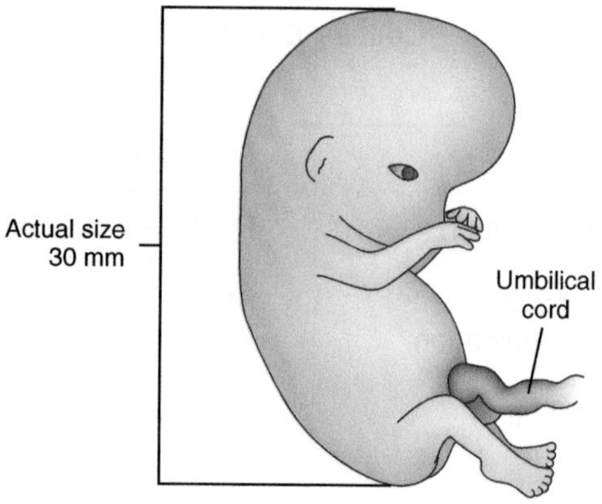

seventh week, rudimentary eyes, ears, nose, jaw, heart, liver, and intestines are visible. At the same time, other systems such as the lungs, kidneys, sex organs, and skeleton are being constructed. By the end of the eighth week (Figure 15.3), the end of what is called the "embryonic period," the embryo is about 30 mm long and weighs about 5 g. The head makes up nearly half its length. Limbs with fingers and toes are present. The tail has disappeared. The embryo is enclosed within two membranes: an inner amnion and an outer chorion. The embryo moves easily within the fluid-filled amniotic sac, anchored to the placenta by the umbilical cord. Amniotic fluid acts as a shock absorber, and the embryo will drink and breathe this fluid.

Weeks 9 to 38

During the next 7 months (the fetal period), the fetus will increase about 12-fold in length and about 650-fold in weight. Development of all the organ systems continues, with the lungs being the last to complete organogenesis.

INTERSECTIONS: EMBRYOLOGY AND EVOLUTION

Ontogeny Does Not Recapitulate Phylogeny

In his monumental work of 1859 entitled *On the Origin of Species*, Charles Darwin proposed that all living species are descended from a single common ancestor. In Darwin's view, early embryonic stages of more advanced organisms are similar to the same embryonic stage of more primitive ones. Evolution, he held, proceeded through a branching life tree of ever increasing complexity (see Chapter 9). This idea ran counter to those circulating in France and Germany from the 1790s that attempted to present a "pattern of unification" in the natural world. The idea of this school of philosophers and embryologists was most clearly articulated in 1866 by the German zoologist, Ernst Haeckel, in his theory of "recapitulation." Its central claim was that the development of an advanced species passes through stages represented by adult forms of more primitive species. His notion that "ontogeny (growth and development of an organism) recapitulates phylogeny (evolutionary history of an organism)" differs profoundly from Darwin's in that it proposes that humans, for example, had adult fish, salamander, turtles, and so on as linear ancestors (see Haeckel's drawing below).

INTERSECTIONS: EMBRYOLOGY AND EVOLUTION—Cont'd

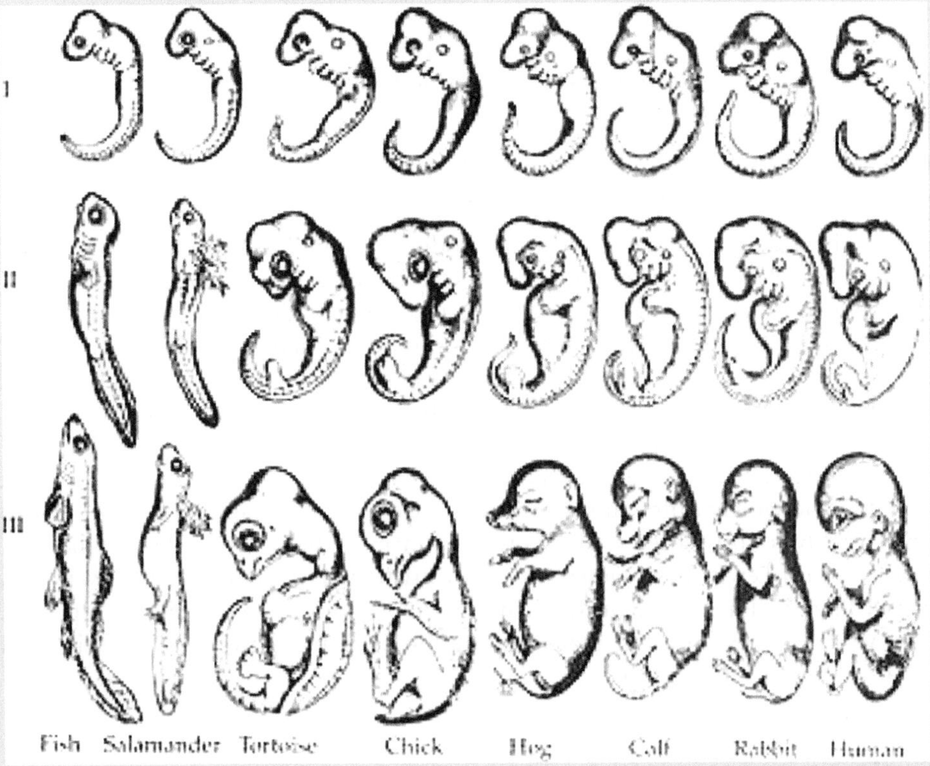

Romane's 1892 copy of Ernst Haeckel's controversial embryo drawings (this version of the figure is often attributed incorrectly to Haeckel).

By the end of the 19th century, the views of Haeckel and other "recapitulationists" were no longer taken seriously. One of the few remnants of this discredited school of thought is the following song of uncertain origin, whose cleverness matches its scientific invalidity (note: amphioxis is a small marine animal):

It's a long way from amphioxis,
It's a long way to us.
It's a long way from amphioxis
To the meanest human cuss.
Goodby fins and gill-slits,
Hello lungs and hair!
It's a long, long way from amphioxis,
But you came from there!

(Sung to the tune *It's a long way to Tipperary*.)

DEVELOPMENTAL GENETICS

Developmental genetics is the field that uses genetics to learn how the fertilized egg of a multicellular organism becomes mature. Most of our information comes from studying drosophila (fruit flies), roundworms, zebrafish, and mice. These "model" organisms can be **mutagenized** (caused to become mutant) experimentally and the results of such genetic manipulation examined in organisms as young as the early embryo and as old as adults. Such study is, of course, unthinkable in humans. But, scientists have been helped greatly by the fact that, during evolution, the genes controlling development have been conserved to a major extent. Specific genes isolated from flies and worms have counterparts (or homologues) in mice and humans. Such conservation in gene structure is supported by studies of function: one can take the human homologue of a particular gene active in fly development and correct a mutation of that gene in the fly. This remarkable conservation says two things:

- some genes controlling development are hundreds of millions of years old;
- what is learned about development in model organisms is often transferable to humans.

Starting in the 1970s, geneticists began to produce mutants in drosophila that interfered with normal development. They found three categories of genes that affected early development in these flies: maternal-effect genes, segmentation genes, and pattern-formation genes. Subsequently, they related the mode of action of these genes with the phenotypes produced by mutating them. These groundbreaking experiments were followed by the experiments of a rapidly expanding group of developmental geneticists. In addition to the fly, this work has concentrated on studying the genes regulating development in worms, zebrafish, and mice. As it became clear that the "rules" of genetic control were conserved in the widely different organisms studied, human genes, too, have been sought and increasingly found. Three scientists who pioneered the study of developmental genetics—Edward Lewis, Eric Wieschaus, and Christiane Nüsslein-Volhard—were awarded the Nobel Prize in Physiology or Medicine in 1995. From a large number of possible subjects, we have selected a few that illustrate the principles discovered and their relationship to human birth defects.

Gene classes

MATERNAL-EFFECT GENES

Genetic control of development is initiated before the zygote is formed. We know this because there is very little transcription of genes during the early cell divisions following fertilization and because there are three classes of mutations in female flies that affect early embryonic development in specific and reproducible ways. As we have said before, the egg contains stores of nutrients vital for sustaining the zygote. In addition, the egg stockpiles ribosomes, nucleotides, enzymes, and spindle proteins required for early rounds of cell division. Among these maternally derived products are proteins that are critical for early development. These proteins are transported into the cytoplasm of the egg and from there into the zygote and early embryo. These maternal-effect gene products regulate the zygotic genes controlling axis formation, making the distinction between head and tail (or anterior and posterior), and top and bottom (dorsal and ventral).

SEGMENTATION AND PATTERN FORMATION GENES: HOMEOBOX GENES

At some point unique to each organism, the embryo's own genes are activated, and they take over from the maternal genes. In humans this shift occurs between the four- and eight-cell stage. Two kinds of events establish the basic body plan. First, the **segmentation genes** interact with one another to subdivide the embryo into a linear, antero-posterior series of progressively smaller segments (Figure 15.4). Then, the pattern-formation genes assign a particular developmental fate to each segment. (These overly simple statements describe profound events: different hierarchies of genes interacting within themselves and with other hierarchies to make possible the myriad events that must occur during embryogenesis, such as cellular proliferation, cellular differentiations, and cellular migration.)

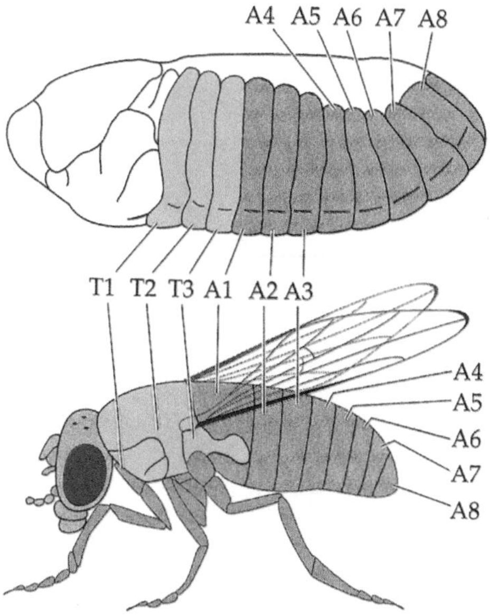

FIGURE 15.4
Early development of *Drosophila melanogaster* embryo: (A) immediately after fertilization; (B) about 9 hours later. Note prominent division of embryo into a linear series of segments. Segmentation results from the action of many genes, which have homologues in humans.

A hierarchy of genes critical to segmentation and pattern formation are the **homeobox genes**. These genes act to specify the structures unique to each segment described above—head structures from the most anterior segment, thoracic features from mid-segments, and so on. They were originally identified by finding mutant flies with striking phenotypes, such as wings emerging from the head region where the antennae should be, or legs sprouting from the head where the eyes should be. Such weird phenotypes led to studies isolating and characterizing these genes. This work led to the recognition of a class of genes called the **homeobox genes** and to mutants of them called **homeotic mutants**. The major conclusions of this work are worth enumerating:

- Multicellular organisms from flies to humans contain a collection of homeobox genes. Flies have 8 of them; humans have 38. The fly genes are abbreviated **HOM**; the mouse genes, **Hox**; the human ones, **HOX**.
- Homeobox genes have arisen by a process of gene duplication and translocation. All HOM genes are located on one fly chromosome. Human HOX genes have been distributed to four chromosomes (numbers 2, 7, 12, and 17) (Figure 15.5).
- There is striking similarity in DNA and amino acid sequences in homeobox genes within a species and between species. The genes contain a **conserved** 180-bp region of DNA coding for a 60 amino acid region of the homeobox protein. This conserved region is called the **homeodomain**.
- The linear order of homeobox genes in one species, such as the fly (Figure 15.5), remains relatively the same in other species as different as humans. Remarkably, these genes are expressed temporally and spatially along the embryo's anterior/posterior axis in the order of their chromosome position. That is, the genes found closest to the 3′ end of the cluster (Figure 15.5), *a1* and *b1* in the mouse) are expressed first. Their expression starts in the most anterior region and extends posteriorly as development proceeds. Then the second genes (*a2* and *b2*) are turned on and are expressed a bit more posteriorly but extending tail-ward. Finally, the most 5′-located gene is expressed, but only in the tail region. Each body segment, then, has its own array (or orchestra) of homeobox gene expression. The HOM, Hox, and HOX gene clusters represent copies of an ancestral gene cluster that likely existed before insects and mammals diverged.
- The proteins encoded by HOM or HOX genes are transcription factors that bind to enhancer or promoter regions of other genes and either turn them on or turn them off. There is still much to be learned about the specific genes modulated by homeobox-derived transcription factors.

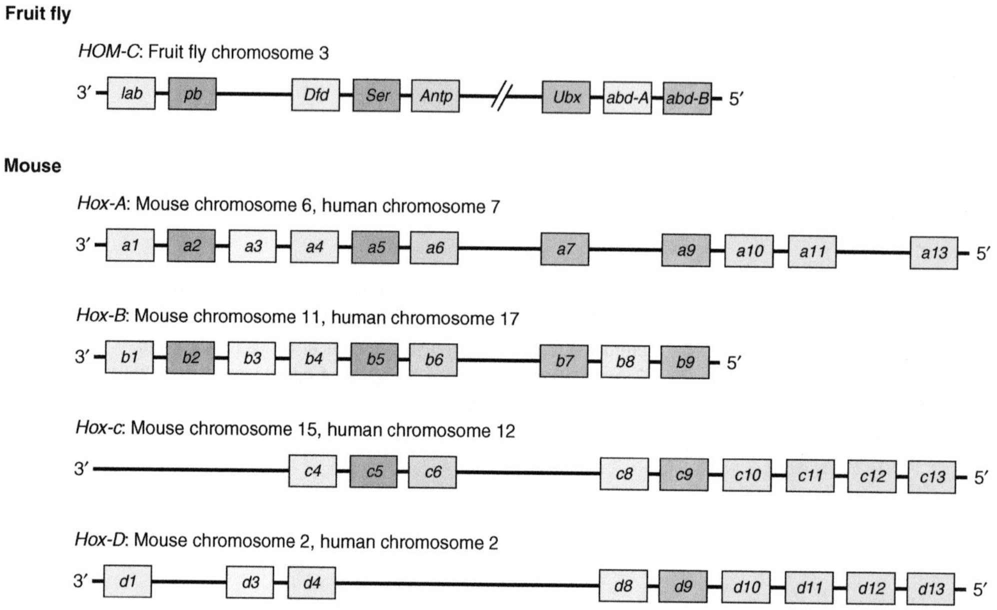

FIGURE 15.5
The homeobox gene family in drosophila and vertebrates. In the fly, eight genes make up the HOM-C family. In mice and humans, 38 genes, distributed to 4 chromosomes, constitute the family called Hox in mice and HOX in humans. Color-coding of the eight HOM-C genes in flies depicts each gene's domain of activity in forming an adult fly. The 3′-to-5′ order of the genes on the chromosome matches the anterior to posterior activity of the genes. In mice and humans, Hox and HOX genes have been dispersed to four chromosomes, but their 3′-to-5′ order has been conserved on each chromosome. See text for additional comment.

- Homeobox genes are involved prominently in embryonic development. Mutational studies have clearly established their role in development of the brain, the limbs, and the vertebrae.
- A rare and unusual human birth defect is caused by a mutation in a single HOX gene (d13). Three families have been described in which hand and foot abnormalities were noted. Fusion and duplication of digits occurred in each affected patient. This condition is inherited as an autosomal dominant trait with incomplete penetrance.

Homeobox genes are but one class of genes regulating segmentation and pattern formation. We discuss them rather than others because more is known about them and because our knowledge about them is representative of the action of other gene classes.

Signaling Molecules

Cells in multicellular organisms must communicate with each other to carry out their many functions. They do so by signaling to each other using chemicals. Hundreds of kinds of signaling molecules have been identified: proteins and peptides, amino acids and fatty acid derivatives, steroids and retinoids, and dissolved gases. Often, these signaling molecules are transported out of one cell and are recognized by an adjacent or distant target cell. After the signaling molecule binds to the target cell—usually by way of a cell surface receptor—a response is initiated. Such cell-to-cell signaling is a vital part of embryonic development. Genes control each aspect of signaling pathways: synthesis of the signal, transport from the cell, receptors on target cells, and metabolic pathways of response to signals. We will discuss two types of signaling that are particularly relevant to humans.

FIBROBLAST GROWTH FACTOR

The human genome contains 10 different **fibroblast growth factor (FGF)** loci encoding proteins that stimulate growth of many cell types. Although they were originally described in fibroblasts, these factors are found in many other kinds of cells. They function in a variety of ways during embryonic development, being required for formation of such structures as the midbrain, spinal cord, facial structures, nerve cell processes (axons and dendrites), and limb buds. In later development they regulate the growth of long bones and closure of skull bones. FGFs signal by binding to FGF receptor (FGFR) proteins on target cells. In turn, the FGFR protein activates pathways leading to growth.

Many mutations of human FGF and FGFR genes have been described. These mutations produce a variety of birth defects specific to the involved locus and each is inherited as a Mendelian trait. Most commonly these mutations produce abnormalities in the face and limbs. Achondroplasia, the most widely recognized form of human dwarfism is caused by mutation of the *FGFR3* gene, and will be discussed in greater detail subsequently. Other mutations of *FGFR3* cause different conditions, some less severe than achondroplasia, some more severe. In all, more than 100 different birth defects result from mutations in one of the four FGFR genes. These mutations often cause serious malformations of the skull, face, and limbs.

SONIC HEDGEHOG GENES

Among the many mutations in flies affecting development is one called **hedgehog** because of its small size and unusual appearance. (It was named after a video game character.) We now know that this gene locus (called *hh*) is required for pattern formation. The product encoded by the *hh* locus is a protein, referred to as a **morphogen**. Morphogenesis means the formation of new structures during development. Morphogens like the hedgehog protein are critical to embryonic development because they diffuse through the tissues of the developing embryo and establish concentration gradients of the protein that causes adjacent cells to assume different fates. The hh family of genes is the best-studied class of morphogen-encoding loci.

Homologues of *hh* occur in many other organisms, including humans. Several different hedgehog gene loci have been identified in humans. The best-studied one is called **sonic hedgehog** (***SHH***). Secretion of SHH protein by cells of the human notochord and neural tube establishes a gradient that induces and organizes cells of the brain and spinal cord. Similarly, secretions of SHH in the developing limb are responsible for the asymmetrical formation of the digits.

SHH mutations in humans are rare. They may produce severe birth defects or barely perceptible ones (Figure 15.6). Those who are severely affected (Figure 15.6A) exhibit failure of mid-face and brain development (called holoprosencephaly). Characteristically, these infants have clefts of the lip and palate, closely spaced eyes, and absence of the forebrain. But the same mutation in the same family may result in a very different phenotype. The woman in

(A) (B)

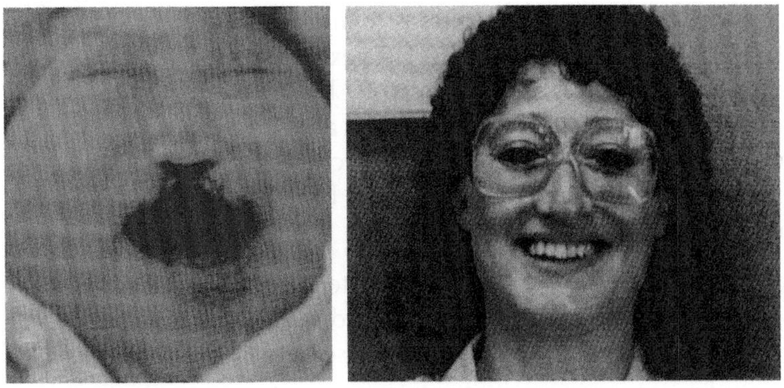

FIGURE 15.6
Variable expression of identical mutations in the SHH gene. (A) An infant has a small head, closely spaced eyes, missing nose bones, and a cleft lip. (B) The mother of the infant shown in (A). She has the same SHH mutation as her child, yet her only clinical manifestation is absence of a single tooth—a central upper incisor.

Figure 15.6B is the mother of the child in Figure 15.6A. Her only manifestation is a missing single central incisor tooth. How can we account for this widely variable expressivity? It seems likely that other genes have modified expression of the SHH mutation, but this supposition prompts the tantalizing and still unanswered question: Which modifying genes, and which actions do they modify?

Programmed Cell Death (Apoptosis)

Integral to the process of cell growth and embryonic development is its opposite, **programmed cell death** or **apoptosis**. Apoptosis is an orderly process, distinct from the one that disposes of cells injured or killed by trauma, infection, or other environmental insults. Apoptosis is systematic, leading to the orderly disassembly of particular molecules, followed by disappearance of the cell marked for destruction. Cells undergoing apoptosis shrink; their nuclear chromatin is broken into fragments and clumps near the nuclear membrane. Finally, neighboring cells or circulating scavenger cells ingest the dead cells.

Apoptosis is a constant feature of human life. Early in development, our hands are shaped like paddles. Then, rows of cells are removed by apoptotic events to form fingers. Heart formation requires remodeling to yield our four-chambered organ, and apoptosis is the mechanism for such remodeling. In the brain, many more cells are formed in the embryo and fetus than exist at birth, with the excess being removed in an orderly fashion. All of this raises the question: How is apoptosis controlled? Not surprisingly, the answer is that cell suicide is controlled by suicide (or cell death) genes. These genes were identified first in *Caenorhabditis elegans* (roundworm), and homologues have been found elsewhere. In the worm, more than a dozen genes act either to promote apoptosis or prevent it. In vertebrates, more than 25 such genes are known. Signals from inside or outside cells affect the action of this "cassette" of death genes. Some of these genes code for proteases that degrade proteins; other products are endonucleases that degrade chromosomal DNA. Still others are transcription factors or cell surface receptors.

It is widely believed that abnormalities in apoptosis underlie such important human conditions as congenital heart defects, limb deformities, autoimmunity (which requires removal of cells signifying "self"), and cancer. One notable example of defective apoptosis involves the *BCL-2* gene. *BCL-2* is an anti-apoptotic gene expressed in B lymphocytes (antibody-reducing cells). Mutations at the *BCL-2* locus lead to a form of B-cell lymphoma because the gene is persistently active, becomes an oncogene, and results in uncontrolled anti-apoptosis and B-cell proliferation. We shall return to this subject in Chapter 16.

TERATOGENS

An intricate network of genes controls embryonic and fetal development. This network is carefully regulated so that the genes involved are turned on when and where they should be. An extra round of cell division may be as lethal as impairment of a single patterning gene or cell signaling molecule. In one way or another, the genes regulating development act on the microenvironment of the developing embryo. They control, for example, the migration of cells and the interaction between cells. Most birth defects are caused by mutations of the genes responsible for development, but a small fraction of birth defects are the result of substances in the external environment that interfere with the finely tuned orchestra of events leading to a healthy newborn. Substances in the environment that produce birth defects or increase their frequency are called **teratogens**.

As noted in Table 15.2, a variety of classes of substances act as teratogens. Nutrients or lack thereof (starvation), hormones, medications, viruses, X-rays, addictive substances, and occupational hazards have the potential to be teratogenic. Here are some general features of teratogens.

TABLE 15.2 Classes and Examples of Human Teratogens

Class	Teratogen Example
Nutrients	Starvation Vitamin A excess
Hormones	Diethylstilbestrol (DES) Adrenal steroids
Medications	Thalidomide Tretinoin (Accutane) Valproic acid
Addictive substances	Nicotine Alcohol
Viruses	Rubella Cytomegalovirus Herpes simplex
Radiation	X-rays
Occupational hazards	Lead Mercury

- Different teratogens often cause specific patterns of birth defects. The risk depends on the mother's genotype, the gestational age at the time the teratogen is encountered, and the magnitude of exposure.
- Most teratogens exert their deleterious effects during the embryonic period; that is, during weeks 3–8 of gestation. This is the period of maximal organogenesis and differentiation and, therefore, of greatest sensitivity to external insults. This concept is illustrated in Figure 15.7. For example, central nervous system development occurs throughout the embryonic period and through much (or all) of the fetal period as well. It is not surprising, then, that the sensitive period of risk to the central nervous system is long, explaining why abnormalities of the brain are so often encountered in teratogen-producing birth defects. In contrast, limb development occurs during a narrow window between weeks 4 and 6. Therefore, exposure to teratogenic substances before or after this sensitive period would not be expected to be harmful to limb development.
- We know very little about which genetic programs are affected by particular teratogens. This ignorance extends to not understanding why certain human teratogens are harmless in mice or monkeys.
- Teratogens are identified empirically by noting a characteristic pattern of birth defects in a cluster of children born to women exposed to a particular substance. This is how all the teratogens shown in Table 15.2 were identified.
- Prevention of birth defects caused by teratogens is accomplished by avoidance of any substance shown to be teratogenic or suspected of being so. This translates into recommending that any woman planning to become pregnant, or believing she has become pregnant, abstain from as many substances conveying a risk as possible.

We will follow these general remarks with a few well-characterized examples.

Thalidomide

In the late 1950s, thousands of newborns in Europe were born with absent or malformed arms and legs. It was discovered that their mothers had been taking a new drug for morning sickness called **thalidomide**. Exposure to the medication occurred during the critical period of limb development, weeks 4–8. When the drug was withdrawn from the market, newborns with these limb defects vanished. The United States was spared this tragedy because an astute physician

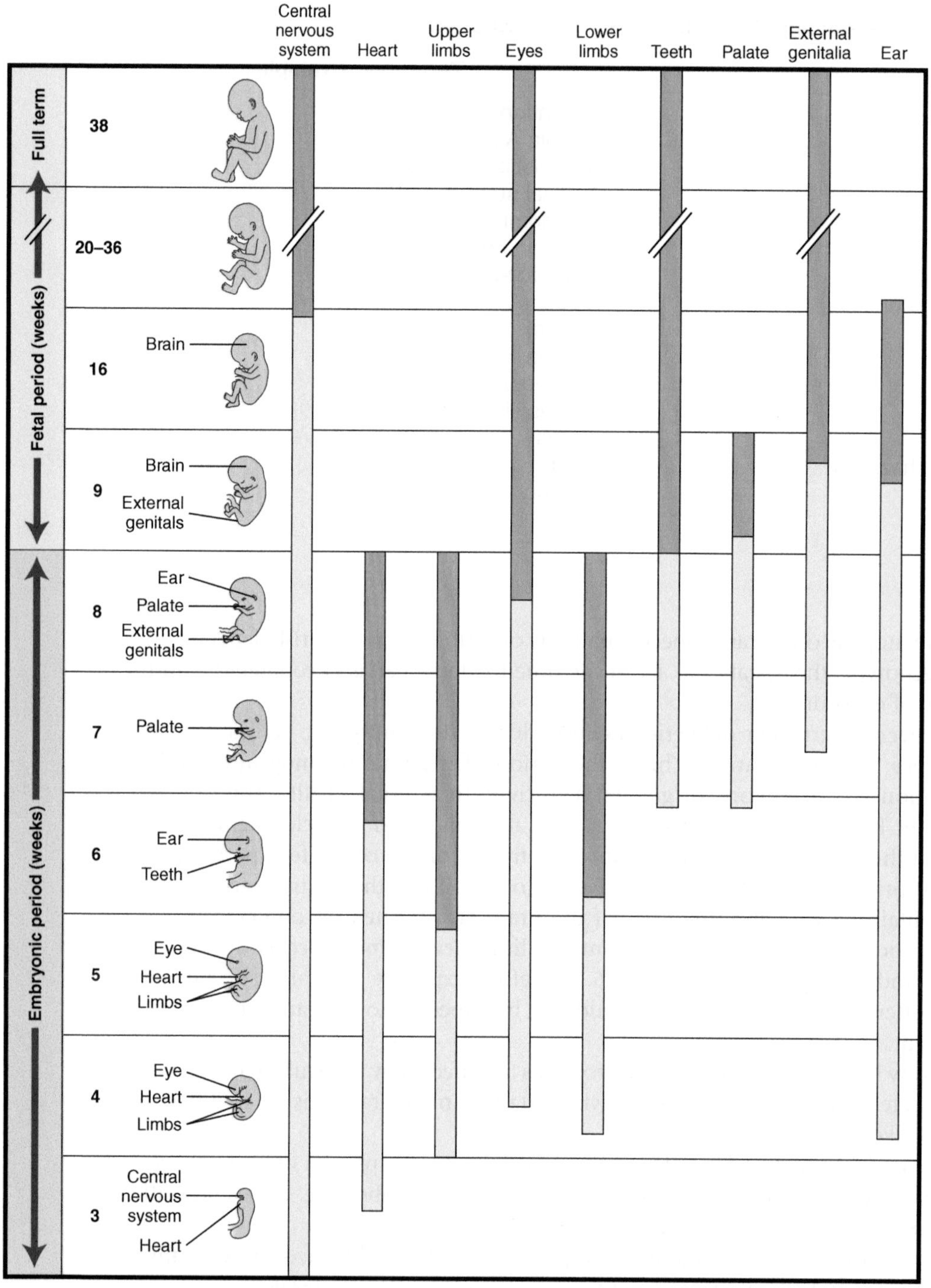

FIGURE 15.7
Critical periods of development in humans. Green bars indicate periods when teratogens may cause major structural abnormalities. Blue bars denote less sensitive periods. For example, exposure during week 3 affects the central nervous system and the heart because these organ systems are developing then. By week 5, the brain, eye, heart, limbs, and ears are in the period of sensitivity to teratogens. By age 16 weeks, only the brain remains highly sensitive to teratogenic effects.

working for the Food and Drug Administration (FDA) barred the marketing of thalidomide there because she was concerned about results in monkeys which suggested that thalidomide might be teratogenic. Subsequently, we have learned that thalidomide's effects on limb formation very likely reflect the effect of the drug on blood vessel growth in the arms and legs.

Alcohol

Imbibing even as little as one alcohol-containing drink per day puts a pregnant woman at risk of having a baby with **fetal alcohol syndrome** (FAS). This syndrome produces more subtle effects than, say, exposure to thalidomide, but is every bit as damaging. Children with FAS have a small head, a flat face, and often other facial features such as a thin upper lip and short nose. More severe than these anatomic features are the effects on the central nervous system, which may include cognitive disability, learning difficulties, and other manifestations of central nervous system dysfunction.

Rubella

German measles (**rubella**) is the best-characterized viral teratogen in humans. In the 1960s, an epidemic of rubella in the US caused 30,000 stillbirths and at least 20,000 cases of birth defects. The most severe problems resulted from exposure to the virus during the first trimester of pregnancy. The most common defects were cataracts, deafness, and abnormalities of the heart. After development of an effective anti-rubella vaccine, rubella essentially disappeared in the US, and with this disappearance went the birth defects caused by the virus. Other viruses, too, may be teratogenic. These include herpes simplex, varicella (chickenpox virus), and cytomegalovirus (CMV).

ILLUSTRATIVE EXAMPLES

Although congenital anomalies occur in only 5–7% of live newborns, they have a profound effect on child health. Twenty percent of infant deaths can be attributed to birth defects. In addition to mortality, birth defects are a major cause of morbidity through their effects on the brain, the heart, the limbs, and other organs and organ systems.

Birth defects have many different causes (Figure 15.8). About 50% show complex inheritance, implying that they result from abnormalities in two or more genes interacting with the environment (see Chapter 13). Examples include cleft lip and palate, congenital heart defects, and neural tube abnormalities. About 25% of birth defects are caused by chromosomal imbalance, particularly trisomy for numbers 13, 18, and 21, and monosomy for the X (see Chapter 11). Single-gene mutations cause about 20% of birth defects. Some are inherited as autosomal dominant traits, others as autosomal recessives or X-linked ones (see Chapter 12). Finally, environmental teratogens are responsible for about 5% of birth defects.

We will now discuss three particular birth defects. One is multifactorial, another is caused by a single-gene mutation, and the third involves an environmental teratogen. They have been chosen to illustrate particular aspects of the cause, mechanism, and management of birth defects.

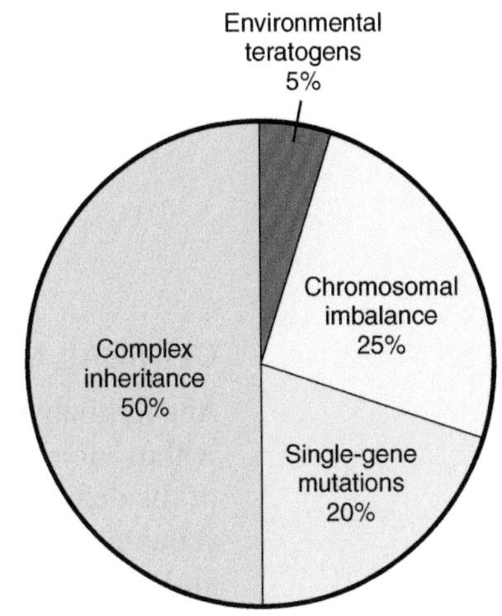

FIGURE 15.8
Relative contribution of multifactorial (complex) traits, single-gene mutations, chromosomal abnormalities, and environmental teratogens as causes of human birth defects.

Neural Tube Defects (NTD)

INCIDENCE AND EPIDEMIOLOGY

Neural tube defects (**NTDs**) are a leading cause of stillbirth, death in early infancy, and disability in children who survive. NTDs occur in about 1 in 500 newborns in the US; in Ireland, the incidence is much higher—about 1 in 100. This difference is unexplained.

PATHOLOGY

As mentioned earlier in this chapter, formation of the neural tube is among the earliest events in human embryogenesis. It occurs during weeks 3 and 4, and the neural tube is closed by the end of week 4. NTDs result from failure of the neural tube to close. As seen in Figure 15.9, NTDs occur in two clinical forms: anencephaly and spina bifida. **Anencephaly** results from failure in closure of the anterior neural tube. As a result, the forebrain, meninges (membranes lining the central nervous systems), skull vault, and overlying skin are absent. In **spina bifida**, the failure in closure occurs posteriorly, usually affecting the lumbar region. Manifestations vary considerably, from an innocent defect in vertebral closure causing no clinical problem on one hand, to protrusion of neural elements and meninges (called meningomyelocele) on the other.

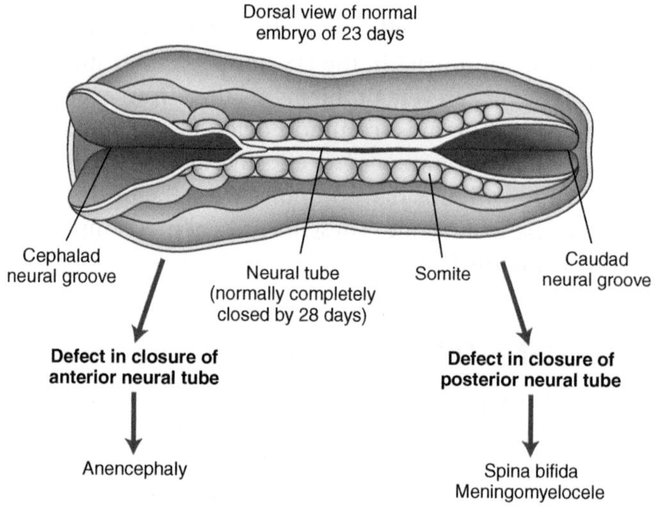

FIGURE 15.9
Neural tube defects resulting from defects in closure. Defect in closure of the anterior neural tube causes anencephaly. Defective closure of the posterior neural tube produces a variety of clinical findings, depending on the location and severity of the lesion. These posterior defects are generally referred to as spina bifida.

CLINICAL MANIFESTATIONS

Anencephaly is a lethal birth defect. Many infants are stillborn; those born alive usually die within a few hours. The clinical course of spina bifida varies greatly depending on the magnitude of the defect in closure. Those individuals with minor abnormalities in the posterior part of vertabrae have no clinical manifestations. Those with open defects affecting neural structures and their covering often have lower limb paralysis and dysfunction of bladder and bowel.

PRENATAL DETECTION

In the presence of an NTD, a protein (**alphafetoprotein**, or **AFP**), made by the liver and circulating in the serum of the developing fetus, leaks into the maternal serum and the amniotic fluid. This provides three possible avenues for prenatal detection of NTDs. The first involves measuring the AFP concentration in maternal serum during the first trimester. Significant elevations increase the likelihood of NTDs markedly. The second avenue involves measuring AFP in the amniotic fluid. The third examines the fetus by ultrasound, looking for skeletal abnormalities characteristic of anencephaly or spina bifida.

ETIOLOGY

In a small minority of infants with NTDs, single-gene mutations, chromosomal abnormalities, or environmental teratogens have been identified. In the large majority, however, identification of susceptibility genes has been lacking. Thus, it came as a great surprise that a single vitamin, **folic acid**, plays a critical role in NTDs. The evidence for this involvement came as follows:

- The risk of NTDs was found to be inversely correlated with maternal serum folate concentration; when serum folate fell below 200 micrograms per liter (μg/l), the incidence of NTDs increased significantly.

- Maternal serum folate concentrations are strongly influenced by dietary folate and may become abnormally low during pregnancy, even with a "normal" intake of 200 mg of folate per day.
- Metabolic studies supported these epidemiologic findings. Folate is a required cofactor in the cycle of reactions by which the amino acid homocysteine is converted to methionine. In the presence of folate deficiency (or of a mutant enzyme methylene tetrahydrofolate reductase, abbreviated MTHFR) that impairs the ability to make methionine from homocysteine, serum homocysteine concentrations rise. Such a rise is seen regularly in women with low serum folate concentrations. Further, mothers of babies with NTDs were twice as likely to be homozygous for MTHFR mutations as were mothers of healthy babies.

These data, taken together, suggested that folic acid deficiency causes NTDs.

TREATMENT AND PREVENTION

The already mentioned data about folic acid led to a true breakthrough in the prevention of NTDs. The incidence of NTDs has fallen by 75% since folate supplements began to be prescribed for women of childbearing potential. Women who intend to become pregnant are urged to supplement their diet with 400–800 mg of folic acid per day, starting 1 month before anticipated conception and continuing for 2 months after conception. The remarkable results obtained with this public health intervention have led some to propose that everyone should consume a diet supplemented with folate, so that by protecting women with unplanned or undetected early pregnancies, an even greater fraction of NTDs can be prevented.

Achondroplasia

FREQUENCY AND EPIDEMIOLOGY

Achondroplasia is the most common cause of human dwarfism. It occurs in 1 in 15,000 to 1 in 40,000 live births, affects both genders equally and is found in all ethnic groups.

PATHOLOGY AND CLINICAL MANIFESTATIONS

People with achondroplasia have a large number of defects. Their skull is large, the forehead prominent, and the midface underdeveloped; the limbs are shortened and the trunk is elongated (Figure 15.10). At birth, height may be near normal, but it falls well behind with age. The skeletal abnormalities often lead to prominent secondary consequences: ear infections, hydrocephalus (abnormal accumulation of fluid within the brain), spinal cord compression, and respiratory dysfunction. Radiographically, the main bones of the arms and legs are tubular in shape, and there is widespread disorganization of skeletal cartilage. Characteristic abnormalities are seen in the skull and vertebral bodies as well.

ETIOLOGY

Achondroplasia is caused by mutations in the fibroblast growth factor 3 receptor (*FGFR3*) gene. Recall that *FGFR3* is one of the family of transmembrane receptor proteins that participate in cellular signaling during embryonic and fetal development. They interact with a number of fibroblast growth factors (FGFs) and thereby participate in blood vessel formation and limb bud and neural differentiation—all through regulation of cell growth.

Virtually all cases of achondroplasia are caused by a single nucleotide substitution in the *FGFR3* gene, which alters one amino acid in the FGR3 protein (Gly380Arg). Other mutations in this gene cause other skeletal disorders in humans. Very few genetic disorders are as mutation-specific as in achondroplasia. In this regard, it resembles sickle cell disease. The Gly380Arg mutation is a gain-of-function one, resulting in activation of the receptor (FGF3) in the absence of fibroblast growth factor.

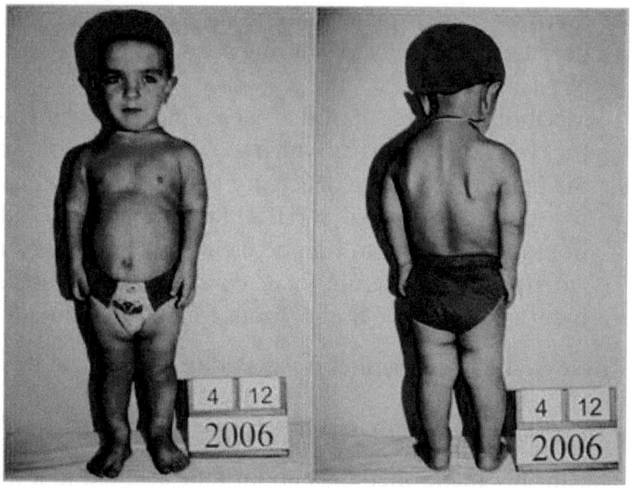

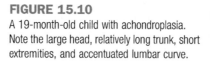

FIGURE 15.10
A 19-month-old child with achondroplasia. Note the large head, relatively long trunk, short extremities, and accentuated lumbar curve.

INHERITANCE

Achondroplasia is inherited as an autosomal dominant trait which is fully penetrant. Offspring of an affected parent have a 50% chance of being affected, but such affected adults usually do not reproduce. Most patients (at least 80%) are born to two normal parents. Why is this so, and why hasn't the incidence of achondroplasia decreased progressively over time? The answer is that a high rate of new mutations occurring in the parental germ cells maintains mutant gene frequency. In a preponderance of cases, the new mutation occurs selectively in the paternal germ cells. Achondroplasia is among a few disorders where this is so. As important, the mutation rate increases with paternal age. These paternal effects reflect the fact that the older a man is, the more rounds of replication have occurred in his germ line and, with that, the higher the frequency of mutations in sperm cells.

In those instances where two achondroplastic dwarfs reproduce together, two-thirds of the offspring are affected and one-third is normal. This deviation from the expected 25 : 50 : 25 proportion of homozygous affected : heterozygous : homozygous unaffected is explained by the fact that homozygosity for the mutation causing achondroplasia is a genetic lethal, meaning that death occurs before birth.

INTERSECTIONS: BIOLOGY AND EPIDEMIOLOGY

Paternal Age and Birth Defects

The association between maternal age and autosomal trisomies is clear. Women who become pregnant over age 35 years have a progressively increased risk of having offspring with trisomy 21 (Down syndrome), trisomy 13, or trisomy 18 (see Chapter 11).

What about older fathers? Does advanced paternal age increase the risk of birth defects in their offspring? These questions have been addressed repeatedly during the past 30 years, but definitive answers have been few. There is general agreement that two autosomal dominant disorders leading to dwarfism demonstrate a paternal age effect. In achondroplasia (discussed in this chapter) and Apert syndrome, most affected patients represent *de novo* mutations because affected adults do not reproduce. Such new mutations are strongly associated with paternal age. These findings are consistent with Penrose's theory, advanced in 1955, in which a "copy error" mechanism leads to the accumulation of mutations in spermatogonia undergoing continuous replication from puberty to the end of life. The older the father, he posited, the greater the likelihood that deleterious mutations would accumulate in male gametes.

INTERSECTIONS: BIOLOGY AND EPIDEMIOLOGY—Cont'd

Based on these results, many epidemiologic studies have been carried out. The results have been conflicting—some reporting positive association between paternal age and a long list of common and rare birth defects, others finding no such association. In the largest and best-controlled retrospective study carried out on more than 5 million births in the United States, paternal age was associated with a slightly increased risk of congenital heart defects, Down syndrome and other chromosomal abnormalities, esophageal and tracheal abnormalities, and some musculoskeletal conditions. Surprisingly, paternal age younger than 20 years was also associated with a modestly increased risk for selected birth defects. A large cohort study conducted in British Columbia supported these findings, including the increased risk in fathers under age 20. In both studies, the associations between paternal age and birth defects were weak, suggesting that paternal age plays a small role in the etiology of birth defects.

What to make of all this? Although males are capable of fathering children into their later years (the oldest recorded being 92), a combination of a well-documented decrease in sperm count and ejaculate volume, plus the associations just mentioned, raises questions for older men contemplating fatherhood. As the average age of becoming a parent rises in both genders in our society, this biological matter and its sociologic trade-offs assume greater relevance.

Accutane Embryopathy (AE)

EPIDEMIOLOGY

Since 1982, thousands of men and women have taken the retinoic acid derivative **isotretinoin** (brand name **Accutane**) for severe acne. Eighty percent of users are women under the age of 30. Survey data indicated that about 35% of pregnant women who took Accutane during the first trimester had spontaneous abortions and that 42% had a living child with birth defects characteristic enough to be called Accutane embryopathy. Neither the incidence nor the severity of birth defects was dose-related.

CLINICAL FINDINGS

Accutane embryopathy (AE) is characterized by a large number of major and minor malformations. Not all abnormalities occur in each affected newborn. The most common birth defects involve the skull and face, the brain, and the heart. Specifically, malformed ears, underdevelopment of the skull and facial bones, a variety of congenital heart defects, and hydrocephalus are prominent findings, as is underdevelopment of the thymus. The diagnosis of AE is based on the history of Accutane used during the first trimester of pregnancy and on the pattern of birth defects found.

ETIOLOGY

Accutane causes birth defects because it is a derivative of **retinoic acid**, an endogenously produced small molecule important in early embryonic development. Retinoic acid, produced from vitamin A, diffuses into cells and binds to gene sequences coding for retinoic acid receptors. Thus, retinoic acid is a transcription factor, with particular specificity for development of the anterior–posterior axis and the limbs. It is known, for instance, to affect HOX expression in mice. Accutane causes birth defects because the carefully regulated effects of retinoic acid on early development are perturbed by the administration of this medication—and the increased tissue concentrations of retinoic acid derived therefrom.

MANAGEMENT AND PREVENTION

Infants born with AE often require surgery to correct or ameliorate problems such as hydrocephalus, cardiac abnormalities, and malformed ears. Prevention takes three forms. Women who find out that they have taken Accutane during their first trimester often choose to abort rather than risk having a child with AE. At a second level, efforts have been undertaken to prevent women of childbearing potential from taking Accutane unless it is certain that they are not pregnant. Strenuous efforts, such as urging prescribing physicians to require use of birth control methods in women before starting Accutane, and, again, 1 month after initiating treatment, have been and are being tried, but infants with AE continue to be born. Of course, all AE could be prevented by removing Accutane from the market. This has been resisted because severe acne is a serious problem, medically and socially.

REVIEW QUESTIONS AND EXERCISES

1. Choose the phrase in the right column that best matches the term in the left column.

a. somites	1. 40-billion-cell result of zygote formation
b. achondroplasia	2. process resulting in formation of three embryonic layers
c. apoptosis	3. programmed cell death
d. maternal-effect genes	4. major class of segmentation genes
e. cell proliferation	5. head-to-tail division of embryo
f. teratogens	6. occurs via mitotic divisions
g. neural tube defects	7. weeks 1–8 of human gestation
h. neonate	8. caused by mutations in fibroblast growth factor receptor
i. segmentation	9. small mesodermal blocks
j. sonic hedgehog	10. control development before zygote is formed
k. differentiation	11. process of cell specialization
l. embryonic period	12. an illustrative morphogen
m. homeobox genes	13. environmental substances causing birth defects
n. gastrulation	14. anencephaly and spina bifida
o. pattern formation	15. assignment of specific function to embryonic segments

2. Why do we know so much more about developmental genetics in drosophila and mice than we do in humans?
3. Discuss the following sentence: All birth defects are congenital but only a fraction of them are inherited.
4. Homeobox genes are conserved from drosophila to humans. What does this tell us about this gene class?
5. Suppose that a particular gene controlled a specific event in morphogenesis of the eye.
   a. What effect(s) might a gain-of-function mutation of this gene have?
   b. What effect(s) might a loss-of-function mutation of this gene have?
   c. Which of the mutations in (a) and (b) would be dominant? Which recessive?
6. Scientists studying limb development in mice using gene-knockout methods often found that knocking out one homeobox gene locus produced mild abnormalities, whereas severe phenotypes required knocking out at least two homeobox gene loci. How can this be explained?
7. Suppose you discovered a newborn with a pattern of birth defects that has not been reported previously.
   a. How would you investigate the cause of this child's defects?
   b. While counseling the healthy parents of this child, one of them says "lightning never strikes twice" with reference to the risk of recurrence. What is your response?
8. Knowing what you do about the value of folic acid supplements in preventing neural tube defects, should all women of child-bearing age be encouraged to supplement their diets with folic acid? Should such supplementation be mandatory?
9. Embryopathy due to accutane has been recognized for more than 30 years, yet new cases continue to be reported.
   a. Why hasn't accutane been taken off the market?
   b. Why doesn't the US Congress pass a law banning use of accutane?

CHAPTER 16

The Genetics of Cancer

CHAPTER OUTLINE

Human Genes and Genomes. DOI: 10.1016/B978-0-12-385212-0.00016-0

CORE CONCEPTS

Cancer is a large group of diseases that result when cells divide out of control and acquire the ability to spread beyond their prescribed borders. Cancer is a genetic disease because it is caused by mutations, but it is rarely inherited because these mutations occur, in most cases, in somatic cells rather than in germ cells. Cancer affects one in three people during their lifetime and is the second leading cause of death in the United States. In women, the most common and most lethal cancers originate in the lung, breast, and colon. In men, cancers of the lung, prostate, and colon predominate. In sharp contrast to cancer, also called a **malignant tumor**, a benign tumor is defined as a collection of cells that grows out of control but does not spread into surrounding tissues.

Fundamentally, cancer results from dysfunction of the repeating pattern of mitotic cell growth and division called the **cell cycle**. This cycle (composed of the four phases G_1, S, G_2, and M) is tightly controlled by the concerted action of hundreds of genes. Through their products, these genes provide the components which make it possible for a single cell to grow, replicate its DNA, prepare to and then divide into two identical daughter cells, and, when necessary, die. Two large families of proteins—the **cyclins** and the **cyclin-dependent kinases**—regulate the transition from one phase of the cell cycle to the next (for instance, from G_1 to S). In addition to these two large classes of proteins, many others—called **checkpoint proteins**—scan and arrest the cell cycle so that any damage that has occurred to the genome (intrinsically or extrinsically) may be repaired. Cancer is the result of some kind of failure of this intricate set of cellular checks and balances to work perfectly.

Cancer cells differ phenotypically from normal cells in many particular ways. Their proliferation is not controlled by such usual phenomena as contact with a neighboring cell, chemical signaling between adjacent cells, and failure to respond to apoptotic signals. Their genomes are unstable, as evidenced by a variety of chromosome abnormalities. They demonstrate immortality, meaning that they no longer observe the usual limits on the number of cell divisions—a process in which **telomeres** and the enzyme that replenishes their length, telomerase, play a critical role. They develop the ability to invade surrounding tissue and to **metastasize** by, in part, stimulating the formation of new blood vessels.

Among genetic diseases, cancer is unique. It is caused not by a single-gene mutation but by a sequence of mutations at different loci in the progeny of a single cell. No one of these mutations by itself produces the cancer phenotype; rather, only their successive nature ultimately topples the cell from healthy to potentially lethal. Cancer is usually sporadic, meaning that it occurs in only one member of a family; this reflects the fact that cancer-predisposing germ-line mutations (which are heritable) are rare, while somatic ones are common. Cancer is the essence of a multifactorial trait in that environmental exposures (such as tobacco, X-rays, bacteria, viruses, dietary substituents) interact with gene mutations in the disorderly process ultimately overwhelming the cell's ability to control its growth.

Mutations in two large classes of genes—**oncogenes** and **tumor suppressor genes**—are the best-studied examples of "cancer genes." Oncogenes are produced by mutations of their normal gene homologue, the **proto-oncogenes**, Once formed, oncogenes are gain-of-function mutations that drive cell proliferation. Oncogenes behave dominantly in that mutation of one of two alleles is sufficient to perturb cellular function. Tumor suppressor genes, in contrast, normally function to inhibit cell division or activate apoptosis. They act recessively in that both alleles must be mutant before they are unable to carry out their usual "braking" function. Hundreds of different mutant oncogenes and mutant suppressor genes have been identified and characterized. Though they vary greatly in structure and function, it is their propensity to arrange themselves in a dysfunctional repertoire that ultimately leads to cancer.

These general remarks will be enlarged upon by discussion of a few illustrative cancers: chronic myelocytic leukemia, retinoblastoma, and breast cancer.

OVERVIEW

The word "cancer" is among the most dreaded in the medical lexicon. With few exceptions, it is more feared by patients and their families than words used to describe any other serious condition. This dread reflects cancer's frequency and lethality.

Frequency

Cancer is a common disease. Over a lifetime, one in three people will develop some form of cancer. Nearly 80% of cases occur in people over age 55, and the incidence rises progressively as people age. Yet cancer is seen at birth and throughout childhood as well. More than 1 million new cases of cancer are diagnosed in the US population each year, and at any one time, 10 million people are being treated.

Lethality

Each year, more than 500,000 Americans die of cancer, making it the second leading cause of death. Mortality from heart disease—particularly coronary artery disease—still exceeds that from cancer, but this may not be true for long. Deaths due to heart disease have declined by nearly 50% in the past 50 years while deaths due to cancer have remained stubbornly constant until the past decade, when they have been inching down. Many experts predict that deaths due to cancer will exceed those from heart disease by the year 2020.

Organs Affected

Cancers develop in every organ, but at widely different rates and with very different consequences. In women, cancers of the lung, breast, and colon account for nearly 60% of new cancers (Table 16.1). In men, cancers of the lung, prostate, and colon account for about the same percentage. Figure 16.1 depicts cancer deaths from four main tumor types, rather than new cases. Again, cancers of the lung, breast, and colon predominate in women (Figure 16.1A), and those of the lung, prostate, and colon in men (Figure 16.1B). Notably, deaths due to lung cancer have increased markedly in women and men during the past half-century, while those due to cancer of the stomach have decreased. These trends reflect, for lung cancer, the rise in the fraction of adults who smoke cigarettes. In children, cancer distribution is quite different, as are its effects. Less than 10% of cancers occur in childhood or adolescence; the most common types of cancer in this age group being acute leukemia, Hodgkin's disease, cancer of the retina (retinoblastoma), and cancer of the peripheral nervous system (neuroblastoma).

TABLE 16.1 Fraction of New Cases of Cancer in the United States According to Tissue of Origin in Men and Women

Male		Female
3%	Melanoma of skin	3%
3%	Oral	2%
14%	Lung	13%
	Breast	32%
2%	Pancreas	2%
10%	Colon and rectum	12%
	Uterus	8%
	Ovary	5%
8%	Urinary	4%
36%	Prostate	
7%	Leukemia and lymphomas	6%
15%	All other	13%

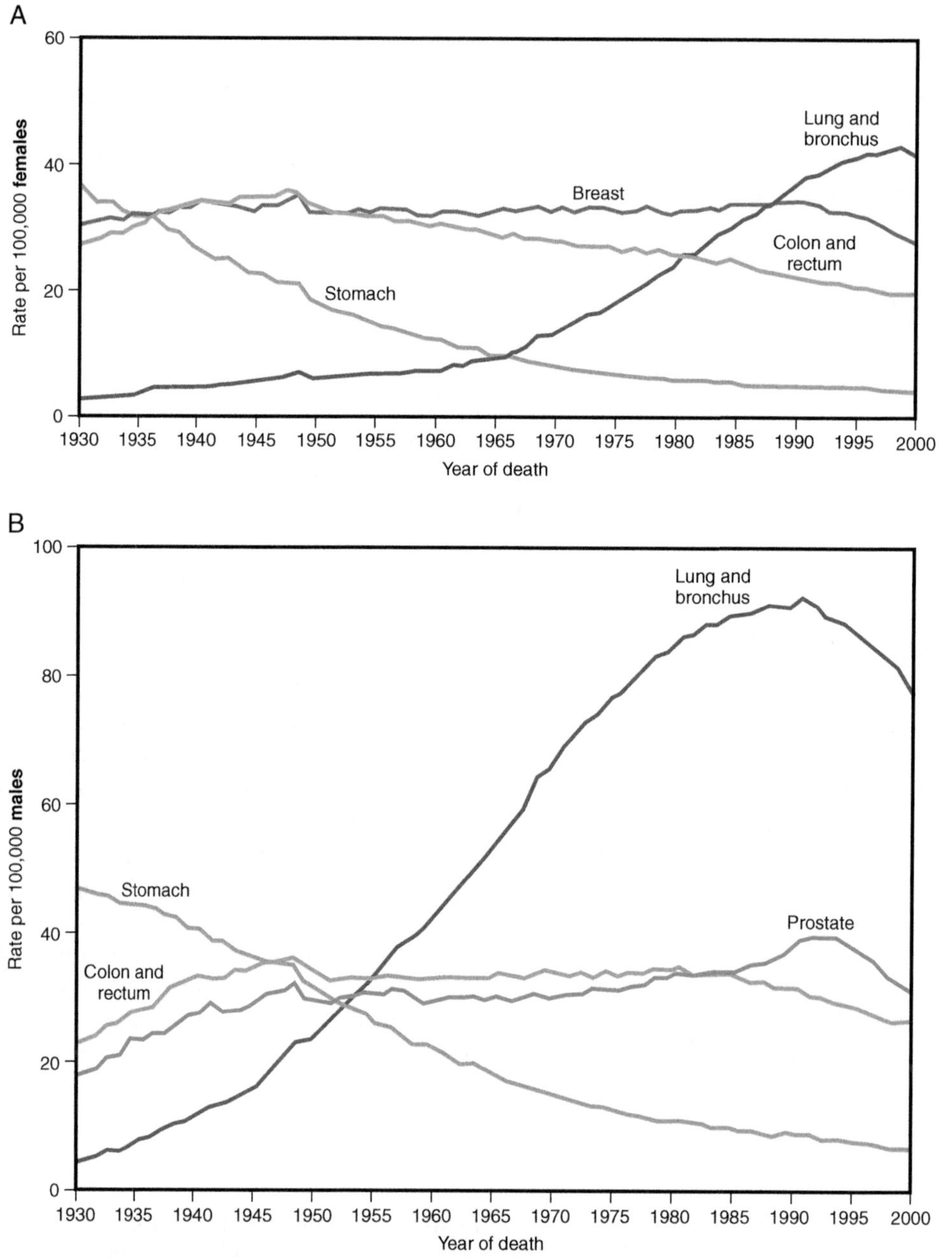

FIGURE 16.1
Rates of US deaths from various cancers in females (A) and males (B) over 70 years from 1930 to 2000. Note the dramatic rise in deaths due to lung cancer (purple line) in both males and females. In contrast, deaths due to breast cancer in women and prostate cancer in men have remained constant. In both sexes, deaths due to stomach cancer have fallen markedly.

Treatment and Prognosis

Three modalities are used to treat cancer: surgery, chemotherapy, and irradiation. Of these, surgery is the "first line" treatment for cancers of solid organs such as breast and lung, while chemotherapy is used first for leukemias and cancers of lymph nodes, such as Hodgkin's disease. The goal of cancer surgery is to remove all cancerous tissue, while the goal of chemotherapy is to kill all of the rapidly dividing cancer cells (while sparing, insofar as possible, healthy cells). Radiation treatment is generally used as an adjunctive modality with either surgery or chemotherapy.

Overall, fewer than 50% of patients with cancer are cured by any or all of the modalities just mentioned. Ninety percent of cancer deaths result from spread of the cancer from its site of origin to other tissues—a process called metastasis. Importantly, cancers that occur most frequently in the first two decades of life, such as leukemia, Hodgkin's disease, and testicular cancer, have a much better prognosis than do cancers of later onset. These cancers in young people are cured more than 80% of the time—a rate considerably higher than that seen in cancers of older adults.

BIOLOGY OF CANCER

Definitions

We must understand a few terms used widely in describing and characterizing the nature of cancer (Table 16.2). **Neoplasia** is a disease process characterized by uncontrolled cellular proliferation. It leads to formation of a **tumor** (or **neoplasm**). **Benign tumors** are circumscribed and do not invade surrounding tissue. Their size and location may not make them "benign" clinically, however, as exemplified by circumscribed tumors of the brain. They must be removed because they interfere with some body process. Cancer (malignancy, or **malignant tumor**) refers to neoplasms that invade neighboring tissues and often spread to more distant sites.

TABLE 16.2 Terms Used to Discuss Cancer

Term	Definition
Neoplasia	A disease process characterized by uncontrolled cellular proliferation
Tumor (or neoplasm)	A growth of cells resulting from uncontrolled cellular proliferation
Benign tumor	A tumor that is circumscribed and does not invade neighboring tissues
Cancer (or malignant tumor)	A tumor that invades neighboring tissues and often spreads to more distant sites

General Types of Cancers

There are three main types of cancer:

- **carcinomas**, which originate in epithelial tissue such as that lining the lung, intestine, or breast;
- **hematopoietic** or **lymphoid**, such as leukemia and lymphoma, which originate in the bone marrow, lymphatic system and peripheral blood;
- **sarcomas**, which originate in bone, muscle, nervous tissue, or connective tissues.

Within each major group, tumors are classified by site, tissue type, and histology.

Cell Cycle Regulation and Dysregulation

Cancer, by definition, results from uncontrolled cellular proliferation. Said another way, cancer is caused by interruption of the normal cell cycle—the series of events that take place between one mitotic division and the next. In Chapter 4, we introduced this cell cycle and briefly described its phases: gap 1(G_1), synthesis (S), gap 2 (G_2), and mitosis (M). We also pointed out that non-dividing cells (such as erythrocytes and brain cells) are in a different phase called G_0. We will now look at the cell cycle in greater detail because its dysregulation is central to understanding the cellular events that go awry in malignancy.

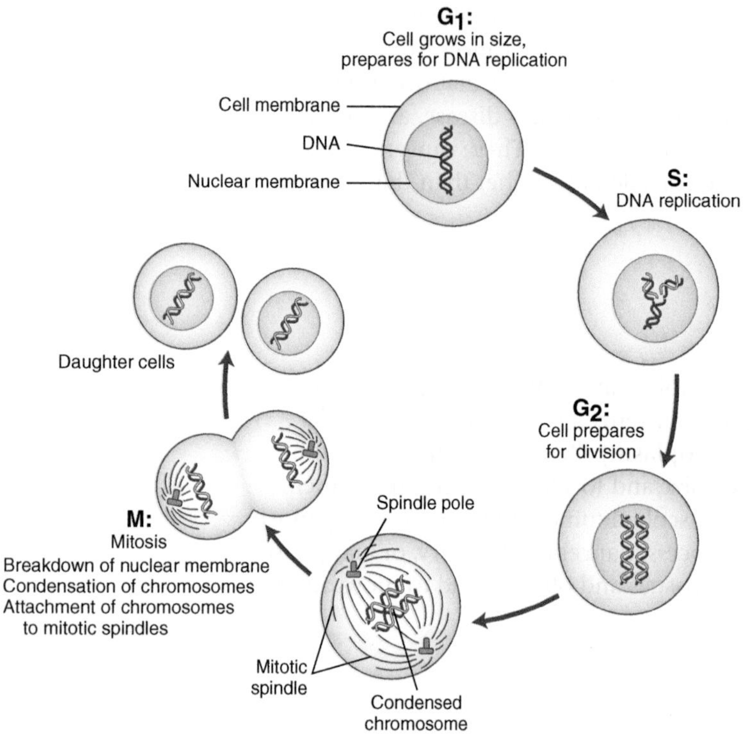

FIGURE 16.2
Details of the cell cycle constituting the events that occur between one mitotic division and the next. The phases of the cycle (G1, S, G2, and M) are discussed in the text, as are the events occurring in each phase.

PHASES

Figure 16.2 describes the four phases of the cell cycle and emphasizes what happens during each phase. After dividing, a cell enters the **first gap phase, G_1**. During G_1, the cell grows in size by making many of the chemicals it needs for maintenance and, in so doing, prepares for S phase. G_1 varies greatly in duration; it is brief in rapidly dividing cells, lengthy in slowly dividing ones, and permanent in non-dividing cells (where it is referred to as G_0).

Synthesis (S) is the phase during which DNA is replicated and chromosomes are duplicated. This duplication produces two identical sister chromatids, which are joined together at the centromere. It is, then, during S that exact copying of the genetic material assures that both daughter cells receive identical copies of chromosomes by the end of mitosis (M). **Gap 2 (G_2)** follows S and is generally of short duration. During G_2, the cell prepares to divide. It grows (but not as much as during G_1) and synthesizes a variety of proteins required for mitosis.

A variety of events occur during the fourth phase of the cycle, **mitosis (M)**: the nuclear membrane breaks down; the chromosomes (each made up of sister chromatids) condense; the mitotic spindle forms; the chromosomes attach to the spindle via their centromeres; the chromosomes segregate to the two poles; and, finally, the cell divides by cytokinesis into two identical daughter cells. During M, many kinds of proteins are needed: those responsible for chromosome condensation; tubulins that make up the mitotic spindle on which chromosomes move; motor proteins that make chromosome movement possible; and proteins that dissolve and reform the nuclear membrane.

How does a cell know when to transition from G_1 to S, or from G_2 to M? This question is fundamental to life. If too little mitotic cell division occurs, organs will not grow or develop and wounds will not be repaired. If too much mitotic division occurs, tumors may result and cancers may grow. Therefore, the cell cycle is monitored and modulated by a large and complex series of events. Hundreds of proteins participate in these events, which we will summarize briefly here.

CYCLINS AND CYCLIN-DEPENDENT KINASES

Two large families of proteins are critical to a cell's passage from one phase of the cell cycle to another. One family is the **cyclin-dependent protein kinases (CDKs)**, enzymes that add phosphate groups to proteins, thereby modifying their activity. The second family is the **cyclins**, proteins that combine with CDKs and determine which proteins a particular CDK will phosphorylate and when. In some instances, phosphorylation will activate a target protein; in others, phosphorylation will inactivate it (Figure 16.3). For instance, one **CDK–cyclin complex** will activate target proteins required for DNA replication at the start of S phase. A different CDK–cyclin complex will activate a different set of proteins needed for chromosome condensation and segregation at the start of M phase. A particular CDK has the ability to

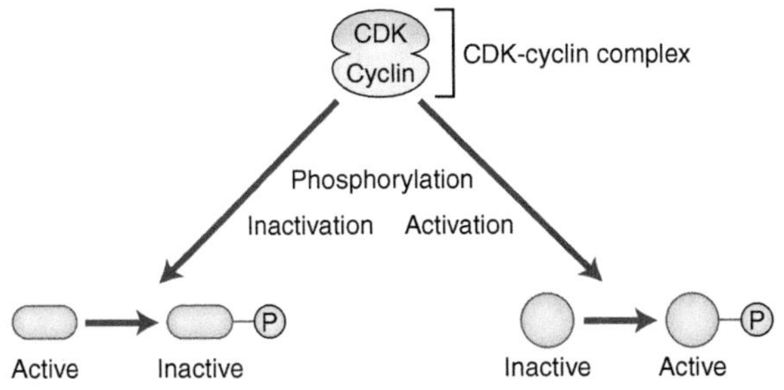

FIGURE 16.3
Cyclin-dependent kinases (CDK) control the cell cycle by phosphorylating or dephosphorylating particular proteins. In some instances phosphorylation activates the target protein, in other instances dephosphorylation is the activating signal.

phosphorylate many different proteins. A particular cyclin serves as a guide, determining which protein targets will be phosphorylated.

Cyclin formation is timed to occur only during that part of the cell cycle when that cyclin is active. Thus, one set of cyclins is synthesized during G_1, binds to particular CDKs that target a particular group of proteins, and then these cyclins are degraded as the cell transitions from G_1 to S. Other cyclins are made and broken down during the G_2 to M transition (Figure 16.4). To simplify, the two critical phases of the cell cycle (S and M) are controlled by two different repertoires of cyclins and CDKs, each activating a large and different set of proteins. As will be discussed subsequently, mutations that interfere with this intricate control of cell cycle phase transitions often lead to cancer.

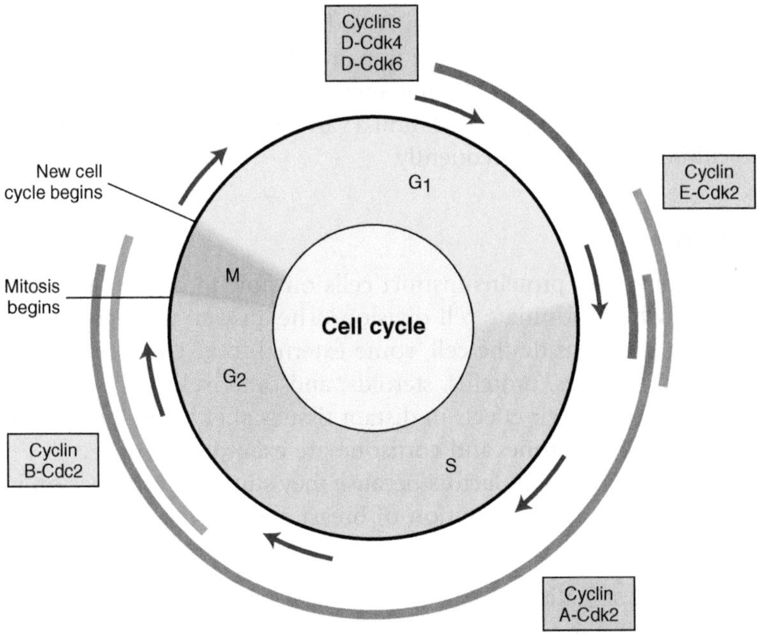

FIGURE 16.4
Temporal expression of different cyclin–CDK complexes during the cell cycle. The cyclin A–CDK2 complex is active throughout S and G_2. In contrast, cyclin B–Cdc2 is active during G_2 and M. Other complexes act during other phases and transitions. More details are provided in the text.

CELL CYCLE CHECKPOINTS

A cell's genome may be damaged by agents in the environment or by random errors occurring during DNA replication and chromosome segregation. An intricate set of reactions has evolved to arrest the cell cycle while repairs caused by these errors are made. These reactions (or controls) are called **checkpoints** because they check the integrity of the genome (and the cell cycle machinery) before permitting the cell to proceed to the next phase.

G_1 to S

When chemical mutagens or radiation damage DNA during G_1, DNA replication is postponed until the damage is repaired. The key molecular species that operates this checkpoint is a protein called **p53**. p53 is a transcription factor that induces the expression of DNA-repair genes. It also acts indirectly to inhibit the activity of the cyclin–CDK complex that normally drives the cell from G_1 to S. Mutations in p53 interfere with its checkpoint function, leading to chromosomal rearrangements and gene amplification (from two copies to hundreds). These rearrangements predispose to cancer. Furthermore, in the presence of p53, irreparably damaged cells commit suicide via apoptosis. In the absence of p53, damaged cells may proliferate, thereby being another predisposing factor toward malignancy.

G_2 to M

Damage to DNA during G_2 calls for repair prior to entering M. As with the G_1 to S checkpoint, many proteins act to halt progression from G_2 to M until the genome has been scanned and necessary repairs made. RAD 9 is a protein critical to this checkpoint. In its presence, many breaks in DNA can be repaired. In its absence, damaged cells will die.

Spindle

A third checkpoint occurs during M phase and is called the **spindle checkpoint**. As chromosomes condense and attach to the mitotic spindle, sometimes one or more chromosomes fail to attach. Under this circumstance, chromosome separation and anaphase movement of chromosomes are delayed until all chromosomes are attached to the spindle.

Cell cycle checkpoints do not determine whether a cell will divide. They do play a major role in ensuring genomic stability. When these checkpoints are impaired by mutation, a variety of changes in chromosome structure may occur—including aneuploidy, translocations, deletions, and gene amplifications. Cancer cells regularly exhibit a variety of these abnormalities in chromosome structure, as will be discussed subsequently.

Initiators of Cell Division

The cyclins, CDKs, p53, and many other proteins instruct cells on how to divide (and how not to). But how do cells know when to initiate cell division? The answer: through a large number of **signaling molecules**, some inside the cell, some external to it. One class of signaling molecules is hormones: peptides, proteins, steroids, and other substances that are made in one tissue (or gland) and exert their effects in distant tissues after being transported in the blood. Thyroxine, estrogen, testosterone, and cortisone are examples of such hormones. Many hormones function as growth factors because they stimulate cell division in their target organs (such as stimulation of proliferation of breast and uterine tissue by estrogen).

Most growth factors carry out their signaling function—that is, deliver their message—by binding to specific receptors found in the cell membrane of the receiving cell. In turn, these activated receptors transmit the signal sent by the external molecule (a hormone, for example) to intracellular proteins, often referred to as signal transducers. The final common "player" in this signal transduction pathway is generally a transcription factor, a protein that binds to particular genes in the nucleus either to promote or inhibit cell proliferation.

Phenotypic Changes in Cancer Cells

Cancer cells differ from normal cells in many ways. The cancer phenotype has four major characteristics: **uncontrolled cell proliferation**, **genomic instability**, **immortality**, and the **ability to disrupt local and distant tissues**. Each of these four deserves detailed description (Table 16.3).

TABLE 16.3 Phenotypic Characteristics of Cancer Cells*

Characteristic	Result
Uncontrolled cell proliferation	• Autocrine stimulation • Loss of contact inhibition • Evasion of apoptosis signals • Loss of gap junctions
Genomic instability	• Defective DNA replication machinery • Chromosomal abnormalities
Cellular immortality	• Loss of limits of cell division • Restoration of telomerase activity
Invasiveness	• Ability to metastasize • Angiogenesis • Evasion of immune surveillance

UNCONTROLLED CELL PROLIFERATION

- Normal cells are directed to divide (or not to divide) by signals from adjacent or distant cells. Cancer cells make their own stimulatory signals (**autocrine stimulation**) and are insensitive to negative signals.
- Normal cells cease to grow and divide when they come into contact with other cells. This is called **contact inhibition** and explains the orderly appearance of healthy tissue. Cancer cells have lost the property of contact inhibition and grow in an often chaotic way.
- Normal cells die according to a carefully regulated pathway referred to as programmed cell death (or **apoptosis**). Apoptosis is a critical mechanism of tissue remodeling and of removal of damaged or senescent cells. Cancer cells are resistant to apoptotic signals.
- Normal cells communicate with their neighbors through tiny pores in their cell membranes called **gap junctions**. Small molecules move through these gap junctions and may be important in cell-to-cell cross talk. Cancer cells have lost these gap junctions.

GENOMIC INSTABILITY

- Normal cells reproduce their genomes faithfully by repairing any damage to the DNA replication machinery that occurs randomly or from environmental exposure. Families of DNA-repair enzymes catalyze this process. Cancer cells are often defective in repairing DNA damage, thereby allowing mutations to accumulate.
- Normal cells reproduce their chromosomes with the utmost fidelity, yielding reproducible karyotypes with 46 chromosomes. Cancer cells display a wide variety of detectable chromosomal aberrations: aneuploidy, translocations, deletions, inversions, and gene amplifications.

CELLULAR IMMORTALITY

- Normal cells have limits on the number of cell divisions they are capable of. This is called **cellular senescence**. Such senescence varies from tissue to tissue. It is observed in cultured cells and in living tissue (*in vivo*). Cancer cells have lost the characteristic of senescence and divide indefinitely.
- Normal cells do not express the enzyme **telomerase**, which facilitates replication of telomeres—the repeating nucleotide sequences found at the tips of chromosomes (discussed in Chapters 6 and 11). Telomeres shorten with each round of cell division in the absence of telomerase. When telomeres shorten to a critical size in normal cells, they are no longer able to divide. Cancer cells regularly express telomerase, giving them the property of maintaining telomere length even in the face of rounds of cell division.

TISSUE INVASION AND METASTASIS

- Normal cells stay within carefully controlled boundaries in their respective tissues. Cancer cells acquire the ability to invade surrounding tissue and, often, to travel through the bloodstream or lymphatic channels to distant sites. The latter property, called **metastasis,** produces secondary tumors that are responsible for 90% of cancer deaths.
- Normally, new blood vessels do not form in the adult except to repair wounds. Cancer cells secrete substances that generate new blood vessels capable of providing nutrients and routes through which cancer cells can metastasize. This property is termed **angiogenesis.**
- The human immune system recognizes a variety of microorganisms as foreign and removes them. Similarly, it can recognize and remove cancer cells before they grow to sufficient size to be dangerous. Cancer cells evade this immune surveillance.

CANCER GENETICS

Cancer is a genetic disease in that the observed disruption of normal cell proliferation is caused by mutations of genes that tell cells when to divide or stop dividing and when to remain viable or die.

Somatic Mutations

The pathway by which a normal somatic cell is transformed into a malignant one begins with a single mutation in a single cell. This mutation is necessary but not sufficient to cause cancer. To develop the cellular phenotype characteristic of cancer cells (Table 16.3), a series of mutations of other genes—perhaps as many as 5 to 10—must occur sequentially in the progeny of this cell, that is, in its cellular descendants or clone (Figure 16.5). Ultimately, this accumulation of mutations impairs a cell's ability to control proliferation, and it becomes a cancer cell. This is why cancer is referred to as a **clonal disorder**.

It is not surprising that mutations occur in somatic cells. Given the extraordinary number of mitotic cell divisions (10^{15}) occurring during the lifetime of an adult, and the low but appreciable number of replication errors occurring per base of DNA per cell division, thousands of genome, chromosome, and single-gene mutations must occur in somatic cells. In the main, these mutations are repaired or the cell loses function and dies. Loss of a single mutant cell is generally of no consequence because this cell is surrounded by a vast number of healthy

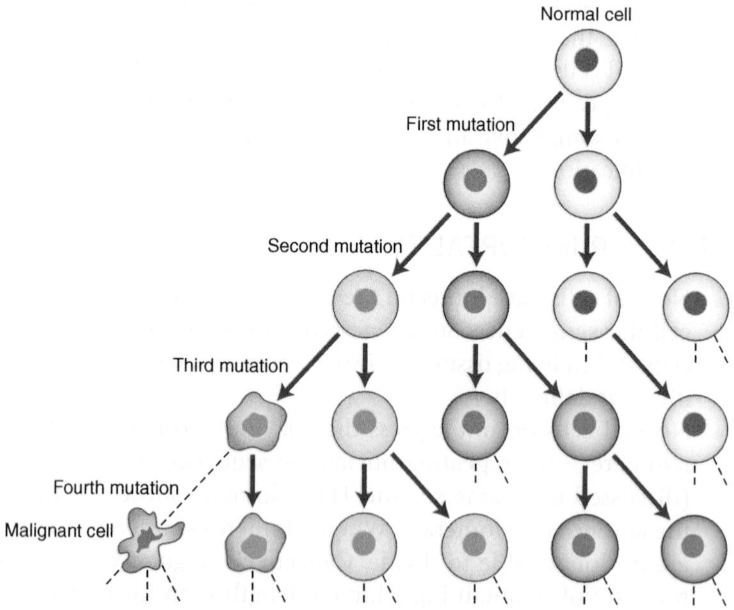

FIGURE 16.5
The clonal nature of cancer. The cancer phenotype results from a series of mutations in the program of a single, normal cell. The first mutation occurs in one gene of a daughter cell; the second in a different one of its progeny, and so on. Each successive mutation weakens the ability of the cell to control its proliferation. Ultimately, a malignant cell is formed.

cells. The genes mutated in cancer are not inherently more prone to mutate than other genes. What distinguishes these **oncogenic** (cancer-causing) mutations is that they occur in genes whose very role is to regulate cell growth and cell death. Because these mutations increase rather than decrease the proliferative capacity of the cell, they are selected for, not against.

Germ-line Mutations

About 95% of cancers occur because of mutations occurring only in somatic cells. The remainder reflect mutations that occur first in the germ cells of affected patients. In some instances such mutations—almost all of which render the individual heterozygous at the involved locus—are both necessary and sufficient to lead to cancer. In other instances, heterozygosity is necessary, but not sufficient. Cancer results when the other allele is mutated in somatic cells. Some oncogenic germ-line mutations, then, are dominant and others are recessive.

Cancer in Families

Perusal of OMIM leads to the recognition that there are about 50 Mendelian disorders with a high risk of cancer. Such disorders are inherited in the typical ways described in Chapters 5 and 12. In addition there are perhaps another hundred Mendelian disorders that predispose to cancer. Although these Mendelian disorders account, overall, for less than 5% of cancer patients, members of such families can often be tested before cancer has appeared or has progressed to become potentially lethal. Therefore, knowledge of the existence of these single-gene causes of cancer has clinical value as well as the scientific value that comes from knowing which specific gene has mutated.

Most forms of cancer are not inherited but are instead sporadic. Yet, it bears saying that most forms of cancer have a higher incidence in relatives of affected patients than in the general population. Overall, first-degree relatives of cancer patients have a two- to three-fold increase in the incidence of cancer, implying that most cancers behave like multifactorial traits in which both genetic and environmental factors play a part (see Chapter 13). Further, in non-Mendelian forms of cancer, concordance of cancer in MZ twin pairs is low and doesn't differ appreciably from that in DZ twins. This suggests that the environment plays a dominant role in the pathogenesis of most cancers compared to that of the germ-line (as distinct from the somatic cell) genome.

Major Questions Posed

Thus far we have said that:

1. cancer results from mutations, most often only in somatic cells;
2. a single mutation in a single cell starts that cell down a dangerous path;
3. a series of other mutations occurring in the clonal descendants of this original cell is required before the cellular phenotype of cancer results;
4. cancer ultimately produces disruption of the mitotic cell cycle that regulates cell proliferation and cell death.

What we've just said prompts several questions:

1. Which genes mutate in cancer?
2. How do mutations in these genes cripple the carefully regulated cell cycle?
3. What environmental factors interact with these mutations and how do they interact?
4. Is the fundamental information being gained about cancer leading to improved treatment of patients with cancer?

Complete answers are not available for any of these questions, but what we do know is substantial and increasing rapidly.

CANCER GENES

The mutant alleles that lead to cancer are often called "**cancer genes**." This widely used term is a deeply ingrained misnomer because all cancer genes are, in fact, mutant alleles of normal genes. Two major classes of cancer genes are known: **oncogenes** and **tumor suppressor** genes. As described in Figure 16.6, oncogenes act dominantly; that is, mutation of one of the pair of alleles is sufficient to lead to uncontrolled cell growth. Tumor suppressor genes act recessively; both alleles must mutate before the cell loses its ability to control cell proliferation.

Gene class	Genotype	Representation	Gene product	Cell proliferation
Protooncogene (dominant)	Homozygous normal		Normal structure and amount	Controlled
	Heterozygous		Normal structure and amount Abnormal structure or excessive amount	Uncontrolled
Tumor suppressor (recessive)	Homozygous normal		Normal structure and amount	Controlled
	Heterozygous		Normal structure and amount Absent or inactive	Controlled
	Homozygous mutant		Absent or inactive	Uncontrolled

FIGURE 16.6
The two major forms of cancer-causing genes. An oncogene acts dominantly; mutation of one allele alters the expression of its gene product. A tumor suppressor gene acts recessively; mutation of both alleles are required before abnormal gene expression is produced. Mutation of one allele only does not disturb function.

Oncogenes

As defined in Table 16.4, a **proto-oncogene** is a normal gene involved in some aspect of cell division, proliferation, or apoptosis. An oncogene is a mutant form of a proto-oncogene that initiates or participates in development of cancer. Oncogenes are gain-of-function mutations because they, directly or indirectly, promote cell proliferation or inhibit apoptosis. Oncogenes may be created by point mutations, gene amplification, or chromosomal rearrangement, any of which may activate it and drive cell proliferation. More than 50 different oncogenes have been described in humans.

TABLE 16.4 Major Classes of Cancer Genes

Class	Definition
Oncogene	A mutant form of a proto-oncogene that initiates or participates in the development of cancer by stimulating cell division or inhibiting apoptosis
Tumor suppressor gene	A gene whose product inhibits cell division and proliferation and activates apoptosis

SEMINAL DISCOVERIES

The first oncogenes were discovered by Michael Bishop and Harold Varmus, in Nobel Prize-winning work carried out in the 1970s. The basis for this work goes back to the turn of the 20th century and to the work of another Nobel laureate, Peyton Rous. He ground up cells from a sarcoma in chickens and injected the tumor tissue into other chickens which also developed sarcomas. The agent responsible for this type of transplantable tumor was identified as a virus, called the Rous sarcoma virus. Decades later it was shown that this virus was composed of four genes. One of these genes, called *src*, was shown to be responsible for its transforming ability; that is, its capacity to turn normal cells into malignant sarcoma cells.

Bishop and Varmus showed that *src* and many other genes capable of causing cancer in animals were homologous to genes (proto-oncogenes) found in normal mammalian cells, including human ones. They identified a large number of these proto-oncogenes and defined the mechanisms by which these normal genes could mutate and become oncogenes. Although viruses rarely cause cancer in humans, they have played a large part in our understanding of oncogenes.

CLASSES

Oncogenes encode proteins that act at many different steps along the pathway controlling cell growth (Table 16.5). Some are components of the cell cycle, such as cyclins and CDKs. Others are growth factors or growth factor receptors. Still others are involved with intracellular signaling. Finally, there are those that inhibit apoptosis or prevent cell senescence. Several examples will serve to describe this variety of oncogenes and their cancer-causing properties.

TABLE 16.5 Classes and Examples of Oncogenes

Class	Example	Types of Cancer
Cell cycle regulator	cyclin D	Esophageal
Growth factor	sis	Glioma
	hst	Stomach
Growth factor receptor	erb-b	Breast
	egfr	Colon, lung, breast
Cytoplasmic signaling		
G-protein	ras	Bladder
Tyrosine kinase	abl	Chronic myelogenous leukemia
Phosphoinositide kinase	pten	Breast, glioma
Serine/threonine kinase	raf	Stomach, melanoma
Transcription factor	myc	Burkitt lymphoma
Cell senescence	telomerase	Many
Anti-apoptotic	bcl2	Chronic lymphocytic leukemia

- **Cyclin D and CDK 4.** Overexpression of cyclin D promotes unscheduled entry of the cell into S phase. Cyclin D overexpression usually results from amplification of the gene (many copies, rather than one). Such amplification is found in about 35% of esophageal cancers and 15% of breast cancers. The gene for cyclin D's partner in the cell cycle, CDK4, is amplified in 12% of glial brain tumors.
- **EGF and EGFR.** Epidermal growth factor (EGF) is a growth factor for many cells. It acts by binding to its cell-surface receptor (EGFR), which activates an intracellular signal transduction pathway that ultimately leads to cell proliferation. The gene coding for EGFR is amplified in a wide variety of cancers, including those of the brain, breast, colon, and ovary.
- **RAS.** One of the first oncogenes discovered, RAS is a small intracellular protein that participates in cell signaling (Figure 16.7). The *RAS* gene encodes a "G-protein"—that is, a protein that binds guanosine triphosphate (GTP) or guanosine diphosphate (GDP).

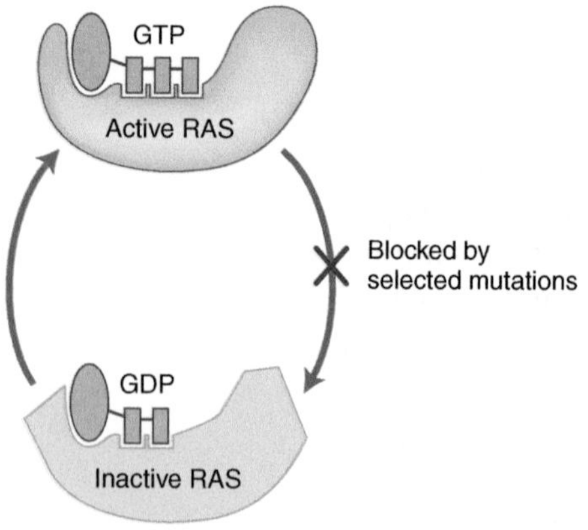

FIGURE 16.7
Regulation of the RAS protein in intracellular signaling. RAS is a G-protein: inactive when bound to GDP; active when bound to GTP. When mutations in RAS "lock" it in its active form, it accelerates cell proliferation. See text for more details.

When bound to GDP, RAS is inactive. When bound to GTP, RAS stimulates downstream signals leading to cell proliferation. Another protein, called GAP, results in conversion of RAS/GTP to RAS/GDP, thereby assuring that RAS/GTP's growth stimulatory property is transient. Point mutations in the *RAS* gene interfere with this cycle of activation and inactivation, causing *RAS* to remain in the "on" position permanently, thereby transmitting a signal for growth. Mutant *RAS* is found in many human tumors.

- **ABL.** ABL is a tyrosine kinase involved in intracellular signaling. Its gene is located on the long arm of chromosome 9. Chronic myelogenous leukemia (CML) is caused by a chromosomal translocation that moves the *ABL* gene to a site on chromosome 22, adjacent to the *BCR* (breakpoint cluster region) gene. The juxtaposition of *ABL* and *BCR* sequences results in formation of a composite (chimeric) protein that is longer than normal ABL and has enhanced tyrosine kinase activity. The enhanced activity drives proliferation of one class of white blood cells, leading to CML. (See the more detailed discussion of CML later in this chapter.)
- **RET.** The vast majority of mutations leading to cancers caused by oncogenes occur in somatic cells only. A familial cancer, called multiple endocrine adenomatosis (MEN), is the exception. In one form of MEN, germline mutations occur in the *RET* gene, which encodes a tyrosine kinase involved in cell signaling. Heterozygotes for these mutations, inherited as an autosomal dominant trait, are at high risk (ranging from 60 to 90%) of developing a form of cancer of the thyroid glands and benign tumors of the adrenal gland.

Tumor Suppressor Genes

The human genome contains dozens of tumor suppressor genes (TSGs)—genes whose normal alleles inhibit cell division or activate apoptosis. If one likens oncogenes to automobile accelerators, TSGs are the brakes. Whereas mutation of one of a pair of alleles of an oncogene is sufficient to drive cell growth toward cancer, both alleles of a tumor suppressor gene must be mutated before their "braking" action is lost. At the cellular level, then, oncogenes are dominant, tumor suppressor genes, recessive.

TSGs affect a variety of reactions in a variety of locations (Table 16.6). Some affect one or more cell cycle checkpoints, others, cell cycle transitions. Some act to promote apoptosis. Still others promote contact inhibition and protect against tissue invasion. TSGs are sometimes divided into two classes: **gatekeepers** that act at steps directly related to the regulation of cell proliferation; and **caretakers** that regulate cell growth indirectly by repairing DNA damage and maintaining genomic integrity. We will expand on this overview with some notable examples.

TABLE 16.6 Sites of Action and Examples of Tumor Suppressor Genes

Site of Action	Example	Cancers Observed when Mutated
G_1 to S transition	Rb	Retina, bone, lung
G_1 to S checkpoint	p53	Many
β-catenin phosphorylation	APC	Colon
DNA mismatch repair	MLH, MSH	Colon, uterus
Excision of DNA damage	XP	Skin
Intercellular contact inhibition	E-cadherin	Many
Pro-apoptotic	Bax	Stomach, colon
	p53	Many

Rb

The **retinoblastoma (*Rb*)** gene product controls the transition from G_1 to S by interacting with the transcription factor E2F, a stimulator of this cell cycle phase transition. In the absence of *Rb* activity, E2F is free to stimulate the enzymes needed for DNA synthesis throughout the cell cycle. Loss of *Rb* function occurs in tumors of the retina (hence its name), melanoma, cancer of the lung, and sarcoma of bone. As mentioned earlier, overexpression of cyclin D and CDK 4 can also effect the same transition. Here, then, is an example in which overexpression of an oncogene or loss of expression of a TSG may produce the same result—excessive and unscheduled rounds of DNA synthesis. Historically, *Rb* was the first TSG identified and will be discussed in more detail later in this chapter.

P53

Sometimes called the "guardian of the genome," *p53* acts at several steps in the reactions controlling cell growth and death. It is a transcription factor acting on several classes of genes: those that stop cell division; those that catalyze repair of damaged DNA; and those that induce apoptosis. Loss of function of *p53* has two major deleterious effects: elimination of the DNA checkpoint between G_1 and S, ultimately allowing cells with damaged chromosomes to proceed into S and M; and failure to stimulate pro-apoptotic signals and inhibit anti-apoptotic signals, thereby allowing damaged cells to survive and divide. Not surprisingly, then, *p53* function is lost in more than half of all cancers, regardless of tissue of origin. Rarely, one *p53* allele is mutated in germ cells, creating a situation in which a single mutation in somatic cells will result in complete loss of *p53* activity. This situation leads to the Li-Fraumeni syndrome, which is characterized by the appearance of many different kinds of cancers in affected family members (Figure 16.8).

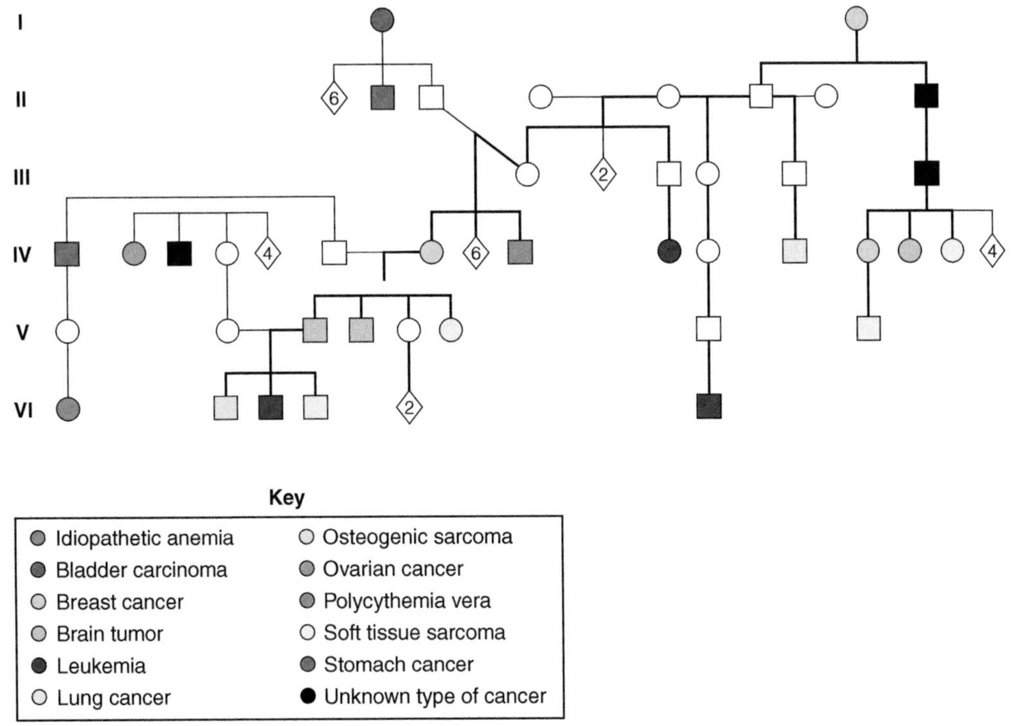

FIGURE 16.8

Pedigree of an extended family with the Li-Fraumeni syndrome. A germ-line mutation in *p53*, followed by a somatic mutation of *p53* in different tissues, leads affected individuals to develop cancers of many different organs.

APC

The *APC* gene product is a protein that signals cells that have come into contact with their neighboring cells to stop growing and dividing. As shown in Figure 16.9, loss of *APC* function leads to the formation of benign tumors (polyps) of the colon. A subsequent oncogenic mutation of *RAS* causes the polyps to enlarge and change their histology. Futher modification of the character of the polyps is caused by deletion of the *DCC* tumor suppressor gene. Then,

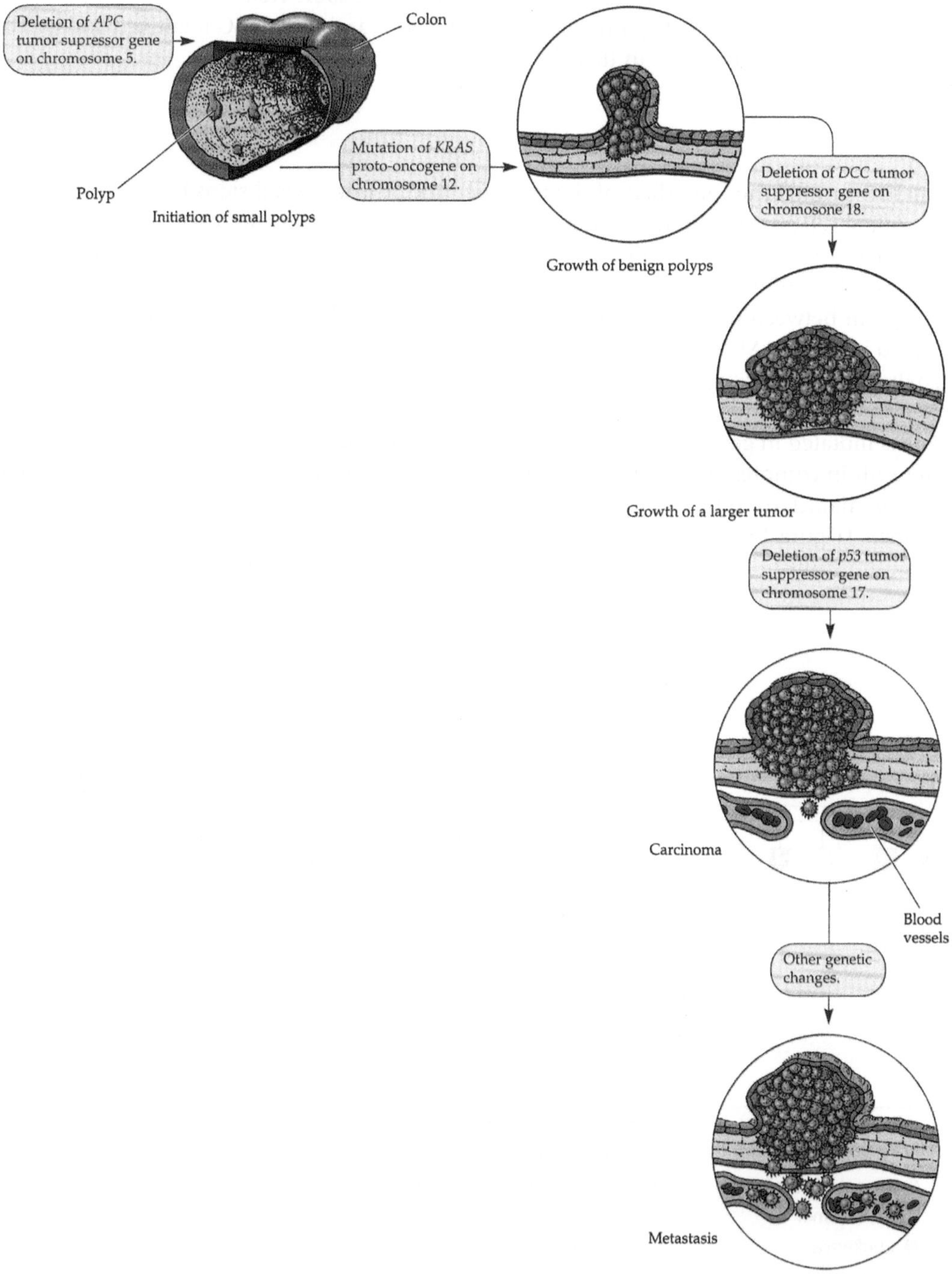

FIGURE 16.9
Sequential mutations leading, stepwise, from normal colon tissure to invasive colon cancer. Loss of activity of the *APC* gene, a tumor suppressor gene (TSG), leads to formation of benign polyps. Subsequent mutations in *ras* (an oncogene) and in *p53* (a TSG) produce the malignant tumor. Other less well characterized mutations promote metastasis.

loss of *p53* function results in conversion of the benign polyps into invasive colon cancer cells. This is the prototype for the general model described earlier, which stated that cancer most often results from a sequence of mutations of different genes, each of which moves the affected cell (or, more properly, the clone of cells) toward cancer. The pathway described here involving *APC, RAS,* and *p53* is but one form of progression. Others exist in which mutations affect different oncogenes and TSGs in different sequential order.

Defective DNA Repair Genes

As discussed in Chapter 6, damage to DNA occurs continuously at a low rate. Such damage would have catastrophic effects if cells did not have effective means of repairing the damage. When these repair systems are defective, mutations accumulate and genomic instability becomes apparent. If such mutations or rearrangements affect oncogenes or TSGs, cancer can result. Such is the case in a small number of human cancer syndromes. For example, xeroderma pigmentosum is an autosomal recessive cancer syndrome in which affected individuals develop many skin cancers when they are exposed to ultraviolet (UV) light (to be discussed further in Chapter 17). These patients lack one or another repair enzyme that functions to replace thymine dimers that form in response to UV light. Defective DNA repair is also proposed as the mechanism for familial breast cancers characterized by mutations in *BRCA1* or *BRCA2*. The precise role these transcription factors play remains uncertain, but current evidence suggests they are TSGs that function to repair double-stranded DNA breaks.

miRNA and siRNA Genes

A fourth class of genes involved with cancer is that composed of genes that are transcribed into RNAs that don't code for proteins. As discussed in Chapter 6, the primary transcripts of these genes (of which there are more than 250 in the human genome) are processed in the nucleus and in the cytoplasm to form small RNAs. These are called **microRNAs (miRNA)** or **small interfering RNAs (siRNA)** and are capable of interfering with gene expression (referred to as **RNAi,** for "**RNA interference**"). They bind to mRNAs and either block translation (as miRNAs do) or produce mRNA degradation (as siRNAs do). siRNAs appear to be mRNA-specific; that is, a single siRNA interferes with a single mRNA. miRNAs are more promiscuous, in that a single miRNA may interfere with the function of up to 200 mRNA targets.

Much remains to be learned about the significance of RNAi in human health and disease, but evidence is accumulating that indicates a role for miRNA in cancer. More than 10% of miRNAs are greatly overexpressed or downregulated in different kinds of tumors. Such miRNAs are termed oncomeres. For example, one of these miRNAs—called miR-21—is overexpressed more than 100-fold in glioblastoma multiforme, a malignant brain tumor with a dire prognosis. The mRNAs targeted by miR-21 have not yet been identified, and the importance of such overexpression has not been defined. Other observations are as intriguing: one particular miRNA-encoding gene is deleted in a particular form of leukemia, and miRNAs have been found at areas of noted chromosome instability. Might miRNAs and siRNAs be another means of modulating oncogenes and TSGs? Might RNAi be a fruitful approach to treatment of cancer? These RNA-encoding genes and their products are among today's most exciting avenues of cancer research, and appropriately so.

Metastasis Genes

Ninety percent of deaths due to cancer are a result of **metastasis**. Once cancer spreads from the tissue in which it arises to other tissues, such as lung, liver, bone, and brain, the likelihood of cure is very small. Although much has been learned in the past three decades about the genes initiating the cancer cell phenotype, very little has been learned about the mechanisms underlying metastasis. Scientists believe that metastasis must reflect some kind of synergistic interaction between cancer cells, the normal cells surrounding them (referred to as stromal cells), and tiny blood vessels formed in response to the molecules secreted by cancer cells.

The identity of the genes that promote metastasis is an area of intense investigation. It seems likely that many genes, working cooperatively or synergistically, will be found to foster metastasis, and that the genes facilitating metastasis to one distant site may be different from those facilitating spread to another. For example, as many as 18 genes may be involved in the metastasis of breast cancer to the lung. Recently, a subset of four of these genes was shown to be capable, when overexpressed, of promoting metastasis. One of these four was a growth factor, two were protein-digesting enzymes, and the fourth was an enzyme that promotes inflammation. Many other studies are underway to confirm and extend this work, to understand how this group of genes promotes invasion and spread, and to define the gene patterns that promote spread to other organs, such as bone or brain. Further, studies aimed at determining whether downregulating these metastasis-promoting genes might prevent the spread of cancer are already underway.

GENOMIC APPROACHES TO CANCER

In recent years the study of cancer genomes has joined with the study of cancer genes in the hope that more can be learned about cancer biology and cancer treatment. Central to these studies is a technique called expression profiling.

Expression Profiling

METHODOLOGY

The elements of this technical approach have been described in Chapters 7 and, in modified form, in Chapter 15. Basically, the technique aims to compare gene expression in a test sample with that in a reference sample. In cancer, for example, it is done like this:

- DNA sequences or oligonucleotides representing all 21,000 human genes are affixed to a plastic microchip (hence, the name "chip" technology).
- mRNA is isolated from a sample of a particular tumor (e.g., breast cancer) and from a representative control sample (e.g., normal breast tissue). The mRNA from the tumor is labeled with one fluorochrome (for example, red); the mRNA from the reference sample is labeled with a different fluorochrome (for example, green).
- The two mRNA samples are mixed together and are then hybridized to the chip-affixed DNAs. If the tumor sample expresses more of a particular mRNA than the control sample does, the corresponding spot on the chip will fluoresce red. If the control sample expresses more of an mRNA than the tumor does, the spot will fluoresce green. If both samples have equal expression, the spot will fluoresce yellow.
- The large amount of data collected from any single experiment (more than 21,000 bits of information per pair of samples) is analyzed using sophisticated statistical and computerized techniques. Using these tools, cluster analysis can be performed in which groups of genes whose expression correlates are collected and portrayed as shown in Figure 16.10. In this way the gene expression "signature" of a tumor is identified.
- From analyses of multiple tumors of the same type, one can determine the signature of the tumor type.

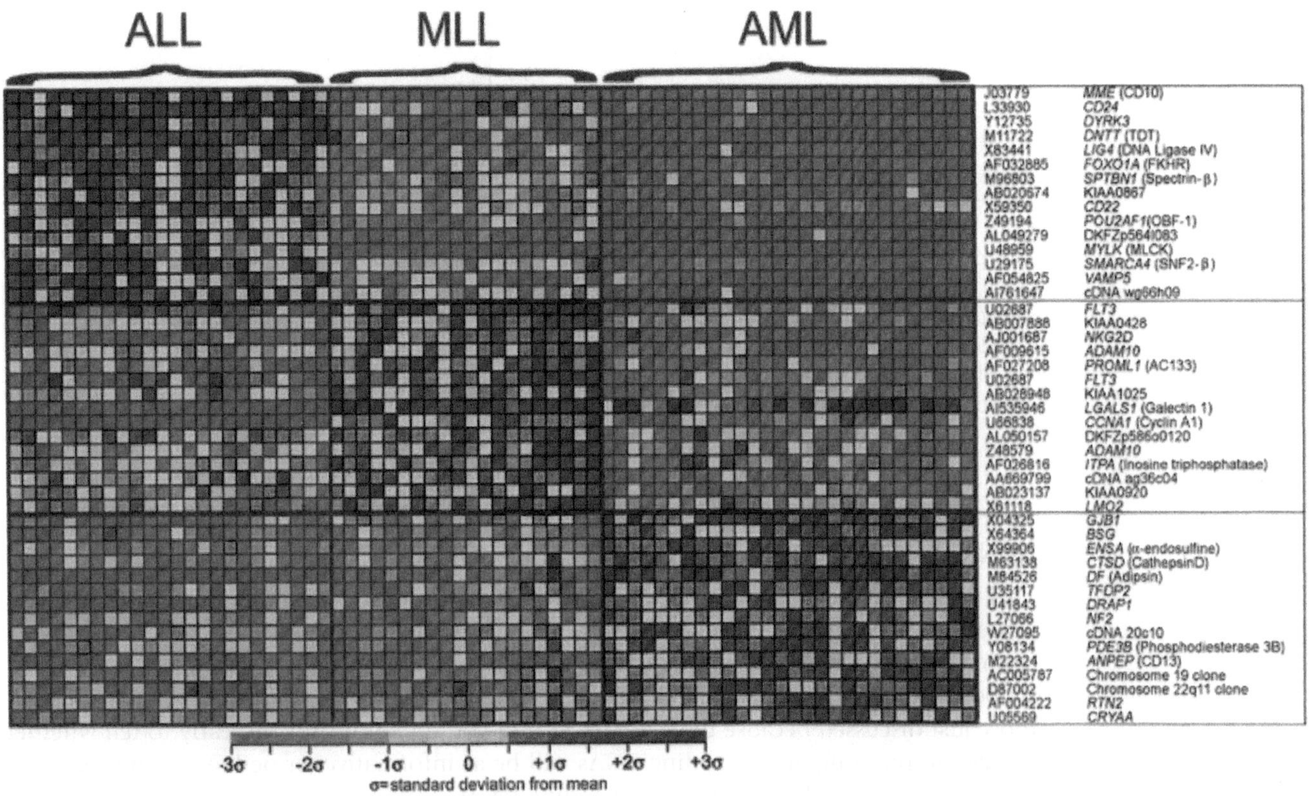

FIGURE 16.10

Expression profiling and cluster analysis. Expression of all 21,000 genes is compared in tumor and normal tissue. Then, those genes overexpressed in tumor tissue, and those underexpressed are grouped together to create expression "signatures." Shown here are expression signatures in acute lymphocytic (ALL), myelo-lymphocytic (MLL) and acute lymphocytic leukemia. Overexpressed genes are shown in red, underexpressed ones in blue. Each leukemia has a distinct pattern.

USE OF EXPRESSION PROFILING

Such expression profiling has been used in two general ways. First, it has been used to discriminate among tumors that have the same histologic appearance and chromosome markers—in other words, to show that tumors that look alike at the tissue or cellular level differ in gene expression. This was the first example of the use of expression profiling in classifying a type of lymphoma.

Second, different signatures have been shown to correlate with different clinical outcomes. For example, expression profiling has been used to determine the probability of survival in people with breast cancer who initially had similar clinical manifestations. In patients with breast cancer in whom the tumor had spread to no more than three lymph nodes, expression profiling divided those into one group with a profile predicting a good prognosis and another group with a poor prognosis (Figure 16.11). Further refining these expression signatures led to increasingly individualized prognostic information. It seems likely that this kind of profiling will educate physicians regarding prognosis and treatment of other types of cancers. We can anticipate the day when every cancer patient will have useful information provided by expression profiling (or, even more definitive, by complete genome sequencing as will be discussed subsequently), and that such information will inform treatment tailored to that patient. Such an individualized approach is not yet in general use, but the direction of the science and its clinical use is clear.

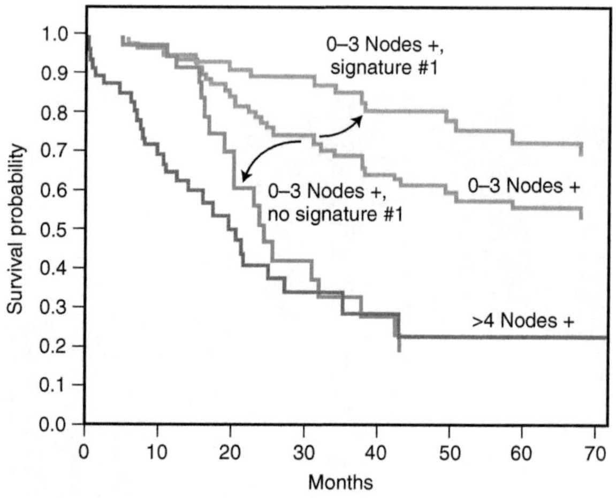

FIGURE 16.11
Use of expression profiling in breast cancer. Survival in patients whose tumor had spread to more than four lymph nodes (pink line) declined more rapidly than it did in patients with zero to three lymph nodes involved (green line). When the latter group's tumors were studied by expression profiling, two subgroups were identified: one with an expression signature predicting a good prognosis (yellow line); one, lacking that signature, predicting a poorer outcome (blue line).

NON-CODING RNA EXPRESSION PROFILING

This discussion of profiling would be incomplete without mentioning non-coding RNA expression. As originally developed, expression profiling catalogued activity of protein-coding genes. Recently, great interest has been shown in expression profiling of genes transcribed into RNA products (so-called non-coding RNAs). These profiles are simpler than those just discussed because there are fewer non-coding RNAs. It is too early to tell whether expression profiling of non-coding RNAs will be as informative, or perhaps even more informative, as profiling expression of coding RNAs.

Complete Genome Sequencing

There is widespread but by no means universal belief that understanding cancer will depend on sequencing the entire genome of cancer tissue from many patients and many tumor types. This view has been translated into the Cancer Genome Project (CGP), which is currently enrolling research centers and patients and collecting cancer tissue. The CGP promises to be as complex and expensive as the HGP, and it is hoped that the basic and clinical findings will justify the effort. As of now, the first complete genome sequences from patients with melanoma, pancreatic cancer, glioblastoma, and acute myelocytic leukemia (AML) have been reported. The appended box provides greater detail on the AML work.

AMPLIFICATION: EARLY LESSONS FROM CANCER GENOMICS

Complete Genome Sequencing in Acute Myelocytic Leukemia

The Cancer Genome Project focuses on defining all the mutations in a cancer cell, believing that such information will speed development of diagnostic tests, prognostic indicators, and, ultimately, new modalities of personalized treatment. Three hundred and fifty cancer genes have been found using a variety of techniques, but experts believe there are many others. There is hope that sequencing complete cancer genomes will help find them. Although still in its early days, such sequencing has yielded unanticipated returns in one type of cancer of the blood—acute myelocytic leukemia (AML).

Timothy Ley and his colleagues at Washington University have carried out complete sequencing in a number of patients with AML. In each case, they have compared the genome sequence of a patient's leukemic cells with that of their normal skin cells, thereby identifying definitively the changes characteristic of the malignant cell line. In brief, here is what they found in the cancer cells of their first patient: 150 somatic mutations spread across the genome, of which 12 occurred in protein-coding or RNA-encoding genes, and 52 were found in non-coding, conserved regions of the genome. Of these

AMPLIFICATION: EARLY LESSONS FROM CANCER GENOMICS—Cont'd

64 mutations, 4 were also found in leukemic cells from more than 180 additional patients: 3 in genes previously associated with AML, 1 in a conserved region. The nine other protein-coding genes had not been previously associated with any human cancer.

This work has been extended to other patients, each time adding additional mutations to the list, some of which have never been seen in any human disorder—for instance, changes in the spliceosome (the RNA/protein complex needed to splice mRNA transcripts) and in connexins (proteins required for proper formation and function of the mitotic spindle).

Up to now, this work has raised more questions than answers. How do these mutations lead to transformation into malignant cells? How many "cancer genes" are there in the genome? Does every patient have a unique set of leukemia-causing mutations? Can annotation of these mutations lead to rules that will enlighten treatment of AML, a leukemia with a generally poor prognosis? Answers to these questions may come from sequencing complete genomes in more patients with AML and from similar studies in patients with other malignancies.

THE ENVIRONMENT AND CANCER

Saying that cancer is a genetic disease and that we have learned much about cancer genes and dysregulation of the cell cycle does not mean that the environment is unimportant in causing cancer. In fact, it has been estimated that 75% of the risk for developing cancer lies with the environment. Although we have little information about the precise mechanism by which environmental agents initiate cancers, and even less about how these environmental agents interact with the genetic and cellular mechanisms focused on in this chapter, it is important to mention these environmental influences, particularly because understanding them is central to preventing cancer.

Microbial Pathogens

HPV

As mentioned earlier, we have known for a century that viruses can cause cancer. Rous demonstrated this clearly in discovering the sarcoma virus in chickens. Despite much effort, very few human cancer viruses have been demonstrated convincingly. One is the **human papilloma virus (HPV)**. HPV is a DNA virus found in more than 90% of cancers of the cervix. It encodes two oncogenic proteins, E6 and E7, whose actions as tumor promoters have been extensively studied: E6 disables p53; E7 inactivates Rb. A recently developed vaccine against HPV has shown remarkable effects in preventing cervical cancer in adolescent girls and young women before they have initiated sexual activity and after.

HEPATITIS B AND C

These two viruses are among the most widespread viral pathogens in the world. Each causes liver disease by leading to inflammation and destruction of the main cells in the liver, called hepatocytes. Although the mechanism is unknown, there is a clearly increased risk for hepatic cell cancer in people infected with either hepatitis B or hepatitis C. As with HPV, widespread use of a highly protective vaccine against the hepatitis B virus has been shown, over many years, to prevent hepatitis and forestall liver cancer.

H. PYLORI

Helicobacter pylori is a bacterium that causes ulcers of the stomach and duodenum. This has been known for more than 20 years. More recently, an association between *H. pylori* infection and cancer of the stomach has been observed. Because *H. pylori* can be eradicated with a 14-day course of antibiotics and because efforts are being made to develop a protective vaccine, it seems likely that prevention of some stomach cancers will follow the path taken in the prevention of cervical and hepatic malignancies.

Chemical Carcinogens

Since the 18th century, and probably before, evidence has been accumulating that chemicals in the environment are carcinogens—probably because these chemicals are mutagens as well. In the 18th century, it was noted that men who worked as chimney sweeps had a much higher incidence of scrotal cancer than did men not exposed to chimney dust and soot. Around this same time, people observed that those who chewed tobacco had a much higher risk for cancer of the mouth and tongue than did non-users. Such epidemiologic correlations have multiplied over time; today there are many chemicals suspected of being carcinogens, but convincing evidence is available in only a few instances, three of which we'll touch on here.

AFLATOXIN B1

The incidence of liver cancer is much higher in people who ingest aflatoxin B1, which is produced by a mold that contaminates peanuts. Aflatoxin is a widespread and potent mutagen. It has been shown to cause a single base change in the critical TSG, *p53*, which results in an amino acid substitution (Arg 249 Ser). This point mutation generates enhanced hepatic cell growth and impaired apoptosis, the likely antecedents of hepatic cell carcinoma. This mutation in *p53* is found in about half of the patients with liver cell cancer worldwide who ingest aflatoxin. There also appears to be synergy between aflatoxin and the hepatitis viruses with respect to risk for liver cancer.

NITRITE

Gastric (stomach) cancer is among the 10 most common epithelial cancers. It is found worldwide, but its prevalence is 5–10 times higher in Japan than anywhere else. There is strong evidence that this difference in prevalence reflects environmental rather than genetic factors. First, the traditional Japanese diet contains large amounts of nitrite, a known carcinogen. Second, decreasing the amount of nitrite in the diet of the Japanese has led to a distinct fall in the prevalence of gastric cancer in Japan over the past 30 years. Third, gastric cancer is three times as common among Japanese living in Japan as it is among Japanese living in the US or Hawaii (where the diet is very different). All of this points to ingestion of nitrite as a dietary chemical predisposing to gastric cancer. How it does this is the subject of current research.

CIGARETTES

People who smoke cigarettes have about a 20-fold greater prevalence of lung cancer than do non-smokers. Smoking increases the risk of developing cancers of other tissues as well: throat, breast, and bladder. Cigarette smoke contains a mixture of polycyclic hydrocarbons that are converted to highly reactive compounds called **epoxides**. The latter damage DNA and are potent mutagens. The precise means by which substances in tobacco smoke interfere with normal cellular proliferation remains under study, but evidence again points to damage to p53. Among the most effective public health efforts conducted in the US is that concerned with curtailing cigarette smoking. In the past 30 years, the number of American men who smoke has been cut in half, and the incidence of lung cancer in them has finally stopped rising (see Figure 16.1B). Disturbing, however, are trends among women: the number who smoke has been increasing for more than 20 years. The incidence of lung cancer in women—once only a small fraction of that in men—has been increasing rapidly and threatens to equal or exceed that in men soon (see Figure 16.1A). It is a testament to the powerful addictive properties of nicotine that 25 million Americans continue to smoke in the face of the high risk cigarettes pose for cancer and such other debilitating conditions as chronic obstructive lung disease and age-related macular degeneration (discussed in Chapter 13).

ILLUSTRATIVE EXAMPLES

The three particular malignancies that we have chosen for more detailed discussion have been touched on earlier as they relate to our information about oncogenes and TSGs. Each of them is of historical, clinical, and societal interest as well.

Chronic Myelocytic Leukemia (CML)

FREQUENCY AND EPIDEMIOLOGY

CML accounts for about 15% of leukemia cases in adults. It affects about 1 in 50,000 people and is somewhat more common in men than in women. It is panethnic and widely distributed geographically.

ETIOLOGY AND PATHOGENESIS

CML is caused by clonal expansion of hematopoietic progenitor cells that give rise to one type of circulating white blood cell called a polymorphonuclear leukocyte. In 1960, two scientists in Philadelphia noticed that virtually all patients with CML had a reproducible, striking karyotypic abnormality in their white blood cells characterized by a tiny chromosome, number 22, which soon became known as the Philadelphia chromosome (Figure 16.12).

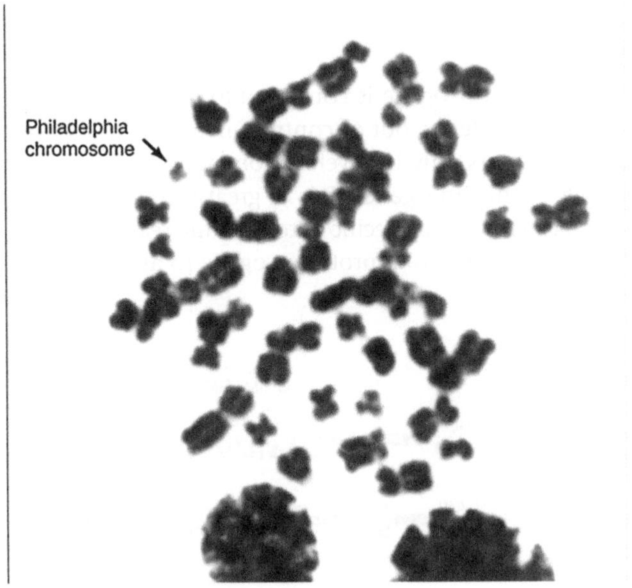

FIGURE 16.12
The Philadelphia chromosome. In this mitotic spread from a white cell of a patient with chronic myelocytic leukemia, the arrow points to a minute chromosome designated the Philadelphia chromosome based on the city in which the discovery was made.

Initially, this chromosome was thought to be a deletion of a portion of the long arm of chromosome number 22. A decade later it was shown that the Philadelphia chromosome was instead produced by a reciprocal translocation between the long arms of chromosomes 9 and 22 (Figure 16.13). As discussed earlier in this chapter, this translocation juxtaposes the *ABL* proto-oncogene normally found on the long arm of chromosome 9 to the *BCR* gene on chromosome 22, thereby creating a chimeric gene, *BCR-ABL* (Figure 16.14).The product of this gene is a hybrid protein composed of amino acid sequences from *BCR* and *ABL*. Critically, this fusion gene (and its fusion protein) has lost its control elements and is continually in the "on"

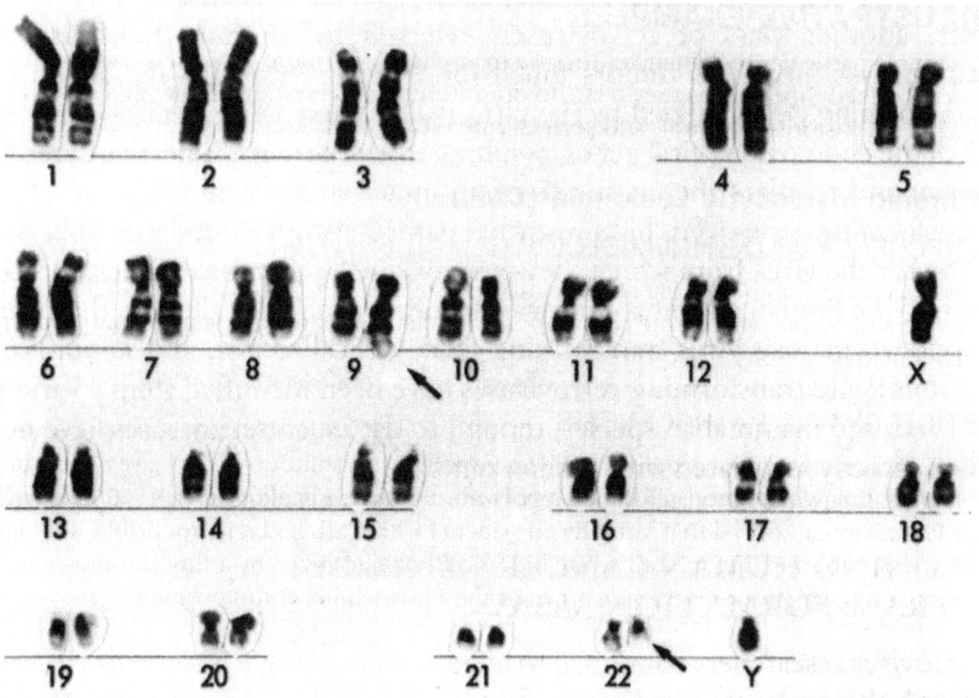

FIGURE 16.13
Karyotype demonstrating that the Philadelphia chromosome is not a deletion, but rather a translocation between chromosomes 9 and 22 (arrows). This discovery led to an understanding of the molecular basis of chronic myelogenous leukemia.

position. Thus, the *ABL* proto-oncogene is converted to the *ABL* oncogene and stimulates myeloid precursor cells to proliferate out of control via its role as a tyrosine kinase enzyme. Because the translocation occurs only in myeloid somatic cells, the cellular expansion affects only hematopoietic cells of this lineage. CML is of great historical importance because it was the first malignancy associated with a specific chromosome rearrangement, and it was the first human cancer in which conversion of a proto-oncogene to an oncogene was understood in molecular terms.

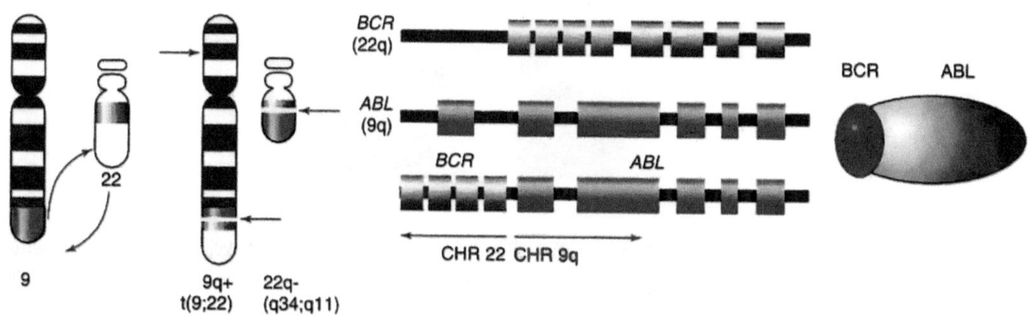

FIGURE 16.14
The chromosomal translocation responsible for chronic myelogenous leukemia (CML). The ABL proto-oncogene, normally found on chromosome 9, is translocated to chromosome 22, where it is fused with the break point cluster (BCR) locus. This chimeric gene encodes a chimeric protein containing BCR and ABL sequences, and lacking sequences normally regulating its activity. The chimeric oncogenic protein drives a class of white blood cell precursors to grow out of control, and produces leukemia.

CLINICAL FEATURES

CML has several phases. It begins insidiously with fatigue, weight loss, and enlargement of the spleen, accompanying the 5- to 40-fold increase in circulating white blood cells (WBCs). Over time, this chronic phase evolves into a more aggressive, accelerated phase and finally to a usually terminal phase in which progressively more immature WBC precursors appear in the peripheral blood. This phase is called a "**blast crisis**," and signals that CML has become an acute leukemia. Nearly 90% of patients are diagnosed during the chronic phase and if untreated, progress to the blast phase within 1–2 years, succumbing soon thereafter.

MANAGEMENT

Up to the year 2000, patients with CML were treated with a combination of cytotoxic (cell-killing) drugs. These were of value for a relatively short time after which blast crisis appeared and, with it, fatal consequences. In a truly revolutionary breakthough, the approach to treating CML changed. **Imatinib** (or **Gleevec®**), a specific inhibitor of the BCR/ABL hybrid kinase, was discovered (Figure 16.15). Oral administration to patients in the chronic phase

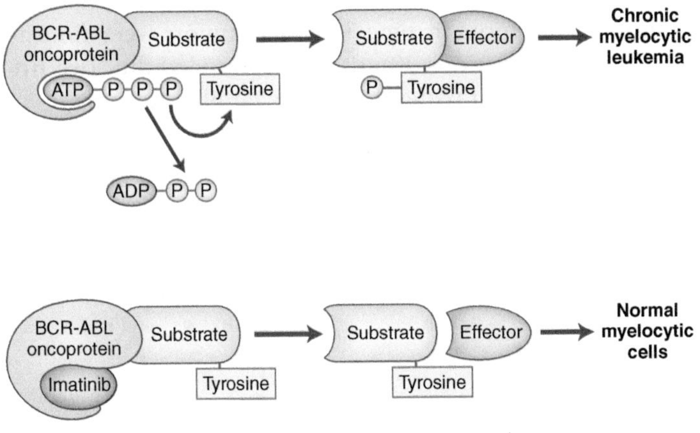

FIGURE 16.15
Mechanism of action of imatinib (Gleevec®) in chronic myelogenous leukemia. (Top frame) In the absence of Gleevec, ATP binds to the BCR-ABL oncogenic protein catalyzing the phosphorylation of a tyrosine residue on its substrate. This generates a signal promoting proliferation of the leukemia cells. (Bottom frame) Gleevec binds to the ATP site, thereby "short-circuiting" the growth signal and preventing proliferation of leukemic cells.

led to normalization of circulating WBC counts in 95% of patients. This dramatic effect was durable, lasting 5 years or more. Further, this treatment resulted in disappearance from the blood of all cells bearing the 9/22 chromosomal translocation. Delayed resistance to imatinib has developed, necessitating development of other kinase inhibitors with specificity for the enzyme escaping imatinib action. Imatinib treatment of CML is the first and best example of a targeted form of cancer treatment in which a pharmaceutical acts by interfering precisely with the molecule responsible for the malignant transformation of cells. It has been a lodestone since, sought with modest success but great intensity for the treatment of other cancers.

Retinoblastoma

FREQUENCY AND EPIDEMIOLOGY

Retinoblastoma is a malignant tumor of retinal cells, with an incidence of about 1 in 20,000 births. It is panethnic and occurs throughout the world.

ETIOLOGY AND PATHOGENESIS

Retinoblastoma is caused by mutations at the *Rb1* locus, which encodes the Rb protein. Rb is a TSG—in fact, the first TSG discovered. The history of its discovery is of unusual interest and instructional value.

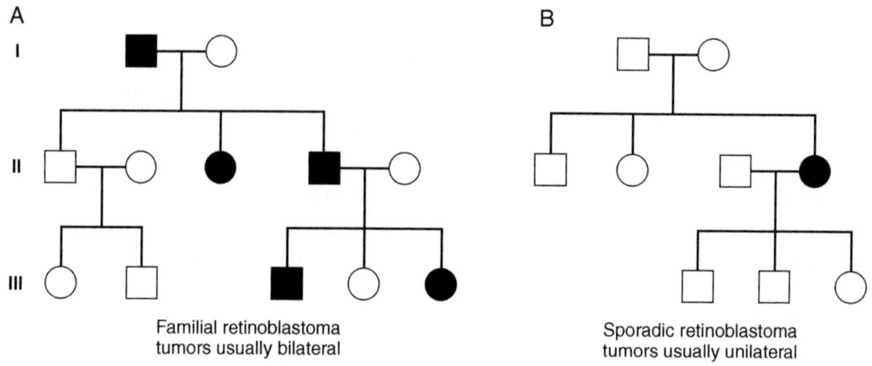

FIGURE 16.16
Typical pedigree observed in retinoblastoma patients. (A) The familial form, inherited as a dominant trait. (B) The sporadic form. Description of these two types of pedigrees (see text for greater detail) led to the formulation of the "two hit" theory of cancer.

In 1970, Alfred Knudson collected a number of families, each with patients who had retinoblastoma. He found two types of families (Figure 16.16): those in which the eye tumor affected only one person (sporadic) and those in which two or more individuals had retinoblastoma (familial). Further, he noted that affected patients with the familial form tended to have more than one retinoblastoma in an eye, or tumors in both eyes. Their tumors presented earlier in life than did those of the sporadic type, in which tumors were almost always unilateral. Based on these clinical observations, Knudson propounded the **"two hit"** theory of cancer (Figure 16.17). He proposed that patients with the familial form of

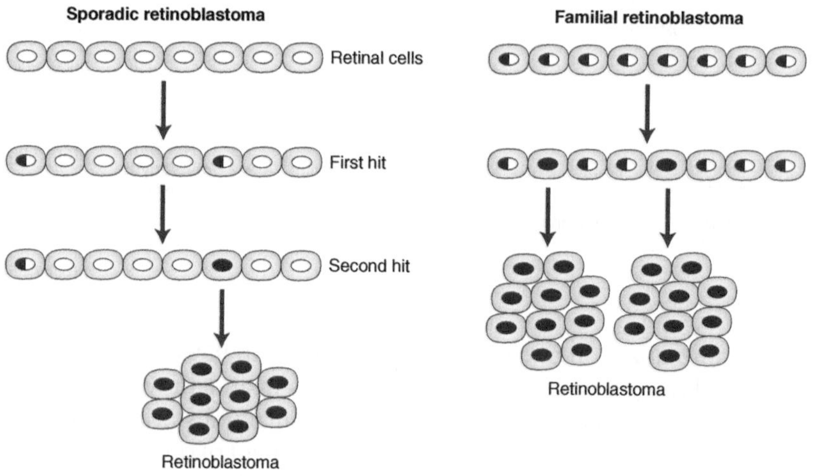

FIGURE 16.17
Schematic representation of the "two hit" theory to explain retinoblastoma. It proposed that the tumor occurs only after both alleles of the *Rb1* gene have been inactivated. In sporadic retinoblastoma, on the left, retinal cells are normal at birth. Both hits (cells marked in red) occur sequentially in a retinal cell. In familial retinoblastoma, one copy of the *Rb1* gene has been inactivated in germ cells, meaning that all retinal cells start out with one "hit." Thus only a single hit in a retinal cell is necessary to produce the tumor. This also explains why tumors in familial retinoblastoma are often multiple and bilateral.

retinoblastoma had a germ-line mutation of one allele at the involved locus (now designated the *Rb1* locus) and developed retinoblastoma after the second allele was damaged in retinal cells by somatic mutation. In the sporadic form, however, both mutations had to occur in somatic cells. This formulation would explain the familial form of the disease in that one "hit" was present at birth, thus only one additional mutation occurring in rapidly proliferating retinoblasts would be required to cause a tumor. Similar reasoning would explain the bilateral tumors in the familial form and the earlier presentation of eye tumors, both phenomena being predicated on the increased probability of mutation of one allele over that of mutations of both. This prescient work preceded current understanding of oncogenes and TSGs. Today we know that Rb is a "gatekeeper" TSG, which regulates the G_1 to S transition of the mitotic cell cycle. If only one allele is inactive, cell proliferation is controlled. If both alleles are inactivated (either by germ-line and/or by somatic mutation) cell division goes out of control.

Rb1 mutations occur throughout the coding and promoter regions. In addition to a variety of point mutations, deletions of the locus on chromosome 13 and epigenetic modification of *Rb1* (resulting from methylation) have been reported.

INHERITANCE

About 40% of cases of retinoblastoma are the inherited form, 60% the sporadic form. As noted in Figure 16.16, the familial form is inherited as a dominant trait, with nearly full penetrance. Yet Rb acts recessively at the cellular level. This apparent paradox is explained by the nature of the disease: in the presence of a germ-line mutation, a second, somatic mutation produces a dominant pattern of inheritance (vertical transmission; males and females affected).

CLINICAL FEATURES

The diagnosis of retinoblastoma is made by seeing a white or yellow reflex when a light is shined at the eye of a young child (Figure 16.18). In more advanced cases, the tumor may

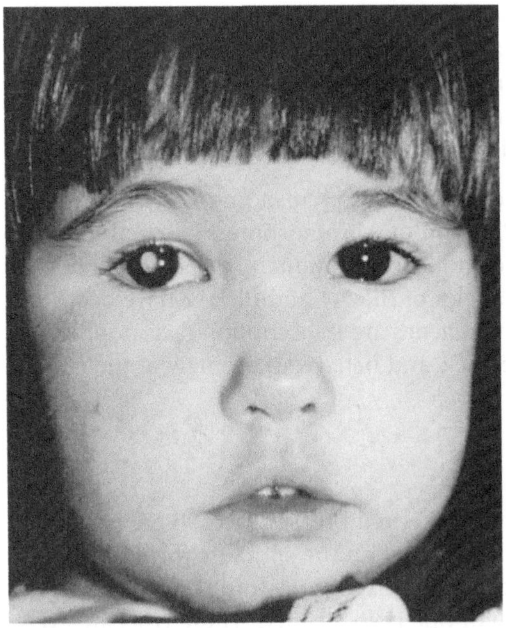

FIGURE 16.18
A young girl with retinoblastoma in her right eye. The tumor is revealed by the white reflex when light is directed through the pupil.

fill the optic chamber and produce a bulging eye. If retinoblastoma goes untreated, it spreads beyond the eyes and is usually fatal. In the inherited form of retinoblastoma, tumors usually appear during the first year of life; in the sporadic form, between 2 and 3 years of age. Sporadic retinoblastoma is most often treated by removing the affected eye. Treatment of the familial form is more complicated. Because tumors are often bilateral, every effort is made to spare at least one eye. This means that radiation and chemotherapy are deployed, along with surgery. Infants with heritable retinoblastoma who survive their childhood eye tumors have a greatly increased risk of developing other tumors later in life, particularly melanoma and osteosarcoma (sarcoma of bone). Given that Rb is expressed in all tissues, why do tumors affect the eye early and these other tissues late? This question has not been answered.

Breast Cancer

It is not possible to do justice to the complexity of **breast cancer** in these pages. There are, however, some issues regarding the genetics of breast cancer that bear discussion.

FREQUENCY AND EPIDEMIOLOGY

Cancer of the breast is among the most common cancers in women—second only to skin cancer. Between 6 and 9% of women in the US and Western Europe will develop breast cancer in their lifetime. Worldwide, the incidence of breast cancer is highest in the US and Western Europe and lowest in East and South Asia. Black women in the US (and in Africa) have a lower incidence than do white women. Breast cancer is the second most common cause of cancer deaths in American women, exceeded only by cancer of the lung.

ETIOLOGY AND PATHOGENESIS

Breast cancer originates most commonly in the inner lining of the milk ducts (**ductal carcinoma**) or the glandular lobules that supply the ducts with milk (**lobular carcinomas**). It results, as all cancers do, from a combination of germ-line and somatic mutations. More than 90% of breast cancers result from somatic mutations only—affecting a large number of different and diverse mutations of oncogenes and TSGs that ultimately lead a single cell to escape its growth-controlling pathways and signals. Beyond this remarkable genetic heterogeneity, however, are a series of dominantly inherited syndromes in which breast cancer occurs. These account for a small fraction of breast cancer overall, but indicate that germ-line mutations occur in this setting (as in other dominantly inherited cancers like retinoblastoma). Regardless of the specific mutations, these families share several features: early onset of cancer, frequent bilateral disease, and frequent association with ovarian cancer.

Among the most interesting germ-line genes where mutations predispose to breast cancer are ***BRCA1***, found on chromosome 17, and ***BRCA2***, found on chromosome 13. Together, mutations at these two loci account for 30–50% of autosomal dominant breast cancer, but less than 5% of breast cancer overall. Many mutant alleles of *BRCA1* and *BRCA2* have been identified. The gene products encoded by these two genes are transcription factors affecting repair of double-stranded DNA breaks. They are TSGs and behave accordingly at the cellular level.

GENETIC CONSIDERATIONS

Beyond the known mutations causing dominantly inherited breast cancer, there is considerable evidence for a genetic component to those with the sporadic form of this common cancer: a woman's risk for breast cancer is increased three-fold if one first-degree relative is affected, and up to ten-fold if more than one is so affected. These risks are even greater if the affected relative developed breast cancer before age 40. Such familial clustering is far more common than the inherited forms mentioned above, but as of now, awaits molecular understanding.

CLINICAL FINDINGS

The most important risk factors for breast cancer are sex, age, childbearing, and breastfeeding. Risk is 100 times greater in women than in men, increases progressively with age, and is much lower in women who have had children and who have breastfed them. Other factors that increase risk include a high-fat diet, alcohol consumption, estrogen replacement, and tobacco use. After diagnosis by physical examination (often aided by mammography), first-line treatment is surgical removal. Then, based on examination of the tumor histologically and functionally, treatment is individualized. Genetic findings play an important role here. If the tumor cells express the estrogen receptor, drugs blocking this receptor will inhibit tumor growth. If the tumor overexpresses **HER2/Neu**, an oncogenic cell surface receptor which acts by stimulating cell proliferation, treatment with a monoclonal antibody (**Herceptin**) against it improves prognosis distinctly. Chemotherapy and radiation complete the armamentarium of modalities that find use in advanced disease.

BRCA SCREENING

In recent years considerable effort has gone into presymptomatic screening of women for *BRCA1* and *BRCA2* mutations in normal skin or white blood cells, in the hope that such screening will identity women before their tumors had grown (or even appeared). Two important issues have arisen, one scientific, and one legal. The scientific issue has to do with the difference between cause and susceptibility (Figure 16.19). In families with highly penetrant mutations of *BRCA* genes, the cumulative risk of developing breast cancer increases to more than 80% in carriers of *BRCA* mutations by age 80 years (compared to 8–9% in the general population). This has led many women in the high-penetrance families to undergo bilateral mastectomy and removal of their ovaries rather than take the risk of developing breast or ovarian cancer. This is a controversial area because the *BRCA* gene mutations are identifying susceptibility, not cause, meaning that many women undergoing preventive mastectomy and removal of the ovaries would not have developed cancer of the breast or ovary.

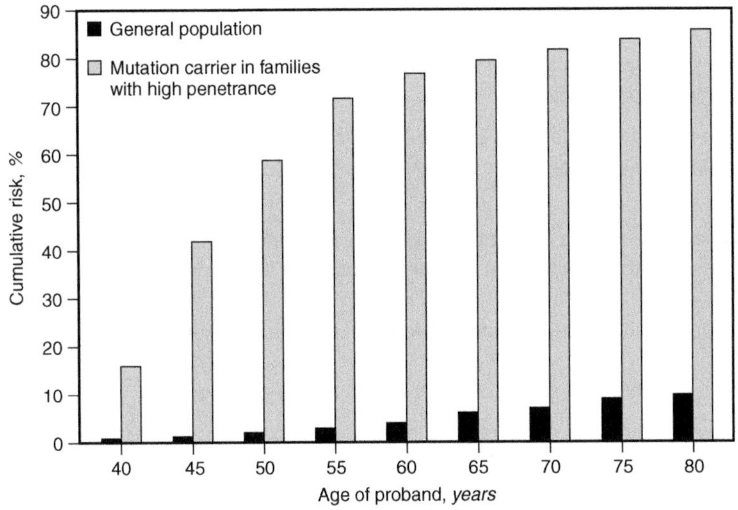

FIGURE 16.19
Cumulative risk of developing breast cancer in women carrying a mutation in *BRCA1* or *BRCA2* from families with high penetrance for these mutations (gray bars) compared to cumulative risk in the general population (black bars). See text for additional details.

The legal issue affects gene patenting. A company named Myriad Genetics, which participated actively (but not exclusively) in the isolation and characterization of the *BRCA* genes, has patented them. This means that Myriad has a monopoly on providing the *BRCA* testing mentioned above. Patients have complained about the high cost of the test (more than $3,000). Others have said that genes are natural substances, thereby prohibiting taking out patents on them. This matter has been hotly contested among scientists and lawyers. Recently, a judge ruled that Myriad's patents are illegal and overturned them. Even more recently, that ruling has been overturned in favor of Myriad. It remains to be seen what will happen when the latter ruling is appealed, as it already has been.

REVIEW QUESTIONS AND EXERCISES

1. Choose the phrase in the right column that best matches the term in the left column

a. checkpoints	1. a growth resulting from uncontrolled cell proliferation
b. human papilloma virus (HPV)	2. tobacco, X-rays, microbes
c. oncogene	3. series of events between one mitotic event and another
d. clonality	4. mutant form of proto-oncogene; drives cell proliferation
e. metastasis	5. variety of chromosomal abnormalities in cancer cells
f. immortality	6. proteins that are required for cell cycle regulation
g. benign tumor	7. being derived from a single cell
h. environmental carcinogens	8. cause of cervical cancer

i. expression profiling
j. tumor suppressor genes (TSGs)
k. cyclin-dependent kinases
l. retinoblastoma gene *(Rb)*
m. genomic instability
n. "two hit" theory
o. tumor (neoplasm)
p. Abelson leukemia gene *(ABL)*
q. contact inhibition
r. cyclins
s. *BRCA1*
t. cell cycle
u. *p53*

9. inhibit cell proliferation and promote apoptosis
10. transcription factor required at several points in cell cycle
11. loss of limits of cell division
12. oncogene responsible for chronic myelocytic leukemia
13. enzymes that add phosphate groups to cyclin proteins
14. the ability of cancer cells to travel to distant sites
15. idea that some cancers require both germ-line and somatic mutations
16. use of microarrays to assess activity of many genes simultaneously
17. first TSG found; implicated in cancer of the eye
18. tumor that is circumscribed and non-invasive
19. reactions that arrest cell cycle so as to permit DNA repair
20. property of normal cells to cease dividing when they touch
21. genes that predispose to breast cancer when mutated

2. What do we mean when we say that cancer is a genetic disease but is rarely inherited?

3. What is the fundamental difference between a benign tumor and a malignant one? Why might a benign tumor histologically be less than benign clinically?

4. A tumor (of human origin) the size of a garden pea contains ~10^9 cells. How many mitotic generations are required to produce a tumor of this size? How long would it take to reach this size?

5. The inherited form of retinoblastoma results from mutations in the *Rb* gene.
  a. What do we mean when we say that this form of cancer is "dominant" at the Mendelian level and recessive at the cellular level?
  b. The *Rb* gene is a tumor suppressor. Describe briefly two kinds of actions of tumor suppressor genes (TSGs).
  c. What clinical observation led to the formulation of the "two hit" theory of cancer?

6. Is there a logical fallacy in saying (a) that all cancers are clonal; and (b) that mutations at several different loci must occur before a cancer develops? Explain briefly.

7. Most inherited forms of cancer are caused by mutations of TSGs, very few by mutations of proto-oncogenes. Why is this so?

8. It seems most unlikely that we will ever find "the" cure for cancer.
  a. Why is this so?
  b. List three technologies that might improve the treatment of histologically similar but prognostically different cancers of a particular tissue (for example, breast or lung).
  c. What do we mean when we say that "targeted" treatment is much more promising than "standard" chemotherapy?

9. During the past 50 years the prevalence of lung cancer has been rising progressively while the prevalence of stomach cancer has been falling progressively.
  a. How can these trends be explained?
  b. What would you recommend as a means to reverse the trend in lung cancer prevalence mentioned above?
  c. We cure only about 20% of patients with lung cancer, compared with a cure rate greater than 50% for breast cancer. Why?

10. Myelocytic leukemia—both acute and chronic—has provided important insights about the genes involved with cancer.
  a. What seminal contribution was made regarding chronic myelocytic leukemia (CML) in 1960?
  b. What milestone was achieved with acute myelocytic leukemia (AML) about 50 years later?

11. Why do some sporadic cancers like retinoblastomas appear during the first few years of life, while most other sporadic cancers appear in adults over age 50 years?

12. Aneuploidy is a regular finding in karyotyping cancer cells. Why is this the case?

CHAPTER 17

Detection and Treatment of Genetic Disorders

CHAPTER OUTLINE

CORE CONCEPTS

Clinical genetics depends on applying the principles of human genetics and genomics to the detection and treatment of the many kinds of genetic disorders discussed previously. Here we present a framework for thinking about the intertwined subjects of **detection** and **treatment**—starting with detection. There are three main reasons to detect (or diagnose) genetic disorders:

- first and foremost, to institute treatment for the affected person;
- second, to provide information to family members regarding their risk for developing the same condition;
- third, to seek answers to scientific questions posed by the disorder, such as its mode of inheritance, its precise cause, and its frequency in different ethnic populations.

Human Genes and Genomes. DOI: 10.1016/B978-0-12-385212-0.00017-2

Detection (or testing) is increasingly being done throughout the cycle of human life: **premarital**, preconceptual, **preimplantation**, **prenatal**, and during **childhood** or **adulthood**. In general, premarital through prenatal testing is carried out to prevent the birth of those with untreatable conditions. In contrast, detection in newborns, children, and adults is performed with the ultimate goal of instituting treatment. An increasingly broad array of biological materials is used in detection systems: blood, urine, cells, proteins, RNA, and DNA. In addition to having to ascertain the analytical and clinical validity of any test employed, testing for genetic disorders sometimes raises complex ethical and social questions that must be discussed with those seeking testing. To assist in answering the range of questions posed, families are referred for **genetic counseling**—a communication process facilitated by a team of professionals, including physicians, nurses, and master degree-trained genetic counselors.

Some genetic disorders may be cured **surgically** (for instance, cleft lip, congenital heart defects, and localized cancers). Some are not treatable (such as Tay-Sachs and α-thalassemia). A larger subset may respond beneficially to a variety of therapeutic strategies and modalities, such as **substance avoidance**, **dietary restriction**, **product replacement**, **cofactor supplementation**, **pharmaceutical administration**, **organ** or **adult stem cell transplantation**, and **gene therapy**. All together, these modalities are beneficial for only a minority of patients with genetic disorders, thereby constituting a major challenge to science and clinical medicine.

Gene therapy and stem cell therapy have been the subject of much attention in recent years. Although *in vivo* and *ex vivo* methods of gene therapy have been worked on for 30 years, clinical progress has been disappointingly slow. Currently, it has proven to be beneficial in less than a handful of disorders. At present, there is no instance in which use of embryonic stem cells has been shown to produce clinical benefit. It seems likely that gene therapy and embryonic stem cell therapy will benefit patients with a variety of conditions in the years and decades ahead.

Some of the disorders used to illustrate the aforementioned generalizations include: Tay-Sachs disease, thalassemia, cystic fibrosis, phenylketonuria, hemophilia, methylmalonic acidemia, polycystic kidney disease, diabetes mellitus, severe combined immunodeficiency, and Leber's congenital blindness.

In this chapter, we will present a framework with which to think about detection and treatment. The rapidly expanding knowledge base about human genes and genomes is creating new opportunities for detecting and treating genetic disorders. Some of these opportunities fit comfortably into the paradigm of clinical medicine. Others do not, instead raising ethical dilemmas and societal controversies of nearly explosive proportions. We will begin this discussion with detection (or diagnosis), and then move on to treatment.

TABLE 17.1 Major Considerations in Detection of Genetic Disorders

Purposes of detection
To treat affected individuals To provide information to people at risk To answer scientific questions
Timing of detection
Premarital Preconceptual Preimplantation Prenatal Neonatal Childhood and adulthood
Special issues regarding detection
Clinical dilemmas Moral judgments and ethical debates

DETECTION

Three words beginning with the letter "W" provide a means for an organized presentation concerning detection of genetic disorders: why, when, and whether. Each will be discussed in turn (Table 17.1).

Why: Purposes of Detection

TO TREAT AFFECTED INDIVIDUALS

This is the most important reason to detect any condition. It is central to the ethos of clinical medicine. There are myriad examples where detection of genetic disorders leads to valuable, even life-saving, treatment: the diagnosis of phenylketonuria (PKU) in neonates leads to use of a low phenylalanine diet that prevents cognitive disabilities; the diagnosis of cystic fibrosis in infants leads to ministrations that prolong life dramatically, and improve the quality of life immeasurably; the diagnosis of chronic myelocytic leukemia leads to treatment with Gleevec®, a medication regularly able to convert this once rapidly fatal disorder to a manageable, chronic condition; the diagnosis of bipolar disorder leads to use of lithium and other mood-stabilizing drugs that often prevent familial strife, professional interruption, and suicide. Notice that we have used in these examples conditions that are single-gene disorders, those that are multifactorial traits, and still others that reflect somatic rather than germ-cell mutations. The goal of detection is the same.

TO PROVIDE INFORMATION TO THOSE AT RISK

This is one of the singular features of clinical genetics and takes many forms: quantifying the risk of Down syndrome (2–3%) in a 40-year-old woman considering pregnancy; estimating the empiric risk of recurrence (3–4%) in a family whose first child has a cleft lip; quantifying the lifetime risk (85%) of breast cancer in women from Ashkenazi Jewish families with *BRCA1* mutations and a highly penetrant family history of breast tumors; quantifying the risk of a second child with sickle cell anemia (25%) in an African American couple with one affected child. Sometimes these risks have a high degree of certainty—as in Mendelian disorders inherited as dominant, recessive, or X-linked traits. In other instances, risk assessment depends on empiric estimates based on information summed from many affected families.

TO ANSWER SCIENTIFIC QUESTIONS

Detection of disorders is often conducted to define the mode of inheritance, or to explain familial or population clustering, or to associate conditions with genotypes. Such work has, as its long-range goal, translation of this science into treatments that are still lacking for so many genetic disorders. Examples of this rationale for detection include:

- learning why Pima Indians have a nearly 50% incidence of type 2 diabetes, compared to less than 10% in the general population;
- learning why folate supplementation has reduced the incidence of neural tube defects by 75%;
- learning why copy number variants (CNVs) are being found in greater than expected numbers in patients with a variety of neurologic and psychiatric disorders.

When: Timing Programs of Detection

Detection of those affected with genetic conditions or those at risk for such conditions is carried out throughout all phases of the human life cycle. The timing of detection is tightly linked to the kind of information being sought. Table 17.2 lists these phases, and some examples for each.

TABLE 17.2 Timing of Detection of Genetic Disorders

Phase Of Life	Examples of Conditions Tested For
Premarital	Carriers of Tay-Sachs disease or of β-thalassemia
Preconceptual	Couples are carriers of Tay-Sachs disease or β-thalassemia
	Female carriers for Duchenne muscular dystrophy or hemophilia
	Males or females at risk for late-onset Huntington or Alzheimer disease
Preimplantation	Couple at risk for familial retinoblastoma or Li-Fraumeni syndrome
	Couple at risk for translocation Down syndrome
Prenatal	Woman pregnant at age 37
	Couple whose prior child has neurodegenerative lipid storage disease
	Couple at risk for neural tube defect
Neonatal	Identification of PKU (and other treatable metabolic disorders)
	Confirmation of diagnosis of Turner syndrome or other kinds of birth defects
Childhood and adulthood	Identification of child with cystic fibrosis
	Identification of adolescents or adults with late-onset disorders (hypertension, bipolar disorder, breast cancer)

PREMARITAL

The vast majority of humans choose their marriage partners based on a complex mix of characteristics: appearance, shared values, sexual attraction, respect, esteem, and mystery (a certain *je ne sais quoi*). Genotypes rarely make the list of marital determinants, at least not explicitly. But this is not always true. Take **Tay-Sachs disease (TSD)** for example. The incidence of TSD is about 1 in 3,000 in Ashkenazi Jews, meaning that about 1 in 30 is a carrier for this autosomal recessive condition. TSD is characterized by progressive neurologic dysfunction, and is almost uniformly fatal by age 3–5 years. An ultra-orthodox sect of Ashkenazi Jews in New York City urges all teenagers in their community to determine whether they are carriers for TSD so that two carriers will be advised not to marry each other. We would be surprised if there are not other examples of such gene-based marital choice that are not publicly disclosed. For instance, would two first cousins be inclined to marry (illegal in 25 states) without genetic testing if a serious autosomal recessive disorder was present in the pedigree? For them, the risk of having an affected child would be much greater than in the general population.

PRECONCEPTUAL

There are a number of clinical settings where detection efforts may be carried out after two people have married but before they become parents. Here are some from a considerably longer list.

Carriers for Autosomal Recessive or X-linked Disorders

As the number of DNA-based tests proliferate, it is increasingly possible for a couple to find out whether they are at risk for having affected children. The two conditions for which such testing has been carried out are TSD in Ashkenazim and β-thalassemia in Italians. For each condition, couples in which both individuals are carriers have a 25% chance—for each offspring—of being affected with these clinically grim disorders. Couples at risk are offered prenatal diagnosis (soon to be discussed) and termination of pregnancies in which fetuses are affected. Using this combined preconceptual and prenatal testing, the incidence of TSD in Ashkenazi Jews in the United States has decreased by 95%, as has the incidence of β-thalassemia in the Italian island of Sardinia. A number of other serious autosomal recessive conditions lend themselves to this pattern: cystic fibrosis, sickle cell anemia, and phenylketonuria to mention but a few.

Preconceptual testing for X-linked traits is equally accessible today. Women at risk for being carriers of Duchenne muscular dystrophy, hemophilia, and ornithine transcarbamylase deficiency (each discussed in Chapter 12), for example, can be tested and their pregnancies monitored, if so indicated and desired.

Identification of People with Autosomal Dominant Traits

Because individuals expressing autosomal dominant traits have a 50% chance, for each pregnancy, of having an affected child, there are numerous examples where preconceptual testing is considered. This is particularly so in settings where incomplete penetrance or variable expressivity pertain. Neurofibromatosis, a condition in which multiple benign tumors of nervous tissue develop, resulting in a variety of consequences, is a good example. Some affected individuals may have only a few pigmented spots on their skin; others may have innumerable large and small skin tumors and growths elsewhere. Individuals known to have the Huntington gene, or who are at high risk for it, are proper candidates for preconceptual testing, with all the dilemmas discussed in Chapter 13 and to be mentioned later in this chapter.

Selective termination of pregnancies is not, by any means, the only option for those being tested preconceptually. For couples both shown to be carriers for autosomal recessive traits, artificial insemination is an option, as is adoption. For X-linked traits, an egg donor may be considered. The options, and the dilemmas associated with them, will be held out to more and more couples as means of detection proliferate.

PREIMPLANTATION

For couples at risk for Mendelian or chromosomal disorders who are not willing to undergo prenatal diagnosis with its attendant use of therapeutic abortion, **preimplantation genetic diagnosis (PGD)** presents another alternative. PGD combines *in vitro* fertilization (IVF) with genetic testing in the following way (Figure 17.1):

- An egg from the woman is obtained by usual IVF technology and fertilized by a single sperm from her male partner;
- The zygote is allowed to undergo several cell divisions *in vitro* (in a Petri dish or test tube), reaching an eight-cell stage after about 48 hours;
- A single blastomere (one of the eight cells) is isolated and tested using DNA-based or cytogenetic methods;
- If the test shows that the early embryo is affected, it will not be implanted; if unaffected, the remaining seven-celled embryo will be implanted and the pregnancy continued.

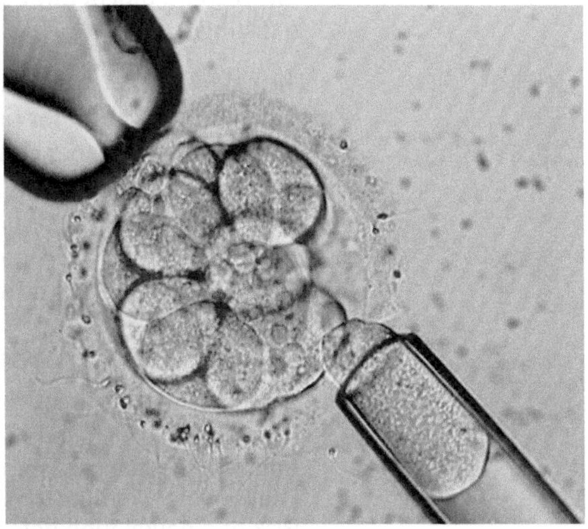

FIGURE 17.1
Removing a single blastomere from an eight-cell preimplantation embryo is the first step in PGD. The blastomere is then tested, usually with a DNA probe. If the test is normal, the remaining seven-cell embryo is implanted in the uterus. If the test reveals an abnormality, the embryo is not transferred to the uterus.

PGD has been used only during the past decade, and its technical success rate is small (in terms of the fraction of implanted embryos that become live-born babies). Further, it is expensive and generally not covered by insurance. Nonetheless, thousands of PGDs have been carried out and thousands of healthy newborns delivered. There does not appear to be any damage to the child caused by the biopsy procedure itself.

PRENATAL

First carried out in 1966, prenatal diagnosis has become an increasingly regular part of prenatal care. In the US each year about 80,000 pregnancies are monitored in this way, and that number is growing. Some couples seek prenatal diagnosis because their family history or carrier-testing puts them at an appreciably higher risk for having a baby with a genetic disorder than the general population. Others seek prenatal diagnosis because of advanced maternal age or because maternal serum tests (to be discussed subsequently) put them at higher risk.

The goal of prenatal diagnosis is to provide couples with information about the risk of having a baby with a genetic disorder and to help the couple understand and manage that risk. For those at risk for a severe genetic disorder, prenatal diagnosis has allowed thousands of couples who otherwise would not have risked conceiving to go on to have healthy babies. For those at risk for severe birth defects, the same is true. In all, more than 95% of the fetuses monitored with prenatal diagnosis are found to be unaffected, and the pregnancy is continued. In this setting, parental anxiety will have been relieved markedly. When fetuses are found to be affected, most families choose to terminate that pregnancy. Some, however, use the information to prepare psychologically for the birth of the affected child, and to arrange for the child's postnatal care.

There are two main types of prenatal diagnosis. One is non-invasive, the other, invasive. The non-invasive category uses **maternal serum screening** and **fetal ultrasonography**. The invasive category employs **amniocentesis** and **chorionic villus sampling**. Before commenting on these methods, we will mention the main indications for prenatal diagnosis.

Indications

The most common reasons couples seek prenatal diagnosis are listed in Table 17.3.

Advanced maternal age is by far the most common indication for prenatal diagnosis because of the increased risk of Down syndrome and other aneuploidies in the offspring of women who become pregnant after age 35. Two other indications involve chromosomal abnormalities: when a couple has had a previous child with *de novo* aneuploidy and when one of the parents has a balanced translocation (see Chapter 11).

Another category of illness often leading to prenatal diagnosis is single-gene disorders of many kinds. In some instances the testing involves enzymatic or other biochemical assays. Increasingly, though, DNA-based testing is becoming the method of choice. For X-linked traits for which no biochemical or DNA-based test is available, fetal sex determination is carried out to distinguish an at-risk male from a female.

The risk of major structural defects, particularly those involving the neural tube, is another important indication for prenatal diagnosis. In the past, such detection was offered to couples that had already had a child with a neural tube defect (NTD). Today's obstetrical practice is very different. Virtually all pregnant women have fetal ultrasound and maternal serum testing done. Such testing identifies women at higher risk of having a child with an NTD. Similarly, women who have been exposed to environmental teratogens known to produce structural defects may opt for ultrasonographic monitoring of pregnancy.

TABLE 17.3 Indications for Prenatal Diagnosis
Advanced maternal age
Previous child with chromosomal aneuploidy
Parent with balanced chromosome translocation
Family history of autosomal recessive, autosomal dominant, or X-linked disorder
Risk of a neural tube defect or other structural abnormality

Methods

Prenatal diagnosis may be carried out in a variety of ways. The non-invasive approaches include measurement of certain biochemical substances in the serum of the pregnant woman—referred to as maternal serum screening (MSS). MSS is a regular part of routine prenatal care (in fact, not offering it may be considered malpractice). During the first trimester, the serum is tested for **alpha-fetoprotein (AFP)** and **human chorionic gonadotropin (HCG).** AFP is made in the liver, HCG in the placenta. Under normal circumstances, these substances are found in small amounts in fetal serum and are then transported to the mother's serum. In the presence of an open NTD, the concentrations of AFP and HCG in maternal serum are distinctly elevated, high enough to identify 70–80% of all open NTDs (such as spina bifida or anencephaly). In contrast, abnormally low concentrations of maternal serum AFP are often found in Down syndrome. When MSS detects abnormal concentrations of AFP or HCG, a second non-invasive approach, **fetal ultrasonography**, is performed. This technique allows direct visualization of the fetus. It is particularly good at seeing the folds behind the neck (nuchal folds), which are often accentuated in Down syndrome, and the neural tube. These non-invasive screening techniques are becoming more sensitive (with fewer false negatives) and more specific (fewer false positives), but they are not diagnostic. They are almost always followed by two invasive tests: **amniocentesis**, and **chorionic villus sampling (CVS)**.

These techniques were discussed in Chapter 11. Amniocentesis is generally carried out at 16–18 weeks' gestation. It involves inserting a needle (under ultrasound visualization) through the abdominal wall and into the fluid-filled amniotic sac. A sample of amniotic fluid is removed, containing fetal cells and non-cellular fluid. Most often, the fetal cells are grown in cell culture for several days, then harvested and tested in a variety of ways. Karyotyping is the most common test performed; DNA-based and protein-based assays are performed for a progressively longer list of single-gene disorders. Measurements are made on the fluid as well, particularly the concentration of AFP, which is usually markedly increased in open NTD. More than 250,000 amniocenteses are performed annually in the US today. The procedure results in miscarriage in about 1 in 1,500 cases, a risk very much lower than the rate of spontaneous abortions occurring at this stage of pregnancy (1–2%).

The other invasive test used for prenatal diagnosis is CVS. It involves inserting a catheter through the vagina and the cervix and into the uterus. The catheter is positioned near the villus area of the chorionic placenta (that is, the fetal part of the placenta). This procedure is also monitored by ultrasound. Once positioned, a biopsy of chorionic villi is performed and cells obtained for analysis. Culturing the cells permits karyotypic, molecular, and biochemical analysis—just as described above with amniocentesis. The main advantage of CVS is that it can be performed as early as 12 weeks' gestation, thereby shortening the interval of anxiety and uncertainty and permitting, when elected, earlier termination of pregnancy. CVS has two drawbacks: it provides no measurement of AFP; and it carries a much greater risk of miscarriage (about 1%) than does amniocentesis.

NEONATAL

Genetic disorders in newborns are detected by individual testing and by population screening. Testing is carried out on a person who is likely to have, or is at risk for, a disorder. It is tailored to the circumstances of the individual—in this case, a neonate. Some examples: a girl with the physical features of Turner syndrome will have a karyotype done to see if she has a single X chromosome; a neonate whose older brother has Down syndrome will be tested by physical examination and karyotyping; a newborn whose sibling has achondroplasia will have combined testing by physical examination and X-rays. Such testing will depend on the family history, prior birth history, and physical features of this individual.

Newborn screening is much more often carried out on whole populations. Its goals are to identify neonates with genetic disorders before signs and symptoms appear and to institute therapeutic measures aimed at preventing damage to the newborn. Newborn screening has been carried out in the United States since 1961, when it was instituted to detect phenylketonuria (PKU). This approach was so successful in preventing the severe cognitive disabilities associated with untreated PKU that it has been expanded progressively over time. Virtually all newborns in the US are screened for PKU, and many others are screened for a longer list of genetic disorders. Because such screening is under the authority of each state, there is no consistency as to the number of conditions screened for. Thus, in some states more than 50 disorders are tested for; in others, many fewer than that. Screening is generally carried out by obtaining a few drops of blood obtained from the baby's heel on the first or second day of life and analyzing the sample by **tandem mass spectrometry**. This method measures the concentration of many substances involved in the metabolism of amino acids, organic acids, and other classes of biochemicals. When an elevated concentration of a particular substance is found, the physician and family are made aware of the need for more definitive testing.

A number of criteria must be met before adding a particular metabolite to a screening panel. The metabolite must be an accurate predictor of the condition being screened for. The chemical test must have sufficient specificity and sensitivity to mitigate misinterpretation. Most important, the disorder being screened for must be one where early institution of treatment has a high likelihood of producing a good outcome for the affected person. Neonatal screening has proven itself to be life-saving, family-preserving, and cost-effective.

CHILDHOOD AND ADULTHOOD

To complete this discussion of the timing for detecting genetic disorders, we must point out that many single-gene disorders and multifactorial traits are not tested or screened for until well after birth. In the main, they are tested for when an individual exhibits signs or symptoms suggestive of a particular condition:

- cystic fibrosis will be tested for in a young child with respiratory or digestive disturbance;
- Duchenne muscular dystrophy will be detected when a young boy shows skeletal muscle weakness;
- familial hypercholesterolemia will be suspected in a young child or a parent with cholesterol deposits in the eye or skin.

Because genetic disorders are so numerous, so diverse, and, together, so common, they are now, and will be even more in the future, conditions that will manifest themselves throughout life and therefore will have to be diagnosed (or detected) by the discerning physician throughout life.

Whether: Special Issues Regarding Detection

Traditional clinical medicine is based on some well-established (though rarely articulated) principles. It is an implied contract between a physician or other healthcare provider and an individual patient. The patient consults the physician because of signs or symptoms of some kind. Diagnostic tests and therapeutic intervention are directed toward ameliorating the clinical disturbance in that patient—and, generally, in no other person. Genetic and genomic medicine differs in three major ways:

- it considers the family as well as the individual;
- it often provides information about risk, rather than certainty (that is, about what may happen, rather than what has already happened);
- it considers the entire human life cycle—from the embryo to the aged.

These differences make up the ethos of genetic and genomic medicine, an ethos that raises important questions and dilemmas not generally encountered in traditional clinical practice. These questions have important implications for detection of genetic conditions. Three criteria for assessing the safety and efficacy of genetic tests will be discussed in turn.

ANALYTIC VALIDITY

Validity has to do with soundness and reliability. A test has **analytic validity** if it is positive when a particular DNA sequence, karyotype, gene product, or metabolic substance is present, and negative when such sequences, karyotypes, or products are absent. Tests with high analytic validity include karyotypic aneuploidy, red blood cell morphology, separation of substances in neonatal blood by mass spectroscopy, and mutations in specific DNA sequences. Analytic validity is proven over time by experience with large numbers of people. In the absence of analytic validity, neither clinical validity nor clinical utility is possible.

CLINICAL VALIDITY

A test demonstrates **clinical validity** if it is positive in people with a given disorder and negative in those without the disorder. Said differently, a positive test predicts disease or susceptibility; a negative test predicts the absence of disease or susceptibility. Some examples of clinical validity include trisomy 21 in Down syndrome, abnormal hemoglobin electrophoresis in sickle cell anemia, and reduced hexosaminidase A activity in carriers of Tay-Sachs disease. Tests demonstrating analytic validity and clinical validity are necessary, but not sufficient, to satisfy the third characteristic of a genetic test—clinical utility.

CLINICAL UTILITY

To be clinically useful, a test's benefits must outweigh its risks. This may sound self-evident but it is not, because genetic testing—unlike other clinical testing—has **clinical utility** in some settings (and for some people) but not in others. For example, PGD and prenatal diagnosis have clinical utility in situations where a couple is prepared not to implant or not to carry to term an affected embryo or fetus. However, for people unalterably opposed to abortion, such testing has risk without appreciable benefit, and hence is of no clinical utility. Similarly, testing for dominantly inherited disorders of the brain with late clinical onset—including Huntington disease and Alzheimer disease—presents excruciating dilemmas for some families. For example, we now have an analytically and clinically valid test for Huntington disease, but the clinical utility varies greatly from family to family. In the absence of any effective treatment, nearly 80% of children of a parent with Huntington disease choose not to be tested rather than deal with the agonizing certainty of Huntington disease decades before its clinical onset. Other members of the same family who are at risk choose to be tested so that they can make informed decisions about having children, or about undergoing prenatal diagnosis. Similar quandaries present themselves in BRCA testing, where a positive test increases the likelihood of breast cancer but not its certainty. In this setting we have analytic validity, but not necessarily clinical validity or utility. As genomic testing proliferates, the health profession, patients and families, and regulatory agencies will confront the kinds of dilemmas just mentioned. Increasingly, understanding such complexity and communicating it accurately and humanely is the province of a relatively new field called genetic counseling.

GENETIC COUNSELING

Genetic counseling is a communication process whose broad aim is to provide information to individuals and families with or at risk for genetic disorders. This process attempts to help consultands (those seeking counseling) to:

- comprehend the medical facts, including those concerned with diagnosis, likely outcome, and medical management;
- appreciate the role of heredity in the disorder, specifically regarding the risk of recurrence in family members;
- understand the options available for dealing with the recurrent risks;
- choose a course of action consistent with their family goals and their ethical and religious principles;
- make the best possible adjustment to the disorder.

TABLE 17.4 Common Indications for Genetic Counseling
Family history of a Mendelian disorder
Child with multiple congenital anomalies or cognitive disability
Advanced maternal age
Recurrent pregnancy loss
Family history of cancer
Teratogen exposure
Consanguinity
Planning for testing regarding late-onset disorders
Follow-up to abnormal prenatal or neonatal test result
Interpretation of direct-to-consumer genomic testing

Indications

Given the diverse (and expanding) nature of genetic disorders, a wide range of indications for genetic counseling exists. The most common of these are shown in (Table 17.4).

Often, the consultands are the parents of a child with a known or suspected genetic condition. They may also be adults concerned about a disorder in family members, or preparing for prenatal diagnosis.

Providers

In the past, genetic counseling was usually provided by a physician whose expertise in medical and clinical genetics made possible the communication of information about the nature of the disorder, the recurrence risks, and the available options. But as the body of knowledge about genetic disorders has expanded geometrically, and as the armamentarium of tests and screens has too, communication with families has required counseling by teams of professionals with different skills. Such teams are usually led by MDs. As important are nurses (who have become expert in clinical genetics by learning on the job) and a cadre of **genetic counselors** who have completed a master's program in genetic counseling at one of about 30 colleges and universities in the United States.

Because these genetic counselors learn the principles of human genetics, the application of such principles to clinical practice, the ability to communicate sometimes difficult and complex information to families, and the sensitivity to the psychosocial aspects of communicating this information, they have become ever more important to the process of genetic counseling—often being the first point of contact with the family and the sustaining link between consultands and those consulted. Initially these counselors worked in general genetics units, interacting with those seeking information regarding the wide range of indications mentioned in Table 17.4. However, now that genetic thinking has permeated many medical specialties, genetic counselors provide services to specialty clinics serving patients with cancer, heart disease, and obstetrical problems, among others. At present, there are only about 3,000 such trained counselors in the US—far fewer than are needed now and in the future.

Process

A rational sequence of events occurs once an individual or family has sought genetic counseling (Table 17.5).

TABLE 17.5 Sequence of Events in Genetic Counseling
Information-gathering
Initial assessment
Counseling
Follow-up
Psychosocial support

- *Information-gathering.* After determining the nature of the question asked by the consultands, a detailed family history is obtained and a pedigree is constructed. Relevant medical history and results of prior tests are also collected.
- *Initial assessment.* Based on a physical examination and laboratory and radiologic assessment, a presumptive diagnosis is established (albeit often without absolute certainty).
- *Counseling.* Here, the nature and consequences of the disorder are presented to the consultands, focusing particularly on the risks of occurrence or recurrence.
- *Follow-up.* Based on the consultands' response to the counseling, there may be referral to other medical specialists, to disease-specific support groups, and to public agencies.
- *Psychosocial support.* Maintaining contact and helping families deal with the often wrenching information is a critical part of the process (usually provided by the genetic counselor described earlier).

The above steps sound straightforward enough, but they are often anything but as clear. Here are some of the kinds of questions that make genetic counseling a delicate process requiring skill and sensitivity.

- *What is the real question being asked?* Some families articulate clearly the kind of information they seek; others are confounded by lack of knowledge, anxiety, and guilt and must be helped and encouraged to identify their reason for seeking help.
- *Who are the consultands?* Usually a married couple seeks counseling together, but not always. Sometimes individuals come alone, or with a friend. Assessing the intellectual and emotional capabilities of the consultands is critical to offering information and advice.
- *How much information should be communicated and at what level?* Here, it is critical that whatever information is being transmitted must be understood by the consultands. This is particularly important when discussing empiric recurrence risks or presenting Mendelian probabilities (for example, 25% or 50%). Because so many lay people do not understand even simple probability or elementary statistics, this matter should not be overlooked.
- *Should counseling be non-directive or directive?* Forty years ago, providers of genetic counseling believed that counseling should be non-directive, meaning that the provider did not offer an opinion or advice but merely presented the facts and options. Over time, it has become clear that many consultands want and need advice from providers of genetic counseling—just as they need advice from other clinical specialists. Rather than a dogmatic "non-directive or directive" rule, every consultand should help define their desire and expectation along this delicate continuum.
- *In what ways will the "genomics era" change genetic counseling?* As knowledge about the structure and function of the human genome explodes, it will cause new issues to be brought to genetic counselors, such as individual variation, susceptibility to common disorders, understanding genetic differences among ethnic groups, and the interpretation of results from direct-to-consumer tests (to be discussed in Chapter 19).

TREATMENT

A physician's overarching goal is to provide a patient with successful **treatment** or **cure**. These terms are not synonymous. Treatment involves the management of a patient's illness with the purpose of combating the disorder. It generally involves long-term intervention. Management of asthma, high blood pressure, or diabetes is illustrative of the term "treatment." Cure, on the other hand, involves restoration of health through a single course of treatment which eliminates the disorder. A course of penicillin often cures a streptococcal sore throat; so does an appendectomy cure appendicitis.

The words "treatment" and "cure" pertain equally well to genetic disorders as to environmental ones, yet this is almost universally misunderstood. When we ask our students annually what fraction of children with genetic disorders are cured—offering three choices: 10%, 1% or 0%—virtually everyone says "0." The correct answer is 10%, because a significant fraction of genetic disorders in children are multifactorial ones such as cleft lip and palate, congenital heart defects, and polydactyly—each curable by surgery. Why this widespread misperception? First, because multifactorial conditions are not routinely thought of as "genetic." Second, because the word "genetic" is mistakenly conflated with "genetic determinism," implying that we are born with our genes—normal and mutant—and there is nothing to be done about that.

The same erroneous misperception holds when students are asked what fraction of adults with genetic disorders are treated successfully—10%, 1%, or 0%. Again, the vast majority of students choose 1% or 0. This answer forgets the large number of common, multifactorial conditions for which we have effective pharmacologic treatment, such as high blood pressure, hypercholesterolemia, asthma, depression, chronic obstructive pulmonary disease, some forms of cancer, and many more. Again, genetic fatalism conditions the response of students.

Having (hopefully) convinced you that "treatment" and "cure" are legitimate words to include in the lexicon about genetic and genomic disorders, it must be said that these words remain the "short straws" in our approach to patients with chromosomal and single-gene disorders. Regarding chromosomal aneuploidies, we know of none that are amenable to successful treatment, implying that, in the main, management is symptomatic and supportive. So, too, is the situation for single-gene disorders (Figure 17.2). In a group of 372 Mendelian disorders, treatment was considered completely successful in only 12%, partly successful in 54%, and unsuccessful in 34%.The results were better in those conditions for which a molecular and biochemical understanding was at hand. This tally makes clear the challenge facing those who care for patients with single-gene disorders. It also explains why families faced with these conditions often seek alternative options such as PGD, IVF, or prenatal diagnosis.

For those single-gene disorders and multifactorial ones in which treatment is successful or partially successful, treatment is a form of prevention. Strictly speaking, the word "prevention" means "to keep from happening" while the word "treatment" deals with intervention after something has happened. Vaccination against polio, rubella, or human papilloma virus constitutes primary prevention because it keeps infection from happening. Secondary prevention aims to prevent the effect of the insult—genetic or acquired—on the person affected. For example, many metabolic disorders, phenylketonuria being a prime example, are treated in the neonatal period to prevent organ damage likely to occur in the absence of treatment. In adults with hypercholesterolemia, treatment with diet and drugs is aimed at preventing coronary artery disease. For many genetic disorders, then, treatment and prevention are not distinct, but rather have a common goal of reducing suffering and maintaining health.

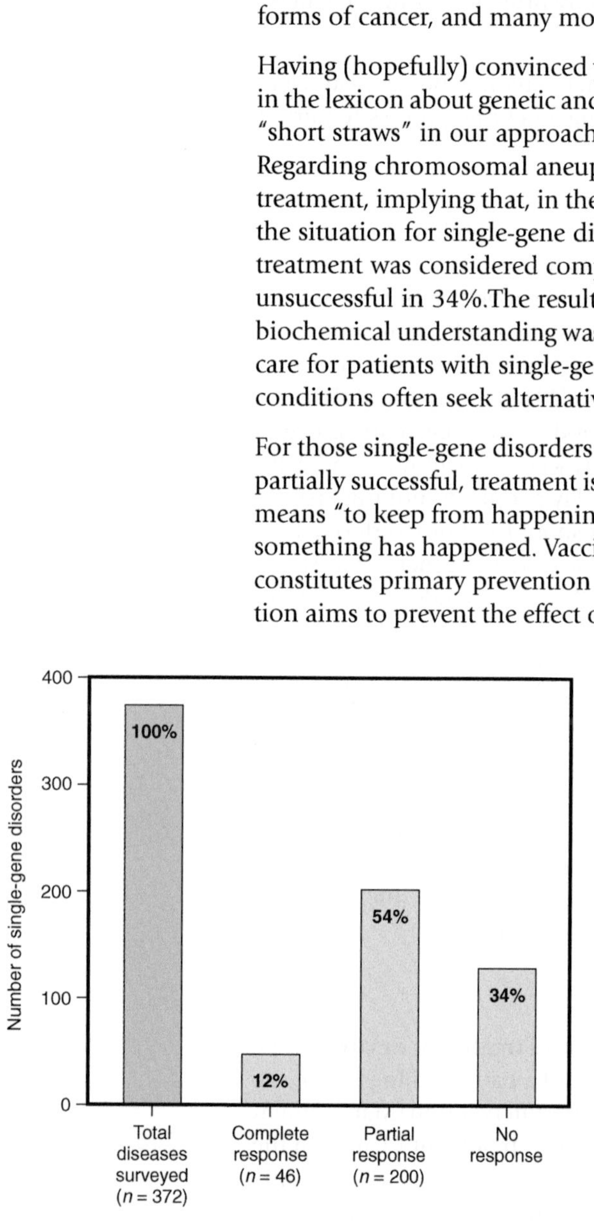

FIGURE 17.2
Effect of treatment in 372 disorders due to single-gene defects, and for which the affected gene or biochemical abnormality is known. The survey was conducted in 1999, but data have not changed appreciably since. Note that complete response was found in only 46 disorders (12%), while partial or no response was found in 88% of the disorders examined.

Timing of Therapeutic Intervention

In the same way that detection of genetic disorders occurs at different periods of human life, so too does treatment—and, as mentioned above, timing is a critical success factor (Table 17.6).

In less than a handful of cases, **prenatal** therapy may be instituted. Here are a few examples: in the face of Rh blood-group incompatibility between mother and fetus, administration of gamma globulin to the mother will prevent fetal damage; in fetuses with congenital defects of their diaphragm muscle, surgical correction may be lifesaving; in a form of methylmalonic acidemia due to defective synthesis of a vitamin B12 cofactor (to be discussed in

TABLE 17.6 Timing of Therapeutic Intervention for Genetic Disorders

Time	Examples
Prenatal	Rh incompatibility Congenital defects of the muscular diaphragm Vitamin B12-responsive methylmalonic acidemia
Neonatal	Phenylketonuria Galactosemia Propionic acidemia Cleft lip Congenital heart defects
Childhood	Cystic fibrosis Sickle cell anemia Type 1 diabetes Asthma
Adulthood	Hypertension Hypercholesterolemia Depression Bipolar disorder Emphysema

detail subsequently), maternally administered prenatal supplements of the vitamin can correct, in part, the metabolic abnormality.

Many genetic disorders are treated in the **neonatal** period. For the 30 or more metabolic diseases that can be detected by neonatal screening, treatment with special diets and/or drugs will regularly limit or entirely prevent delayed manifestations such as cognitive delay or disability, liver damage, or abnormalities in maintaining acid/base balance. In the neonatal period, too, surgical correction of birth defects is often curative (as in cleft lip and palate, congenital heart defects, polydactyly, or obstruction of outflow from stomach to small intestine, called pyloric stenosis).

A growing number of multifactorial conditions and single-gene defects are treated during **childhood** and **adolescence**. Examples of common single-gene defects approached this way include cystic fibrosis and sickle cell anemia. Type 1 diabetes and asthma are but two examples of multifactorial conditions that require treatment during this critical period of growth and development.

Finally, the majority of common disorders with complex inheritance, including hypertension, emphysema (obstructive pulmonary disease), unipolar and bipolar depression, and hypercholesterolemia, reveal their chemical or clinical abnormalities during **adulthood** and are treated then.

Therapeutic Strategies and Modalities

STRATEGIES

As noted in Table 17.7, the treatment of genetic disorders occurs over a uniquely broad number of levels—from the clinical phenotype to the gene—with several plateaus between.

TABLE 17.7 Levels at which Treatment is Offered

Level	Examples
Clinical phenotype	Surgical correction of cleft lip, congenital heart defects Blood transfusion for sickle cell anemia or β-thalassemia
Metabolic abnormality	Phenylketonuria Galactosemia Methylmalonic acidemia
Mutant gene product	Type 1 diabetes Hemophilia SCID
Mutant mRNA	DMD
Mutant gene	SCID X-linked leukodystrophy

Surgical correction of congenital heart defects and blood transfusion in the thalassemias exemplify intervention at the phenotypic level. Gene therapy for severe combined immunodeficiency disease (SCID) or Leber's congenital blindness typify treatment available at the level of the gene. Between these poles, some diseases are successfully treated by addressing the metabolic abnormalities (as in phenylketonuria and many other disorders of amino acid or organic acid metabolism), the mutant or absent protein (as in type 1 diabetes or hemophilia), and potentially the mutant mRNA (as in Duchenne muscular dystrophy).

MODALITIES

Table 17.8 identifies the many **therapeutic modalities**, medical and surgical, that are in use for genetic disorders.

TABLE 17.8 Modalities Employed in Treating Genetic Disorders

Medical	Surgical
Avoidance	Excision
Restriction	Correction
Replacement	Organ transplantation
Supplementation	
Drug administration	
Cell therapy	
Gene therapy	

Medical Modalities

Avoidance Although not usually considered a therapeutic modality, avoiding particular chemical or physical agents constitutes an important therapeutic intervention. In xeroderma pigmentosum (Figure 17.3), for example, avoiding sunlight prevents to some degree the formation of skin cancers in patients lacking the enzyme required to repair DNA damage resulting from exposure to ultraviolet light. Similarly, avoiding the anesthetic succinyl choline is critical to those lacking the enzyme required to break down this widely used anesthetic agent. Avoidance of cigarettes is clearly beneficial in a number of rare disorders, including alpha-1-antitrypsin deficiency, in which lung damage is worsened by smoking; in

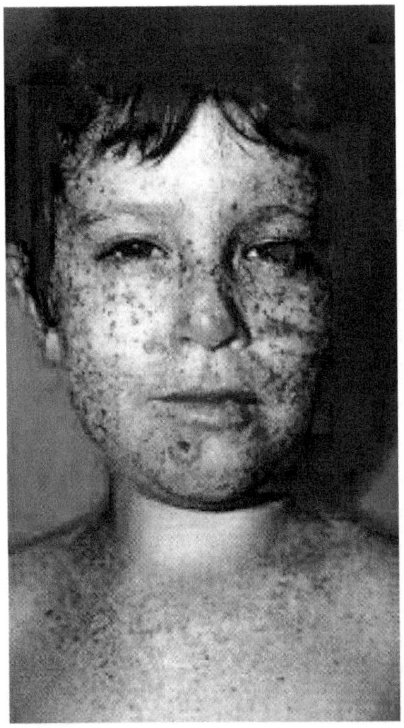

FIGURE 17.3
A patient with xeroderma pigmentation whose face and chest are studded with numerous skin cancers. Exposure to sunlight results in formation of these tumors, because the patient lacks an enzyme required for DNA repair following UV exposure.

several common cancers (lung, breast, pancreas); and in emphysema resulting from chronic obstructive lung disease.

Restriction Marked dietary restriction of single small molecules represents the "classical" therapeutic approach to treating inborn errors of metabolism. Diets low in phenylalanine have been shown definitively to prevent cognitive disability in patients with phenylketonuria (PKU), who lack the enzyme required to break down the essential amino acid phenylalanine. This modality, in use for more than 40 years, has been extended to many other disorders of amino-acid, organic-acid, and sugar metabolism. The particular substance restricted in the diet depends on knowing the step "blocked" by the mutation and the metabolite that accumulates because of that block.

Replacement A considerable number of genetic disorders can be treated effectively by replacing a molecule rendered deficient by mutation. For example, if a mutation blocks intestinal absorption of vitamin D, parenteral administration of the vitamin will prevent rickets. If synthesis of a hormone is interfered with (as when thyroid hormone or cortisone formation is blocked), replacing the hormone is vital. If a circulatory protein (such as hemoglobin, clotting factor VIII, or gamma globulin) is rendered deficient by mutation, protein replacement is widely successful. If an intracellular enzyme is defective, enzyme replacement has been shown to be effective in rare instances (Gaucher's disease, SCID due to adenosine deaminase deficiency) where the administered enzyme can find its way to its normal location of intracellular activity.

Supplementation We now recognize more than 20 disorders that respond beneficially to administration of supplements of a particular vitamin. This approach "works" for one of two reasons, each reflecting the fact that most vitamins act as cofactors needed to activate one or more enzymes. In certain instances in which vitamin-requiring enzymes are impaired, but not absent, vitamin supplementation will restore enzymatic activity by stabilizing the enzyme in one of several ways. In other instances, a mutation interferes with conversion of the vitamin to its coenzyme form, but doesn't block synthesis entirely. Here, vitamin supplementation "drives" sufficient coenzyme synthesis and is thereby beneficial.

AMPLIFICATION: SERENDIPITY IN SCIENCE

Inside the Hospital and the Laboratory: The Case of Robby

In 1969, Robby was admitted to the pediatric intensive care unit at Yale-New Haven Hospital in coma at the age of 8 months. Dr Rosenberg found that he had ketoacidosis, a metabolic state associated with elevated concentrations of acetate, acetoacetate, and ketone bodies in the blood. His blood pH was 6.95—several points lower than the normal value of 7.4, and barely compatible with life.

Judicious residents saved Robby's life by correcting his acidosis with intravenous solutions of bicarbonate, but the cause of his coma was difficult to determine. Usual causes, including diabetes mellitus or ingestion of some poison, were ruled out. Playing a hunch, Dr Rosenberg assayed Robby's urine for methylmalonic acid (MMA), previously detected in large amounts in urine from only one baby with lethal ketoacidosis living in the United Kingdom. Huge amounts of MMA were found in Robby's urine—more than that excreted by all the hospital's employees combined.

MMA is an intermediate formed in the breakdown of several amino acids, some fatty acids, cholesterol, and pyrimidines. MMA is converted, in turn, to succinate, which enters the tricarboxylic acid cycle—a set of reactions which ultimately generates energy in the form of ATP. The critical enzyme (methylmalonyl mutase) that catalizes the breakdown of MMA requires a coenzyme form of vitamin B_{12}. Reasoning that the cause of Robby's metabolic block might lie in the interaction between the mutase enzyme and its B_{12} coenzyme, Dr Rosenberg played a second hunch and gave Robby megadoses of vitamin B_{12}, 1,000 times the normal daily requirement. Urinary MMA fell by 80%, only to rise again when the B_{12} supplements were withdrawn!

From these clinical and laboratory clues, a long series of investigations ensued, leading to the following conclusions. Robby suffered from a genetic defect that disrupted vitamin B_{12} metabolism such that the required coenzyme was formed in insufficient amounts. The vitamin B_{12} supplements were effective because they drove formation of sufficient coenzyme to overcome the block in MMA breakdown. Subsequent elucidation of the complex pathway of cellular B_{12} coenzyme synthesis led to the discovery of at least eight other mutations in children, some responsive to B_{12} supplements, others not.

Today, urine of most newborns is screened for MMA. Babies excreting it in excess are given high doses of vitamin B_{12} supplements briefly as a test. Those responding chemically are maintained on supplementary B_{12} indefinitely. (Robby has been so treated for 40 years.) Children with B_{12}-responsive methylmalonic acidemia have a far better long-term prognosis than do those who fail to respond.

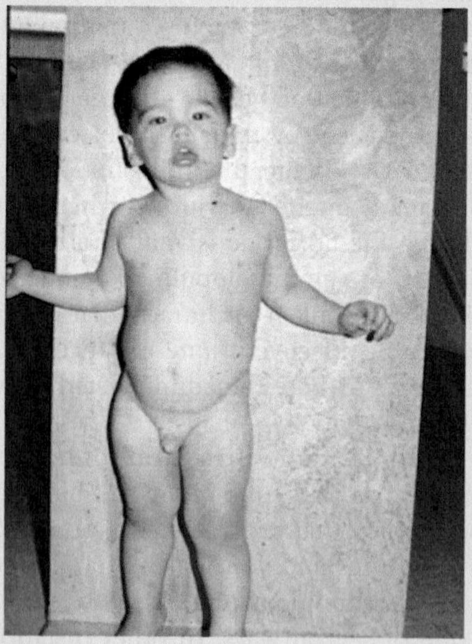

(Continued)

AMPLIFICATION: SERENDIPITY IN SCIENCE—Cont'd

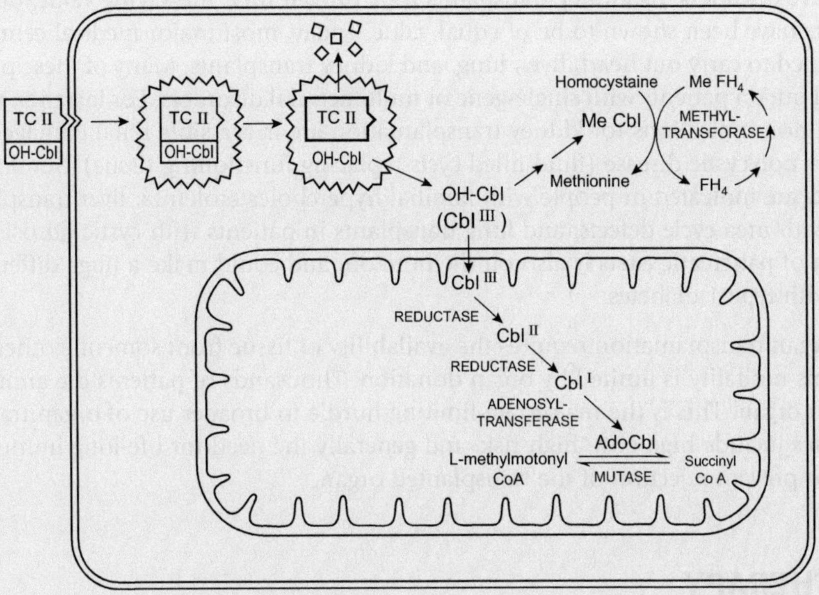

Drug Administration A growing list of inherited disorders respond beneficially to administration of pharmaceutical agents. As with everything we've discussed about treatment previously, efficacy depends on understanding the nature of the disorder. A few examples:

- **Wilson's disease**, a disorder of copper metabolism, responds to administration of chemical chelators that bind copper and enhance its excretion
- **Statins** interfere with intracellular synthesis of cholesterol, and reduce its serum concentration
- As discussed in Chapter 16, **Gleevec®** interferes with the ABL oncogene, thereby exerting a striking effect in chronic myelocytic leukemia (CML).

mRNA Modification Already proven in principle but not yet in practice, recent work suggests that it may be possible to affect mutant mRNA molecules and thereby the products they code for. For example, a particular deletion mutant in **Duchenne muscular dystrophy (DMD)** has been approached by administration of modified oligonucleotides that hybridize with mRNA and allow ribosomes to "skip" past the deleted mRNA segment and restore some mRNA (and protein) function. In another example, a modified oligonucleotide has led to skipping a premature stop codon in one form of cystic fibrosis. Use of tailored miRNA molecules represents another potential avenue of therapy carried out at the RNA level.

Surgical Modalities

As highlighted in Table 17.8, surgical treatment of genetic disorders offers much to affected patients. Three approaches deserve mention.

Excision In the many forms of polydactyly or syndactyly, removal of the extra digits or the interdigital bands restores normal structure and function.

Correction More often, as in cleft lip or congenital heart defects, surgery aims to correct the anatomic problem. Most such surgical procedures are performed early in life so that normal structure and function are restored. Because birth defects like cleft lip and heart defects are so common (about 0.1% of live births for each category), corrective surgery is by far the most commonly used surgical modality.

Organ Transplantation For 50 years kidney transplantation has been carried out in patients with irreversible renal failure brought on by diabetes and a variety of inflammatory and degenerative disorders. As kidney transplants have proven their life-saving value, other organ transplants have been shown to be of equal value. Today, most major medical centers have teams trained to carry out heart, liver, lung, and kidney transplants. Many of these procedures are carried out on patients with single-gene or multifactorial disorders. For instance, two of the most common indications for kidney transplantation are irreversible renal damage from diabetes or polycystic disease (fluid-filled cysts replacing functioning tissue). Similarly, heart transplants are indicated in people with familial hypercholesterolemia; liver transplants in children with urea cycle defects; and lung transplants in patients with cystic fibrosis. Transplantation of pancreatic islets is also on the horizon, and could make a huge difference to patients with type 1 diabetes.

Because organ transplantation requires the availability of tissue from someone other than the patient, this modality is limited by organ donation. Thousands of patients die annually for want of an organ. This is the major rate-limiting hurdle to broader use of organ transplants. Other issues include high cost, high risk, and generally the need for life-long immunosuppression to prevent rejection of the transplanted organ.

GENE THERAPY

For more than 40 years, we have told students that one day we would treat disorders with genes. This prediction was neither hubris nor fantasy. It reflected two things: that treatment for most single-gene disorders was absent or inadequate and that the ability to isolate and propagate genes was proceeding at a sometimes remarkable rate. Fast forward to the present. Disappointingly, only a handful of disorders have been treated successfully at the level of the gene. Nonetheless, progress is being made, and there is increasing hope that gene therapy will become available for a substantial list of currently untreatable conditions.

We will approach this subject by trying to answer the following questions:

- What does the term "gene therapy" mean?
- How is gene therapy carried out?
- What steps are required for development of successful gene therapy?
- Why has progress been slow—and sometimes halting?
- In which conditions has gene therapy been proven to be successful?
- What is the likely future of gene therapy?

Definitions

Human gene therapy gene is defined as the process by which a normal human gene is introduced into a patient's cells for the purpose of treating or curing that patient's disease (Table 17.9).

TABLE 17.9 Defining Terms Concerning Gene Therapy

Human gene therapy	The process by which a normal human gene is introduced into a patient or a patient's cells for the purpose of treating or curing that patient's disease
Somatic cell gene therapy	That form of gene therapy in which the normal gene is introduced to tissues other than germ cells
Germ line gene therapy	That form of gene therapy in which a normal gene is introduced into male or female gametes

In theory, two kinds of cells are potential targets for gene therapy. **Somatic cell gene therapy** is that form of gene therapy in which the normal human gene is introduced to tissues other than germ cells; hence, this kind of gene therapy is not heritable. **Germ line gene therapy** is that form in which the normal gene is introduced into male or female gametes. Thirty years ago the scientific community urged that germ line gene therapy be avoided until it was absolutely clear that this approach, heritable by its very nature, was proven to be safe for future generations. To this day, germ line gene therapy in humans remains "off limits," and properly so.

Types of Somatic Cell Gene Therapy

As shown schematically in Figure 17.4, two types of somatic cell gene therapy are possible—***ex vivo*** and ***in vivo***. In the *ex vivo* form, the normal gene is introduced to cells removed from the patient. Then, the treated cells are returned to the patient via the circulation. *In vivo* therapy differs in that the normal gene is introduced directly into the patient—generally into the tissue requiring treatment (such as muscle, liver, or brain). In both types, the normal gene must be incorporated into a vector (usually a non-pathogenic virus) and propagated to obtain a sufficient number of new gene copies to make treatment clinically feasible.

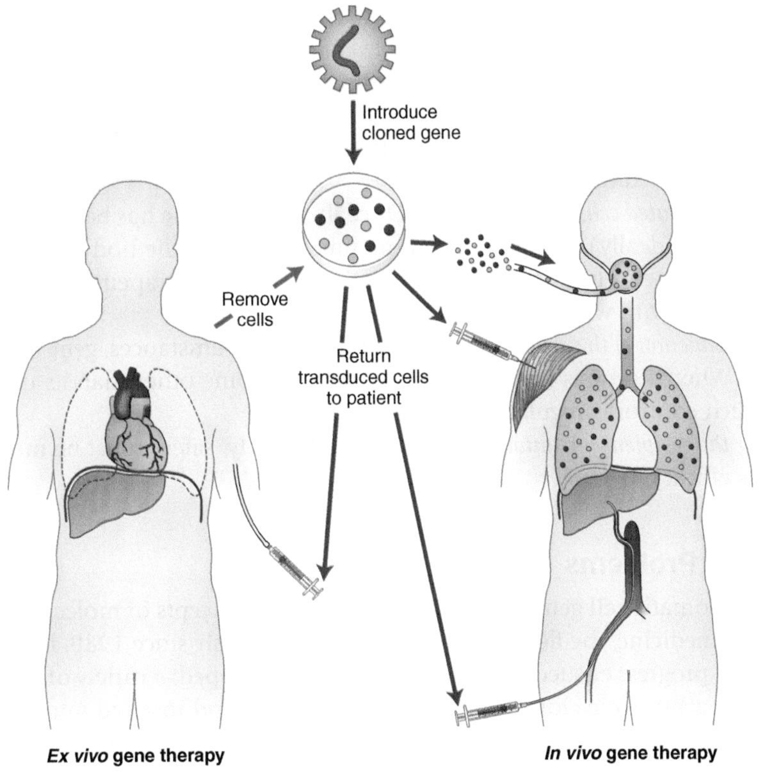

FIGURE 17.4
Distinctions between *in vivo* and *ex vivo* gene therapy. In the *in vivo* approach the normal gene packaged in a viral vector is introduced directly into, for example, the patient's muscle, liver, or respiratory tract. In *ex vivo* therapy, the normal gene is packaged in a vector, transduces the patient's cells in culture, and then is administered. Additional details are provided in the text.

Steps Required for Development

Table 17.10 lists a long (and daunting) list of steps toward the development of *ex vivo* somatic cell gene therapy. A brief comment about each follows.

- *Choice of clinical indication.* Given the limited progress of gene therapy, it is crucial that a risk–benefit assessment favors its use. If other modalities are effective, most clinical geneticists believe that trying gene therapy is currently unwarranted.

TABLE 17.10 Sequential Steps Required for Development of Successful *ex vivo* Gene Therapy

- Ascertaining that risk–benefit assessment favors gene therapy
- Isolating and cloning the normal gene
- Defining the regulatory DNA sequences necessary for expression of the normal gene
- Packaging the normal gene into a suitable vector
- Transducing the patient's cells with the vector carrying the normal gene
- Propagating the transduced cells in culture
- Administering the transduced cells into the patient
- Scrutinizing the patient for untoward effects from the treatment

- *Isolating and cloning the normal gene.* When gene therapy was dawning, this step constituted a formidable hurdle. Recombinant DNA technology has changed that. Ironically, today this is the easiest step toward gene therapy.
- *Defining necessary regulatory sequences.* For some genes, expression requires only the nucleotide sequence coding for the gene product. For others, expression demands that both coding and regulatory sequences be provided. (This matter has bedeviled development of gene therapy for sickle cell anemia and β-thalassemia.)
- *Integrating the cloned gene into recipient cells via an exogenous vector.* This step is, in actuality, usually two steps. For *ex vivo* gene therapy, the cloned normal gene is inserted into a damaged virus, where it propagates. Then, the viral vector is used to transfect the recipient cells, where it integrates into the genome. In the *in vivo* approach, the gene propagated in the vector is returned to the patient directly.
- *Propagating the treated cells in the recipient.* Once the normal gene has been introduced to the patient, it must (ideally) be stable (i.e., not eliminated from the body) and direct production of proteins to sufficient concentration to have a therapeutic effect. There are no established rules with which to predict success at this step.
- *Regulating production of the gene product.* Under normal circumstances, gene activity is tightly controlled. When a gene is inserted at a site in the genome other than its usual one, however, such control is regularly interrupted.
- *Scrutinizing the recipient for untoward effects.* Stated simply, safety must be monitored as carefully as efficacy.

Progress and Problems

Because human somatic cell gene therapy melds tools and concepts of molecular biology with those of clinical medicine, the field has been pursued vigorously since 1980. During the 1980s much preclinical progress ensued: animal models were developed; a variety of viral vectors was identified; many genes were cloned, packaged into vectors, and inserted into the genome of human cells in culture; and expression of such inserted genes was shown.

By 1990, the stage was set for clinical trials in humans. In that year the first such trial was conducted in two children with a form of SCID caused by deficiency of adenosine deaminase. Its gene was inserted into circulating lymphocytes. The results were equivocal, but (importantly) no safety concerns were noted. Subsequently, hundreds of clinical trials were initiated in a large number of single-gene disorders and in cancer. The results were very disappointing. Not a single example of clinical efficacy was observed.

In 1999, disaster struck the field. An 18-year-old man named Jesse Gelsinger, who had a mild version of ornithine transcarbamylase (OTC) deficiency, volunteered to participate in an early phase, *in vivo* gene therapy trial. After receiving a high dose of the OTC gene inserted into an adenoviral vector, he collapsed from a massive immunologic reaction to the virus, developed multiple organ failure, and died. This catastrophe was made worse by a lengthy inquiry showing that the investigators had not received informed consent from the patient, had not followed experimental protocols put in place by the FDA and the National Institutes of Health (NIH), and almost surely had financial conflicts of interest. The ensuing uproar, which included congressional hearings, halted all American gene therapy trials for several years.

Ironically, concurrent with this calamity, the first genuine cures produced by gene therapy were reported from France. A number of children with a form of SCID, produced by deficiency of a cell surface receptor required to initiate a cellular immune response, underwent *ex vivo* gene therapy in which the normal gene was transduced into their bone marrow cells. In each case, the cellular and clinical manifestations of their SCID disappeared. The exuberance produced by this breakthrough was muted when two of the children developed a form of (treatable) leukemia because the insertion of the normal gene had occured immediately adjacent to an oncogene. The insertion activated the oncogene and drove overproduction of white blood cells.

Between 2000 and 2010, *ex vivo* gene therapy trials resumed in the US and at least two more unquestioned successes were noted—in adenosine deaminase deficiency and in X-linked adrenal leukodystrophy (which produces failure of the adrenal gland). In Leber's congenital blindness, *in vivo* gene therapy in which the normal gene was packaged and inserted directly into the retina has provided impressive restoration of vision.

INTERSECTIONS

Successful Gene Therapy for an Inherited Form of Blindness: A Dog, A Virus, and a Gene

Leber congenital amaurosis (LCA) is a rare, inherited disorder that leads to severe and progressive loss of vision and results in blindness. Mutated genes prevent cells in the eye from producing a particular protein (RPE65) that is necessary for vision. This disease appears at birth or in the first few months of life. It was first described by a German ophthalmologist, Theodore Leber, in 1869, and occurs in about 1 in 800,000 people.

Although LCA has long been recognized as inherited, there has been no known treatment for it. It occurred to two scientists, Jean Bennett and Albert Mcguire, that if they could somehow insert corrective genes into the cells of the eye, these genes might facilitate the production of the missing chemical and thereby treat the disease.

Although inherited eye diseases had been identified as early as the time of Aristotle (384–322 BCE), it was not until 1989 that scientists at the NIH (National Institute of Health) identified the first gene that caused such eye disease in humans. The specific gene that caused LCA was identified in 1997. In 1998, it was recognized that this same mutation caused blindness in a particular breed of dog, the Briard.

A Briard.

(Continued)

INTERSECTIONS—Cont'd

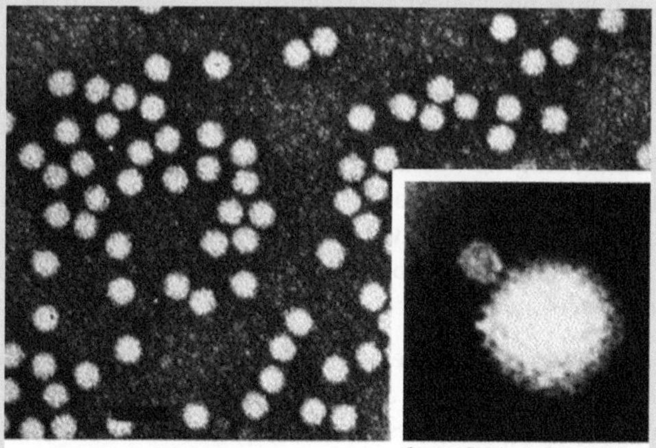

Adeno-associated viruses.

Now Bennett and Mcguire had a non-human subject on which they could test their theory of gene injection into the eye. The next question, however, was how to get this gene into the cells of the eye so that they could produce the necessary protein, RPE65, which would enable vision.

Bennett and Maguire decided to take a modified virus that could no longer produce any deleterious changes in the eye, add to it the missing gene, and infect the cells of the blind dogs' eyes with it. They hoped that the normal gene would be expressed in the retinal cells of the dog's eye and cause the production of RPE65. The modified virus was injected through the white of the eye to the retina, the light-sensitive lining of the inner surface of the eye. The trials with the blind Briards were successful, restoring some vision in more than 90% of them with no side-effects.

In 2008, Bennett and Macguire injected eight human patients with the modified virus. Within 3 weeks of surgery, there was noticeable improvement in the vision of these patients. For the first time, *in vivo* gene therapy had produced safe, positive results. Subsequently, more than 30 patients with LCA have been treated using this form of gene therapy. Initial results have been most encouraging, but longer-term follow up is required to gauge the clinical utility of this promising modality.

Whither Gene Therapy?

Many scientists have conjectured about the reasons for the slow progress just described. There are certainly multiple factors at play: random insertion of the cloned gene into recipient cells results in insertion that may not be efficacious, and may be detrimental; development of suitable viral vectors has been vexingly slow; inadequate attention to regulatory regions needed along with the normal gene has been inadequate; and difficulties are always encountered when any new form of treatment is studied. We believe that successes with somatic cell gene therapy will continue to occur, but at a rate remarkably slower than many had predicted—and all have hoped for.

STEM CELL THERAPY

In a number of ways, the questions we've just posed about gene therapy are relevant to another potentially exciting modality—the therapeutic use of **stem cells**. In Chapter 4, we introduced the subject of stem cells by defining them as unspecialized cells capable of self-renewal and of giving rise to specialized cells. We pointed out that there are two general types of stem cells: **embryonic stem cells** found in the inner cell mass of the blastocyst, that are capable of specializing into differentiated cells of virtually every organ and tissue—a property referred to as being **pluripotent**; and **adult stem cells** found in each organ or tissue and capable of differentiating into all of the specialized cell types of that tissue, but not other tissues—a property denoted as **multipotent**. After briefly reviewing some of the fundamental science associated with stem cells, we will introduce the subject of stem cell therapy with either adult stem cells (ASCs) or embryonic stem cells (ESCs).

Physiologic Properties

Stem cells divide by mitosis to yield two daughter stem cells, or one stem cell and one progenitor cell. These progenitor cells divide to yield progressively more differentiated daughter cells. Stem and progenitor cells are generally described in terms of their potential—the number of possible fates of their daughter cells. ESCs are pluripotent because their daughter cells can differentiate into the more than 250 different cell types found in humans. Accordingly, ASCs are multipotent in that their daughter cells can give rise to the more specialized cellular repertoire of say the heart, kidney, or brain.

"**Stemness**," defined as the basis for stem cell behavior, is a reflection of differential gene expression. Different cassettes of genes accord different properties. For instance, a small number of genes are transitorily active in ESCs but nowhere else. A different cassette is turned on in bone cells, but not muscle cells. In turn, these stemness genes lead to the production of mineral binding proteins in bone, of beating myocytes in the heart, and of axons and dendrites capable of transmitting signals in the brain.

Stem cells are required for the growth and development of the organism. ESCs seed each organ with daughter cells unique to it. ASCs are required for organogenesis during the embryonic period and persist in virtually all tissues thereafter. ASCs may be activated by illness or injury to replace damaged or dead cells with healthy ones. Although stem cells have been the object of intense scientific inquiry for a half-century, important scientific questions remain—for example: What signals turn genes on and off as stem cells give rise to differentiated ones? What signals an ASC in a particular tissue to become a cancer stem cell?

Although ASCs were identified and isolated decades before ESCs, the latter are much easier to study. They are easier to obtain than ASCs. They grow in culture indefinitely, whereas ASCs have a finite life span. Nonetheless, ASCs were shown to be clinically useful 40 years ago whereas ESCs are in their earliest phase of clinical testing. The reasons for this are scientific and societal.

Clinical Utility of ASCs

ASCs are a time-honored, Nobel Prize-recognized modality in the treatment of many human diseases. These ASCs are obtained from bone marrow, peripheral blood, and umbilical cord blood. Following isolation of stem cells, they are administered to patients with leukemia, lymphoma, sickle cell anemia, thalassemia, aplastic anemia (failure of the marrow to produce red blood cells), and other conditions. These stem cell transplants (formerly called bone marrow transplants) regularly restore normal hematologic function and are an important form of life-saving clinical practice. Unfortunately, ASCs obtained from other tissues such as heart muscle, skeletal muscle, or brain have not proven to be efficacious. Use of ASCs from bone marrow or blood represent a proof of principle that stem cells from one person can proliferate and differentiate in the marrow of another person. Having said that, events beginning in 1998 shifted the emphasis and excitement regarding the therapeutic potential of stem cells from ASCs to ESCs.

Isolation of Human ESCs: Scientific Opportunity and Societal Cacophony

In 1998, James Thomson reported the first isolation of human ESCs from frozen embryos left over from attempts at *in vitro* fertilization (IVF). Although ESCs had been isolated from mouse embryos nearly 20 years earlier, finding a way to isolate ESCs from humans had been difficult, and therefore their derivation was widely acclaimed by the scientific community. Scientists believed that such cells presented enormous opportunities to study the control of differentiation, the identification of new targets for drug discovery, the testing of drugs for toxicity in human rather than animal cells, and the means for expanding tissue and organ transplantation.

The clinical promise of human ESCs was seen as exuberant. The ability to turn ESCs into any differentiated cell desired raised the possibility of preparing pancreatic islet cells for type 1 diabetes, replacing nerve cells in spinal cord injury or Parkinson disease, and using cardiac myocytes in people who had sustained myocardial infarcts (heart attacks). This led to a new term, "regenerative medicine," defined as treatments in which stem cells are induced to differentiate into specific cell types required to repair damaged or depleted cell populations or tissues.

The societal clamor concerning human ESCs began almost immediately. Despite public opinion polls showing large majorities in favor of laboratory and clinical research using ESCs, and despite a strongly positive report from a presidential commission appointed by President Clinton, a passionate minority decried human ESC research from the start. Because, they argued, living embryos must be destroyed to obtain the ESCs, it is morally wrong and ethically unacceptable to isolate ESCs regardless of their clinical potential. This is the same argument used by those who want to prohibit RU-486 and prenatal diagnosis, or who want to overturn Roe *v.* Wade and prohibit abortion. The subsequent societal battle over ESCs has gone on longer, has been more intense, and has risen to the highest reaches of government in a way that no prior medical research issue has.

Here is some of the evidence for that statement. Soon after taking office, President George W. Bush issued an executive order, in August 2001, that restricted use of federal funds for human ESCs to a small number of previously derived cell lines (later shown to be grossly inadequate and often contaminated). During the Bush presidency, Congress twice voted to lift this restriction, but the legislation was vetoed by the president. In the 2004 presidential election, President Bush and John Kerry took polar positions on ESC research in a presidential debate (the first time in memory that a medical research issue had been so debated by candidates for the presidency). In 2008, Barack Obama campaigned for ESC research and John McCain against it. After being elected in 2008, President Obama lifted the moratorium on ESC research, but a federal judge ruled not long after that federal funds cannot be used to support such research because prior law forbids federal funding on any research carried out with embryos. At the time of publication of this book, President Obama's executive order has been sustained by higher judicial review, but there is no reason to think that the matter has been resolved.

This governmental cacophony has had many effects: it has allowed research conducted with private funds to proceed much faster than that with public funds; it has led a few American scientists doing ESC work to leave the US for the United Kingdom, Australia or South Korea, where no restrictions exist; and it has slowed dramatically the work designed to test the promise articulated ardently by many patients and families. As late as the year 2010, only a single preliminary clinical trial had been initiated. Patients with paraplegia due to spinal cord transection have received neuritic cells derived from human ESCs. In late 2011 this study was discontinued for unstated reasons, suggesting that it had been unsuccessful.

Induced Pluripotential Stem Cells (iPSCs)

In 2008, stem cell research was galvanized by work from Shinya Yamanaka in Japan. He reported that he could induce adult skin cells growing in tissue culture to become pluripotential stem cells by adding just four isolated genes to the culture medium. The work was quickly confirmed, making work with iPSCs of great interest. Subsequently iPSCs have been obtained by supplementing the growth medium with only three genes previously shown to be

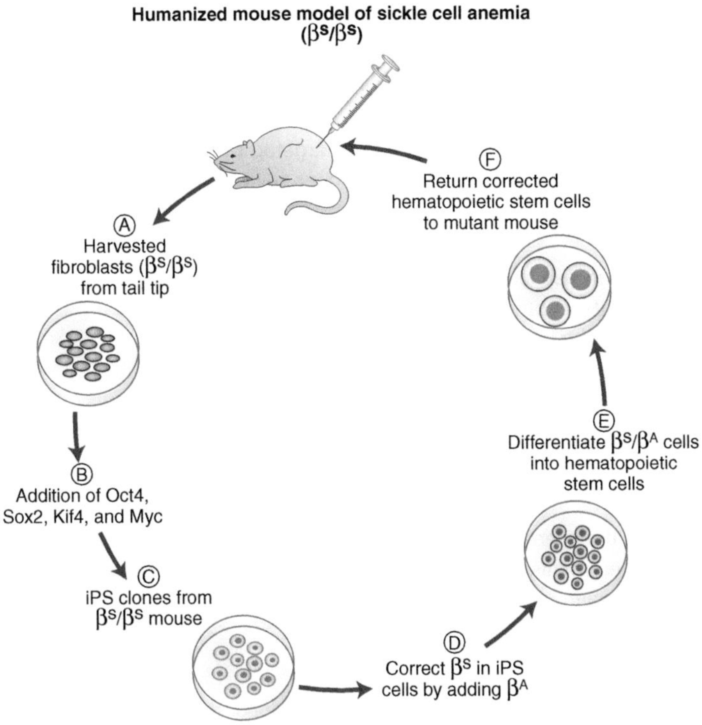
Humanized mouse model of sickle cell anemia
(β^S/β^S)
(A) Harvested fibroblasts (β^S/β^S) from tail tip
(B) Addition of Oct4, Sox2, Kif4, and Myc
(C) iPS clones from β^S/β^S mouse
(D) Correct β^S in iPS cells by adding β^A
(E) Differentiate β^S/β^A cells into hematopoietic stem cells
(F) Return corrected hematopoietic stem cells to mutant mouse

FIGURE 17.5
Schematic representation of combined use of induced pluripotential stem cells (iPSC) and gene therapy to cure a mouse model of sickle cell anemia. (A) Fibroblasts from animal are grown in cell culture. (B, C) Four genes are used to create pluripotential cells. (D) Normal β-globin gene is inserted into iPSCs. (E) (F) After the iPSCs have been differentiated into hematopoietic stem cells, they are reintroduced into the same animal from whom the original cells were removed.

required for stemness behavior. As shown in Figure 17.5, the clinical potential of this discovery is great, and it could work as follows:

- Culture skin fibroblasts from a patient with, for example, sickle cell anemia;
- Prepare iPSCs from these cells using the technique mentioned above;
- Prepare differentiated hematopoietic cells from the iPSCs (this is now straightforward);
- Insert the normal β-globin gene into the differentiated cells;
- Reintroduce the cells into the patient, hoping to cure the genetic disorder.

It is premature even to hazard a guess as to the likelihood of success with this kind of approach, but at best it marries stem cell therapy and gene therapy in a remarkably logical way.

REVIEW QUESTIONS AND EXERCISES

1. Choose the phrase in the right column that best matches the term in the left column.

a. advanced maternal age	1. first inherited disorder treated by dietary restriction
b. gene therapy	2. test positive in those affected and negative in those unaffected
c. Jesse Gelsinger	3. died in ill-advised gene therapy trial
d. induced pluripotent stem cell (iPSC)	4. two forms of gene therapy
e. Preimplantation diagnosis	5. Most common indication for prenatal diagnosis
f. clinical validity	6. multipotent cells that differentiate into lineages of single organ
g. genetic counseling	7. testing single blastomere
h. alpha-fetoprotein	8. being tried in Duchenne muscular dystrophy
i. *ex vivo*/*in vivo*	9. insertion of normal gene into cells with defective gene
j. embryonic stem cell (ESC)	10. employed in Tay-Sachs disease
k. prenatal diagnosis	11. protein measured in maternal serum screening
l. exon skipping RNA	12. communication process that conveys genetic risk information
m. phenylketonuria	13. derived from somatic cells with ESC-like properties
n. severe combined immunodeficiency (SCID)	
o. neonatal screening	

p. preconceptual diagnosis
q. imatinib (Gleevec[R])
r. adult stem cell (ASC)

14. first successful use of gene therapy
15. legally mandated detection of metabolic disorders
16. carried out using chorionic villus sampling and amniocentesis
17. first major success in targeted treatment of cancer
18. pluripotent cell from inner cell mass

2. List three general reasons to detect genetic disorders, and give an example showing the utility of each.
3. Carrier detection for Tay-Sachs disease in Ashkenazi Jews has been widely employed.
   a. Why has such testing been done in Jews but not everyone else?
   b. Who is tested and what is done with the information?
   c. How effective has this program been?
   d. Will this program lead to the disappearance of the Tay-Sachs disease in Ashkenazi Jews?
4. One in seven marriages in the US is now interracial, and this trend is increasing.
   a. If all marriages become interracial, what effect would that have on the prevalence of genetic disorders?
   b. Identify two other countries where the interracial marriage rate far exceeds that in the US.
   c. Why shouldn't federal or state governments encourage or mandate interracial marriage for the purpose of improving the public's health?
5. There is a big difference between treatment and cure.
   a. What is the difference?
   b. Discuss briefly four factors that severely limit our ability to treat chromosomal and single-gene disorders today.
6. Human gene therapy has been worked on for more than 30 years but its successes have been disappointingly few.
   a. Discuss briefly five reasons for this slow progress.
   b. What is the difference between *in vivo* and *ex vivo* gene therapy?
   c. Two major events concerning gene therapy occurred in 1998–1999. One was a major success, the other a disastrous failure. What were these events, and why did the disaster attract so much more attention than the success?
7. Most single gene disorders are called "orphan diseases."
   a. What does that designation mean?
   b. Discuss briefly three practical implications of being an "orphan disease."
   c. Why have pharmaceutical companies generally not tried to develop medicines for such conditions?
8. For nearly 15 years some scientists and clinicians have talked about "regenerative medicine."
   a. What is meant by this term?
   b. What will be required to make the idea of regenerative medicine a reality?
   c. What do stem cells have to do with this concept?
   d. An important form of regenerative medicine has been practiced for 40 years. What is it and how is it carried out?
   e. Discuss briefly four reasons why use of embryonic stem cells (ESC) has been halting, slow, and frustrating.
   f. Now that we know how to induce skin and other somatic cells to become pluripotent stem cells, should we cease working with ESC? Why or why not?
9. Genetic counseling is an increasingly important tool in communicating information about genetic disorders.
   a. Optimally, all clinicians should provide such counseling. Discuss briefly two major reasons why this is not the case.
   b. What are the rate-limiting steps for such counseling by providers?
   c. What are the rate-limiting steps for those seeking counseling?
   d. How can we increase the number of genetic counselors with masters degree education?

PART 3

Populations and Individuals

CHAPTER 18

Population and Evolutionary Genetics

CHAPTER OUTLINE

CORE CONCEPTS

Population genetics is the quantitative study of the distribution of **genetic variation** in a population and of how the frequencies of its genotypes, alleles, and phenotypes are maintained or changed. It seeks answers to such practical questions as why the frequency of PKU in Caucasians is so much greater than in Japanese, or why the frequency of the sickle cell allele varies markedly in people from different West African countries. The mathematical cornerstone of population genetics is the **Hardy-Weinberg law** or principle. The law has two parts. First, it states that in a large, randomly mating population with two alleles at a locus (for example, A and a), there is a simple relationship between these allele frequencies (frequency of $A = p$; frequency of $a = q$) and the genotype frequencies (p^2, $2pq$, or q^2) they define. Second, it holds that this relationship between allele and genotype frequencies, constructed simply on the binomial expansion of $(p + q)^2$, does not change from one generation to the next. When a population conforms to this two-part law, it is in Hardy-Weinberg equilibrium. In such populations, the law is of great value in showing why dominant traits do not increase in frequency from one generation to the next and why recessive traits do not decrease. Further, the

Human Genes and Genomes. DOI: 10.1016/B978-0-12-385212-0.00018-4

law is regularly used in genetic counseling settings where estimates of genotype, allele, and carrier frequencies are calculated from limited phenotypic information in small families, such estimates then being employed to estimate specific genetic risk.

Hardy-Weinberg equilibrium is never fully realized in human populations because it is perturbed by one or more deviations. First, individuals do not usually mate randomly. Mating is more often **assortative** (mate choice depends on geographic proximity), **stratified** (within an ethnic subset), or **inbred** (among relatives or a small group). Second, allele frequencies do not remain constant for a number of reasons: random or chance events producing major changes in population size and composition (called "**genetic drift**"); migration of individuals from one population to another, followed by mating between the populations, referred to as **gene migration**; **new mutations** that occur at a low rate constantly; and **natural selection** in which some genotypes are better suited to reproduce and thrive (called "**fitness**") and therefore give rise to a disproportionate share of offspring. A particular form of such selective advantage occurs when gene–environment interaction leads to the situation in which the fitness of heterozygotes for a particular genetic condition exceeds that in either homozygote. This is referred to as **heterozygote advantage**, and has been best studied in the relationship between sickle cell anemia and malaria.

Such examination of single-gene frequencies and perturbations is now being complemented and supplemented by genome-wide studies employing SNPs and CNVs. These genomic approaches have revealed that most genetic variation occurs within a population rather than between two populations—adding additional complexity to the meaning of the word "race" and making it clear that such population categories as European, Asian, African, and Hispanic are in no way genetically distinct.

As we understand more about the structure of genes and genomes, that information informs our ideas about the evolution of populations. **Molecular evolution** is concerned with determining how the study of genomes, chromosomes, genes, and proteins helps us account for the evolution of our species—and other species as well. Molecular evolution employs many techniques (DNA hybridization, chromosome banding, amino acid sequences in proteins, and whole-genome sequencing), all aimed at providing more precise estimates of the timing of evolutionary events (molecular clocks) and of the relationship between our species and that of others near or distant from our own (ancient DNA).

TERMINOLOGY

To this point we have discussed, for the most part, genetic variation in individuals and families. Now, we broaden our lens to consider genes in populations. To do this, we need to define a few new terms (Table 18.1).

Human population genetics, broadly speaking, encompasses the evolutionary ideas of Darwin and Wallace, the laws of Mendel, the insights of molecular biology, and the

TABLE 18.1 Terms Used in Population Genetics

Term	Definition
Population	A group of interbreeding individuals of the same species inhabiting the same space at the same time
Gene pool	The sum total of all of a population's alleles
Population genetics	The quantitative study of the distribution of genetic variation in a population, and of how the frequencies of its alleles are maintained or changed
Genotype frequency	The proportion of all individuals in a population that are of a particular genotype
Allele frequency	The proportion of all copies of a given gene in a population that are of a particular allele
Phenotype frequency	The proportion of all individuals in a population with a particular phenotype

contributions of the genome project. Using more specific examples, population genetics seeks answers to questions such as:

- Why does the frequency of PKU vary widely in different populations—from 1 in 2,600 Turks, to 1 in 10,000 US whites, to 1 in 119,000 Japanese?
- Why does the frequency of sickle cell anemia vary so much in different parts of Sub-Saharan Africa (Figure 18.1)?
- Why are 90% of Asian Americans unable to digest milk while 90% of white Americans can?
- Why does the vast majority of genetic variation exist within races rather than between them?

Today, rather than being abstract or arcane, population genetics is critically important to many fundamental and applied aspects of human genetics, including gene mapping and sequencing,

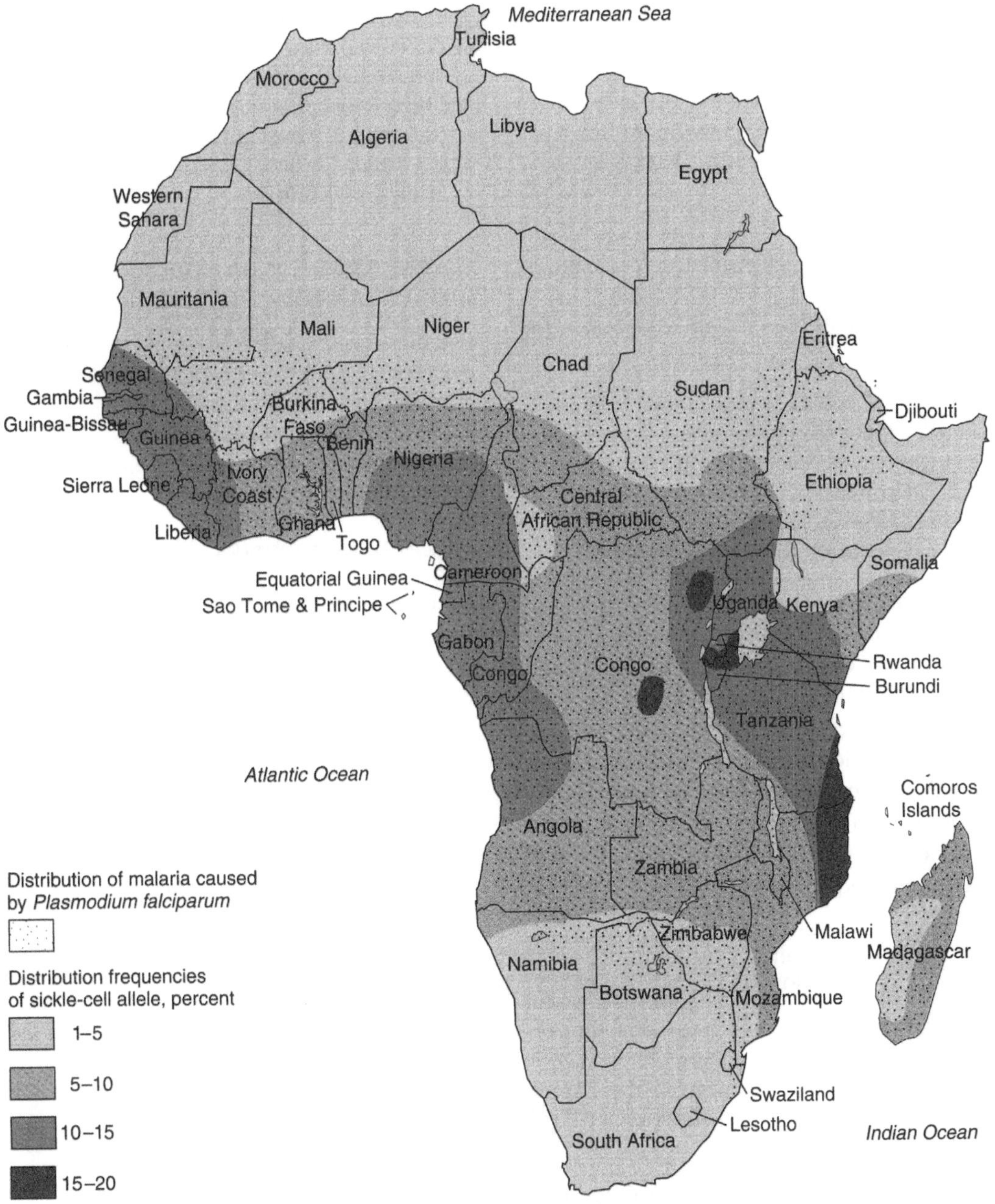

FIGURE 18.1
Frequency of the sickle-cell allele in Africa, superimposed on regions where malaria is endemic.

genetic screening and counseling, DNA-based fingerprinting and forensics, and for comprehending biological evolution.

THE HARDY-WEINBERG LAW

In 1908, Geoffrey Hardy, an English mathematician, and Wilhelm Weinberg (Figure 18.2), a German physician, formulated independently the law (or principle) that bears their names—and which constitutes the cornerstone of population genetics. Rather than merely stating "**the law**," we'll start with a particular example.

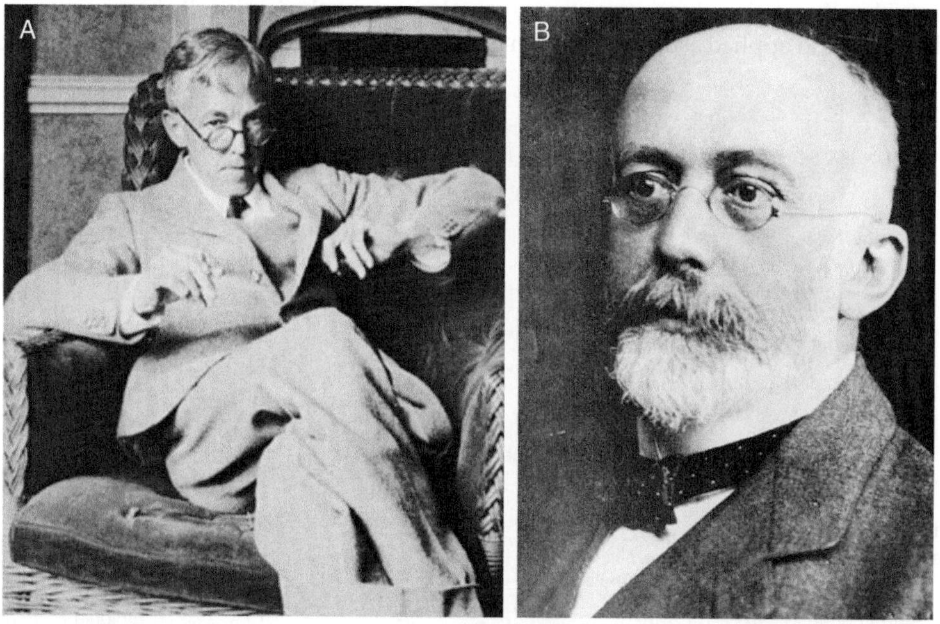

FIGURE 18.2
(A) Geoffrey Hardy and (B) Wilhelm Weinberg. Hardy was a mathematician, Weinberg a physician.

Consider the gene *CCR5* that encodes a cell surface receptor required by the human immunodeficiency virus (HIV) to gain entry into human lymphocytes and infect them. A 32 base-pair deletion of *CCR5*, called Δ*CCR5*, creates an allele encoding a non-functional protein because of a frame shift and premature termination. The Δ*CCR5* allele was discovered in people whose behavior placed them at high risk for developing HIV/AIDS but who failed to develop the disease. Individuals homozygous for *CCR5* are susceptible to HIV/AIDS; those homozygous for Δ*CCR5* are resistant; heterozygotes (*CCR5*/Δ*CCR5*) are partially protected. Loss of function of *CCR5* has no deleterious effect in homozygotes or heterozygotes with the Δ*CCR5* allele. These clinical observations led to the study of the *CCR5* gene pool in a population we'll now discuss. **Genotype frequencies** and **allele frequencies** were determined as shown in Table 18.2.

TABLE 18.2 Counting Genotypes in a Model Human Population Using CCR5/ΔCCR5 as an Example

Population	1,000 people
Genotypes	795 *CCR5/CCR5*
	190 *CCR5*/Δ*CCR5*
	15 Δ*CCR5*/Δ*CCR5*
Genotype frequencies	*CCR5/CCR5* = 795/1,000 = 0.795
	CCR5/Δ*CCR5* = 190/1,000 = 0.19
	Δ*CCR5*/Δ*CCR5* = 15/1,000 = 0.015
Allele frequencies	*CCR5* = 795 + 795 + 190 = 1,780/2,000 = 0.89
	Δ*CCR5* = 15 + 15 + 190 = 220/2,000 = 0.11

Because molecular analysis can readily distinguish between the *CCR5* and the *ΔCCR5* alleles, it was straightforward to identify both kinds of homozygotes and the heterozygote by counting them: in a sample of 1,000 French people, 795 were *CCR5/CCR5*, 190 were *CCR5/ΔCCR5*, and 15 were *ΔCCR5/ΔCCR5*. The genotype frequencies, then, were 795/1,000 (or 0.795) for *CCR5* homozygotes, 190/1,000 (or 0.19) for heterozygotes, and 15/1,000 (or 0.015) for the *ΔCCR5* homozygotes. Allele frequencies were determined by counting all the *CCR5* and *ΔCCR5* alleles: they totaled 2,000 (2 for each of the 1,000 people); *CCR5* alleles equaled $795+795+190=1{,}780$; *ΔCCR5* alleles equaled $15+15+190=220$. The allele frequencies then were $1{,}780/2{,}000=0.89$ for *CCR5* and $220/2{,}000=0.11$ for *ΔCCR5*. Note that genotype frequencies add up to 1.0, just as do allele frequencies. This must be the case, because all of the genotypes and alleles were counted.

In the *CCR5* example, each of the genotypes was distinguishable, enabling direct determination of allele frequencies. But in most cases clinical geneticists encounter two phenotypes—normal and variant. How, then, can allele and gene frequencies be determined? Hardy and Weinberg answered the question. The key to the answer lay in establishing a quantitative relationship among genotype, phenotype, and allele frequencies within and between generations. Hardy, believing that this relationship was obvious, wrote in his brief paper, "I am reluctant to intrude in a discussion concerning matters of which I have no expert knowledge, and I should have expected the very simple point I wish to make to have been familiar to biologists." Before articulating the Hardy-Weinberg law, we will present five simplifying assumptions on which its derivation depends:

1. The population is large (such that fluctuations in genotype frequencies cannot be explained just by chance).
2. Individuals mate at random (in the sense that each individual's genotype at the locus in question does not influence his or her choice of mate).
3. No new mutations enter the gene pool.
4. No migration of individuals from a population with allele frequencies very different from the original population has occurred.
5. No genotype-dependent difference in the ability to survive and reproduce exists.

A population satisfying all these assumptions is said to be in Hardy-Weinberg equilibrium. In such a population, a simple mathematical relationship, called the **Hardy-Weinberg law**, permits calculation of genotype frequencies from allele frequencies. No actual population is at true Hardy-Weinberg equilibrium: every population is finite, some are small; mating is not always random; mutations occur constantly; migration into and out of a population is common; and many genotypes affect the ability to survive and reproduce. Despite such deviation from one or more of the simplifying assumptions, the Hardy-Weinberg law is remarkably robust at providing estimates of genotype frequencies.

Deriving the Law

The Hardy-Weinberg law has two critical components (Table 18.3). The first is that a simple relationship exists between genotype frequencies and allele frequencies in a population.

TABLE 18.3 The Hardy-Weinberg Law

1. Consider a single autosomal locus with two alleles: A and a
   In male and female gametes, the frequency of allele $A=p$; the frequency of allele $a=q$
2. In a population at equilibrium, the progeny have the following genotypes and the following genotype frequencies:
   Genotypes, AA Aa aa
   Genotype frequencies, p^2 $2pq$ q^2

Because there are only two alleles, $p+q=1$ and $p^2+2pq+q^2=1$

Suppose that a given autosomal locus has two alleles, A and a, and that, in sperm and eggs, the frequency of allele A is p and the frequency of allele a is q. As shown in Table 18.4, the likelihood that two A alleles will pair up in a zygote is $p \times p$, or p^2. The likelihood that two a alleles will pair up is $q \times q$, or q^2. The likelihood that the zygote will have the genotype Aa is 2pq (because either allele could be inherited from either parent). The frequencies of the three genotypes (AA, Aa, aa), then, are p^2, 2pq, and q^2. Since these genotype frequencies are the only ones in the population, they must sum to 1. Thus, the law states that the frequency of the three genotypes is given by the terms of the binomial expansion of $(p+q)^2$. In the absence of any difference in fitness, the genotype frequencies in the adults who develop from the zygotes just described will be identical to those in the zygotes.

TABLE 18.4 Schematic Derivation of the Hardy-Weinberg Law

		Paternal gametes*	
		Allele A **Frequency p**	**Allele a** **Frequency q**
Maternal gametes*	**Allele A**	**AA**	**Aa**
	Frequency p	p^2	pq
	Allele a	**Aa**	**Aa**
	Frequency q	pq	q^2

After mating, the genotype frequencies of the progeny in this population will be p^2 for **AA, 2pq for **Aa,** and q^2 for **aa**.*

A second component of the law is that the genotype frequencies in a population will remain constant from generation to generation if the allele frequencies (p and q) remain constant. This is referred to as **Hardy-Weinberg equilibrium**. Such equilibrium states that if genotype frequencies in one generation are in the proportion $p^2 : 2pq : q^2$, these same relative proportions will exist in the next generation and the one after that. An important corollary of this equilibrium follows. Let us say that allele A is dominant and allele a is recessive. In the absence of differences in fitness, the genotype frequencies for those homozygous for the dominant allele will not increase from generation to generation, and the frequency of homozygotes for the recessive allele will not decrease.

Table 18.5 presents data proving that genotype frequencies do not change. It shows gene frequencies in the offspring of individuals produced by the five possible mating types. Algebraic treatment of the genotype frequencies yields allele frequencies of $p^2 : 2pq : q^2$.

TABLE 18.5 Frequency of Different Genotypes in Offspring of Each Mating Type

		Offspring		
Mating Type	**Frequency**	**AA**	**Aa**	**aa**
AA × AA	p^4	p^4		
AA × Aa	$4p^3q$	$2p^3q$	$2p^3q$	
AA × aa	$2p^2q^2$		$2p^2q^2$	
Aa × Aa	$4p^2q^2$	p^2q^2	$2p^2q^2$	p^2q^2
Aa × aa	$4\ pq^3$		$2pq^3$	$2pq^3$
aa × aa	q^4			q^4

AA offspring $= p^4 + 2p^3q + p^2q^2 = p^2(p^2 + 2p + q + q^2) = p^2(p + q)^2 = p^2(1)^2 = \mathbf{p^2}$
Aa offspring $= 2p^3q + 4p^2q^2 + 2pq^3 = 2pq(p^2 + 2pq + q^2) = 2pq(p + q)^2 = 2pq(1)^2 = \mathbf{2pq}$
aa offspring $= p^2q^2 + 2pq^3 + q^4 = q^2(p^2 + 2pq + q^2) = q^2(p + q)^2 = q^2(1)^2 = \mathbf{q^2}$

Let us return to the *CCR5* example used earlier. The relative frequencies of the two alleles in the gene pool of 1,000 people were 0.89 for *CCR5* and 0.11 for Δ*CCR5*. The Hardy-Weinberg law states that the relative proportions of the three combinations of alleles (in other words, the genotype frequencies) are as follows: $p^2 = 0.89 \times 0.89 = 0.79$ for *CCR5* homozygotes; $2pq = 2 \times 0.89 \times 0.11 = 0.19$ for those heterozygous; and $q^2 = 0.11 \times 0.11 = 0.01$ for Δ*CCR5* homozygotes. When these genotype frequencies calculated by the Hardy-Weinberg law are applied to the population of 1,000 individuals, the derived numbers of people with the three genotypes (795 + 190 + 15) are identical to the numbers observed in the population of 1,000 people shown in Table 18.2.

Clinical Implications

FREQUENCY OF HETEROZYGOUS GENOTYPES

An important implication of the Hardy-Weinberg law is that for a rare disorder (such as is the case in most disorders due to single-gene mutations), the frequency of heterozygotes far exceeds that of homozygotes for the rare allele. For example, if the frequency of the rarer of two alleles is 0.1 (the value for q), then the ratio of heterozygous genotypes to homozygous genotypes for the rare allele is

$$2 \times 0.9 \times 0.1/0.1 \times 0.1 = 18$$

That is, heterozygotes carry 18 times as many of the rare allele copies as do homozygotes. If the frequency of the rare allele is 0.01, this ratio is about 200. This general relationship has major implications for genetic screening and counseling.

USE OF THE HARDY-WEINBERG LAW IN GENETIC COUNSELING

Autosomal Recessive Conditions

The major practical application for the Hardy-Weinberg law is in genetic counseling for autosomal recessive disorders. In cystic fibrosis (CF), for example, 1 in 2,500 whites of northern and central European backgrounds are affected with this serious disorder. They are homozygous for the mutant allele(s). Heterozygotes for CF are not identifiable clinically, but their frequency can be estimated using the Hardy-Weinberg law: frequency of q then = square root of 1/2,500 or 1/50 (or 0.02); because $p + q$ must equal 1, frequency of heterozygous carriers (2pq) $= 2 \times 0.98 \times 0.02$ or 0.039 (which approximates 0.04 or about 4%). Thus, 1 in 25 individuals in this population is a carrier for CF. This calculation is summarized in Table 18.6.

TABLE 18.6 Estimating Carrier Frequency in Autosomal Recessive Disorders Using Cystic Fibrosis as an Example

Disease frequency	1/2,500 Caucasians of northern and central European origin
Mutant allele frequency	$q = \sqrt{1/2{,}500}$ or 1/50
Carrier frequency	$2pq = 2 \times 49/50 \times 1/50 = {\sim}1/25$

Application to a typical counseling problem is depicted in Figure 18.3. The woman identified as II2 wants to know the likelihood that her unborn baby (III1) will have CF. Her brother (II1) has the disorder. No one else in her family or her husband's (II3) family has CF. To answer her question, we first estimate the likelihood that II2 is a carrier. Given that both her parents must be CF carriers (because they have an affected son), the likelihood that she is a carrier is 2/3 (since she is not affected, Mendel's first law says that the likelihood is 2/3, not 2/4). Next, we estimate the likelihood that II3 is a carrier. From the paragraph above, we know that to be 1/25. Therefore, the likelihood that III1 will have CF is $2/3 \times 1/25 \times 1/4 = 1/150$. The 1/4 term comes from knowing that any two carriers have a 1 in 4 chance that each will pass the mutant allele to their offspring. Counselors would put this into perspective by pointing out that their

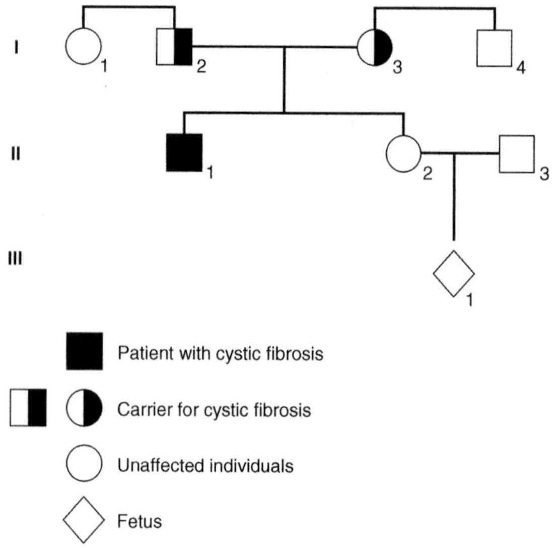

FIGURE 18.3
Using the Hardy-Weinberg law in genetic counseling: cystic fibrosis as an example.

risk of having a child with CF (1/150) is more than 16-fold greater than that for two people from the general population with no history of CF in the family.

If the consultands were African Americans, in whom the frequency of carriers for CF is about 1/65, the above calculation would yield an estimate of 1/390 for the probability of having an affected child ($2/3 \times 1/65 \times 1/4 = 1/390$). Obviously, then, knowledge of ethnicity is an important factor in providing accurate counseling using the Hardy-Weinberg law.

Returning to Figure 18.3, what would be the probability that II2 and II3 would have an affected child if they were first cousins on her mother's side? The answer is as follows: I3 is an obligate carrier; the likelihood that any one of her siblings (I4, for example) is a carrier is 1/2; the likelihood that such a sib would transmit the mutant allele to II3 is $1/2 \times 1/2$, or 1/4. Therefore, the likelihood that II2 and II3 would have an affected child is $2/3 \times 1/4 \times 1/4$, or 1/24. In other words, consanguinity in this setting increases the probability of having an affected child from 1/150 to 1/24, or about six-fold. The rarer the frequency of the mutant allele, the greater is the effect of consanguinity on the risk calculation.

X-Linked Traits

Genes on the X chromosome are present twice in females, but only once in males. Therefore, the equilibrium frequencies under the Hardy-Weinberg law must be stated separately for the two sexes. If we use hemophilia as an example and let H represent the normal allele and h the hemophilia-causing one, Table 18.7 shows genotypes and genotype frequencies for males and females separately.

Among females, the gene frequencies are just as they are for an autosomal gene ($p^2 : 2pq : q^2$). A male's single allele, though, is either H or h. This means that genotype and phenotype frequencies are the same as allele frequencies (p and q). For example, hemophilia occurs in about 1 in 10,000 males, who have the genotype h–. Since q equals the frequency of h, it equals 1 in 10,000, or 10^{-4}. The Hardy-Weinberg law reveals why the trait is very rare among females: an affected female (h/h genotype) must have a gene frequency for q of q^2 or $1/10{,}000 \times 1/10{,}000$, or 1/100,000,000. This means that one would expect to find only three to four female hemophiliacs in the population of the United States. A few

TABLE 18.7 The Hardy-Weinberg Law In X-Linked Conditions: Hemophilia as an Example

Gender	Genotypes*	Genotype Frequencies
Females	H/H	p^2
	H/h	2pq
	h/h	q^2
Males	H–	p
	h–	q

*H, normal allele; h, hemophilia allele.

such individuals, born from hemizygous affected fathers and carrier mothers, have indeed been reported.

FACTORS THAT PERTURB HARDY-WEINBERG EQUILIBRIUM

As discussed earlier, the Hardy-Weinberg law holds in the presence of several simplifying assumptions: an infinitely large population; random mating; and no mutation, migration or selection. Because these conditions are almost never met in human populations, Hardy-Weinberg equilibrium is an approximation. Deviation from any of its underlying assumptions will result in small or large increases or decreases in allele frequencies from one generation to the next. Some deviations skew allele frequencies more than others. Violating the assumption of random mating can cause large deviations in the frequency of autosomal recessive conditions. However, changes due to mutation, migration, or selection usually produce smaller effects on Hardy-Weinberg equilibrium because they tend to affect allele frequencies over many generations.

Exceptions to Random Mating in Large Populations

The principle of random mating holds that for any locus, an individual with a given genotype has an absolutely random probability of mating with a person of any other genotype—the proportions of these matings being determined only by the relative frequencies of the genotypes in the population. In human populations, however, mating is seldom random. Non-random mating may occur as a result of three closely related but distinct phenomena: **assortative mating**, **stratification**, and **consanguinity**.

ASSORTATIVE MATING

When one chooses a mate based on some particular trait possessed by that person, mating is assortative. It is estimated that 80% of people the world over marry individuals born within 10 miles of each other. Because we tend to choose mates similar to ourselves in one or more of several ways (skin color, intelligence, native language, artistic talent, athletic ability), such assortative mating is often positive, its overall genetic effect being to increase the frequency of homozygotes at the expense of heterozygotes. But in certain clinical settings, the effects may be negative. For example, if two people with blindness resulting from homozygosity for the same mutant allele mate, all of their children will be blind. Another example: if two individuals with achondroplasia (an autosomal dominant form of dwarfism) mate, their offspring homozygous for the mutation have a lethal form of dwarfism not observed otherwise. This kind of assortative mating may have little effect on the entire population, but its effects on a specific family may be large.

STRATIFICATION

A population in which subgroups have remained relatively (or absolutely) genetically separate over generations is said to be stratified. The US population, for example, is stratified into many subgroups—whites, Latinos, African Americans, Asians, native Americans. Stratification exists in many other parts of the world as well. When mate selection in a population is restricted to one subgroup, the result is an excess of homozygotes in the population as a whole for any locus with more than one allele.

Suppose, for instance, that a population contains a minority subgroup constituting 10% of the population. Suppose further that a mutant allele for an autosomal recessive trait has an allele frequency of 0.05 (denoted $q_{min} = 0.05$). In 90% of the population—that is, the majority—the frequency of $q_{maj} = 0$. A close approximation of this situation is the mutant allele for sickle cell anemia in the African American population. In the total population, then, the overall frequency of the disease allele q_{pop} is 0.05/10 or 0.005, and the overall frequency of sickle cell anemia would be $0.005 \times 0.005 = 0.000025$ if mating were purely random. If members of the minority group marry exclusively within their group, then the frequency of sickle cell anemia in the minority group will be $q^2_{min} = (0.05)^2 = 0.0025$. Because this group makes up 10% of the population, the true frequency of sickle cell anemia in the total population, $(0.0025/10 = 0.00025)$ is 10-fold higher than one would expect from applying the Hardy-Weinberg law to an unstratified total population.

CONSANGUINITY AND INBREEDING

Mating between close relatives (**consanguinity**) or among members of genetic isolates (**inbreeding**) brings about an increased frequency of autosomal recessive disorders because such matings increase the frequency with which carriers of such disorders mate. Consanguineous mating not only increases the frequency of rare recessive disorders (Chapter 12); it also increases the susceptibility toward common multifactorial traits, the magnitude determined by the coefficient of relatedness between the mating pair: uncle/niece; first cousins; etc.

Inbreeding has had a particularly large effect in Ashkenazi Jews. Some 15 different conditions, many very serious, have a much higher frequency in this group than in the total population. For example, 1 in 30 Ashkenazim is a carrier for Tay-Sachs disease, compared to 1 in 300 non-Ashkenazim. The frequency of the disease is 100 times as common in Ashkenazim (1 in 3,600) as in non-Ashkenazim (1 in 360,000). Because Tay-Sachs is common and almost always lethal by age 5, the Ashkenazi community instituted a program to detect carriers and affected fetuses that has resulted in a decrease in disease frequency among Ashkenazi of more than 90% in recent decades (see Chapter 17). Other disorders, often referred to as "Ashkenazi Jewish diseases," include Gaucher's disease, familial dysautonomia, and Fanconi anemia. The *BRCA1* and *BRCA2* mutant genes, which increase susceptibility to breast and ovarian cancer, also have a considerably higher frequency in Ashkenazi Jews, who are by no means the only inbred subgroup in North America. Others include the Old Order Amish, the Mennonites, and the French Canadians in Quebec.

Exceptions to Constant Allele Frequencies

GENETIC DRIFT

Chance or other unpredictable events can produce large changes in allele frequencies over a single generation. This is called genetic drift, and it usually has a much more profound effect on small populations than large ones. Drift often occurs because a newly formed, isolated group is not representative of the original population. Events like natural catastrophes (hurricanes) or epidemics may be responsible for creating such a genetic isolate. For example, the Pingelapese people occupy a group of Micronesian islands in the eastern Caroline Islands. This group has a frequency of 1 : 10 for an autosomal recessive form of blindness that occurs in other parts of that region with a frequency of 1 : 20,000 to 1 : 50,000. This situation resulted from a typhoon in 1780 that killed all but 9 males and 10 females among the Pingelapese. This catastrophe created what is referred to as a **founder effect** and a **population bottleneck**. A founder effect may lead to an unusually high frequency of a genetic disorder in a small isolated population. The catastrophe just described also created a "population bottleneck," in which a population's sample of alleles is changed by chance when the population is reduced in size suddenly.

MIGRATION AND GENE FLOW

Migration can change allele frequency by a process known as gene flow (defined as the slow diffusion of genes across a geographic or cultural barrier). When groups or populations (and thus their genomes) move from place to place and then mate with the indigenous people, allele frequencies in both the migrant and the indigenous populations may change. This usually occurs slowly over many generations. Let us consider again the *CCR5* gene discussed earlier in this chapter. As shown in Figure 18.4, the Δ*CCR5* allele frequency is approximately 10–15% in western Europe and Russia. Its frequency is less than 5% in the Middle East and Asia, and essentially zero in Africa. This suggests that the mutation originated in whites and diffused into more easterly populations.

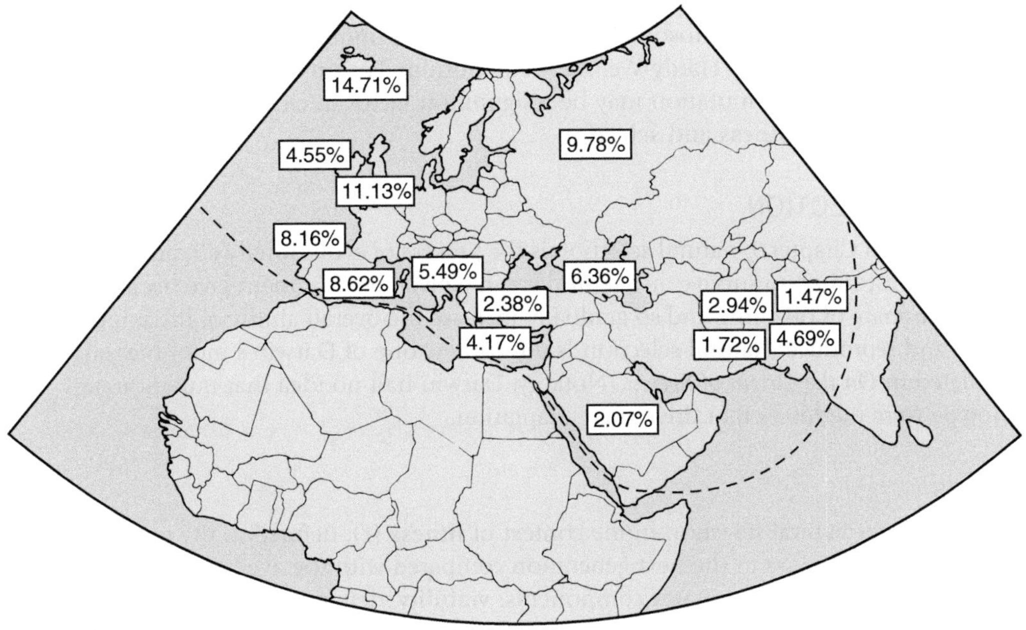

FIGURE 18.4
Migration of Δ*CCR5* allele from origin in northern Europe to numerous countries and geographies.

Another example of gene flow between population groups involves particular mutant alleles for PKU. These mutations have their highest frequency in those of Celtic origin. As the Celts migrated, these PKU alleles moved into the populations they bred with. Witness: the frequency of PKU that is 1 in 4,500 in Ireland is less prevalent in other parts of Europe and is found at a frequency of 1 in 109,000 in Japan (where gene flow from the Celts has been minimal).

MUTATION

As discussed in detail in Chapter 8, mutation is defined as a permanent, heritable change in the genomic DNA sequence. Mutations occur all the time—usually resulting from inherent errors in the DNA replication process, in addition to ionizing radiation, chemical mutagens, and other ill-defined environmental agents. Most mutations are identified and repaired by the several DNA repair systems in our cells. Some mutations, however, are not repaired and become established. What is the frequency of such mutations, and what is their effect on Hardy-Weinberg equilibrium?

Determining the mutation rate is simple in experimental organisms where changes in a large stable population can be observed directly. This situation, of course, does not pertain in humans. We rely on observations made in populations that are as large as possible, using phenotypes that are as distinct as possible. Because the vast majority of mutant alleles for any given autosomal recessive trait are carried in heterozygotes, such conditions are of no value in estimating human mutation rates. Certain autosomal dominant conditions, though, lend themselves to such estimates. For rare autosomal dominant traits, affected individuals who have two normal parents have, by definition, a new mutation. Its rate can be calculated by the equation $M = n/2N$, where M is the mutation rate, n is the number of affected patients with unaffected parents, and N is the total number of births in the population. (The denominator is

2N because the mutation could occur in either allele at the autosomal locus.) The disorders used in such estimates include achondroplasia, neurofibromatosis, and Duchenne muscular dystrophy (DMD). Mutation rates estimated this way range from 10^{-4} to 10^{-6} mutations per gene per generation. Given that we have about 21,000 genes in our nuclear genome, this means that 0.02 to 2.0 new phenotypically detectable mutations occur somewhere in the genes of each gamete each generation. Because the estimates of M are small compared to the mutant allele frequencies for most autosomal recessive traits, new mutation does not appear to be a major factor in perturbing Hardy-Weinberg equilibrium. For some autosomal dominant and X-linked traits, however, mutation may be a significant factor in changing allele frequencies because of effects on **fitness** and **selection**.

NATURAL SELECTION

As presented in Chapter 9, natural selection is the process of evolutionary adaptation in which genotypes best suited to survive and reproduce in a given environment give rise to a disproportionate share of offspring and so gradually increase the overall ability of the population to survive and reproduce. Natural selection is the cornerstone of Darwin's monumental work articulated in *On the Origin of Species*. Notably, Darwin had no idea that mutation and genotype were the forces that drive such adaptation.

Fitness

We speak about natural selection in the context of **fitness (f)**, defined as the probability of transmitting one's genes to the next generation compared with the average probability of the population. Fitness has two major components: **viability**, defined as the probability that a newly formed zygote will survive to reproductive age; and **fertility**, which is the average number of offspring produced by an individual of a specified genotype. If a mutant allele has just as much likelihood of appearing in the next generation as does the normal allele, f is 1; if the mutant allele causes death or sterility, f is 0. For example, zygotes homozygous for the Tay-Sachs mutant allele have a fitness of zero because they die in early childhood before reaching reproductive age. Zygotes with Turner syndrome due to monosomy for the X chromosome also have $f = 0$ because they are sterile. It must be stressed that fitness has no connotation of superiority or inferiority except as related to reproduction.

Selection and Hardy-Weinberg Equilibrium

Under certain circumstances, natural selection has the potential of altering Hardy-Weinberg equilibrium because it produces deviation from the assumption that each genotype in the population has the same probability of reproducing ($f = 1$). It turns out that this exception has little effect on the use of the Hardy-Weinberg law for genetic counseling concerning rare autosomal recessive traits (like cystic fibrosis or phenylketonuria). For even severe autosomal recessives, selection against affected homozygotes doesn't change genotypic frequencies much because most mutant alleles are carried (or "hidden") in heterozygotes whose $f = 1$. Even if $f = 0$ in homozygotes, such conditions will not die out; rather, they would be expected to decrease in frequency in tiny decrements over many generations.

The inefficiency of selection against rare recessive alleles has two practical and important implications. First, it opposes the (mistaken) belief that treatments that save the lives of those with rare recessive disorders will affect negatively the human gene pool because those who carry the deleterious genes will reproduce. This belief is fallacious because the proportion of those with homozygous genotypes is so small compared to the proportion of heterozygotes that reproduction by homozygous affected persons will change allele frequencies only negligibly. Similarly misguided are the views of eugenicists who propose to improve (or "cleanse") the human gene pool by preventing the reproduction of affected persons. Given that persons with severe genetic disorders rarely reproduce, they have essentially no effect on allele frequency when they do reproduce because the main reservoir of harmful recessive alleles is found in the genomes of phenotypically normal carriers.

For autosomal dominant disorders, however, heterozygous carriers express the condition, thus their dominant alleles are directly exposed to selective forces. One would expect that a dominant disorder with $f = 0$ (that is, a genetic lethal) would disappear in a single generation. This does not happen because new mutations occur at a rate such that they replace the alleles lost by negative selection. This means that affected offspring with such conditions regularly have two unaffected parents. This has important implications for genetic counseling: parents of a child with an autosomal dominant, genetically lethal condition have a low recurrence risk because the condition would require another new mutation in subsequent progeny—a statistically unlikely event.

Heterozygote Advantage

As just stated, in many autosomal recessive traits homozygotes for the mutant allele have reduced fitness ($f < 1$ to $f = 0$) compared to heterozygotes or those homozygous for the normal allele. There are a few disorders where environmental conditions result in the fitness of heterozygotes being greater than that for either homozygous genotype. This is termed **heterozygote advantage**. It is important because even a slight heterozygote advantage may act to increase the frequency of the mutant allele in the population—even if the mutant allele causes major reduction in fitness in homozygotes in that population. Those situations in which natural selection acts, at the same time, to maintain a deleterious allele in the gene pool and to remove it from the pool are termed **balanced polymorphisms**.

Sickle cell anemia is the best-known example of heterozygote advantage and balanced polymorphism in humans. As discussed in Chapter 12 and again in Figure 18.1, the frequency of heterozygotes for hemoglobin S (genotype AS) and the frequency of homozygotes for the mutation (SS) is higher in areas where malaria occurs than in non-malarious areas. This is explained as follows. In regions where malaria is endemic (as in West Africa), homozygotes (AA) for the normal allele are susceptible to malaria; many become infected and some die, leading to reduced fitness. SS homozygotes are even more reproductively disadvantaged because of their severe hematologic disease, their fitness approaching zero. Heterozygotes (AS), however, have red blood cells that resist infection by the malaria parasite (*Plasmodium falciparum*) and burst more quickly when infected, thereby leading to the death of the parasites before they infect other cells. In this situation, the fitness of heterozygotes is greater than that for either homozygote. Accordingly, heterozygotes reproduce at a higher rate. Over time the S allele has reached a frequency of 20% (or more) in some areas of West Africa (meaning that as many as one in three people are carriers for the mutant allele). This gene frequency is much higher than can be accounted for by mutation alone. Under these circumstances, Hardy-Weinberg equilibrium is significantly perturbed. When black Africans move to countries like the United States where malaria no longer exists, one would predict that the frequency of the S allele would decline over time. It may already be beginning to do so.

Heterozygote advantage has been shown to exist for other genotypes affecting the red blood cell. Deleterious alleles for the thalassemias, certain other hemoglobinopathies, and glucose-6-phosphate dehydrogenase (G6PD)—an enzyme erythrocytes use to extract energy from glucose—are all thought to be maintained at higher than expected frequencies because they provide protection against malaria. Heterozygote advantage may also explain the high frequency of CF in whites and of a number of lipid storage diseases, including Tay-Sachs, in Ashkenazi Jews. The environmental forces responsible in these situations remain unclear.

ANCESTRY AND DISEASE IN THE GENOMIC ERA

The laws, assumptions, and exceptions discussed thus far depend on examining genes in populations. Today, however, we must add genomic data to that provided by examining populations one gene at a time. As the capability to perform genotyping using SNPs, copy number variations, and complete genomic sequencing in more individuals from diverse populations increases rapidly, we are beginning to gain a broader view of variations among different populations.

Genetic Variation in Populations

As discussed in Chapter 10, two of the most important means, currently, to detect human genetic variation are SNPs and CNVs. SNPs are single-nucleotide base pairs that differ among individuals' DNA sequences (for example A–T in one individual, G–C in a second). CNVs are larger contiguous blocks of DNA sequence (usually larger than 1 kb) that vary in copy number from individual to individual (a block could be duplicated in one person but deleted in another). At the SNP level, any two humans differ at about 1 in 800 base pairs (or 0.1% of genomic DNA). CNVs contribute an additional 0.4% difference between individuals because they comprise much larger segments of DNA. Taking SNPs and CNVs together, any two humans are about 99.5% identical in DNA sequence. This reflects the relatively recent origin of our species. We must emphasize that the 0.5% non-identity (15 million nucleotides) provides ample information for individual variation.

SNPs AND CNVs BETWEEN AND AMONG POPULATIONS

Most common SNPs (that is, those for which the prevalence of the rare allele exceeds 5%) are shared among populations in different continents. This commonality reflects continued migration and gene flow among populations consistent with our common origin in Africa. Going a step further, the great majority of human genetic variation (85–90%) can be found within any human population (for example, people from Spain or from Senegal); only the remaining 10–15% is produced by variation between populations. In other words, persons of different ancestries are genetically similar.

On occasion, a SNP or CNV is relatively common in one population but absent (or very rare) in others. The *ΔCCR5* allele which protects against HIV/AIDS is an example (Figure 18.4). It appeared first in northern Europe, but has not had time to spread to populations in Africa or Asia. In other cases, natural selection is the force behind prevalence differences. The balanced polymorphism between the S allele and malaria is one such example. Another one involves the ability to digest milk, which depends on the enzyme, lactase, that breaks down lactose (milk sugar). Lactase activity, which disappeared in most human populations, persists in many European populations where milk consumption beyond early childhood confers a selective advantage.

When the prevalence of a particular SNP or CNV is compared among populations, there is a strong correlation between geographic proximity and genetic similarity—in keeping with the idea that populations close to one another geographically are more likely to share migrants. Conversely, African populations have relatively more genetic diversity, and the genetic diversity found outside Africa tends to be a subgroup of African variation. Such comparisons support a common African origin of *Homo sapiens*.

INDIVIDUAL ANCESTRY

When large numbers of SNPs are analyzed, the ancestry of an individual's genome can often be inferred. The larger the number of SNPs employed, the more precisely such ancestry can be determined. Few (perhaps no) variants are found in all members of one population and in no members of another. Thus, it is possible to make only statistical conclusions about an individual's ancestry from such studies.

Another limitation to determination of individual ancestry derives from growing understanding that population categories are not distinct. African Americans, for example, are estimated to have about 20% European ancestry, whereas European Americans have as substantial recent African genetic ancestry. The greater the number of populations genotyped, the more difficult it is to define population boundaries. What, then, is a race or an ethnic group? Increasingly, these words are social rather than biological constructs.

EVOLUTION IN POPULATIONS

As discussed in Chapter 9, all life forms on earth are descendants of a single cell that formed about 4 billion years ago. Mutation and natural selection are the main forces that have driven adaptive evolution of that first cell through an incalculable number of organisms—first prokaryotic, then eukaryotic—eventually to include humans. Although RNA was probably the earliest informational macromolecule, it gave way to DNA. In turn, the pathway from DNA to mRNA to protein is common to organisms as simple as *E. coli* and as complex as humans. It is this commonality that has permitted the study of molecular evolution.

The foregoing fundamental concept (or central pillar) of biology is buttressed by many kinds of observations from many fields:

- the fossil record of past ages, showing temporal sequences of more or less continuous life forms;
- the geographic distribution of past and present species;
- the retention of developmental stages of remote ancestors during the embryology of current species;
- the similar underlying structures of some functionally dissimilar body parts;
- the greater similarity of cellular and molecular structures among more closely related species;
- the nearly universal genetic code among all living creatures.

Using data from the abovementioned kinds of inquiry, it is possible to construct a life tree for our species (see Chapter 9). **Hominoids**, animal ancestors to apes and humans, appeared in Africa only 20 to 30 million years ago. About 7 million years ago the hominoid tree began to branch. By 5 million years ago, **hominids** appeared. The first hominids were called ***Australopithecus***. Their evolution continued toward ***Homo***, who appeared about 2 million years ago. Several species of *Homo* (*habilis, erectus*) led ultimately to ***Homo sapiens***, who appeared about 200,000 years ago.

INTERSECTIONS: EVOLUTION AND LANGUAGE

To Speak is to be Human (or Neanderthal?)

The capacity for complex speech and language skills belongs only to our species, *Homo sapiens*. Is there one gene or are there multiple genes that control this ability? How can we find out about them? When, during evolution, did the remarkable ability we refer to simply as "speech" appear?

One way to approach this problem is to study individuals in whom mutation adversely affects their ability to speak. The KE family, in England, was brought to the attention of the scientific community in about 1990. Over three generations, about half of this family's members suffered from severe difficulty in speaking, to such an extent that their speech was largely unintelligible. As a substitute for speech, they were taught sign language as children. Their condition was complicated and included impairment of broader intellectual and physical abilities, but language impairment was most prominent. Examination of the family's pedigree indicated that the disorder was caused by a single-gene defect inherited as an autosomal dominant trait.

Could mutation of a single gene lead to disruption of a human trait as fundamental and complicated as complex language? In the 1990s, scientists carried out a linkage study that localized the mutation in the KE family to a short piece of chromosome 7. In 1998, additional studies uncovered a single point mutation in a single gene: a G to A substitution that changes the amino acid arginine to histidine. This mutation was found in all the affected individuals in the KE family and in none of their unaffected relatives, or in any controls in a normal population of about 360 people.

(Continued)

INTERSECTIONS: EVOLUTION AND LANGUAGE—Cont'd

The gene was named *FOXP2*. It encodes a transcription factor, meaning that it has the potential to bind to DNA and affect the expression of one or more genes—in this case a potentially large number of genes that are important in the development of the brain, generally, and linguistic skills, specifically. Thus, *FOXP2* may act as the "conductor" of a whole orchestra of genes that affect linguistic skills.

In addition to humans, *FOXP2* is present in many animals that have been studied, including crocodiles, birds, mice, and primates. The protein *FOXP2* codes for is identical in the chimpanzee, gorilla, and rhesus monkey, and differs from the human version by only two amino acids.

When did the mutations occur that differentiated the language skills of humans from other primates? A fascinating study in 2008 involved extracting DNA from Neanderthal bones found in a cave in northern Spain. The Neanderthal *FOXP2* gene was found to be identical to that in humans, suggesting that Neanderthals may have had the ability to form intelligible speech. This work implies that the human *FOXP2* gene evolved more than 400,000 years ago, before humans and Neanderthals diverged, and that there was likely a selective advantage afforded by the mutation(s) in *FOXP2* characteristic of these most advanced *Homo* species.

Molecular Evolution

The field concerned with determining and comparing genomes, chromosomes, genes, and proteins among different animal species is called **molecular evolution**. Its basic assumption is that DNA and amino acid sequences change over time due to mutation and selection. Its underlying hypothesis is this: the fewer the differences between a DNA or amino acid sequence between two species, the more closely related they are and the more recently they diverged from a common ancestor.

METHODS OF STUDY

As shown in Table 18.8, a variety of increasingly sophisticated methods have been used to study molecular evolution.

TABLE 18.8 Methods Used to Study Molecular Evolution

- DNA hybridization
- Chromosome banding
- Protein sequences
- Gene structure and function
- Whole-genome sequences

DNA Hybridization

The oldest method is DNA hybridization, shown schematically in Figure 18.5. Double-helical DNA from two species is unwound (denatured), cut, and mixed. Complementary pieces bind (called **reannealing**), forming some **hybrid DNAs**—one strand from each species. Then the temperature is raised and the rate of interspecies double-helix separation determined. The higher the temperature needed to separate the hybrid DNA, the more similar the DNA sequences must be. Such hybridization studies led to the conclusion that humans differed from chimpanzees at slightly more than 1% of their DNA sequences, compared to more than 2% between humans and gorillas. As we will see, the common statement that our DNA differs from that of chimpanzees by only 1% is a considerable underestimate.

Chromosome Bands

The basic idea is the same: the more similar the chromosome structure and banding pattern of two organisms, the more closely related they are evolutionarily. Chimpanzees and the other great apes have 48 chromosomes, compared to 46 in humans. This difference in chromosome number is accounted for by observing that human chromosome 2 is formed by end-to-end joining of two independent chimpanzee chromosomes. After adjusting for this single change in ploidy, the banding pattern of chimpanzee chromosomes is identical to that in humans.

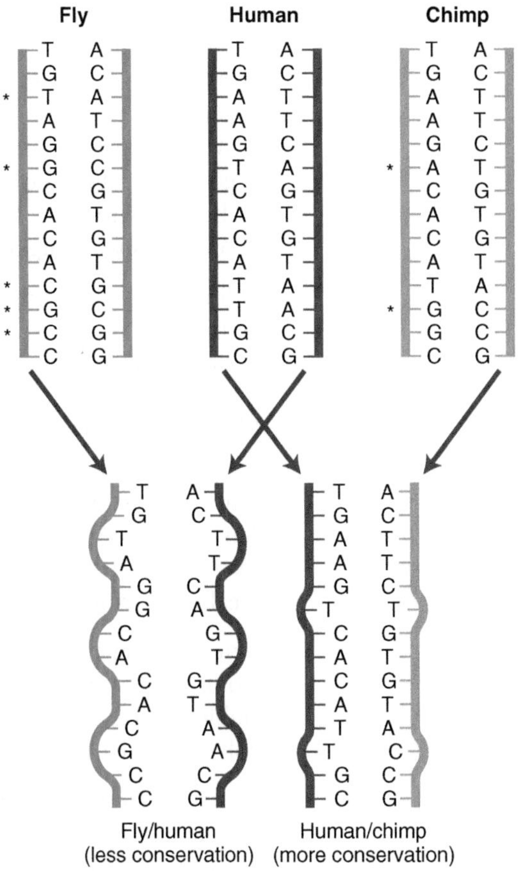

FIGURE 18.5
Schematic representation of information obtained from interspecies DNA hybridization. See text for details.

Protein Sequences

The universal triplet genetic code argues for common ancestry of all life on Earth. So, too, does examination of amino acid sequences in a protein conserved through eons of time—from yeast to humans. Cytochrome C, a protein required for electron transfer in mitochondria and found in all eukaryotes, is an apt protein to examine (Table 18.9).

The amino acid sequence of yeast cytochrome C differs from that of human at 42 amino acid residues out of 104 total. As one examines the sequence of cytochrome C in progressively more complex multicellular organisms, the number of amino acid differences decreases progressively (24 in drosophila, 10 in the cow, 1 in the rhesus monkey). The amino acid sequences in chimpanzees and humans are identical. Other proteins have been examined this way as well, with similar results. This is powerful molecular evidence in support of descent with modification.

TABLE 18.9 Evolution of Cytochrome C

Organism	Number of Amino Acid Differences from Human
Yeast	42
Wheat germ	37
Fruit fly	24
Bullfrog	20
Pigeon	12
Cow	10
Rabbit	9
Rhesus monkey	1
Chimpanzee	0

Gene Structure and Function

Among the many human traits that distinguish our species from those of other primates—including chimpanzees—are larger frontal lobes in the brain, opposable thumbs, spoken language, and the amount and distribution of body hair. We are beginning to be able to account for such differences by looking at single genes. Here are three examples:

- *Why are chimpanzees so much hairier than humans?* The gene for formation of keratin, a major protein in hair, is expressed in chimpanzees but silenced by a stop codon in humans. Perhaps relative hairlessness allowed for greater body cooling when humans left the trees.

- *Why are chimpanzees not capable of intelligible speech?* Remarkably, this all-important human trait is controlled by a single gene called *FOXP2*, as discussed above. The *FOXP2* gene in chimpanzees differs from its human counterpart by two amino acid residues. This seemingly small difference explains, in part, why humans have complex speech, and chimpanzees don't.
- *Why do some primates lack fetal hemoglobin?* That is, why are they unable to switch off the gene for embryonic hemoglobin and turn on the gene for fetal hemoglobin? It appears that the ability to make fetal hemoglobin correlates with the length of gestation, which in turn correlates with larger brain volume.

Additional study of single-gene function and gene expression will likely shed additional light on the many differences between humans and our biologically and genetically nearest animal species, the chimpanzee.

Whole-genome Sequences

As complete genome sequences from hundreds of animal species have appeared, interest in comparing them has soared. Earlier we mentioned that humans and chimpanzees are 99% identical at the SNP level. When indels and CNVs are taken into account, however, this identity falls to 96%. Beyond such organism-to-organism comparisons, DNA sequences have been revealing by comparing human DNA sequence with itself.

There is strong evidence that the human genome underwent duplication since diverging from a common vertebrate ancestor 500 million years ago. The human genome may even have doubled twice. Such extensive duplication distinguishes humans even from more recently evolved simpler primates. Here are a few examples:

- Our smallest chromosome, number 22, is densely packed with genes and contains eight large duplications.
- Much of chromosome 2's short arm is duplicated and makes up 75% of acrocentric chromosome 14.
- Nearly 50% of chromosome 20 repeats itself in rearranged form on chromosome 18.

Such duplications and rearrangements are almost surely responsible, in part, for what makes us human, but there remain huge gaps in our understanding.

APPLICATION OF TOOLS USED TO EXAMINE MOLECULAR EVOLUTION

The methods (or tools) used to examine evolution in molecular terms are beginning to provide exciting information about human origin, migration, and ancestry. Here is a small sample of such applications.

Ancient DNA: Neanderthals

Under special conditions of little or no oxygen, DNA can withstand a broad range of changes in humidity, pressure, and temperature and remain relatively intact for up to many thousands of years. In this way, DNA has become part of the fossil record so useful to anthropologists and paleobiologists. Such ancient DNA was first recovered from amber, a hardened resin from pine trees. Chemicals in amber acted as drying agents and preservatives for biological specimens that entered it before it hardened. More recently, ancient DNA has been extracted from fossilized bones, where the mineral matrix preserved the biological material found in its marrow and interstices.

Not surprisingly for our species, the most provocative and exciting studies have involved Neanderthals (Figure 18.6), an archaic people that coexisted with the ancestors of modern humans for about 200,000 years—from 250,000 years ago until 30,000 years ago, at which point the Neanderthals abruptly disappeared. Aided by use of the polymerase chain reaction (PCR), tiny pieces of Neanderthal DNA have been analyzed since 1997, starting with mitochondrial DNA (mtDNA). When mtDNA obtained from Neanderthal bones was compared to that of humans, three times as many nucleotide differences were observed as were seen with human samples from disparate populations. This was interpreted to mean that Neanderthals diverged from humans before humans left Africa. The people destined to become Neanderthals migrated to Europe; the remainder of humans migrated there and to all other parts of the Earth.

Although it was not considered possible in the 1990s, nuclear DNA from Neanderthals has been isolated and sequenced since 2007. A complete sequence has not yet been reported, but from the approximately 60% that has been described, it has been suggested that humans and Neanderthals interbred after humans left Africa. The data supporting this remarkable conclusion have been questioned, as has the idea that such interbreeding occurred in the Middle East, not in Europe where Neanderthals and humans coexisted for many thousands of years. This controversy will be resolved by collection of additional DNA sequence data.

FIGURE 18.6 Artistic rendering of a Neanderthal. This population group co-existed with *Homo sapiens* for over 200,000 years.

Mitochondrial DNA: An Evolutionary Clock

Humans are continually fascinated about their origins, seeking answers to questions such as: How old is our species? Where did *H. sapiens* originate? How does one account for our presence in North America? By melding evidence from the fossil record and mtDNA studies, scientists have, over the past 20 years, been able to construct an increasingly accurate evolutionary clock.

Recall that all mitochondria are maternally inherited and that each human contains a large number of them (perhaps 10^{15}). Each mitochondrion contains many copies of a single chromosome whose DNA mutates much faster than nuclear DNA does, undergoes no recombination, and doesn't repair the mutations that occur. These characteristics lend themselves to the use of mtDNA in constructing an **evolutionary clock**. The central thesis of such work is this: mtDNA can be used as such a clock if its nucleotides are changed at a known and constant rate. For example, if mtDNA changes at a rate of 7 bases per million years, then two species differing in mtDNA at 21 bases diverged 3 million years ago ($21/7 \times 1$ million).

Beginning in the late 1980s, mtDNA sequences were analyzed from more than 300 people, including Africans from several parts of the continent, Europeans, Asians, New Guineans, African Americans, and Australians. This analysis found greater mtDNA sequence differences among Africans—particularly sub-Saharan Africans—than among Europeans or Asians. Because mutations accumulate over time, they concluded that the African population has had the longest time to evolve, and that Africa is the place where modern humans originated. Using computer-aided estimates, they went on to suggest that humans originated in east Africa about 200,000 years ago, in good agreement with the oldest *H. sapiens idaltu* fossils dating to 170,000 years ago. Because mitochondria are exclusively maternal and are inherited generation after generation in DNA, it is possible to identify the first woman, a proverbial "Eve," who lived in east Africa at the time of our appearance as a species. Subsequent work with Y-chromosome DNA and other nuclear DNA fragments supports these suggestions of where and when we appeared as a species.

Studies using mtDNA have also been used to determine what happened to human populations after they appeared. This has led to the **replacement theory** (dubbed the "out of Africa" theory), which proposes that humans migrated to Asia, Europe, the Americas, and elsewhere, and eventually replaced the archaic hominid groups living in these places by outcompeting

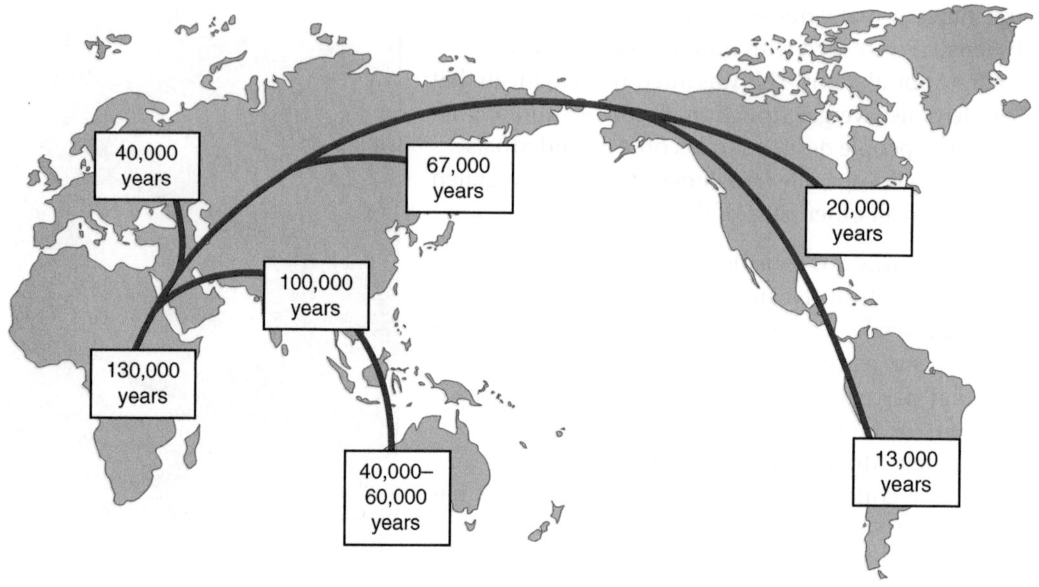

FIGURE 18.7
The migration of human populations from sub-Saharan Africa beginning about 130,000 years ago. The dates are based on fossil and archaeological evidence, and on mtDNA results.

them. As shown in Figure 18.7, the prevailing view is that humans left Africa about 130,000 years ago, migrating first to the near East, then the far East, then to Europe and Australasia, then to North America, and finally to South America. Another theory, referred to as the **regional continuity** theory, proposes that humans evolved in many parts of the world (rather than one), where they interbred with archaic peoples. Current evidence strongly favors the replacement theory. However, there is no agreement as to whether humans migrated out of Africa once or multiple times.

SOCIETAL DEBATES ABOUT EVOLUTION: SCIENCE VERSUS RELIGION

As stated in Chapter 9 and again in this chapter, the theory of evolution proposed by Darwin and Wallace maintains that the diversity and complexity of living organisms are the result of natural selection of pre-existing variations, which we now know are encoded by the genome. **Creationism** maintains instead that a divine being created the world and all the diversity in it. Whereas evolution is a scientific theory which has been tested in countless ways and supported in those tests, creationism is a religious belief whose hypothesis cannot be tested.

A pitched battle between evolution and creationism has been ongoing for 150 years and still rages in the US. In the famous "monkey trial," a schoolteacher named John Scopes was sued by the state of Tennessee in 1925 for teaching evolution in the science classroom. The lawyer for the plaintiffs, whose argument was based on creationism, was William Jennings Bryan. The counselor for the defense, Clarence Darrow, based his arguments on the scientific evidence for Darwinian evolution. The state of Tennessee won the case, but the decision was later overturned.

More recently, evolution has become an integral part of the biology curriculum, but the controversy has not abated. Despite a Supreme Court ruling in 1987 that struck down a Louisiana law requiring that creationism be taught alongside evolution in the science classroom, a new challenge has appeared in the form of **Intelligent Design (ID)**. According to its proponents, life is too complex to have been created by evolution and required an intelligent designer—taken to be God. When a school board in Dover, Pennsylvania ordered that ID be taught alongside evolution in the science classroom, some parents brought suit. In 2009, US District Judge John Jones said the following in ruling against the school board:

> [ID] is a religious view, a mere re-labeling of creationism, and not a scientific theory. [The board's action] singles out the theory of evolution for special treatment [and] misrepresents its status in the scientific community ... and instructs students to forego scientific inquiry in the public school classroom and instead to seek out religious instruction elsewhere.

At heart, the controversies just cited indicate that there is an unwillingness in American society to believe that evolutionary science and religion can coexist. Witness that fewer than 50% of college-educated people in the US believe that evolution alone explains human origins. This fraction is much lower than that found in many countries in Europe, in Japan, or in China. The authors are puzzled by this. Science, after all, seeks to explain the visible (external) world, religion, the spiritual (internal) world. We see no reason for this unremitting conflict unless one simply doesn't believe in science, or, conversely, that one doesn't believe in religion, or that belief in one is an absolute refutation of the other. Coexistence between these two value systems has, thus far, been impossible to achieve in the United States. Hopefully, future generations will be more tolerant than current and past ones have been.

REVIEW QUESTIONS AND EXERCISES

1. Choose the phrase in the right column that best matches the term in the left column.

a. replacement theory	1. the capacity of a population to survive and reproduce
b. allele frequency	2. human migration out of Africa
c. consanguinity	3. studied in amber and fossilized bones
d. molecular evolution	4. the proportion of all individuals in a population with a particular genotype
e. gene migration	5. a group of interbreeding individuals
f. population genetics	6. a major form of inbreeding
g. balanced polymorphism	7. $p^2 + 2pq + q^2 = 1$
h. *Australopithecus*	8. choice of mate based on a particular characteristic
i. assortative mating	9. earliest hominid
j. Scopes *v.* Tennessee	10. quantitative study of genetic variation in a population
k. gene pool	11. population in which subgroups have remained genetically separate
l. fitness	12. proportion of all copies of a gene in a population that are of a specific allele
m. ancient DNA	13. all the genes in a population
n. Hardy-Weinberg law	14. challenged teaching of evolution
o. genetic bottleneck	15. chance events producing large change in allele frequency
p. heterozygote advantage	16. fitness of heterozygote greater than that for either homozygote
q. population	17. compares genomes, genes, and proteins among different species
r. genetic drift	18. population's allele frequencies change when population size is reduced
s. genotype frequency	19. change in allele frequency due to slow diffusion across a geographic or cultural barrier
t. stratification	20. maintaining a deleterious gene in a gene pool as well as removing it

2. A particular human gene has two alleles, A and a. In a population of 1,000 people, 500 have genotype AA, 400 have genotype Aa, and 100 have genotype aa.
   a. Calculate the allele frequencies for A and a in this population.
   b. Is this population in Hardy-Weinberg equilibrium? Why or why not?
   c. List five assumptions underlying derivation of the Hardy-Weinberg law.

3. Investigators typed 5,000 people in a population at the MN blood groups locus (which has two co-dominant alleles, M and N). They found the following numbers of individuals for each of the three genotypes: 2,300 MM; 2,200 MN; 500 NN.
   a. What are the allele frequencies for M and N in this population?
   b. Is this population in Hardy-Weinberg equilibrium?

4. Consider an autosomal recessive disorder with a population frequency of 1 : 10,000 Asians. In the accompanying four-generation pedigree, individual I1 is Asian, I2 is not.

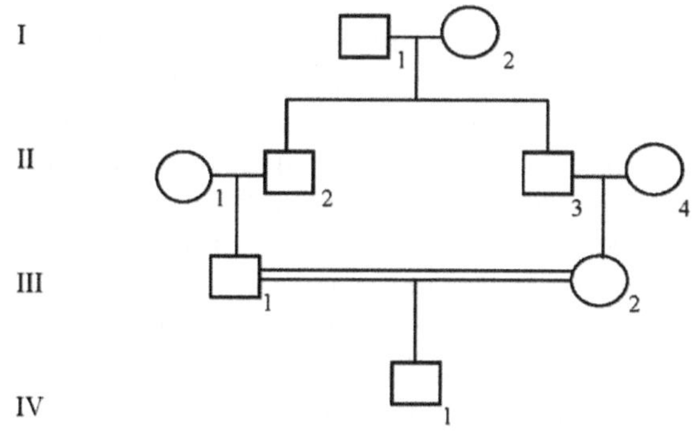

   a. What is the likelihood that I1 is a carrier for this disorder?
   b. Assuming that I1 is a carrier, what is the probability that IV1 would be affected with this disorder?
   c. A neighboring Asian population has a 10-fold greater frequency of the disorder. List four possible reasons for the differences in population frequency.

5. Homocystinuria is an autosomal recessive disorder causing dislocated optic lenses and cardiovascular disease. In the United States the prevalence of this condition is 1:250,000.
   a. Assuming Hardy-Weinberg equilibrium, what is the frequency of the mutant allele?
   b. What fraction of the American population is a carrier for this condition?
   c. What is the ratio of carriers to homozygous affected in the U.S.?
   d. Given that patients with homocystinuria have a fitness of zero, why hasn't this condition disappeared?

6. In a screening program to detect α-thalassemia in Sri Lanka, the carrier frequency was 4%. α-Thalassemia is recessively inherited.
   a. Calculate the frequency of the α-thalassemia allele, assuming only one α-thalassemia mutant allele in this population.
   b. Calculate the fraction of matings that could produce an affected child.
   c. Calculate the incidence of affected newborns in the population (assuming, hypothetically, that all affected are born alive).
   d. Calculate the incidence of α-thalassemia in the offspring of couples both of whom are carriers.

7. Phenylthiocarbamide (or PTC) is an organic compound that tastes bitter to some people and is tasteless to others. The ability to taste PTC is an autosomal trait: the taste allele is abbreviated T; the non-taste allele t. Imagine that you sampled 215 individuals taking this course and determined that 150 detected the bitter taste of PTC and 65 did not.
   a. Calculate the allele frequencies of each allele.
   b. Calculate the number of heterozygotes you would predict in the class.
   c. What would happen to the allele frequencies for T and t if the professor teaching the course said that next year only PTC tasters could enroll?

8. Well before completion of the human genome project, mitochondrial DNA (mtDNA) was used as an "evolutionary clock."
   a. What is an "evolutionary clock" and why could mtDNA be employed as one?
   b. Describe briefly one major limiting factor concerning use of mtDNA as a clock.
   c. What other techniques have been deployed as evolutionary clocks? Why?

9. Why do 90% of adults in countries such as Japan and Sweden accept Darwin's views about evolution as proven, whereas fewer than 50% of American adults do? Does this disparity affect other aspects of these countries' views about genetics?

CHAPTER 19

Personalized Genetics and Genomics

CHAPTER OUTLINE

CORE CONCEPTS

Each of us is unlike anyone else who is alive today or who has ever lived. This uniqueness is a product of our genes and our environment. Among the many goals of the study of human genes and genomes is to help understand this individual uniqueness. Although we have barely scratched the surface of such understanding, useful signposts are being constructed. We will discuss three of them: **DNA fingerprinting**, **pharmacogenetics**, and **direct-to-consumer genomic testing**.

DNA fingerprinting (or **profiling**) is based on the nearly 50% of our genome that is composed of a variety of repetitive DNA sequences. In so-called **microsatellites**, the repeating units are 2–10 bp in length; in **minisatellites** the repeats are 10–100 bp long. Microsatellites and minisatellites are extremely polymorphic, meaning that each of us has a multitude of alleles at these loci. One form of these alleles is called **variable number of tandem repeats** (**VNTRs**), and it is in their identification that DNA fingerprinting is

Human Genes and Genomes. DOI: 10.1016/B978-0-12-385212-0.00019-6

possible. Such fingerprints are developed by obtaining a sample of DNA from viable biologic material, cutting it with a particular bacterial restriction enzyme, running the DNA out using polyacrylamide gel electophoresis, hybridizing it with a large set of VNTR probes, and developing the profile of fragments using radioautography or fluorescence. The extraordinary power of this approach is shown by estimating that if 24 different minisatellite probes are used, there is only a 1 in 700 trillion chance of identity between any two individuals (except for identical twins). The many uses of DNA fingerprinting include forensic identification of the guilty and the innocent; proof of paternity and maternity; identification of victims of catastrophic events; and confirmation of the identity of long deceased persons—famous and infamous. Whereas unequivocal identification of an individual is available from DNA fingerprints, we are nowhere close to achieving such results in any other sphere.

For instance, we all take many medicines. Sometimes these medicines are efficacious, other times, toxic. Optimally, we would like to know how to predict efficacy and toxicity for any medicine in any person. No profiling exists with which to do this. We know that genes form part of the basis for differences in response to medicines (a field called pharmacogenetics). In some instances, genes have been identified that control the absorption, distribution, and metabolism of pharmaceuticals. In other situations, specific genes regulate the target of the pharmaceutical. Certain enzyme systems, such as the cytochrome P450 family, play a central role in metabolism of administered drugs, and their variation predicts accurately both efficacy and toxicity. In a slowly growing list of examples, we are able to understand the pharmacogenetic basis for response to such important medicines as opiate analgesics, antiplatelet agents, antibiotics, and antimalarials. Extending this small body of information will depend on use of genomic (as well as genetic) information and on major modifications in the systems by which pharmaceutical companies make medicines and practicing physicians use them.

A third approach toward understanding genetic uniqueness is a new industry—direct-to-consumer genetic testing (DTC genome testing). It depends on SNP profiling of an individual's DNA and comparing it to profiles from many thousands of other people's. The most robust of the small companies providing DTC genome testing now provides information on susceptibility to more than 120 serious and harmless traits, response to nearly 20 medicines, carrier status for more than 20 single-gene disorders, and ancestry. It is almost certain that DTC genome testing will burgeon in coming years in the form of more tests for more conditions, more medicines, and more disorders.

In Chapter 10 we discussed, as a core concept, the idea of genetic uniqueness and the halting understanding of it. Early work on inborn errors was followed by studies of protein polymorphisms. As we enter the genomic era, information from SNPs, CNVs, and a small (and rapidly growing) number of whole-genome sequences has directed us toward (but only a trivial distance toward) fathoming phenotypic individuality in biochemical or genomic terms. Nonetheless, some scientists and physicians have already begun to talk about a "genomic medicine" in which disease prevention, treatment, prognosis, and cure will be predicated on knowledge of one's own genome. Genomic medicine is a growing reality to a few, a dim, distant beacon to many people, but a frightening kind of illumination to others.

In this final chapter of our book, we will illustrate some of what we know as "personalized" genetics and genomics using three quite different cases: **DNA fingerprinting**; **pharmacogenetics** and **pharmacogenomics**; and **direct-to-consumer genomic testing**.

DNA FINGERPRINTING

Any of you who have arrived in the US at one of its international airports will have seen evidence that taking a person's fingerprints is still a time-honored (even if scientifically crude) way of identifying an individual with a high degree of certainty. As discussed in Chapter 13, each of us has a fingertip ridge count built on a pattern of whorls, arches, and loops. This is a pure polygenic trait which, though still mysterious regarding the genes that account for it, is absolutely private (in that no one has a set of prints exactly like our own). Over the past 20 years DNA fingerprints have become as or more widely used than dermal ones, because they have broader applicability.

TABLE 19.1 DNA Satellites Used in Fingerprinting

Satellite Class	Length of Repeat	Number of Repeating Units	Example
Microsatellite	2–10 bp	15–100	TATATA ... AAATAAAT ...
Minisatellite	10–100 bp	100–1,000(s)	ATGCAGCTTA ...

Molecular Basis

Recall from Chapter 6 that about 50% of total genomic DNA consists of repeating nucleotide sequences. These sequences do not code for proteins, and they come in two general classes, as shown in Table 19.1.

Microsatellites are stretches of DNA consisting of 2–10 bases (for example, TATATA or AATAAT ...) repeated between one and a few dozen times. For any microsatellite locus, an individual has two alleles: they may each contain the same number of repeats (homozygous) or a different number (heterozygous). Because the number of repeats varies widely, and because a microsatellite locus has many alleles in the population, they are, by definition, polymorphic and are referred to as **short tandem repeat polymorphisms (STRP)**.

The other class of sequence polymorphisms results from the presence, in tandem, of large numbers (hundreds to thousands) of copies of a DNA sequence 10 to 100 base pairs in length. These are called **minisatellites**. This class has many more alleles than does the STRP variety; its alleles are called **VNTR**, for **variable number of tandem repeats**. At any minisatellite locus, an individual has two alleles—that is, two sizes of VNTRs (Figure 19.1). Because there are so many minisatellite loci scattered along the genome, because there are so many alleles at each locus, and because the VNTRs can be distinguished from one another by size, there is vanishingly little likelihood that any two individuals will have exactly the same pattern of VNTRs. Therefore, VNTRs are the molecular substrate for a DNA fingerprint. For example, if 24 different unlinked minisatellite loci are examined simultaneously, the chance of any 2 individuals having identical DNA fingerprints is 1 in 700,000,000,000,000. Given a total human population of about 7 billion, there is essentially no chance that two individuals other than monozygous twins would have identical DNA fingerprints when assessed this way. (DNA profiling or DNA typing would be a more appropriate name for this technique, because it has nothing to do with fingers. Nonetheless, we'll use the term DNA fingerprinting because that is what Sir Alec Jeffreys—the man who pioneered its applicability—called it.)

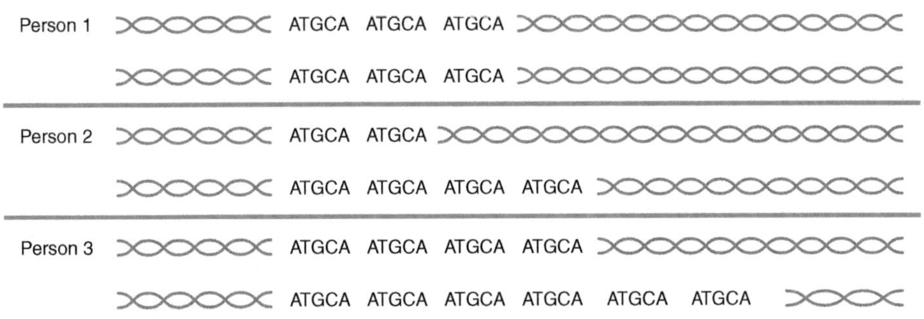

FIGURE 19.1
Schematic representation displaying variable number of tandem repeat (VNTRs) alleles in three different individuals. Person 1 is homozygous for the ATGCA repeat. Persons 2 and 3 are heterozygous.

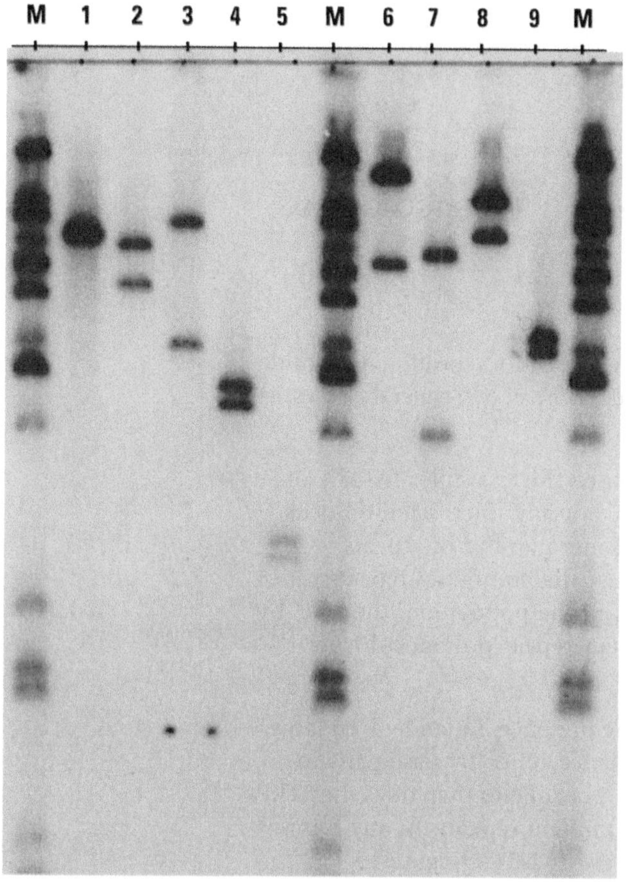

FIGURE 19.2
Genetic variation in VNTRs analyzed using a single locus minisatellite probe. Individual 1 is homozygous, individuals 2–9 are heterozygous. No two people have the same allelic pattern. M, molecular weight markers. Additional details in text.

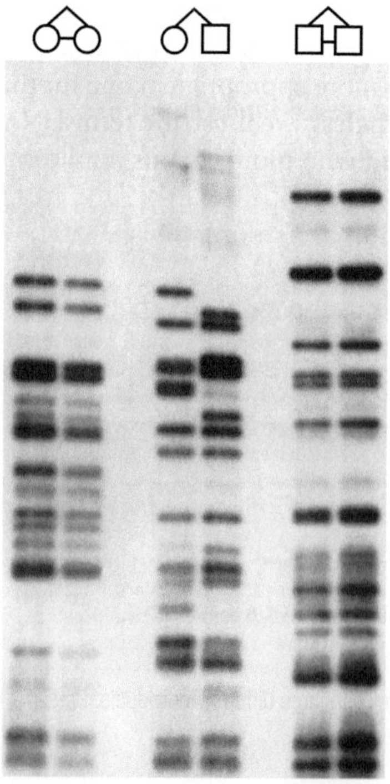

FIGURE 19.3
DNA fingerprinting of three sets of twins using a multilocus probe that detects many VNTRs. Each pair of lanes examines DNA from a single twin pair. As shown at the top, pairs 1 and 3 are identical twins; pair 2, non-identical.

Methodology

DNA fingerprinting employs several of the techniques and reagents described in Chapter 6: restriction enzymes, gel electrophoresis, Southern blotting, hybridization, and radioisotopic or fluorescent labeling. In brief, here is how the fingerprinting is done:

- A sample of cell-containing biological material (blood, hair, skin, cheek lining, semen, vaginal fluid) is obtained, from which DNA is isolated.
- The DNA is "cut" into pieces of varying length using a bacterial restriction enzyme (usually EcoR1). The restriction enzyme is chosen because it cuts between the minisatellite loci, not within them—thereby preserving the length of the polymorphic alleles.
- The sheared DNA is applied to a solid support—usually a polyacrylamide gel—and electrophoresed (DNA is negatively charged because of its phosphate groups; therefore, it migrates toward the positive electrode when current is applied). This separates the DNA pieces by size, the largest fragments travelling more slowly than the smaller ones.
- The DNA is then transferred to (usually) a nylon membrane and probed with a radiolabeled or fluorescent-labeled piece of DNA containing one or more minisatellite sequences. If one uses a monolocus probe (a probe that recognizes a single minisatellite), each person has a pattern showing one or two bands depending on whether he or she is homozygous or heterozygous for the number of repeats at that locus. However, if a multilocus probe is used, a larger number of different-sized bands is observed (Figure 19.2). The more minisatellites represented in the probe, the larger number of bands observed. In this example, none of the four individuals have identical fingerprints.

Figure 19.3 further illustrates this point. Three sets of DNA fingerprints are shown in twin pairs—two pairs of identical twins, one pair of non-identical twins. The DNA pattern is identical in each member of an identical twin pair but very different from those in the other pair. In the non-identical twins, one twin's pattern is clearly distinguished from the other twin's.

It must be pointed out that there are several caveats to DNA fingerprinting: identical twins will have identical patterns; first-degree relatives will have patterns that are more alike than those found in unrelated individuals; the DNA must not have been degraded before or during use; contamination must be avoided. But when properly carried out and interpreted, DNA fingerprinting has found many uses, and its utility continues to expand.

Utility

FORENSICS

DNA fingerprints have received wide and visible use to answer this question: does the DNA pattern of a suspect match that found at a crime scene? The first use of the technique occurred in England in 1986, when Jeffreys' method was instrumental in solving a perplexing rape/murder case. Since then, thousands of criminal cases each year have employed DNA fingerprinting. Figure 19.4 illustrates the approach: DNA is isolated from a victim of rape and from semen found on her external genitalia or in her vagina; DNA is also obtained from one or more men suspected of being the rapist. The DNA fingerprints from each person are then examined. A perfect match between the semen and a suspect leaves virtually no doubt as to the rapist. Negative results are every bit as important. More than 100 men convicted of murder or rape have been exonerated by DNA fingerprinting—often years after being imprisoned.

Today, the FBI uses a system called **CODIS** (combined DNA index system). It uses a large number of polymorphic microsatellite probes from different chromosomes and is said to reduce the likelihood that any two individuals would have identical patterns to 1 in 250,000,000,000,000. More than 3 million convicted felons have their DNA fingerprints stored in this FBI database.

ESTABLISHING PARENTAGE

Until recently, disputes about parentage focused on ascertaining the identity of a child's father (for, as the Swedish playwright August Strindberg wrote, it is obvious who one's mother is). In

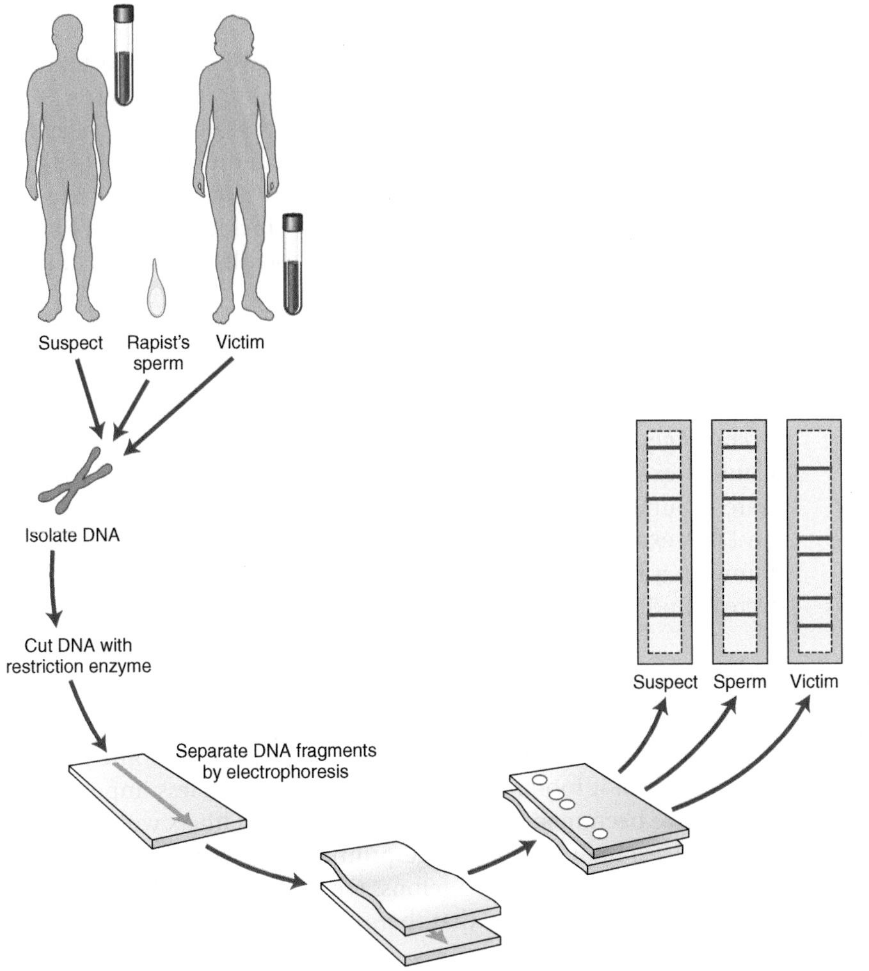

FIGURE 19.4
Schematic representation of forensic use of DNA fingerprinting. Blood samples are obtained from suspect and victim, sperm sample from rapist. DNA fingerprints are developed from three samples: only a subset of bands is shown (compared to the many actually identified, as in Figure 19.3)

today's world of egg donors, IVF, and surrogacy, however, disputes about a mother's identity surface from time to time. In such settings DNA fingerprinting may be useful because the DNA profiles of parent and child—although not identical—are similar enough to be convincing (see Figure 19.3 showing the patterns in non-identical twins). Probes specific for the Y chromosome may be particularly useful in assessing father–son pairs. That Thomas Jefferson was the father of one or more of Sally Hemings' children (she was one of his female slaves and half-sister to his deceased wife) was made much more plausible when Y-specific probes were used to examine DNA in male heirs of Jefferson and Hemings.

Similarly, mtDNA probes have shown their value. Because cells contain so much mtDNA and because it is more resistant to degradation than is nuclear DNA, mtDNA has been helpful in establishing maternity. One very poignant case from the 1970s involves Chilean women who wanted to identify their children from among many murdered by the ruling junta and deposited in mass graves. mtDNA profiling was instrumental in returning the remains of some of these children to their rightful mothers.

IDENTIFYING VICTIMS OF CATASTROPHES

Unfortunately, the past few decades have witnessed catastrophes in which many people have died. Some of these are the explosion of an airplane over Lockerbie, Scotland, in 1988, the destruction of the World Trade Center in 2001, and the Indonesian tsunami in 2004. In each of these events, the remains of those who died were either so burned or decayed that identification was not possible using other means. In some instances, but by no means all, DNA fingerprinting has made identification and proper burial possible.

IDENTIFYING PROMINENT PEOPLE

Two recent, celebrated cases demonstrate how DNA fingerprints have been used to identify famous people. One concerns Czar Nicholas II and his family, who were executed in 1918 during the Bolshevik revolution. Geneticists carried out DNA fingerprinting on several people unearthed in a grave in the city of Ekaterinburg who were suspected of being the Romanov Czar and his family. Their patterns were compared with those of living relatives of the Romanovs—including Prince Philip, consort to Queen Elizabeth II. This work established the identity of those in the grave as the Czar and his family.

The second instance concerns the astronomer Nicolaus Copernicus (Figure 19.5), who changed the world by espousing the idea that the Earth revolves around the sun rather than the other way around. Accordingly, in 1543 he was judged a heretic by the Catholic Church, executed, and buried secretly. In 2008, remains of a body were exhumed from under a church. Fingerprinting was carried out on DNA extracted from teeth of the exhumed person and from hair found in a book known to belong to Copernicus. Perfect matches were found!

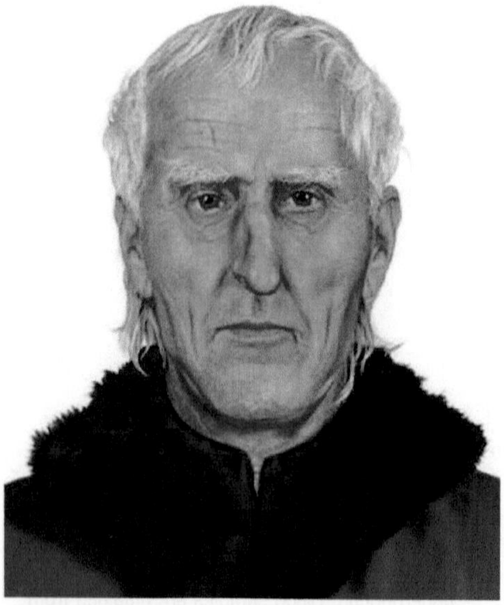

FIGURE 19.5
Copernicus, who first proposed that the Earth orbited the sun, rather than the other way around.

Controversy

As has been the case with virtually every technical advance in human genetics, DNA fingerprinting has found its detractors. Some point out that it is not foolproof and that a few, nonetheless important, errors have been made in its court-related use. Others worry that any database built on DNA fingerprinting invades the privacy of individuals—even convicted felons. On the other hand, proponents of DNA fingerprinting say that there should be a national database containing profiles of everyone, arguing that such a database would help identify perpetrators of serious crimes and would be mercifully useful in the event of catastrophes. These opposing views and the very different risk–benefit assessments they demonstrate—will not likely be resolved any time soon.

GENETICS AND MEDICINES

A second important arena where individual uniqueness demonstrates itself concerns one's response to prescription or over-the-counter medicines. Although evidence abounds concerning this crossroad of genes and the environment, understanding is woefully lacking—except for a few dozen single-gene examples, which chart the direction science and clinical medicine must travel.

The magnitude of the problem is indicated by the following statistics:

- More than 10,000 medicines are in current use or have been employed in the past.
- More than 3 billion prescriptions are filled in the US annually (an average of about 10 per person).
- More than 50% of people in the United States take two or more prescription or over-the-counter medicines daily.

Such widespread use is not, itself, a problem. It is the efficacy and safety of these medicines that are problematic. Again, a few statistics:

- Only about 50–60% of patients for whom medicines are prescribed to treat major, common disorders are helped by these pharmaceuticals (these include such conditions as asthma, depression, hypercholesterolemia, and osteoarthritis).
- More than 2 million patients annually have an **adverse drug reaction** (ADR) to prescription drugs; such ADRs are usually mild and reversible, but not always.
- A recent analysis combining data from 39 prospective studies (called a **meta-analysis**) indicated that 6–7% of all hospitalized patients have serious ADRs—and that such ADRs prove fatal in 0.3% of hospitalized patients who die.
- Fatal ADRs may cause as many as 100,000 deaths in the US annually, putting this category into the top 10 causes of demise.

These troubling statistics reflect two disturbing realities. First, companies that develop pharmaceuticals test them on general populations using the unstated but clearly fallacious assumption that all patients will respond to a pharmaceutical in the same way—that is, that all users are the same. Second, practicing physicians accept the modest efficacy and significant side effects of prescribed medicines as an unavoidable hazard of clinical practice rather than an issue to be avoided or overcome.

This does not mean that discovering, developing, regulating, and prescribing medicines is haphazard or careless. It is anything but.

National Pharmaceutical Policy

The national system (or enterprise) used to bring needed medicines to the marketplace has evolved over a century. Its three major participants are **academia**, **industry**, and **government**—each with its own rights and responsibilities.

Scientists in academia are expected to carry out the basic and applied research that defines mechanisms of health and disease, thereby pointing to targets for new drugs to address unmet medical needs. Those academicians doing clinical research also lead many of the clinical trials required to test new drugs.

The pharmaceutical and biotechnology industries are expected to discover and develop new drugs through sequential (biological) target validation, screening, preclinical testing, and sponsorship of clinical trials before and after approval to sell and market them.

The federal government is expected to provide the legislative framework and the regulatory oversight required to serve and protect the public. On the legislative side, government funds the **National Institutes of Health (NIH)**, the largest single sponsor of medical research in the world; encourages technology transfer from academia to industry; establishes a patent policy for the protection of intellectual property; and conducts inquiries where legal and ethical norms have been violated. On the regulatory side, the federal government delegates responsibility to (principally) the **Food and Drug Administration (FDA)** by charging it with the awesome responsibility of assuring the safety and efficacy of new drugs, devices, and tests, both before they are approved for marketing and while they are in use.

Despite its limitations, the unstated but well-understood social contract just described has made the US the major source of new medicines throughout the past 40 years. Hundreds of new pharmaceuticals have been discovered, developed, and deployed for the treatment of a wide range of conditions, including AIDS, arthritis, asthma, bacterial infections, cancer, coronary artery disease, depression, diabetes, gout, hypertension, osteoporosis, Parkinson disease, schizophrenia, and seizures.

Pharmaceutical Research and Development (R&D)

A deeper glance at **R&D** within the pharmaceutical industry (sometimes called the biopharmaceutical industry) is needed here. Such R&D is generally subdivided into two main components—discovery and development (Figure 19.6)—which make up the "pipeline" of industry. Scientists responsible for the discovery component choose the disease targets, conduct experiments to validate targets, conduct chemical screening to identify attractive lead compounds, and nominate moieties for development. The development units take the output from discovery and test candidate drugs. Initially, they conduct many preclinical studies assessing toxicology, pharmacology, and metabolism of the potential medicine. Then, in another "hand off," **clinical trials** are initiated in three main phases. In phase I, the drug is administered to a small number of (usually healthy) individuals with the expressed goal of establishing its safety. In phase II, the drug is used in a larger number of patients with the disease to be treated to test for preliminary efficacy. In phase III, the drug is used in a very large

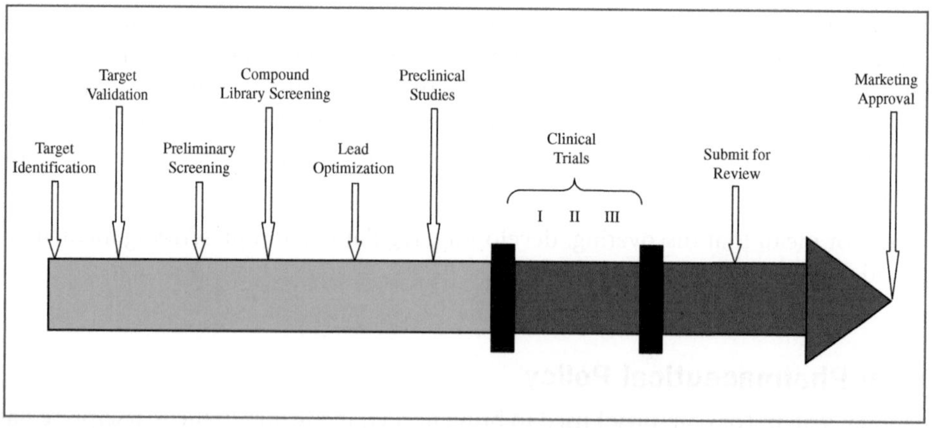

FIGURE 19.6
Diagram of the research and development pipeline used in the biopharmaceutical industry. The discovery activities are shown in blue, the development ones in red. Individual steps are depicted by narrow vertical arrows, the clinical trials by the bracket containing the three phases (I, II, and III). The steps in the blue section are target identification, target validation, preliminary screening, compound library screening, load optimization, and preclinical studies. The steps in the red section are clinical trials (I, II, and III), submit for review, and marketing approval. The first vertical bar denotes submission to the FDA of an investigational drug application (IND); the second bar denotes submission to the FDA of a new drug application (NDA).

patient population—generally in randomized, placebo controlled, double-blind, and prolonged studies to establish efficacy and safety in a statistically validated way. The FDA is involved at several places in this pipeline: it often provides guidance during the discovery phase, it must approve an **Investigational New Drug Application (IND)** before clinical testing is begun, and it examines the voluminous application to market a new drug called an **NDA** (for **New Drug Application**) after completion of the phase III trials. Only after FDA approval can the drug be marketed and sold in the US.

This lengthy pipeline is replete with risk, delay, extraordinary cost, and disappointment. Generally, it takes 8–12 years to move through it from idea generation to drug approval. Only about 1 in 10 drugs that go into preclinical testing enters clinical trials. Only about 1 in 8 drugs that start clinical trials makes it to market. Only about 1 in 2 drugs that enter phase III is finally approved. This explains why companies in the biopharmaceutical industry estimate that it costs them more than 1 billion dollars for every new drug that comes to market.

Emergence of Genetic "Thinking"

Because pharmaceutical R&D is, by definition, applied research, it has generally followed by some years the path blazed by basic and disease-oriented scientists in academia. For example, the remarkably productive antibiotic and antiviral era in industry followed by about 20 years the work done in academia to identify the biology of pathogenic organisms. So, too, did understanding of cholesterol metabolism prefigure the development of statins and identifying the importance of serotonin in depression lead to the development of the selective serotonin reuptake inhibitors (SSRIs). Looked at in this way, it is not surprising that the revolution in understanding human genetic disorders in the 1960s through the 1980s did not influence pharmaceutical R&D until well into the 1990s. But much has changed since then. Every large company has ramped up its expertise in genetics and genomics—particularly since the HGP provided the entire sequence of the human genome in 2003. Now, it is widely held that information about genes will fundamentally change how drug targets are identified, how they are tested, how clinical trials will be conducted, and for whom new medicines will be marketed. In other words, genetic thinking is a critical route to competitive advantage in industry and to personalized medicine—often reduced to the slogan: "the right medicine for the right person at the right dose."

PHARMACOGENETICS

The concept of individual uniqueness carries with it the idea that each person will respond to a pharmaceutical in his or her own way—not shared exactly with anyone else. This view, born from Garrod's notion of chemical individuality more than a century ago, gave rise to the field of **pharmacogenetics** in the 1950s and 1960s. Pharmacogenetics is defined as the study of the genetic basis for differences in response to pharmaceuticals. Implicit in this definition is the view that genes influence all aspects of one's response to drugs: absorption and distribution, metabolism and excretion, effect on a target, and balance between efficacy and toxicity.

The field of pharmacogenetics can be divided into two parts: genetic factors that control **pharmacokinetics**, defined as the rate at which the recipient's body absorbs, transports, metabolizes, and excretes the drug; and **pharmacodynamics**, defined as the effect of the drug on its target, such as an enzyme, receptor, transcription factor, or oncogene. We'll illustrate the distinction between pharmacokinetics and pharmacodynamics using the widely used anti-epilepsy drug phenytoin (Dilantin®). As shown in Figure 19.7, many steps affect the pharmacokinetics of orally used phenytoin: its absorption from the GI tract; its distribution in blood; its metabolism and degradation in the liver; its passage through the blood–brain barrier; and its elimination (or, more precisely, the elimination of its metabolite) via the kidney. Phenytoin's pharmacodynamic features are referable to its therapeutic targets in the brain—sodium and potassium channels that regulate flow of electrical current across brain cell

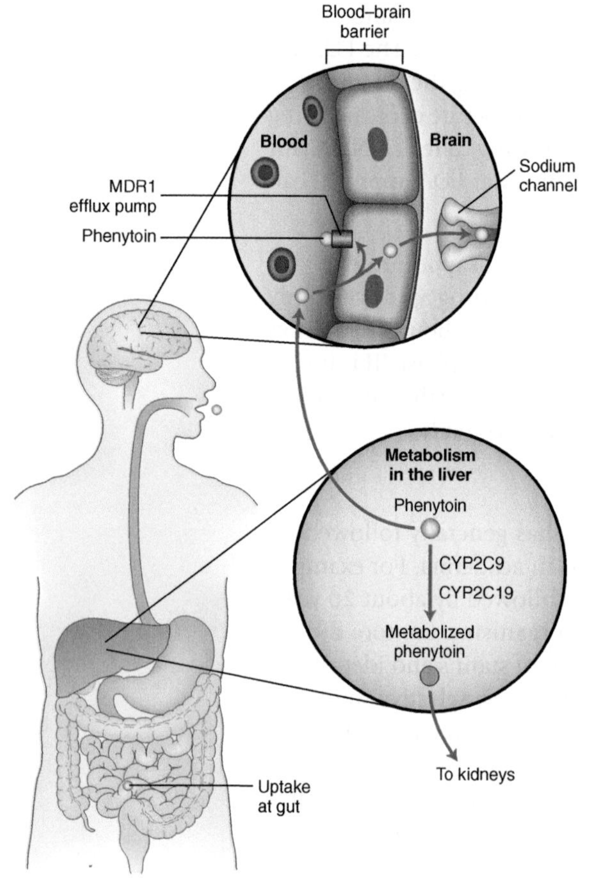

FIGURE 19.7
Schematic representation of metabolic pathway for phenytoin from ingestion to pharmacologic effect in brain. Critical steps are: absorption from intestine; transport in blood; metabolism in liver; passage through blood–brain barrier; and action on cerebral sodium channels. Abbreviations: CYP2C9, CYP2C19—cytochrome P450 enzymes; MDR1—multiple drug resistance efflux pump.

membranes. Phenytoin's efficacy as an antiseizure medication depends on integrating all of its kinetic and dynamic properties in a beneficial way and avoiding the toxicity (highlighted by cardiac arrhythmias or central nervous system depression) not infrequently encountered. A complete understanding of the genes regulating the many steps identified above would assist neurologists in their use of phenytoin, but such understanding is not yet at hand. We know how phenytoin is carried in blood (bound to serum albumin), and the enzymatic basis for its metabolism in the liver (by two cytochrome P450 enzymes to be discussed subsequently). We lack, however, a paradigm that would enable more effective use of phenytoin at the level of the individual patient.

As of this writing, we recognize between 40 and 50 situations in which particular genes (and mutations of them) are responsible for clinically important variation in response to medications. A subset of these is presented in Table 19.2, illustrating that variation in drug transport, metabolism, and target effect has been validated in one or many situations.

Each example shown here affects a minority of individuals, but, as is true with inborn errors discussed in Chapter 12, frequency is not of interest to an individual with special needs. Two examples: possessing the TT variant of the gene coding for the *MDR* gene required for the transport of phenytoin across the blood–brain barrier enhances the efficacy of phenytoin; and deficiency of the thiopurine methyltransferase enzyme required for the breakdown of the anticancer drug, methotrexate, can result in fatal bone marrow suppression due to prolonged and elevated concentrations of the drug. From the long list of genetic variants producing clinically significant effects, we will discuss two: one involving pharmacokinetics and the other, pharmacodynamics.

TABLE 19.2 Some Clinically Significant Pharmacogenetic Variants

Gene	Allele	Phenotype
Multiple Drug Resistance-1	C3435T	TT epileptics respond better to phenytoin
Pseudocholinesterase	Several	Prolonged apnea following succinylcholine
CYP2D6	Several	Dyskinesia on antipsychotics
	Several	Poor analgesic effect of codeine
CYP2C19	Several	Poor antiplatelet effect of clopidogrel
N-acetyltransferase2	Several	Neuropathy during treatments of tuberculosis with isoniazid
Thiopurine methyltransferase	Several	Bone marrow suppression during treatment with methotrexate
Bradykinin receptor B2	C58T	T allele predisposes to cough associated with use of an angiotensin converting enzyme (ACE) inhibitor
Serotonin transporter	Promoter indel	Long promoter associated with better response to Prozac
SLC01B1 anion transporter	SNP	Predisposed to skeletal myopathy during treatment with statins
G6PD	Many	Hemolytic anemia from antimalarial drugs

Cytochrome P450s

The **cytochrome P450s (CYPs)** are the most important class of drug-metabolizing enzymes in humans. The CYPs are a family of 56 different enzymes, each encoded by a different gene. All CYPs are heme-containing proteins whose most regular effect is to add hydroxyl (OH^-) groups to other molecules, including pharmaceuticals. Such addition, usually the first step (or phase I) of drug metabolism, is generally followed by other steps of activation or inactivation (phase II). The CYPs are grouped into 20 families according to amino acid sequence homology. Three of these families will be discussed further.

CYP1, CYP2, AND CYP3

Each of these gene families contains enzymes that act on multiple different substrates (accordingly said to be **promiscuous**). They metabolize a wide array of substances that enter the body from the outside, including pharmaceuticals. Six genes (*CYP1A1, CYP1A2, CYP2C9, CYP2C19, CYP2D6*, and *CYP3A4*) are unusually important pharmacogenetically because the six enzymes they code for catalyze phase I metabolism of more than 90% of commonly used medications.

CYP1, CYP2, and CYP3 enzymes do not affect drug metabolism equally. As shown in Figure 19.8, CYP3A4 is involved in phase I metabolism of more than 40% of all commonly used drugs, the remaining percentage being accounted for by a relatively small number of other CYPs. Another point of importance concerns the population distribution of *CYP* alleles and its clinical significance. *CYP2D6* has a slow-metabolizing phenotype that is present in 1 in 14 whites but rare in Asians and virtually absent in Native Americans and Pacific Islanders. Further, slow-metabolizing phenotypes for *CYP2C19* are found in 16% of Asians but only 3% of whites. Such differences surely affect the use of globally distributed and globally used medicines.

Many of these CYP genes have multiple alleles (that is, are polymorphic) with varying effects on enzyme activity. Some result in increased activity (copy number), others in decreased or no activity (missense, nonsense, frameshift, splicing). Accordingly, this allelic variation produces quantitative effects on drug metabolizing ability. *CYP2D6* is a good example, because it accounts for phase I metabolism of more than 70 drugs. The *CYP2D6* locus has 26 alleles that affect enzymatic activity variably—some increasing it and others decreasing it. As shown in Table 19.3, different pairs of alleles produce different classes of metabolizer phenotypes, ranging from ultrafast through normal to poor. Those people homozygous for the normal (or wild-type) allele are said to be normal metabolizers. Ultrafast metabolizers having one wild-type allele and one gain-of-function one may be at risk of being undertreated because they break down or excrete the medicine so rapidly that it doesn't reach the needed concentration. Conversely, poor metabolizers (that is, those partially or completely deficient in the required CYP) are at risk of drug toxicity because the drug will accumulate to dangerously high concentrations.

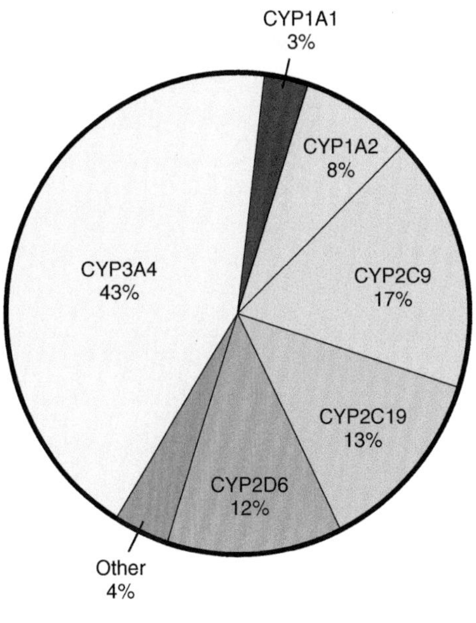

FIGURE 19.8
Contribution of various cytochrome P450 enzymes in phase I metabolism of commonly used medicines. See text for additional details.

TABLE 19.3 Metabolizer Phenotypes Arising from Various Combinations of CYP2D6 Alleles

	Allele on one chromosome			
Allele on other chromosome	**Wild-type**	**Reduced**	**Absent**	**Increased**
Wild-type	Normal			
Reduced	Normal	Poor		
Absent	Normal	Poor	Poor	
Increased	Ultrafast	—	—	—

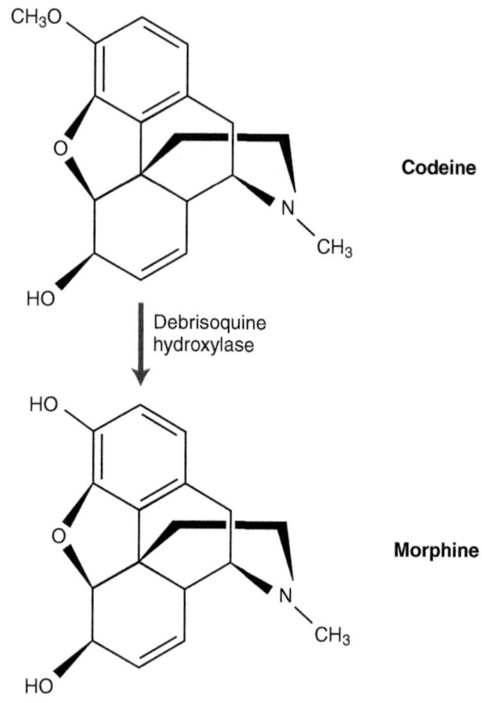

FIGURE 19.9
Conversion of codeine to morphine is catalyzed by debrisoquine hydroxylase, a CYP enzyme.

CYP2D6 AND OPIOID ANALGESIA

A good example of the pharmacogenetic significance of the CYPs concerns the opioids. Opioids are widely prescribed to relieve pain. Codeine, perhaps the most widely prescribed opiate for moderate to severe pain, is not active as ingested. It must be converted to morphine before exerting its analgesic properties. The enzyme that converts codeine to morphine (Figure 19.9) is debrisoquine hydroxylase, a member of the CYP2D6 family. The poor-metabolizer phenotype for *CYP2D6* (found in as many as 10% of North Europeans) results in failure to convert codeine to morphine, thereby producing little pain-relieving effect of codeine. Conversely, the ultrafast-metabolizer phenotype leads to excess morphine formation, thereby risking dangerous or even lethal toxicity. This example underscores that whereas CYPs generally act to detoxify medicines, in some cases they serve to activate ingested drugs.

CYP2C19 AND CLOPIDOGREL

Another example of drug activation by CYPs has to do with clopidogrel (Plavix®). Clopidogrel, the second most widely sold drug worldwide, is an antiplatelet agent shown to be effective in preventing heart attacks and strokes. Clumping of platelets is generally the first step in formation of an intra-arterial thrombus (clot). It has been known for some time that a significant minority of patients is resistant to the beneficial effects of clopidogrel. The mechanism for this clinical failure involves CYP2C19 (Figure 19.10). Clopidogrel is a prodrug that is converted to its active moiety in the liver; CYP2C19 is the enzyme catalyzing this conversion. Loss-of-function alleles of this *CYP* are responsible for poor-metabolizer phenotypes resulting in little conversion of clopidogrel to its active form—thus negating the value of the drug.

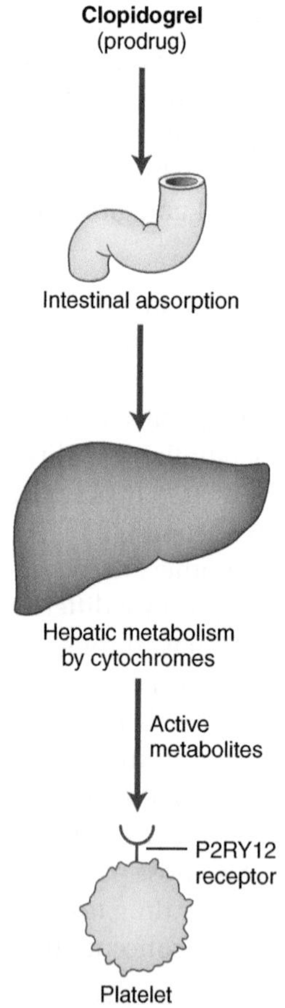

FIGURE 19.10
Pathway of metabolism of clopidogrel, a widely used antiplatelet agent. Clopidogrel is a prodrug which must be converted to its active form in the liver. It then acts on a receptor protein (P2RY12) found on the cell surface of platelets.

CYP GENOTYPING: AMPLICHIP™

Given that CYP enzymes are involved so often in the first, and often rate-limiting, step in the metabolism of myriad commonly used medicines, it seems logical that safe and effective use of such medicines should take CYP genetics into account. With notable exceptions, however, this was not the case until recently. A major pharmaceutical company (Roche) began to market AmpliChip in 2004. Using the kind of array technology described in Chapters 6, 13, and 16, AmpliChip distinguishes among 27 different variants of *CYP2D6* and three variants of *CYP2C19*. The goal is to classify individuals as slow, intermediate, fast, or ultrafast metabolizers of many drugs. CYP genotyping among different populations gave interesting results. Using *CYP2D6*, the frequency of slow metabolizers was higher in whites than in Africans or Asians. Using *CYP2C19*, the frequency order of slow metabolizers was reversed—highest in Asians, intermediate in Africans, lowest in whites. Use of AmpliChip by practicing physicians has been limited to date, largely because the FDA has not validated its claims. When and if it does so, it seems likely that this test, or one like it, will become a routine part of laboratory assessment of children and adults.

Glucose-6-phosphate Dehydrogenase (G6PD)

The first and still most interesting example of pharmacogenetic variation affecting pharmacodynamics (rather than pharmacokinetics) concerns the enzyme **glucose-6-phosphate dehydrogenase (G6PD)**. This enzyme catalyzes a critical step in an alternate pathway of glucose metabolism called the **hexose monophosphate shunt**. G6PD deficiency affects more than 400 million people worldwide, making it the most prevalent single-gene defect known. The story of its scientific and clinical unraveling is worth telling.

THE MILITARY AND MALARIA

During World War II, American soldiers fighting in Africa and Asia were beset by malaria, which caused great health problems for these troops. American scientists were tasked by the federal government with developing new drugs to combat malaria, and they did so. One of these drugs—a very effective one—was primaquine (a close chemical relative of the most widely used antimalarial at the time, chloroquine). During the Korean War in the 1950s, malaria was again a problem for American troops. Accordingly, primaquine was routinely prescribed to prevent (and treat) malaria. It soon became apparent that more than 10% of African American soldiers suffered ill effects from primaquine. They developed acute, generally **self-limiting** (cured by itself, without treatment) anemia due to rapid destruction of their RBCs (a condition referred to as **hemolytic anemia**). A smaller number of white soldiers (mostly Italian and Greek) also developed hemolytic anemia. The government instituted a crash program to investigate the cause of this worrisome problem and to define a means of dealing with it. The result: a brilliant story of medical detective work with worldwide ramifications.

BIOCHEMICAL UNDERSTANDING

When RBCs from healthy soldiers and those developing hemolytic anemia were studied enzymatically, deficiency of G6PD activity was found in the affected soldiers. As shown in Figure 19.11, G6PD catalyzes the conversion of glucose-6-phosphate to 6-phosphogluconate. One of the products of this complex reaction is the formation of nicotinamide adenine dinucleotide phosphate (NADPH). NADPH is an important intracellular cofactor that, in RBCs, protects the cell membrane from oxidative damage by regenerating reduced glutathione from oxidized glutathione. In G6PD-deficient RBCs, oxidant drugs like primaquine deplete the cell of reduced glutathione, leading to oxidative damage to the RBC membrane and to hemolysis. Other substances (sulfa drugs, mothballs) damage RBCs from G6PD-deficient individuals in the same way.

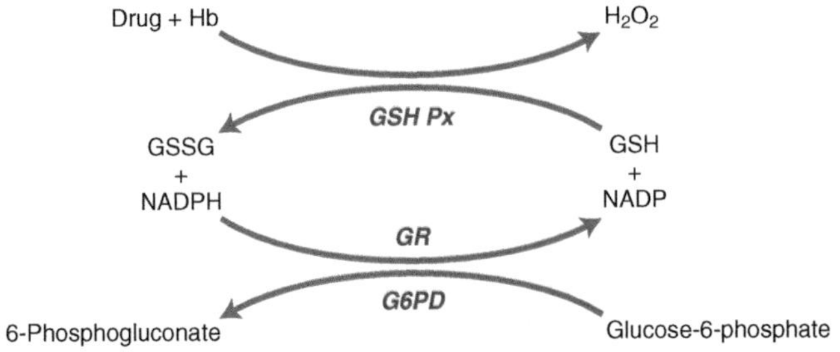

FIGURE 19.11
Role of glucose-6-phosphate dehydrogenase (G6PD) in protecting the red blood cell from oxidative damage caused by drug administration. NADPH produced by G6PD catalysis increases intracellular concentration of reduced glutathione (GSH), which in turn protects the cell membrane from oxidative damage resulting from formation of peroxide (H_2O_2).

In addition to reduced activity, the G6PD from those with G6PD deficiency had an altered electrophoretic mobility (called A^-) that distinguished it from the normal form (called B). Why were the effects of G6PD deficiency limited to RBCs? The answer: because RBCs lack a nucleus and ribosomes, they cannot repair or replace proteins as cells and other tissues can. Why was the anemia self-limited? The answer: because newly formed RBCs continuously produced in the bone marrow have sufficient reduced glutathione to protect them from destruction by oxidant drugs. Why was the hemolytic anemia observed in Greek or Italian soldiers more severe—even lethal—than that in African American ones? The answer: their mutant G6PD (designated "the Mediterranean variant") was a different protein with distinctly less residual activity. (Herodotus foreshadowed this finding in the 5th century BC when he recorded an untoward response of some Greeks after ingesting fava beans. Fava beans will cause oxidant injury to RBCs in the absence of G6PD.)

GENETICS

Family studies conducted in the wake of the Korean War experience revealed that G6PD deficiency was an X-linked trait. The gene for G6PD was subsequently mapped to the proximal portion of the long arm of the X chromosome. The *G6PD* gene is 19 kb long, contains 19 exons, and encodes a protein of 514 amino acids. The A^- variant results from a missense mutation causing a single amino acid substitution. A different missense mutation is responsible for the Mediterranean variant. By now, more than 400 different allelic variants of G6PD have been described—some of clinical significance, many not. Not surprisingly, different populations have different frequencies of these G6PD variants: 60% of Kurdish Jews have G6PD deficiency; 30% of some Sardinian villagers do, as well. All told, more than 400 million people worldwide have G6PD deficiency, reflecting the presence of the many variants just mentioned.

HETEROZYGOTE ADVANTAGE

If one looks at a world map showing frequencies of G6PD deficiency (Figure 19.12), it resembles that shown for sickle cell anemia, and for the same reason. RBCs from

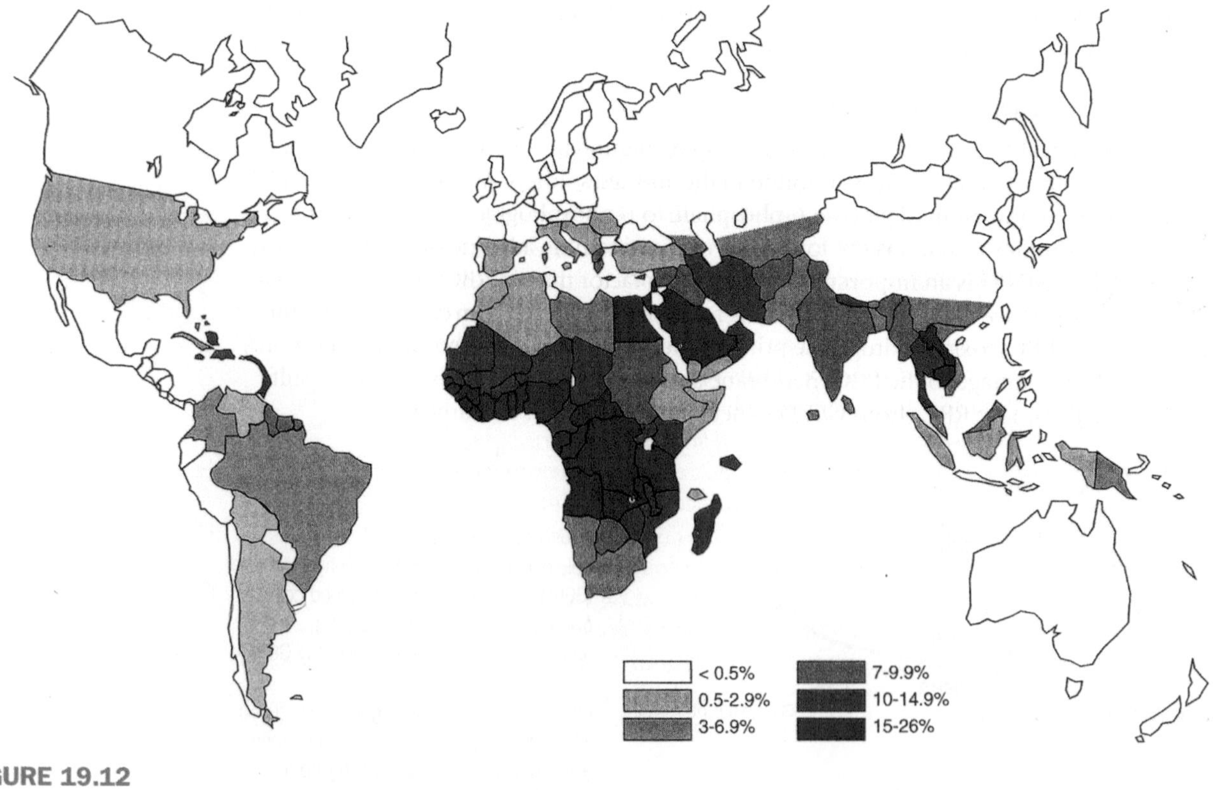

FIGURE 19.12
Worldwide distribution of G6PD deficiency, showing estimated population frequencies.

G6PD-deficient people are inhospitable to the plasmodia causing malaria. Hence, a balanced polymorphism exists in which resistance to malaria in heterozygous females maintains a high frequency of mutant alleles in the entire population.

Limited Application

From the foregoing descriptions, it should be clear that, in a small number of instances, understanding of pharmacogenetic principles can and does affect use of medicines. That it so rarely does so, however, reflects a number of forces: the slow penetration of genetic principles into pharmaceutical R&D; the slow recognition by the FDA that genetic variation should become a regular part of defining the safety and efficacy of new drugs; and the lack of understanding by the lay public that their genetic makeup will influence their individual response to a prescribed medicine. We now turn to a more recent version of pharmacogenetics—pharmacogenomics—that may expand the utility of genetic principles into medical practice.

PHARMACOGENOMICS

In 1997, a new term appeared in the scientific literature as the HGP accelerated toward completion. The term: **pharmacogenomics**. The simplest definition of this word is to relate it most closely to the discipline from which it is derived—pharmacogenetics. Accordingly, pharmacogenomics is the study of the genomic basis for differences in response to drugs. Rather than focusing on single genes, as pharmacogenetics has from its inception, pharmacogenomics uses genomic technologies such as SNPs to correlate particular variants with phenotypic responses to drugs. Pharmacogenetics has almost always been retrospective in approach. That is, it has begun with some observed untoward response to a drug in a small group of patients and then attempted to explain that response by examining candidate genes implicated in the drug's transport, metabolism, or effect on a target. Pharmacogenomics tries to answer the question differently. Generally, it identifies a group of people with an untoward response to a drug and compares their SNP profiles (or mitochondrial DNA sequences) with the SNP profiles in individuals who have not exhibited untoward effects from the drug. In essence, it uses the same approach as that to identify genes conferring susceptibility to common disorders. The long term goal of pharmacogenomics—like pharmacogenetics—is to identify genetic variants that can be used prospectively to predict who will respond beneficially to a particular medicine, and who won't; who will develop an ADR, and who won't.

In the heady days following completion of the HGP, pharmacogenomics was touted as the essence of genomic medicine—meaning that it would soon revolutionize medical practice. That has proven to be wildly hyperbolic for several reasons.

- Pharmaceutical companies are interested in "blockbuster" drugs with sales of more than 1 billion dollars annually, but pharmacogenomic approaches will, by definition, segment the market.
- Clinical trial design would have to be totally re-engineered such that genotypic assessment would precede sorting of the clinical population to be treated into those with a "susceptibility" profile from those without it.
- R&D costs would rise initially because the technology needed for genomic testing would need to be purchased, validated, and added to steps used in drug research and development.

As of now, there are less than a handful of examples where the pharmacogenomic approach has been shown to be successful. Two of these are worth examining because they foreshadow events that will surely take place during the coming decade and beyond.

Statins and *SLCO1B1*

In 2008, the first genomic study demonstrating a statistically significant association between a genetic locus and toxicity of a drug was reported. This was the clinical setting. Statins are perhaps the most widely used drugs in the world today, with proven efficacy in preventing heart attacks and strokes by lowering serum LDL cholesterol (see Chapter 12). Fewer than 5% of statin-treated patients develop severe skeletal muscle pain in the arms and legs (called skeletal myopathy), necessitating discontinuation of the drug. An even smaller fraction has a much more severe reaction: widespread breakdown of skeletal muscle which may even be lethal.

A study was conducted using 85 statin-treated patients who had developed myopathy, and 90 statin-treated controls with no adverse response to the drug. A genome-wide association study was performed using 300,000 SNPs per patient and control. A single informative SNP was found on chromosome 12. This SNP was shown to be within the coding sequence for the *SLC01B1* gene—an anion (negatively charged) transporter regulating statin uptake by the liver. This suggested that the variant form of *SLC01B1* resulted in statin accumulation in liver and muscle, thereby causing or predisposing to the myopathy.

Subsequently, GWAS were performed in a much larger group of statin-treated patients with myopathy and in controls. The association with *SLC01B1* was confirmed. Further, those homozygous for the variant developed more severe myopathy than did heterozygotes. From these studies, it would seem sensible to test for the variant before initiating treatment with statins. We are not aware that clinically useful tests for this variant have been developed, but the example makes clear the direction that will surely be followed for this and other drugs.

Aminoglycoside Antibiotics and Mitochondrial DNA

Young children with life-threatening bacteremia (bloodstream infection) are often treated with aminoglycoside antibiotics particularly effective against gram-negative organisms (those which fail to retain crystal violet dye in the traditional Gram-staining protocol). Hearing loss, even complete deafness, and kidney damage are common side effects of these antibiotics. Following complete sequencing of the human mitochondrial DNA sequence in 2008, small studies employing complete mitochondrial sequencing suggested that a missense mutation at nucleotide position 1555 (adenine to guanine, abbreviated *A1555G*) was strongly associated with toxicity to the aminoglycosides.
In a subsequent study, 9,371 children between ages 7 and 9 years were tested for the *A1555G* variant. A population prevalence of 0.19% (that is, 1 in 520 children) was found. Based on this work, it seems prudent to genotype children before treating them with aminoglycosides.

DIRECT-TO-CONSUMER GENOMIC TESTS

In 2006, a new industry was born. Some refer to it as "personal genome services," others as **direct-to-consumer genome testing**. We will use the latter name to describe this commercial enterprise. In the past, genetic testing was either mandated by law (neonatal screening) or ordered by physicians as part of an individual's medical care. In both instances, results were interpreted with professional guidance. DTC genome testing differs fundamentally in that the consumer (read: patient) subscribes to the service online and receives the results online without any intervention by a healthcare provider. The stated goal of DTC genome testing is to inform individuals of their genetic risks. More than 100,000 people have purchased DTC genome testing thus far—a sufficient number to allow for a progress report.

Testing Protocol

Of the more than 20 companies that provide DTC genome testing, a small minority sequence all or a part of a customer's genome, at a cost of many thousands of dollars. The majority of the companies, in contrast, base their assessments on SNP-based data using 300,000 to over 1 million SNPs covering the entire genome. In all instances, the protocol proceeds as follows:

- The consumer subscribes to a company's services and pays a fee (generally 200—1,000 dollars for SNP-based DTC genome testing).
- The consumer then submits a sample (saliva or cheek swab) from which the company extracts DNA.
- The DNA is then tested, and the results are returned to the consumer in the form of risk assessments (increased risk of some trait or disorder, typical risk, or decreased risk).
- In some instances the company provides interpretation of test results by a genetic counselor; in other instances no professional interpretation is offered.

Information Provided

All of the companies performing DTC genome testing emphasize that they are assessing risk or susceptibility, rather than making diagnoses. For instance, they may say that a person has a 50% risk of developing age-related macular degeneration compared to a population risk of 10%. As noted in Table 19.4, the individual companies differ considerably in the range of services offered. Three companies are profiled as examples.

Each tests for disease susceptibility—the number of conditions ranging from 28 to 92. Two provide pharmacogenetic information on a range (12—18) of medicines in wide use. Two supply information about benign traits (such as eye color, baldness or character of ear wax). One tests for carrier status in 24 autosomal or X-linked disorders. Two provide information about ancestry. In all cases, subscribers are provided an extensive general review of human and medical genetics.

TABLE 19.4 Some Companies Offering Direct-to-Consumer Genomic Testing

	Services Offered*				
Company	**Disorders**	**Traits**	**Carrier**	**Drug Response**	**Ancestry**
A	54	4	—		Yes
B	28			12	No
C	92	44	24	18	Yes

**Numbers indicate number of conditions tested for in each category.*

IMPLICATIONS: DIRECT-TO-CONSUMER GENOMIC TESTING

In Her Own Words: Diane's Story

I was quite excited about the prospect of having my "genes done." Several companies had started up, offering to determine one's ancestry, chances of having certain phenotypic traits, susceptibility to drugs, or the likelihood of having certain diseases, using SNPs from DNA collected from one's saliva. I ordered a kit that arrived promptly and spent half an hour spitting into a test tube to fill it to the demarcated line. The chance of looking both into the past and into the future appealed to my life-long desire to "know myself."

Several weeks later, I received an email saying my results had arrived. I knew something about my parents' ancestry—my mother's family had come from Wales in about 1850, and my father's from Russia in the 1880s. Now, however, there was information filling in thousands and thousands of years.

(Continued)

IMPLICATIONS: DIRECT-TO-CONSUMER GENOMIC TESTING—Cont'd

Of course, everyone's earliest ancestors came from Africa. But what excitement to find out that my mother's ancestors were probably Vikings! Because I did not possess a Y chromosome, I was provided no information about my father's ancestors. I promptly gifted my brother with a kit, a somewhat self-interested gift for me, as his results would work for me as well as him. Our father's male ancestor had come from North Africa and Southern Europe.

When it came to my likely phenotypic traits, there were few revelations. Yes, I was most likely to have brown eyes and yes, I have dry ear wax rather than wet. I was more likely a sprinter than a long-distance runner. There was one disagreement with actuality—preliminary research on one marker indicated that I would probably have a lower than average "good" cholesterol level. Since I have an abnormally high level of good cholesterol, I concluded that this marker gave meaningless information.

Almost all my disease results were OK. I wasn't particularly susceptible to most cancers and I had almost average risk of coronary heart disease. I was more susceptible to the drug Warfarin, a blood thinner, than average. There was one unusual result that caught my eye: I did possess a SNP for factor V Leiden (FVL). What was this? After reading in more detail, I discovered that in 1994, scientists in the Netherlands published the first paper describing a change in the gene encoding the blood clotting protein, Factor V, which can increase the risk of blood clots forming in veins. As a result, I have more than three times the average risk for deep vein thrombosis (DVT) or pulmonary embolism (PE), where a blood clot forms in a vein deep in the legs and may travel to the lung. Not good.

The phrase "economy class syndrome" has been coined to describe the occurrence of deep vein thrombosis in people who have spent long hours in cramped airline seats. To prevent such an occurrence, it is strongly suggested that such people—particularly those with Factor V Leiden, like me—wear tight stockings to help with circulation of the blood in the legs when traveling in airplanes. Also, I've been told to get up and move about frequently. I promptly bought three pairs of tight, knee-length stockings in different colors for flights. Though it is a disturbing condition to have, I am grateful for the knowledge. Also, I have a good reason to upgrade to business class.

Present Impact of DTC Genome Testing

BEST CASE

Under optimal circumstances, a person would learn about his or her genetic susceptibilities and take steps to mitigate personal risks. Some examples: a woman with a mutation in clotting factor V that predisposes to venous thrombosis and pulmonary embolism shouldn't take estrogen and should wear pressure stockings when flying; a man with a significantly increased risk for age-related macular degeneration (AMD) should not smoke, because smoking is a well-established risk factor for AMD; a young married couple each found to be a carrier for cystic fibrosis should be aware that they have a one in four chance with each pregnancy of having a child with that disorder. Further, the explanatory information about genetics provided on each company's website offers a valuable "primer" concerning terms and concepts. Finally, the information can (and usually should) be shared with one's physician or genetic counselor to avoid misunderstanding and erroneous action.

WORST CASE

There are several reasons why DTC genome testing could be harmful rather than helpful.

- If the test is analytically invalid, the customer will be misled, thereby leading to unnecessary anxiety or needless environmental modification.
- If the customer is unable to interpret the result, emotional harm may be caused. For example, a relative risk of 1.5 may be trivial for a condition with a frequency of 1 in 20,000. Proper interpretation requires knowledge of probability.
- If the customer's physician is not conversant with the condition about which the customer has received information, he or she may doubt or dismiss genuine increased risk.

Prospects

As we learn more and more about the genes that cause and predispose to disorders, the demand for information will surely increase. This behooves all involved parties to take steps to ensure that the information provided by DTC genome testing is valuable, not injurious. The federal government, through the FDA, should see to it that all companies providing this kind of service are certified through the Clinical Laboratories Improvement Amendments (CLIA) rather than being unregulated because they claim to provide only "information," not "service." Physicians should become aware of the DTC genome testing industry so that they can help their patients make informed decisions and avoid unnecessary stress. The public should take every opportunity to educate themselves about the role of genes in health and disease, so that they can make optimal use of DTC genome testing (just as they are expected to do with direct-to-consumer information about medical devices and pharmaceuticals). It is toward these ambitious goals that DTC genome testing should reach. To the extent that this book facilitates the pursuit of and arrival at these goals, and of the many other bits of information seeking to account for the genetic basis of individual uniqueness, it will have been deemed successful by the authors.

REVIEW QUESTIONS AND EXERCISES

1. Choose the phrase in the right column that best matches the term in the left column.

a. pharmacokinetics	1. regulates marketing of pharmaceuticals
b. Czar Nicholas; Copernicus	2. family of enzymes responsible for metabolism of most drugs
c. genomic medicine	3. opened field of pharmacogenetics
d. cytochrome P450 (CYP)	4. employs SNPs to identify individual susceptibilities
e. DNA fingerprinting	5. studies genetic basis for differences in response to pharmaceuticals
f. Amplichip™	6. skeletal remains identified by DNA fingerprinting
g. direct-to-consumer genomic testing	7. examines effect of drug on target
h. FDA	8. idea that medical diagnosis and treatment will rely on genomics
i. minisatellites	9. among 10 leading causes of death
j. pharmacodynamics	10. chromosomal regions containing many polymorphic repeats
k. adverse drug reactions (ADR)	11. tool to genotype CYPs
l. G6PD deficiency	12. employs VNTRs and electrophoresis
m. pharmacogenetics	13. examines rate of drug absorption, transport and metabolism

2. Minisatellites are employed widely in DNA fingerprinting.
   a. Why are minisatellites preferred over microsatellites?
   b. Describe briefly the difference between STRPs and VNTRs.
   c. Why have some victims of the tragic events occurring on September 11, 2001 not been identified with certainty, while Copernicus (who died in the 15th century) has been identified?

3. Below are shown gel patterns for a single VNTR polymorphism in three families (1–3) each containing a father (F), mother (M), and a child (C).

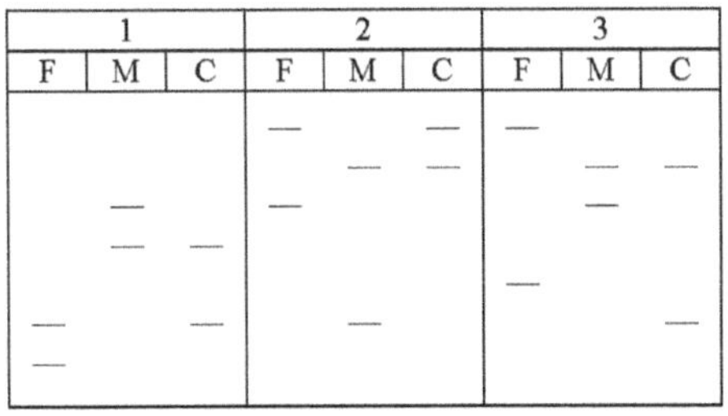

a. Why does the DNA from each person have two bands?
b. Would you be surprised to find that some other people had a single band?
c. If each of the fathers in these families said the respective children were not their biologic offspring, what would you tell them?

4. A VNTR polymorphism detects four different alleles, each with a frequency of 0.25. What fraction of people would be expected to be heterozygous at this locus?

5. Suppose that a young man previously in excellent health develops severe anemia soon after starting a newly marketed antihistamine used to treat hay fever.
   a. How would you determine whether the drug was responsible for the anemia?
   b. If you determined that this man was having an ADR, what would you do with this information?
   c. What would you say to the parents of this young man if they planned to sue the prescribing physician for malpractice regarding use of this antihistamine?

6. Cytochrome P450 enzymes (CYPs) figure prominently in any discussion of drug metabolism.
   a. What makes CYPs so important?
   b. Do they participate in pharmacokinetic or pharmacodynamic properties of pharmaceuticals?
   c. How could a single CYP be responsible for producing a drug overdose of two different pharmaceuticals by two different mechanisms?
   d. CYPs are evolutionarily old; pharmaceuticals are young. What does this tell us about CYPs?

7. SNPs and GWAS are proposed to be the leading edge in the development of "personalized medicine."
   a. What is "personalized medicine"?
   b. Some people complain that we are still far away from personalized medicine even a decade after completion of the HGP. What is your response to them?
   c. What does direct-to-consumer genomic testing (DTC genome testing) have to do with personalized medicine?

8. Some people say that DTC genome testing is beneficial, others say it is harmful.
   a. What is DTC genome testing, and how is it done?
   b. List three possible benefits of DTC genome testing.
   c. List three possible harms from DTC genome testing.
   d. Would it be unethical to offer DTC genome testing for Alzheimer disease? For diabetes mellitus? For Hodgkin's disease?

GLOSSARY

Acceptor splice site The boundary between the 3′ end of an intron and the 5′ end of the following exon.
Accutane embryopathy (AE) The pattern of birth defects that may be caused in an embryo after exposure to Accutane during pregnancy.
Acetylation A reaction that introduces an acetyl function into a chemical compound.
Achondroplasia A disease of the skeletal system inherited in humans as an autosomal dominant trait and characterized by insufficient growth of the long bones.
Acrocentric chromosome A chromosome where the centromere is found near one end.
Adaptation The evolutionary process whereby a population becomes better suited to its habitat over many generations.
Adjacent segregation Segregation of homologous or non-homologous centromeres during meiosis in a carrier of a reciprocal or Robertsonian translocation such that unbalanced gametes with duplications or deficiencies are produced.
Adrenocortical stimulating hormone (ACTH) Pituitary hormone that stimulates the secretion of adrenal cortical steroids and the induced growth of the adrenal cortex.
Adult stem cells (ASCs) Undifferentiated cells found among differentiated cells in a tissue or organ that can differentiate to yield some or all of the major specialized cell types of that tissue or organ.
Age-related macular differentiation (AMD) A degenerative condition of the retina which usually affects older adults and results in a loss of vision in the center of the visual field.
Alkaptonuria A rare, inherited disorder of phenylalanine and tyrosine metabolism where a person's urine turns a dark brown to black color when exposed to the air. It was the first autosomal recessive trait discovered in humans.
Allele frequency The proportion of a given allele among all the different forms of a gene in a population.
Allele An alternative form of a given gene.
Alpha-fetoprotein (AFP) A plasma protein produced by the yolk sac and liver during fetal development in humans, encoded by the *AFP* gene.
Alpha-satellite DNA Highly repetive DNA sequences found in mammalian centromeres.
Alpha-thalassemia A recessively inherited blood disorder characterized by reduced synthesis of α-globin chains, and caused by mutations of the A1 and A2 globin genes.
Alternate segregation One of two patterns of segregation resulting from the normal disjunction of homologues during meiosis I.
Alternate splicing Production of different mature mRNAs from the same primary transcript by joining different combinations of exons.
Alu repeats DNA sequences, each about 300 bases long, that are dispersed throughout the genome and constitute nearly 10% of total genomic DNA.
Alzheimer disease The most common form of dementia, usually diagnosed in people over the age of 65, which is progressive, incurable, and ultimately lethal.
Amniocentesis A medical procedure in which a sample of amniotic fluid containing fetal cells is taken from a pregnant woman in order to detect abnormalities in the fetus.
Amplification Multiple copies of a gene or DNA sequence in a chromosome.
Amyloid precursor protein An integral membrane protein expressed in many tissues and concentrated in the synapses of neurons, which is the primary component of plaques found in the brains of Alzheimer disease patients.
Anabolism The metabolic process by which simple substances are synthesized into the complex materials of living tissue.
Anencephaly A catastrophic disorder that results when the neural tube fails to close during embryonic development, resulting in the absence of a major portion of the brain, skull, and scalp.
Aneuploidy Any chromosome number that is not an exact multiple of the haploid number for the species.
Angelman syndrome A neuro-genetic disorder characterized by intellectual and development delay, sleep disturbance, seizures, jerky movements, and frequently laughter or smiling.
Angiogenesis The physiological process involving the growth of new blood vessels from pre-existing vessels, a normal and vital process in growth and development and wound healing, as well as a fundamental step in the transition of tumors from a dormant state to a malignant one.
Anomalies Birth defects resulting from malformations, deformations, or disruptions.
Antibody A blood plasma protein produced in response to a specific antigen and capable of binding that antigen.
Anticodon A three-base unit in transfer RNA that is complementary to a codon in mRNA.
Apolipoprotein E (APOE) A type of protein that combines with fats or cholesterol to transport lipids through the lymphatic and circulatory systems. It binds to specific receptors on hepatic and peripheral cells.

Apoptosis Programmed cell death; an orderly process in which certain cells are eliminated at certain stages of development, controlled by specific cell death genes.

Archaea A group of single-celled microorganisms with no cell nucleus or any other membrane-bound organelles within their cells.

Association studies Studies of populations to determine if a marker (such as a single nucleotide polymorphism) is found more frequently in a group of patients than in matched controls.

Assortative mating Non-random mating between unrelated individuals with similar characteristics.

Autosomal dominant A pattern of inheritance in which an affected individual has one copy of a mutant gene and one copy of a normal gene on a pair of homologous autosomal chromosomes.

Autosomal recessive A pattern of inheritance in which an affected individual has two copies of a mutant gene on a pair of homologous autosomal chromosomes.

Autosome Any chromosome that is not a sex or a mitochondrial chromosome.

Bacteria Single-celled, prokaryote microorganisms; one of the three major kingdoms of living things.

Bacterial artificial chromosome (BAC) A stable cloning vector made by ligating a large fragment of DNA into the DNA of a bacterial cell.

Bacteriophage A virus that infects bacteria.

Balanced polymorphism A polymorphism maintained in the population by heterozygote advantage, allowing an allele, even a deleterious one when homozygous, to persist at a relative high frequency in the population.

Banding A characteristic staining pattern produced by several techniques which allows identification of individual chromosomes and their structural abnormalities.

Barr body A darkly staining body found in the interphase nucleus of certain cells of female mammals, consisting of a condensed, inactive X chromosome.

Base pair Two complementary nucleotide bases, as found in double-stranded DNA.

Benign tumor A tumor that does not invade its surrounding tissue, and lacks the ability to metastasize.

Biochemical genetics The branch of biology that deals with the formation, structure, and function of macromolecules essential to life, such as nucleic acids and proteins, and especially with their role in cell replication and the transmission of genetic information.

Bipolar disorder (BPD) A psychiatric mood disorder in which episodes of mania or hypomania (abnormally elevated energy levels and grandiosity) alternate with episodes of depression (pervasive sadness and loss of interest in life).

Birth defect Any defect present in a baby at birth, irrespective of whether the defect is caused by a genetic factor or by prenatal events that are not genetic.

Bivalent A pair of juxtaposed homologous chromosomes, as seen at metaphase in meiosis I.

Blastocyst An early stage of animal development where the initial ball of cells derived from the zygote secrete fluid and form a fluid-filled internal cavity within which is a separate group of cells called the inner cell mass.

Blastomere An early embryonic cell.

Blood group Genetically determined antigens on red blood cells.

B lymphocyte A white blood cell that matures in the bone marrow and strongly responds to antigens by differentiating into antibody-producing plasma cells.

Cancer A group of conditions in which cells grow inappropriately without restraint, lose their normal form and function, invade nearby tissue, and spread to remote sites.

Candidate gene analysis A method whereby a gene is mapped to a region known to be linked to a specific disease, the gene being known to code for a protein that is plausibly associated with the disease phenotype.

Carcinoma A solid cancer derived from epithelial cells.

Carrier An individual heterozygous for a particular mutant allele.

Catabolism The metabolic breakdown of complex molecules into simpler ones, often resulting in a release of energy.

cDNA Complementary DNA; DNA which has a base sequence complementary to that of its mRNA template, and contains no introns.

Cell communication Any of several ways in which living cells of an organism communicate with one another, whether by direct contact between cells or by means of chemical signals carried by neurotransmitter substances, hormones, and cyclic AMP.

Cell cycle regulator A gene whose product regulates the cell cycle.

Cell cycle The sequence of events from one mitotic cell division to the next, including all stages of interphase and mitosis.

Cell differentiation The process by which a cell becomes specialized in order to perform a specific function, as in the case of a liver cell, a blood cell, or a neuron.

Cell migration The programmed movement of cells during development.

Cell senescence The phenomenon by which normal diploid cells lose the ability to divide.

Cell surface receptors Specialized integral membrane proteins that take part in communication between the cell and its surroundings.

Cellular immortality When cells divide beyond their normally controlled limit.

Centrioles Tiny, self-reproducing structures located just outside the nuclear membrane in animal cells that replicate during interphase, the new daughter centrioles migrating to opposite spindle poles during nuclear division.

Centromeres Indented regions of metaphase chromosomes that divide them into two arms, and the region where spindle fibers attach during mitosis or meiosis.

Chaperones Proteins that assist in the three-dimensional folding of other proteins.

Charcot-Marie-Tooth Disease (CMT) A group of inherited disorders of nerves (neuropathy), characterized by loss of motor and sensory function predominantly in the feet and legs, but also affecting the hands and arms in the advanced stages of disease.
Checkpoints Locations in the cell cycle at which the cell determines whether to proceed to the next stage of the cycle (as in G_1 to S, or G_2 to M).
Chorea An abnormal involuntary movement characterized by brief, quasi-purposeful, irregular contractions that are not repetitive or rhythmic, but appear to flow from one muscle to the next.
Chorion villus sampling (CVS) A procedure used for prenatal diagnosis at about 12 weeks' gestation where bits of chorionic tissue are removed from the developing fetal placenta.
Chromatids Two parallel identical strands of a single chromosome after DNA synthesis, connected at the centromere.
Chromatin The complex of DNA and proteins of which chromosomes are composed.
Chromatin-remodeling complexes (CRC) Chemical complexes containing enzymes, other proteins and chromatin that make coiled nucleosome cores chemically accessible.
Chromosomal abnormalities Abnormalities of chromosome number or structure.
Chromosome banding Staining techniques that give rise to a unique pattern of lateral bands along the length of each chromosome and thus provide a means of identification and analysis.
Chromosome mutations Changes in the genetic material at the chromosome level.
Chromosomes Self-replicating structures in eukaryotic nuclei that consist of coiled strands of DNA and proteins, and contain sets of linked genes, regulatory elements, and other nucleotide sequences.
Chronic myelocytic leukemia (CML) A cancer of the white blood cells characterized by the increased and unregulated growth of predominantly myeloid cells in the bone marrow and the accumulation of these cells in the blood.
Cis-acting elements Short DNA sequences (6–15 base pairs long) that constitute the control elements adjacent to genes that bind to transcription factors, thereby controlling initiation of transcription.
Clone A cell line derived from a single ancestral cell, or an animal identical to its parent.
Clopidogrel An oral antiplatelet agent used to prevent the formation of blood clots in coronary artery, peripheral vascular, and cerebrovascular diseases.
Clustered repeats Locations in a genome where a short nucleotide sequence is organized in repeating units.
Codon Three bases, referred to as a "triplet," in a DNA or RNA molecule that specify a single amino acid.
Colchicine A chemical toxin derived from plants of the genus *Colchicum* that prevents the formation of the spindle during mitotic division. It is used in the treatment of gout.
Co-linearity The parallel relationship between the sequence of bases in DNA and the sequence of amino acids in its corresponding polypeptide.
Comparative genomic hybridization (CGH) A molecular technique, usually employing microarrays, which aims to determine the number of copies of a particular genomic sequence in a test sample compared to a reference one.
Complementarity The nature of base-pairing in DNA (A to T; G to C).
Concordance A state in a family where a pair of relatives both have a certain qualitative trait or have values of a quantitative trait that are similar in magnitude.
Congenital Present at birth.
Consanguinity A relationship between individuals who have a common ancestor.
Copy number variation (CNV) A variation in DNA sequence defined by the presence or absence of a segment of DNA, or where alleles have tandum duplications of two to four, or more, copies of a DNA segment.
Corona radiata Two or three layers of follicular cells that surround an ovum or unfertilized egg cell, are attached to its outer protective layer, and supply vital proteins and offer protection to it.
Coronary artery disease (CAD) The condition in which there is accumulation of atheromatous plaques within the walls of the coronary arteries that supply the muscle of the heart with oxygen and nutrients.
Creationism The religious belief that humanity, life, the Earth, and the universe are the creation of a supernatural being.
Crossing over The exchange of chromosome parts between the chromatids of synapsed homologues during prophase I of meiosis.
Cyclin-dependent protein kinases (CDK) A family of kinases that, once activated by cyclin, regulates the cell cycle by adding phosphate groups to a variety of protein substrates that control processes in the cycle.
Cystic fibrosis A disorder caused by mutations in a chloride transport channel protein that is inherited as an autosomal recessive trait. It affects principally the lung and the pancreas.
Cytochrome P450s A large and diverse group of enzymes that catalyze the oxidation of organic substances.
Cytogenetic map Diagrammatic representation of a chromosome.
Cytogenetics The study of chromosomes.
Cytoskeleton The protein scaffolding present in all cells within the cytoplasm, containing such structures as flagella and cilia, which plays an important role in both intracellular transport of vesicles and organelles, as well as in cellular division.
Deletion The subtraction of a block of one or more nucleotide pairs from a DNA molecule.
Dementia A serious loss of cognitive ability in a previously unimpaired individual, beyond what might be expected from normal aging.
Demethylases Enzymes that remove methyl (CH_3-) groups from proteins and other substances.
Depurination DNA alteration in which the hydrolysis of a purine base, either A or G, from the deoxyribose-phosphate backbone occurs.

Developmental genetics The use of genetics to study how the fertilized ovum of a multicellular organism becomes an adult.

Diabetes mellitus (DMS) A group of metabolic diseases in which a person has high blood sugar, either because the body does not produce enough insulin or because cells do not respond to the insulin that is produced.

Dicer A cytoplasmic enzyme in the RNAse family that recognizes double-stranded RNA duplexes and trims off both ends to create duplex products 21–24 base pairs long that contain a *miRNA* strand and a complementary miRNA* strand that are passed to the RNA-induced silencing complex (RISC).

Differentiation The process whereby the pattern of expression of genes and proteins of a cell becomes tissue-specific, and the cell develops a characteristic phenotype.

Diploid Cells containing two copies of each of the chromosomes characteristic of that species (ZN).

Discordant twins A twin pair in which one member exhibits a certain trait and the other does not.

Disease gene Any gene with one or more alleles that increase the risk of occurrence of a disease phenotype in an individual carrying such an allele.

Disjunction Separation of homologous chromosomes to opposite poles of a division spindle during anaphase of a mitotic or meiotic nuclear division.

Dispersed repeats A variety of simple di-, tri-, tetra-, and penta-nucleotide tandem repeats that are dispersed in the euchromatic regions of most chromosomes.

Dizygotic twins Non-identical (fraternal) twins that arise from two different zygotes, that is, two separate eggs fertilized by two separate sperm.

DNA fingerprinting Electrophoretic identification of individuals by the use of DNA probes for highly polymorphic regions of the genome. The genome of virtually every individual exhibits a unique pattern of bands.

DNA hybridization A technique that measures the degree of genetic similarity between pools of DNA sequences, used to determine the genetic distance between two species.

DNA methylase The enzyme that facilitates the addition of a methyl residue to the 5 position of the pyrimidine ring of a cytosine base in DNA to form 5-methylcytosine.

DNA methyltransferase (DMT) An enzyme that promotes the transfer of a methyl group to a specific nucleotide base in the molecule of DNA.

DNA polymerase An enzyme that forms new DNA by linking together a string of deoxyribonucleotides using single-stranded DNA as a template.

DNA replication The copying of a DNA molecule.

DNA sequencing Any method or technology that is used for determining the order of nucleotide bases in a molecule of DNA.

DNA Deoxyribonucleic acid, the macromolecule, usually composed of two polynucleotide chains in a double helix, that is the carrier of genetic information in all cells and many viruses.

Dominance A relationship between two variant alleles of a single gene in which one of the alleles masks the expression of the other in expressing a trait.

Donor splice site The nucleotide sequence recognized by the splicing machinery in the process of intron removal.

Dosage compensation The mechanism that accounts for the observation that the amount of product formed by two copies of a gene is the same as that formed by having only one copy of that gene, as in the expression of genes in XX females and XY males.

Drusen Tiny yellow or white accumulations of extracellular material that build up in the membrane of the eye.

Duchenne muscular dystrophy (DMD) A X-linked recessive form of muscular dynstrophy which results in muscle degeneration, difficulty in walking and breathing, and death.

Duffy blood group A genetically determined, immunologically distinct group of human erythrocyte antigens.

Dysmorphology The area of clinical genetics concerned with the diagnosis and management of congenital anatomic abnormalities.

Ectoderm One of the three primary germ layers of the early embryo that begins as the layer farthest from the yoke sac and ultimately gives rise to the nervous system, skin, and neural crest derivatives.

Electrophoresis A technique used to separate molecules on the basis of their different rates of movement in response to an applied electric field, typically through a gel.

Embryology The science that studies the development of an embryo from fertilization of the ovum to the fetus stage.

Embryonic period The period between the fertilization of the ovum and the fetus stage.

Embryonic stem cells (ESCs) Cells in the blastocyst that give rise to the body of the embryo that, when cultured, continue to divide without differentiating, and when reintroduced into the inner cell mass of the blastocyst, can repopulate all the tissues of the embryo.

Empiric risk studies Studies of the probability that a familial trait will occur in a family member, based on observed numbers of affected and unaffected individual in family studies rather than on knowledge of the causative mechanism, Mendel's laws, or genetic linkage.

Endoderm One of the three primary germ layers of the early embryo, the one closest to the yolk sac, that ultimately gives rise to the gut, liver, and portions of the urogenital system.

Endonuclease An enzyme that breaks internal bonds of DNA.

Endoplasmic reticulum (ER) An eukaryotic organelle that forms an interconnected network of tubules, vesicles, and cisternae within cells, where proteins, lipids, and steroids are synthesized, carbohydrates and steroids are metabolized, and calcium is regulated.

Enhancers DNA elements that can regulate transcription from nearby genes by acting as binding sites for transcription factors.

Epicanthal folds Skin fold of the upper eyelid covering the inner corner of the eye.

Epigenetic Any factor that can affect gene function without affecting gene structure.

Epigenome The sum of all the factors that affect genomic function without affecting genomic structure.

Epistasis Any type of interaction in which the genotype at one locus affects the phenotypic expression of the genotype at another location.

Erythrocytes (RBCs) Red blood cells, the most common type of blood cell, which deliver oxygen to the body tissues via the blood flow through the circulatory system.

Euchromatin The most abundant portion of chromosomes. It stains lightly during interphase upon Giemsa treatment.

Eukarya One of the major kingdoms of living organisms, in which the cells have a true nucleus and divide by mitosis or meiosis.

Eumelanin A black to brown variety of melanin pigment.

Euploid A cell or an organism having a chromosome number that is the exact multiple of the haploid number.

Evolution The change over time in one or more inherited traits found in populations of individuals.

Exons The sequences in a gene that are retained in the messenger RNA after the introns are removed from the primary transcript. These sequences are found both in a gene's DNA and in the corresponding mature messenger RNA (mRNA).

Expression profiling A quantitative assessment of the mRNAs present in a cell type, tissue, or tumor, often in comparison to the expression profile of another cell type, tissue, or tumor.

Expressivity The extent to which a genetic defect is expressed, often from mild to severe.

Familial hypercholesterolemia An autosomal semi-dominant genetic disorder characterized by elevated serum concentrations of low density lipoprotein cholesterol (LDLC), and causing early cardiovascular disease.

Familial A trait that is more common in a relative of an affected individual than in the general population.

Fertility The natural capability of giving life.

Fetal alcohol syndrome (FAS) A pattern of mental and physical defects that can develop in a fetus in association with high levels of alcohol consumption during pregnancy.

Fetal period Stage of interauterine development of the fetus from weeks 9 to 40.

Fetal ultrasonography The application of medical ultrasonography in order to visualize the fetus in its mother's uterus.

Fibroblast growth factor (FGF) One of a family of growth factors involved in angiogenesis, wound healing, and embryonic development, a key player in the processes of proliferation and differentiation of a wide variety of cells and tissues.

Fitness The probability of an individual's genes being transmitted to the next generation compared with the average probability for the population.

Fluorescent in situ hybridization (FISH) A physical mapping approach that uses fluorescent tags to detect hybridization of nucleic acid probes with chromosomes.

Frameshift mutations A nutational event caused by the insertion or deletion of one or more nucleotide pairs in a gene, resulting in a shift in the reading frame of all codons following the mutational site.

Functional cloning Identifying and cloning a gene starting from knowledge of the protein product of the gene.

G banding A technique used in cytogenetics to produce a visible karyotype by staining condensed chromosomes.

Gain-of-function mutation A mutation that increases the normal activity of its encoded protein.

Gametes Specialized haploid cells (eggs or sperm) that carry genes between generations via reproduction.

Gap junctions A specialized intercellular connection between a multitude of cell types where the cytoplasm of two cells is connected, allowing various molecules and ions to pass freely between cells.

Gel electrophoresis A process used to separate DNA fragments, RNA molecules, or polypeptides according to their size by passing an electrical current through gel, which forces molecules to migrate into the gel at different rates dependent on their sizes.

Gene amplification A process in which certain genes undergo differential replication either within the chromosome or extra-chromosomally, increasing the number of copies of the gene.

Gene density The number of genes per million bases of chromosome.

Gene flow Exchange of genes among populations resulting from either dispersal of gametes or migration of individuals.

Gene map The characteristic arrangement of the genes on the chromosomes.

Gene pool The totality of genetic information in a population of organisms.

Gene therapy Treatment of a disease by introducing a normal gene into cells bearing a defective one.

Gene A hereditary unit or segment of DNA that occupies a specific site on a chromosome and contains genetic information that is replicated, transcribed into messenger, transfer, or ribosomal RNA, and translated into a functional product, often a polypeptide chain making up a protein.

Genetic code The 64 triplets of bases that specify the 20 amino acids found in proteins.

Genetic drift Random fluctuation of allele frequencies in a population.

Genetic heterogeneity A single phenotype produced by more than one different mutation.

Genetic lethal A genetically determined trait that results in the failure to reproduce.

Genetic screening Testing a population to identify those with an inherited disorder or at risk for a disorder.

Genome The complete DNA sequence, containing the entire genetic information, of a gamete, an individual, a population, or a species.

Genome-wide association studies (GWAS) An approach involving scanning markers across the entire genome of many patients with a particular condition and in age-matched controls to find genetic variations associated with the condition.

Genomic imprinting Differential expression of genetic information depending on whether it was inherited from the father or the mother.

Genomic instability An increased tendency of the genome to acquire mutations when various processes involved in maintaining and replicating the genome are dysfunctional. It is the driving force behind cancer development.

Genomics The field of scientific study concerned with structure and function of the genome.

Genotype frequency The proportion of those in a population with a particular genotype.

Genotype The genetic constitution of an individual or the alleles at a particular genetic locus.

Germ layer One of the three layers of cells (ectoderm, mesoderm, and endoderm) that are formed in the early embryo.

Germ-line gene therapy A genetic engineering technique that modifies the DNA of germ cells that are passed on to progeny.

Germ-line mutation A mutation in a cell from which gametes arise, as opposed to a somatic mutation.

Glucokinase An enzyme that facilitates phosphorylation of glucose to glucose-6-phosphate.

Glycosuria Excretion of glucose into the urine.

Golgi apparatus An organelle found in most eukaryotic cells that processes and packages macromolecules, such as proteins and lipids, after their synthesis and before they make their way to their destination.

Growth factor receptor A receptor that binds to growth factor.

Growth factor A naturally occurring substance capable of stimulating growth, proliferation, and cellular differentiation, usually a protein or a steroid hormone.

Haploid The chromosome number (N) of a normal eukaryotic gamete, containing one repesentative copy of each homologue; in humans, 23.

Haplotype A unique array of closely linked alleles that is usually inherited as a unit.

HapMap A large set of haplotypes, defined by SNPs, distributed throughout the genome.

Hardy-Weinberg law The formula that defines the relationships between genotype and allele frequencies within a generation and from one generation to the next, where there is no selection, mutation, or migration.

Helicase An enzyme that is capable of unwinding the double strands of the deoxyribonucleic acid helix at a replication fork.

Helicobacter pylori Bacteria that can inhabit various areas of the stomach, particularly the antrum, that causes a chronic, low-level infection of the stomach lining.

Heme molecule An iron-containing molecule that binds as a cofactor or prosthetic group to form hemoglobin, myoglobin, and the cytochromes.

Hemizygote A gene present in only one dose, such as the genes on the X chromosome in XY males.

Hemoglobin The oxygen-carrying protein in red blood cells.

Hemoglobinopathies Genetic defects that result in abnormal structure of one of the globin chains of the hemoglobin molecules.

Hemophilia A group of hereditary genetic disorders that impair the body's ability to control blood clotting or coagulation.

Hepatocytes Cells in the main tissue of the liver that are involved in protein synthesis; protein storage; transformation of carbohydrates; synthesis of cholesterol, bile salts, and phospholipids; and detoxification, modification, and excretion of exogenous and endogenous substances; as well as initiating the formation and secretion of bile.

Hereditary neuropathy with pressure palsies (HNPP) A disorder that affects peripheral nerves, characterized by repeated pressure neuropathies such as carpal tunnel disease and peroneal palsy with foot drop.

Heritability (h^2) A measure of the fraction of total variance of a quantitative trait in a population that is caused by genes, and that can be modified by selection.

Heteromorphism A normal variant of a chromosome.

Heteroplasmy The presence of more than one type of mitochondrial DNA in the mitochondria of a single cell of a single individual.

Heterozygote advantage The situation where heterozygotes have a higher fitness than either homozygote—a possible explanation for the survival of recessive genetic diseases in a population.

Heterozygote An individual or genotype with two different alleles at a given locus on a pair of homologous chromosomes.

Histones Small DNA-binding proteins with a preponderance of the basic, positively charged amino acids lysine and arginine that are the fundamental protein components of nucleosomes.

Histopathologic disease A disease that can be studied by microscopic examination of a tissue in order to study the manifestations of that disease.

Holoprosencephaly A cephalic disorder in which the prosencephalon fails to develop into two hemispheres.

Homeobox genes A sequence of 180 nucleotide base pairs found within many developmental genes that encodes a DNA-binding protein which acts as a gene regulator.

Homeotic mutants Disorders in the developmental genes that regulate pattern formation within body segments, resulting in extra or misplaced body parts.

Hominids A taxonomic, bipedal primate family that includes four extant genera—chimpanzees, gorillas, orangutans, and humans—as well as their ancestors.

Hominoids A kind of primate with two groups, the lesser hominoids (gibbons and siamangs) and the great hominoids (orangutans, champanzees, bonobos, gorillas, and humans) that never have tails.

Homogentistic acid The molecule, formally called alkapton, that is excreted in the urine and turns black upon oxidation.

Homologues Chromosome pairs that match in size, shape, and banding and contain the same linear gene and nucleotide sequence, each derived from one parent.

Homoplasmy The presence of only one type of mitochondrial DNA in the mitochondria of a single individual.

Homozygote An individual or genotype with identical alleles at a given locus on a pair of homologous chromosomes.

Human leukocyte antigen (HLA) Any group of antigens present on the surface of nucleated body cells that are coded for by the major histocompatibility complex of humans and thus allow the immune system to distinguish between self and non-self.

Human papilloma virus (HPV) Any of various strains of papilloma virus that cause warts, especially on the hands, feet, and genitals, with some being responsible for cancers of the cervix, vagina, vulva, and penis.

Huntington disease An inherited neurodegenerative disorder that is inherited as an autosomal dominant trait and leads to muscle dyscoordination, cognitive decline, and dementia. Its clinical onset is usually in midlife, and it is uniformly fatal.

Hybrid An organism produced by the mating of genetically unlike parents.

Hybridization probes A fragment of DNA or RNA of variable length that is used to interact with DNA or RNA samples to detect the presences of nucleotide sequences that are complementary to the sequence in the probe, thereby identifying them.

Hydrocephalus A medical condition in which there is an abnormal accumulation of cerebrospinal fluid in the ventricles or cavities of the brain.

Hyperglycemia A condition where excessive amounts of glucose circulate in the blood plasma.

Hypotonia A state of low muscle tone, often involving reduced muscle strength.

Ideogram A diagram of chromosome morphology, especially the banding pattern, that is used to compare karyotypes of different cells, individuals, or species.

Imprinting The difference in expression of alleles depending on the parent of origin.

In vitro fertilization (IVF) A reproductive technology in which sperm are allowed to fertilize an egg in tissue culture, followed by introduction of the fertilized egg into the uterus to allow implantation and development.

Inborn error of metabolism A genetically determined biochemical disorder in which a specific enzyme defect produces a metabolic block that may have pathologic consequences.

Inbreeding The mating of close relatives; one of the factors disturbing Hardy-Weinberg equilibrium in a population.

Incomplete dominant A trait that is inherited in a dominant manner but is more severe in a homozygote than in a heterozygote.

Incomplete penetrance The fraction of individuals with a given genotype who do not have the expected phenotype. This gives rise to "skipped generations" for dominant traits.

Incomplete recessive The result of expression of alleles that are neither dominant nor recessive. In this situation, a trait that is inherited as a recessive shows phenotypic differences between heterozygotes and affected homozygotes.

Indels A polymorphism defined by the presence or absence of a segment of DNA, ranging from one base to a few hundred base pairs.

Inherited metabolic diseases (Synonymous with "inborn errors of metabolism.")

Inner cell mass The cluster of cells within a blastocyst that are destined to become the embryo.

Insertion (1) A structural chromosomal abnormality in which part of the material from one chromosome is inserted into a non-homologous chromosome. (2) Addition of one or more bases into a gene.

Intelligent design (ID) The proposition that "certain features of the universe and of living things" require the action of some positive force, such as God.

Interphase The stage in the cell cycle between two successive cell divisions.

Interstitial deletion A chromosomal abnormality in which a portion of a chromosome—not involving its terminal ends—is removed.

Intron A segment of a gene that is initially transcribed into RNA but is then removed from the primary RNA transcript by splicing together the exon sequences on either side of it to form a mature RNA.

Invasiveness The ability of malignant tumor cells to infiltrate surrounding tissue.

Inversion A structural aberration in a chromosome in which the order of several genes is reversed from the normal order; it is pericentric when the centromere is contained within the inverted region and paracentric when it is not.

Isochromosomes Abnormal chromosomes in which one arm is deleted and the other arm is duplicated so that two arms of equal length are formed with the same loci in reverse sequence.

Karyotype The chromosome constitution of an individual, also, a photomicrograph of the chromosomes of an individual systematically arranged according to their lengths and positions of their centromeres.

Lineage The progeny of a cell or of an individual.

Linkage map A depiction of the relative positions of genes and other DNA markers on a chromosome, as defined by linkage studies.

Linkage marker A gene, nucleotide sequence, protein, chromosome appearance, or disease useful in conducting linkage studies.

Linkage study A genetic mapping technique that seeks to identify co-inherited genes.

Locus The position occupied by a gene on a chromosome.
Loss-of-function mutation One leading to reduced or complete loss of function of a protein.
Low-density lipoprotein cholesterol (LDL) A complex of lipids and proteins that transports cholesterol in the blood. High levels are associated with an increase risk of atherosclerosis and coronary heart disease.
Lysosomes Cellular organelles that contain acid hydrolase enzymes which break down waste materials and cellular debris.
Magnetic resonance imagining (MRI) A medical imaging technique that depends on the properties of some atoms to alter systematically their alignment in a magnetic field, and thereby allow visualization of internal organs and structures.
Major depressive disorder A mental illness characterized by pervasive sadness, low self-esteem, and loss of interest or pleasure in normally enjoyable activities.
Major histocompatability complex (MHC) The complex of human leukocyte antigen (HLA) genes on the short arm of chromosome 6 that is sufficiently polymorphic to be a useful marker for linkage studies.
Mania A state of abnormally elevated or irritable mood, arousal, and energy levels, generally the opposite of depression.
Maternal effect genes Genes in the mature ovum that are maternal in origin, whose products influence early embryonic development.
Maternal inheritance The transmission of genetic information exclusively through the mother.
Maternal serum screening (MSS) A blood test for pregnant women in the first or second trimester of pregnancy that helps determine the risk of certain genetic abnormalities that may affect the fetus.
Maturity onset diabetes of the young (MODY) A group of heredity disorders which produce diabetes mellitus by interfering with the action of insulin.
Meiosis The process of nuclear division in germ cells of sexually reproducing organisms during which gametes containing the haploid chromosome number are produced from diploid cells.
Meiotic non-disjunction The failure of chromosome pairs to separate properly during meiosis that causes the formation of a cell with an imbalance of chromosomes.
Melanin The intracellular pigment found in most organisms; in humans, a derivative of the amino acid, tyrosine.
Melanocortin-1 receptor (MRC-1) The protein found on the surface of specialized cells, including melanocytes, to which pituitary hormones bind.
Melanocyte stimulating hormone (MSH) A pituitary hormone that stimulates melanocytes to produce melanin.
Melanocytes Melanin-producing cells located in the bottom layer of the skin's epidermis, the middle layer of the eye, the inner ear, the meninges, bones, and heart that are primarily responsible for the color of the skin.
Melanosomes Organelles in the cell containing melanin.
Menarche The first menstrual cycle in female mammals.
Mendelian Patterns of inheritance that follow Mendel's classic laws.
Mesoderm The middle layer of the three germ layers in the very early embryo.
Messenger RNA (mRNA) An RNA species transcribed from DNA that forms the template on which polypeptides are synthesized from amino acids.
Metabolic pathway A set of chemical reactions that take place in a definite order to convert a starting molecule into one or more specific products.
Metabolism The set of physical and chemical reactions that are involved in the maintenance of life.
Metacentric chromosome A chromosome or chromatid whose centromere is centrally located.
Metaphase In mitosis, meiosis 1, or meiosis II, the stage of nuclear division where the centromeres of the condensed chromosomes are arranged in a plane between the two poles of the spindle.
Metastasis Spread of malignant cells from their original site to other sites in the body.
Methylation The process by which a methyl group is added to a substrate. This process is used to regulate gene expression and to protect DNA from some types of cleavage.
MicroRNA (miRNA) A class of non-coding RNA molecules that are processed in the nucleus into short, double-stranded species that interfere with mRNA structure or function in the cytoplasm.
Missense mutation A single DNA base substitution that results in a codon specifying a different amino acid.
Mitochondria A membrane-enclosed organelle found in most eukaryotic cells that generate most of the cell's energy in the form of adenosine triphosphate (ATP), and which have their own, independent genome.
Mitochondrial DNA (mtDNA) The DNA present in the circular chromosome of the mitochondria, which is inherited only through the female line since only the mitochondria present in the ovum's cytoplasm are passed on to mature cells (no mitochondria are contributed by the sperm cell).
Mitosis The process of cellular division in which the replicated chromosomes divide and the daughter cells have the same chromosome number and genetic composition as the parent cell.
Modifier gene A gene that alters the phenotype produced by mutations in a non-allelic gene.
Molecular cloning The process by which a single DNA fragment is purified from a complex mixture of DNA molecules and then amplified into a large number of identical copies.
Molecular evolution A process of evolution at the scale of DNA, RNA, and proteins.
Monosomy A condition in which one of a pair of chromosomes is missing, as in monosomy X (Turner syndrome).
Monozygotic Twins arising from a single zygote which divides into two genetically identical embryos.
Morphogen A substance that defines different cell fates in a concentration-dependent fashion.
Morula The embryo at an early stage of embryonic development, consisting of cells in a solid ball contained within the zona pellucida.

Mosaic A tissue or individual with at least two cell lines of different genotype.
Multifactorial trait or condition A trait or condition determined by the combined action of many factors, typically some genetic and some environmental.
Multipotent When applied to stem cells, cells that have the ability to self-renew and to differentiate into specialized cells of a particular tissue or organ.
Mutagen Any substance capable of producing changes in the sequence of bases in DNA.
Mutation A heritable alteration in a gene or chromosome.
Myelin A complex substance that forms a layer, the myelin sheath, usually around neuronal axons.
Natural selection The process by which genotypes best suited to survive in a given environment reproduce more successfully than others and gradually increase the overall ability of the population to thrive.
Neoplasia An abnormal mass of tissue that forms from an abnormal proliferation of cells, whose growth exceeds and is not coordinated with that of the cells around it. Such growths may be benign or malignant.
Neural tube defect (NTD) An opening in the spinal cord or brain that occurs very early in human development; among the most common human birth defects.
Newborn screening A process by which blood samples from neonates are tested for a list of disorders that may be treatable before they cause clinical disturbance.
Non-allelic homologous recombination (NAHR) A form of homologous recombination that occurs between two lengths of DNA that have high sequence homology, but are not alleles. It is responsible, in most instances, for gene duplication and gene deletion.
Non-disjunction The failure of two members of a chromosome pair to separate during meiosis I, or of a pair of chromatids to separate during meiosis II or mitosis.
Nonsense mutation A mutation that changes a codon specifying an amino acid into a stop codon, thereby resulting in premature polypeptide chain termination.
Non-synonymous mutation A single nucleotide substitution in a codon that changes the amino acid specified by the codon.
Notocord A flexible, rod-shaped body found in embryos of all chordates, the cells of which are derived from the mesoderm, and defines the primitive axis of the embryo.
Nucleosome A fundamental DNA packaging unit, composed of DNA wrapped around a core of histones and histone-associated proteins.
Nucleotide A molecule composed of a nitrogenous base, a five-carbon sugar, and a phosphate group. A polymer of many nucleotides strung together is called a nucleic acid.
Nucleus A membrane-enclosed organelle in eukaryotic cells which contains (most of) the cell's genetic material complexed with a large number of RNAs and proteins.
Oligonucleotide A short, nucleic acid polymer, usually no more that 50 base pairs long, for use in DNA sequencing as a probe or for use in the polymerase chain reaction.
Oncogene A gain-of-function mutation in a cellular gene whose normal function is to promote cellular proliferation or inhibit apoptosis, often associated with tumor production by uncontrolled cell division.
Oncomere A chromosomal locus harboring a cluster of mini-genes coding for miRNAs or siRNAs. They may be involved in promoting or protecting against cancer.
Ontogeny The developmental history of an organism.
Oogenesis The creation of an ovum in the female as the result of gametogenesis.
Opioids (Opioid analgesia) Narcotic drugs that depress the central nervous system and are generally prescribed to manage pain.
p53 A tumor suppressor protein that, in humans, regulates the cell cycle, maintains genomic stability, and promotes apoptosis.
Paracentric inversion An inversion that does not include the centromere.
Pattern formation genes Genes that create a spatially ordered and differentiated embryo from a seemingly homogenous egg cell.
Pedigree A diagram representing the familial relationships among relatives.
Pericentric inversion An inversion that includes the centromere.
Pharmacodynamics The effects of a drug or its metabolites on physiological function and metabolic pathways.
Pharmacogenetics The area of biochemical genetics concerned with the impact of genetic variation on drug response and metabolism.
Pharmacogenomics The application of genomic information or methods to pharmacogenetic problems.
Pharmacokinetics The rate at which the body absorbs, transports, metabolizes, or excrete a drug or its metabolites.
Phenotype frequency The proportion of individuals in a population that are of a particular phenotype.
Phenotype The observed biochemical, physiological, and morphological characteristics of an individual as determined by his or her genotype and the environment in which it is expressed.
Phenylalanine One of the 20 common amino acids used to form proteins, an excess of which in the human infant causes phenylketonuria.
Phenylketonuria (PKU) A genetic disorder, inherited as an autosomal recessive trait, that results in the inability to metabolize phenylalanine and which can be controlled by limiting the amount of phenylalanine in the diet.
Pheomelanin Yellow to red-brown pigment produced by the melanocytes.
Phosphodiester bonds Strong covalent bonds between a phosphate group and two five-carbon ring sugars, creating two ester bonds.

Phosphorylation The addition of a phosphate group to a protein or other organic molecule, which activates or deactivates many protein enzymes.

Phytohemagglutinin A lectin found in plants, especially legumes, that can affect cell metabolism by inducing mitosis and which agglutinates red blood cell types. It can act as a toxin.

Plasma membrane The two-layered membrane that surrounds the cell and separates it from the outside environment, and which is selectively permeable to ions and organic molecules.

Pluripotent Describes an embryonic cell that is capable of giving rise to different types of differentiated tissues or structures, depending on its location and environmental influences.

Point mutation Mutation caused by the substitution, deletion, or addition of a single nucleotide pair.

Polar body A cell produced by meiosis I or meiosis II during oogenesis that does not become the primary or secondary oocyte.

Poly-A tail The sequence of adenines added to the 3′ end of many eukaryotic mRNA molecules in processing.

Polygenic trait The determination of a trait by alleles of two or more different genes.

Polymerase chain reaction (PCR) Repeated, polymerase-catalyzed cycles of DNA denaturation, renaturation with primer oligonucleotide sequences and replication, resulting in exponential growth in the number of copies of the DNA sequence located between the primers.

Polymorphism The presence in a population of two or more relatively common forms of a gene, chromosome, or genetically determined trait.

Polyploid Any multiple of the haploid chromosome number other than the diploid one.

Polyuria A condition usually defined as excessive or abnormally large production and/or passage of urine.

Population genetics The quantitative study of the distribution of genetic variance in a population and of how the frequencies of its genotypes, alleles, and phenotypes are maintained or changed.

Population All the organisms that both belong to the same species and are inbreeding and live in the same geographical area at the same time.

Positional cloning The process that allows identification and cloning of a gene starting from the map position of the gene without any prior knowledge of its protein product or function.

Prader-Willi syndrome A rare genetic disorder in which seven genes on chromosome 15 are deleted or unexpressed on the paternal chromosome, resulting in people that are obese, have reduced muscle tone and mental ability, and have sex glands that produce small amounts of or no hormones.

Preimplantation genetic diagnosis (PGD) A method of determining certain genetic traits of the embryo prior to its implantation in the uterine wall on the basis of one cell extracted from an eight-cell embryo.

Pre-mRNA Precursor mRNA; a single strand of messenger ribonucleic acid synthesized from a DNA template in the cell nucleus by transcription that comprises the bulk of heterogeneous nuclear RNA.

Prions A protein infectious agent in a misfolded form that can transmit disease.

Proband The person through whom a family is ascertained (also referred to as the propositus).

Progenitor A biological cell that, like a stem cell, has a tendency to differentiate into a specific type of cell but is already more specific than a stem cell.

Prokaryote A simple unicellular organism, such as a bacterium, that lacks a separate nucleus.

Proliferation The process by which cells multiply and form tissues or organs.

Promoter A sequence of nucleotide bases at the 5′ end of a gene at which RNA polymerase binds and initiates gene transcription.

Protease Any enzyme that conducts proteolysis; that is, begins protein catabolism by hydrolysis of the peptide bonds that link amino acids together in the polypeptide chain.

Protein sequencing A technique to determine the linear order of amino acids in a protein.

Proteosome Large multiprotein complexes in the cytoplasm of eukaryotic cells that degrade proteins tagged with ubiquitin.

Proto-oncogene A normal eukaryotic gene that functions to promote cellular proliferation or inhibit apoptosis, and which can mutate into a cancer-causing oncogene.

Pseudogene An inactive form of an active gene, produced by mutation of an earlier version of that gene.

Pseudohypertrophy An increase at the site of an organ or a part of it that is the result of an increase in the size or number of some other tissue, not of the specific functional elements.

Q banding A laboratory technique for staining a karyotype with quinacrine dye.

Qualitative trait A trait that an individual either has or doesn't have, such as a cleft palate or a congenital heart defect.

Quantitative trait A trait that varies continuously over a range of measurements from one extreme to the other with no discernable breaks in between, such as height or blood pressure.

Quaternary structure The three-dimensional configuration of a protein made up of more than one polypeptide subunits.

R banding A technique for staining a karyotype using Giemsa stain that effects the reverse of G banding.

RAD9 A DNA damage-dependent checkpoint protein required for cell-cycle arrest and DNA damage repair.

Random mating Selection of a mate without regard to that mate's genotype.

Reannealing Reassociation of dissociated single strands of DNA to form a duplex molecule.

Recessive An allele or corresponding phenotypic trait expressed only in homozygotes.

Reciprocal translocation Interchange of parts between non-homologous chromosomes.

Recombinant DNA A hybrid DNA molecule of two or more pieces, formed by the fusion of pieces of DNA that are not normally in physical proximity and are often derived from different species.

Recombinant A chromosome or individual that has a new combination of alleles not found in either parent.

Recombination The reassortment of parental genes to form new combinations of alleles in the offspring as a result of independent assortment or crossing over.

Regional continuity theory A theory that proposes that humans evolved simultaneously from several ancestral populations.

Regulatory gene A gene that codes for an RNA or protein that affects the expression of one or more other genes.

Relative risk ratio The prevalence of a disorder in the relatives of an affected person divided by the prevalence of the disease in the general population.

Renal glomeruli The tiny clusters of looped blood vessels at the ends of the branches of arteries that enter the kidney. They act as filters for water and plasma constituents that ultimately become urine.

Replacement theory A theory that *Homo erectus* originated in Africa, evolved into *Homo sapiens*, and dispersed throughout the rest of the world, replacing archaic predecessors.

Replication fork Y-shaped area consisting of the two unwound DNA strands branching out into unpaired but complementary single strands during replication.

Restriction enzymes Proteins made by bacteria that recognize specific, short nucleotide sequences and cut DNA at those sites.

Restriction fragment length polymorphism (RFLP) A difference in polymorphic sequences of DNA among individuals that can be recognized by restriction enzymes.

Reticulocytes Immature red blood cells that develop in the bone marrow and circulate in the blood stream before developing into mature red blood cells.

Retinoblastoma A cancer originating in cells of the retina.

Retinoic acid A metabolite of vitamin A that is required for normal growth and development.

Retinopathy A term that refers to some form of non-inflammatory damage to the retina of the eye.

Reverse transcriptase An RNA-dependent DNA polymerase that synthesizes DNA strands complementary to an RNA template.

Ribonucleic acid (RNA) A nucleic acid formed on a DNA template, containing ribose instead of deoxyribose.

Ribosome A cytoplasmic organelle composed of ribosomal RNA and proteins on which polypeptides are synthesized using an mRNA template.

RNA polymerase An enzyme that makes RNA by copying the base sequence of a DNA strand.

RNAi (Interference RNA) A system within living cells that takes part in controlling gene expression by degrading or silencing mRNA.

RNA-induced silencing complex (RISC) A large enzymatic complex in the cytoplasm of all eukaryotic cells that binds to a miRNA and performs sequence-specific RNA interference.

Robertsonian translocation A chromosomal aberration in which the long arms of two acrocentric chromosomes become joined to a common centromere.

rRNA (Ribosomal RNA) RNA molecules that are components of the ribosome.

Rubella Commonly known as German measles, a disease caused by the rubella virus capable of producing congenital defects in infants born to mothers infected during the first 3 months of pregnancy.

Sarcoma A cancer that arises from transformed cells in one of a number of tissues that develop from embryonic mesoderm, i.e., bone, cartilage, fat, muscle, vascular, and hematopoietic tissues.

Schizophrenia A common, devastating mental illness affecting thought and behavior, most often characterized by delusions and hallucinations.

Segmentation gene Any of a group of genes that determines the subdivision of the body into an array of body segments arranged in an anterior to posterior order.

Segregation The disjunction of homologous chromosomes during meiosis.

Selection The action of environmental factors on a particular phenotype, and hence its genotype, based on differences in biological fitness.

Selective serotonin reuptake inhibitors (SSRI) A class of compounds typically used as antidepressants in the treatment of depression, anxiety disorders, and some personality disorders.

Semi-conservative replication The usual mode of DNA replication in which each strand of a duplex molecule serves as a template for the synthesis of a new complementary strand, resulting in the daughter molecules being composed of one old (parental) and one newly synthesized strand.

Short tandem repeat polymorphism (STRP) A locus with repetitive units sometimes used in DNA fingerprinting, each unit being 2–10 nucleotides in length.

Shotgun sequencing The DNA sequencing approach in which the overlapping fragments to be sequenced have been randomly generated from large insert clones, from shearing the whole genome with sound waves, or from digestion with restriction enzymes.

Sickle cell anemia An autosomal recessive disorder of human hemoglobin resulting from a missense mutation in the beta chain, and characterized by red blood cells that assume an abnormal, rigid, sickle shape when deoxygenated.

Silencer A DNA sequence that interferes with transcription of a nearby gene, or an RNA sequence that downregulates or abolishes the translation of an mRNA.

Silent mutation A point mutation that changes a codon, but not the amino acid specified. Accordingly, it has no phenotypic effect.

SINEs Pieces of foreign DNA that have been inserted into the chromosomes of a multicellular organism by foreign agents such as retroviruses, useful for ascertaining phylogenetic relationships.

Single-gene defects Disorders caused by mutant alleles at a single genetic locus, the phenotypic consequences of which vary from none to lethal.

Single-nucleotide polymorphism (SNP) A DNA marker in which a single nucleotide pair differs in the DNA sequence of homologous chromosomes and in which each of the alternative sequences occurs relatively frequently.

siRNA (Short interfering RNA) An RNA molecule, 21–24 bases long, that interrupts translation by degrading mRNA.

SNP chips A type of microarray that enables the SNP genotype of an individual to be determined with great accuracy.

Somatic cell gene therapy The introduction of a normal gene into somatic cells to replace or supplement an abnormal gene.

Somatic cell Any cell in an organism except gametes (germ cells) and their precursors.

Somatic mutation A mutation occurring in a somatic cell, thus not being transmitted to offspring.

Somites Small mesodermal blocks that are found on either side of the developing neural tube in an early embryo.

Sonic hedgehog (SHH) A morphogenic gene whose protein product plays a crucial role in the positioning and growth of the brain, skull, limbs, fingers, and toes during embryonic development.

Southern blot analysis A nucleic acid hybridization method where, after electrophoretic separation, denatured DNA is transferred from a gel to a membrane filter and then probed with pieces of labeled or fluorescent DNA or RNA to locate the DNA fragments according to their sizes.

Speciation The evolutionary process by which new biological species arise.

Spermatogonia Intermediary male germ cells in the production of spermatozoa.

Spina bifida A birth defect caused by the incomplete closure of the embryonic neural tube, usually in the lumbar or sacral area.

Spindle A structure composed of fibrous proteins on which chromosomes align during metaphase in mitosis and meiosis, and then move to opposite poles during anaphase.

Spliceosomes An RNA-protein particle in the nucleus in which introns are removed from primary RNA transcripts.

Splicing A modification of an RNA after transcription in which introns are removed and exons are joined.

Statins A class of drugs used to lower blood cholesterol levels that act by inhibiting the rate-limiting enzyme in cholesterol biosynthesis, HMG-CoA reductase.

Stem cell A type of cell that is capable of reproducing itself and of differentiation into other body cells.

Stop codon One of three codons, UAA, UAG, and UGA, found in mRNA, that do not specify any amino acid, but instead stop the translation process in such a way that the nascent polypeptide can be released from the ribosome.

Stratification A population in which subgroups have remained genetically separate over generations.

Stromal cells Connective tissue cells of any organ that support the framework of the functional cells of that organ.

Submetacentric Describes a chromosome whose centromere lies close to, but not at, its middle.

Synapsis Close pairing of homologous chromosomes in prophase of the first meiotic division.

Syndactyly A condition in which two or more digits are fused together.

Synonymous A mutation in which substitution of a single nucleotide in a codon does not alter the amino acid specified by the codon.

Tandem duplication When two or more identical or closely related DNA sequences are adjacent and arranged in a direct head-to-tail succession along a chromosome.

Tay-Sachs disease (TSD) A deadly autosomal recessive disorder that affects the nervous system of young children almost exclusively of eastern European (Ashkenazi) Jewish descent.

TDF (Testis determining factor) The product of the *SRY* gene on the Y chromosome. In its presence, the internal and external genitalia are those of a male.

Telocentric Chromosomes in which the centromere is terminally placed so that there is only one chromosomal arm.

Telomerase An enzyme composed of protein and RNA that adds hexamers of nucleic acid bases to the telomeric ends of chromosomes.

Telomere The end of each chromosome arm.

Telophase The stage of cell division that begins when the daughter chromosomes reach the poles of the dividing cell, and that lasts until the two daughter cells take on the appearance of interphase cells.

Teratogens Substances that are found in the environment that can cause or increase the incidence of a birth defect.

Terminal deletion A deletion involving the terminal part of the short or long arm of a chromosome.

Termination codon One of three triplets (UAA, UAG, UGA) that stop the ribosomal synthesis of a polypeptide.

Tetraploidy The presence in a cell of four complete sets of chromosomes.

Trans-acting elements Genes that code for transcription factors.

Transcription factor One of a large number of proteins that binds to regulatory regions of genes, and regulates transcription by forming complexes with other transcription factors and RNA polymerase.

Transcription The process by which RNA polymerase directs the synthesis of a strand of RNA from a DNA template using complementary base pairing.

Transformation The process by which a normal cell becomes a cancer cell.

Transition A mutation in which one purine is substituted for another purine or one pyrimidine is substituted for another pyrimidine.

Translation The process by which a polypeptide is synthesized from an mRNA template on cytoplasmic ribosomes.

Translocation The interchange of parts between two or more non-homologous chromosomes.

Transposable elements DNA segments that, regardless of the mechanism, move about in the genome.

Transversion A mutation in which a purine replaces a pyrimidine, or a pyrimidine replaces a purine.

Trinucleotide expansion An increase in the number of tandem copies of a trinucleotide repeat during DNA replication.

Triploidy A cell with three complete copies of each chromosome.

tRNA (Transfer RNA) A small molecule of RNA that brings amino acids into position along the mRNA template, thereby facilitating the translation from mRNA sequence to protein sequence.
True-breeding When two organisms with a particular, inheritable phenotype produce only offspring with that same phenotype.
Tumor suppressor gene (TSG) A gene that normally controls cell proliferation or that activates apoptosis. Loss-of-function mutations lead to initiation or progression of cancer.
Tyrosine One of the 20 amino acids that are used by cells to synthesize proteins.
Ubiquitination The process by which a small molecule, ubiquitin, binds to a protein and acts as a signal for the protein's destruction by signaling the protein transport machinery to ferry the protein to the proteasome for degradation.
Ultrasound Ultrasonic waves used for diagnostic or therapeutic purposes, specifically to image an internal body structure, monitor a developing fetus, or generate localized deep heat to the tissues.
Uncontrolled cell proliferation Continuous cell division with no interruption, as in tumor formation.
Uniparental disomy Homozygosity for a particular chromosome segment or chromosome due to the inheritance of two chromosomes from one parent.
Variable number of tandem repeats (VNTR) Head-to-tail repetition of short DNA sequences at many defined locations throughout the genome that form the basis of DNA fingerprinting procedures.
Vascular endothelial growth factor (VEGF) A signal protein produced by cells that stimulates blood vessel formation during embryonic development, after injury, following exercise, or to bypass blocked vessels. It also creates new blood vessels that nourish solid cancers.
Vector A virus or other particle in which DNA is packaged for the purpose of being delivered to another cell or tissue.
Viability The probability of survival to reproductive age.
Whole-genome sequencing A laboratory process that determines the complete DNA sequence of an organism's genome, including both its chromosomal and its mitochondrial DNA.
Wild type The normal allele or the normal phenotype.
Wilson disease An autosomal recessive genetic disorder in which copper accumulates in tissues, resulting in neurologic or psychiatric symptoms and liver disease.
X-chromosome inactivation Inactivation of genes on one X chromosome in somatic cells of female mammals.
X-linked dominant A dominant allele carried on the X chromosome.
X-linked recessive A recessive allele carried on the X chromosome.
X-linked traits Traits produced by genes found on the X chromosome.
XXX syndrome A condition in which human females carry an extra X chromosome. In most cases, this leads to no phenotypic abnormalities.
XXY syndrome (Klinefelter syndrome) A condition in which human males have an extra X chromosome, often leading to poor testicular development and mild mental retardation.
Y linkage Genes on the Y chromosome or traits determined by such genes.
Zygote A fertilized ovum.

Additional Reading

Chapter 1

Bredenoord, A. L., Kroes, H. Y., Cuppen, E., et al. (2011). Disclosure of individual genetic data to research participants: the debate reconsidered. *Trends Genet., 27*, 41–47.

Collins, F. S. (2006). No longer just looking under the lamppost. *Am. J. Hum. Genet., 79*, 421–426.

Hudson, K. L. (2011). Genomics, health care, and society. *N. Engl. J. Med., 365*, 1033–1041.

Kitscher, P. (1996). *The Lives to Come.* New York, NY: Simon & Schuster.

Korobkin, R. (2008). The Genetic Information Nondiscrimination Act—a half-step toward risk sharing. *N. Engl. J. Med., 359*, 335–357.

McGuire, A. L., & Lupski, J. R. (2010). Personal genome research: what should the participant be told? *Trends Genet., 26*, 199–201.

Slaughter, L. M. (2007). Your genes and privacy. *Science, 316*, 797.

Chapter 3

Bearn, A. G. (1994). Archibald Edward Garrod, the reluctant geneticist. *Genetics, 143*, 1.

Berg, P., & Singer, M. (2003). *George Beadle: an uncommon farmer—the emergence of genetics in the 20th century.* New York, NY: Cold Spring Harbor Laboratory Press.

Judson, H. F. (1996). *The Eighth Day of Creation: The Makers of the Revolution in Biology.* New York, NY: Cold Spring Harbor Laboratory Press.

Orel, V. (1996). *Gregor Mendel: The First Geneticist.* New York, NY: Oxford University Press.

Watson, J. D. (1968). *The Double Helix: A Personal Account of the Discovery of the Structure of DNA.* New York, NY: Touchstone.

Chapter 4

Annas, G. J. (2011). Assisted Reproduction—Canada's Supreme Court and the Global Baby. *N. Engl. J. Med., 365*, 459–463.

Ferguson-Smith, M. A., & Trifinov, V. (2007). Mammalian karyotype evolution. *Nat. Rev. Genet., 8*, 950–962.

Handel, M. A., & Schimenti, J. C. (2010). Genetics of mammalian meiosis: regulation, dynamics and impact on fertility. *Nat. Rev. Genet., 11*, 124–136.

Hegreness, M., & Meselson, M. (2007). What did Sutton see? Thirty years of confusion over the chromosomal basis of Mendelism. *Genetics, 176*, 1939–1944.

Marston, A. L., & Amon, A. (2004). Meiosis: cell cycle controls shuffle and deal. *Nat. Rev. Mol. Cell. Biol., 5*, 983–997.

Ohlsson, R. (2007). Widespread monoallelic expression. *Science, 318*, 1077–1078.

Trask, B. (2002). Human cytogenetics: 46 chromosomes, 46 years and counting. *Nat. Rev. Genet., 3*, 769–778.

Willard, H. F. (2009). William Allan Award Address: Life in the sandbox—unfinished business. *Am. J. Hum. Genet., 86*, 318–327.

Chapter 5

Guttmacher, A. E., Collins, F. S., & Carmona, R. H. (2004). The family history—more important than ever. *N. Engl. J. Med., 351*, 2333–2336.

Hartl, D. L., & Fairbanks, D. J. (2007). Mud sticks: on the alleged falsification of Mendel's data. *Genetics, 175*, 975–979.

Lyon, M. F. (2002). X-Chromosome inactivation and human genetic disease. *Acta Paediatr. Suppl., 91*, 107–112.

McKusick, V. A. (2007). Mendelian inheritance in Man and its online version, OMIM. *Am. J. Hum. Genet., 80*, 588–604.

Mendel, G. (1966). Experiments in plant hybridization [transl.]. In C. Stern & E. Sherwood (Eds.), *The Origins of Genetics: A Mendel Source Book*. New York, NY: W.H. Freeman.

Stern, C. (1965). Mendel and human genetics. *Proc. Am. Philos. Soc., 109*, 216–226.

Chapter 6

Armanios, M., Alder, J. K., Parry, E. M., et al. (2009). Short telomeres are sufficient to cause the degenerative defects associated with aging. *Am. J. Hum. Genet., 85*, 823–832.

Bartel, D. P. (2004). MicroRNAs: genomics, biogenesis, mechanism, and function. *Cell, 116*, 281–297.

Cordaux, R., & Batzer, M. A. (2009). The impact of retrotransposons on human genome evolution. *Nat. Rev. Genet., 10*, 691–703.

ENCODE Project Consortium. (2007). Identification and analysis of functional elements in 1% of the human genome by the ENCODE Pilot Project. *Nature, 447*, 799–816.

Feero, W. G., Guttmacher, A. E., & Collins, F. S. (2011). Genomic medicine—an updated primer. *N. Engl. J. Med., 362*, 2001–2011.

Ferguson-Smith, A. C. (2011). Genomic imprinting: the emergence of an epigenetic paradigm. *Nat. Rev. Genet., 12*, 565–575.

Gresham, D., Dunham, M. J., & Botstein, D. (2008). Comparing whole genomes using DNA microarrays. *Nat. Rev. Genet., 9*, 291.

Hunter, D. J., Khoury, M. J., & Drazen, J. M. (2008). Letting the genome out of the bottle—will we get our wish? *N. Engl. J. Med., 358*, 105–107.

International Human Genome Sequencing Consortium. (2003). Finishing the euchromatic sequence of the human genome. *Nature, 431*, 931–945.

Marshall, E. (2011). Waiting for the revolution. *Science, 331*, 526–529.

McGuire, A. L., Caulfield, T., & Cho, M. K. (2008). Research ethics and the challenge of whole genome sequencing. *Science, 358*, 105–107.

Venter, J. C., Adams, M. D., Myers, E. W., et al. (2001). The sequence of the human genome. *Science, 291*, 1304–1351.

Chapter 7

Feinberg, A. (2007). Phenotype plasticity and the epigenetics of human disease. *Nature, 447*, 433–440.

Hartl, F. U., & Hayer-Hunt, M. (2002). Protein folding: molecular chaperones in the cytosol: from nascent chain to folded protein. *Science, 295*, 1852–1858.

Maniatis, T., & Reed, R. (2002). An extensive network of coupling among gene expression machines. *Nature, 416*, 499–506.

Miyoshi, N., Barton, S. C., Kaneda, M., et al. (2006). The continuing quest to comprehend genomic imprinting. *Cytogenet. Genome Res., 116*, 6–11.

Szyf, M. (2009). Epigenomics and its implications for medicine. In H. F. Willard & G. S. Ginsburg (Eds.), *Genomic and Personalized Medicine*. New York, NY: Elsevier Inc., Ch. 5.

Chapter 8

Antonarakis, S. E. (1998). Recommendations for a nomenclature system for human gene mutations. Nomenclature Working Group. *Hum. Mutat., 11*, 1–3.

Crow, J. F. (2000). The origins, patterns and implications of human spontaneous mutation. *Nat. Rev. Genet., 1*, 40–47.

Mills, R. E., Bennett, E. A., Iskow, R. C., et al. (2007). Which transposable elements are active in the human genome? *Trends Genet., 23*, 183–191.

Chapter 9

Darwin, C. (1969). *The Autobiography of Charles Darwin, 1809–1882 [with original omissions restored; edited with appendix and notes by his grand-daughter.* New York, NY: Nora Barlow. W.W. Norton.

Davies, R. (2008). *The Darwin Conspiracy: Origins of a Scientific Crime.* London, UK: Golden Square Books.

Gibbons, A. (2011). A new view of the birth of. *Homo sapiens. Science, 331*, 392–394.

National Academy of Sciences. (2008). *Science, Evolution, and Creationism.* Washington, DC: National Academy Press.

Wallace, A. R. (1908). *My Life: A Record of Events and Opinions.* London, UK: Chapman & Hall.

Chapter 10

Garrod, A. (1909). *Inborn Errors of Metabolism.* London, UK: Oxford University Press.

Harris, H. (1980). *The Principles of Human Biochemical Genetics* (3rd ed.). Amsterdam, The Netherlands: Elsevier.

Chapter 11

Bahado-Singh, R. O., Choi, S. J., & Cheng, C. C. (2004). First- and mid-trimester Down syndrome screening and detection. *Clin. Perinatol., 31*, 677–694.

Carrel, L., & Willard, H. F. (1999). Heterogeneous gene expression from the inactive X chromosome: an X-linked gene that escapes X inactivation in some human cell lines but is inactivated in others. *Proc. Natl Acad. Sci. USA, 96*, 7364–7369.

Driscoll, D. A., & Gross, S. (2009). Prenatal screening for aneuploidy. *N. Engl. J. Med., 360*, 2556–2562.

Epstein, C. J. (1986). *The Consequences of Chromosome Imbalance: Principles, Mechanisms, and Models.* New York, NY: Cambridge University Press.

Jiang, Y. H., Bressler, J., & Beaudet, A. L. (2004). Epigenetics and human disease. *Annu. Rev. Genomics Hum. Genet., 5*, 479–510.

Ledbetter, D. H. (2008). Cytogenetic technology—genotype and phenotype. *N. Engl. J. Med., 359*, 1728–1730.

Rosenberg, L. E. (1982). *The Congressional Record: Hearings Before the Subcommittee on Separation of Powers of the Committee on the Judiciary United States Senate—The Human Life Bill.* Washington, DC: US Government Printing Office. pp 48–52

Speicher, M. R., & Carter, N. P. (2005). The new cytogenetics: blurring the boundaries with molecular biology. *Nat. Rev. Genet., 6*, 782–792.

Chapter 12

Antonarakis, S. E., Chakravarti, A., Cohen, J. C., et al. (2010). Mendelian disorders and multifactorial traits: the big divide or one for all? *Nat. Rev. Genet., 11*, 380–384.

Beaudet, A. L., Scriver, C. R., Sly, W. S., et al. (2001). Genetics, biochemistry, and molecular bases of variant human phenotypes. In C. R. Scriver, A. L. Beaudet & W. S. Sly (Eds.), *The Metabolic and Molecular Bases of Inherited Disease* (8th ed.). New York, NY: McGraw-Hill. Chapter 1.

Campbell, A. G. M., Rosenberg, L. E., Snodgrass, P. J., & Nuzum, C. T. (1973). Ornithine transcarbamylase deficiency, a cause of lethal neonatal hyperammonemia in males. *N. Engl. J. Med., 288*, 1–6.

Pauling, L., Itano, J. A., Singer, S. J., et al. (1949). Sickle cell anemia, a molecular disease. *Science, 110*, 543–548.

La Spada, A. R., & Taylor, J. P. (2010). Repeat expansion disease: progress and puzzles in disease pathogenesis. *Nat. Rev. Genet., 11*, 247–258.

Rosenberg, L. E. (2008). Legacies of Garrod's brilliance. One hundred years—and counting. *J. Inherit. Metab. Dis., 31*, 574–579.

Sankaran, V. G., Xu, J., Byron, R., et al. (2011). A functional element necessary for fetal hemoglobin silencing. *N. Engl. J. Med., 365*, 807–814.

Short, E. M., Conn, H. O., Snodgrass, P. J., et al. (1973). Evidence for X-linked dominant inheritance of ornithine transcarbamylase deficiency. *N. Engl. J. Med., 288*, 7–12.

Schechter, A. N. (2008). Hemoglobin research and the origins of molecular medicine. *Blood, 112*, 3927–3938.

Weatherall, D. J. (2004). Thalassaemia: the long road from bedside to genome. *Nat. Rev. Genet., 5*, 1–7.

Chapter 13

Barroso, I. (2005). Genetics of type 2 diabetes. *Diabet. Med., 22,* 517–535.

Cirulli, E. T., & Goldstein, D. B. (2010). Uncovering the roles of rare variants in common disease through whole-genome sequencing. *Nat. Rev. Genet., 11,* 415–425.

Dixon, M. J., Marazita, M. L., Beaty, T. H., et al. (2011). Cleft lip and palate: understanding genetic and environmental influences. *Nat. Rev. Genet., 12,* 167–178.

Fanous, A. H., et al. (2009). Bipolar disorder in the era of genomic psychiatry. In H. F. Willard & G. S. Ginsburg (Eds.), *Genomic and Personalized Medicine.* New York, NY: Elsevier Inc., Ch. 106.

Hirschhorn, J. N. (2009). Genome-wide association studies—illuminating biologic pathways. *N. Engl. J. Med., 360,* 1699–1701.

Mackay, T. F. C., & Anholt, R. R. H. (2007). Ain't misbehavin'? Genotype–environment interactions and the genetics of behavior. *Trends Genet., 23,* 311–314.

Manolio, T. A. (2010). Genome-wide association studies and assessment of the risk of disease. *N. Engl. J. Med., 363,* 166–176.

Nousbeck, J., Burger, B., Fuchs-Telem, D., et al. (2011). A mutation in a skin-specific isoform of SMARCAD1 causes autosomal-dominant adermatoglyphia. *Am. J. Hum. Genet., 89,* 302–307.

St George-Hyslop, P. H., & Petit, A. (2005). Molecular biology and genetics of Alzheimer's disease. *C.R. Biol., 328,* 119–130.

Smith, D. J., & Lusis, A. J. (2009). Genomic approaches to complex disease. In H. F. Willard & G. S. Ginsburg (Eds.), *Genomic and Personalized Medicine.* New York, NY: Elsevier Inc., Ch. 3.

The Wellcome Trust Case Control Constorium. (2007). Genome-wide association study of 14,000 cases of seven common diseases and 3,000 shared controls. *Nature, 447,* 661–678.

Visscher, P. M., Hill, W. G., & Wray, N. R. (2008). Heritability In the genomics era—concepts and misconceptions. *Nat. Rev. Genet., 9,* 255–266.

Chapter 14

Eichler, E. E., Flint, J., Gibson, G., et al. (2010). Missing heritability and strategies for finding the underlying causes of complex disease. *Nat. Rev. Genet., 11,* 446–450.

Knight, H. M., Pickard, B. S., Maclean, A., et al. (2009). A cytogenetic abnormality and rare coding variants identify ABCA13 as a candidate gene in schizophrenia, bipolar disorder, and depression. *Am. J. Hum. Genet., 85,* 833–846.

Lee, C., Hyland, C., Lee, A. S., et al. (2009). Copy number variation and human health. In H. F. Willard & G. S. Ginsburg (Eds.), *Genomic and Personalized Medicine.* New York, NY: Elsevier Inc., Ch. 9.

Lupski, J. R. (2009). Genomic disorders ten years on. *Genome Med. 1.* Article 42.

Lupski, J. R., Reid, J. G., Gonzaga-Jauregui, C., et al. (2010). Whole-genome sequencing in a patient with Charcot-Marie-Tooth neuropathy. *N. Engl. J. Med., 362,* 1181–1191.

Redon, R., Ishikawa, S., Fitch, K. R., et al. (2006). Global variation in copy number in the human genome. *Nature, 444,* 444–454.

Sebat, J., Levy, D. L., & McCarthy, S. E. (2009). Rare structural variants in schizophrenia: one disorder, multiple mutations; one mutation, multiple disorders. *Trends Genet., 25,* 528–535.

Stankiewicz, P., & Lupski, J. R. (2009). Structural variation in the human genome and its role in disease. *Annu. Rev. Med., 61,* 437–455.

Stranger, B. E. (2007). Relative impact of nucleotide and copy number variation on gene expression phenotypes. *Science, 315,* 848–853.

Zhang, F., Carvalho, C. M. B., & Lupski, J. R. (2009). Complex human chromosomal and genomic rearrangements. *Trends Genet., 25,* 298–307.

Chapter 15

Epstein, C. J., Erickson, R. P. & Wynshaw-Boris, A. J. (Eds.). (2004). Wynshaw-Boris. *Inborn Errors of Development: The Molecular Basis of Clinical Disorders of Morphogenesis.* New York, NY: Oxford University Press.

Gilbert, S. F. (2003). *Developmental Biology* (7th ed.). Sunderland, MA: Sinauer Associates.

Mitchell, L. E., Adzick, N. S., Melchione, J., et al. (2004). Spina bifida. *Lancet, 364,* 1885–1895.

Chapter 16

Chaffer, C. L., & Weinberg., R. A. (2011). A perspective on cancer cell metastasis. *Science, 331*, 1559–1564.

Croce, C. M. (2009). Causes and consequences of microRNA dysregulation in cancer. *Nat. Rev. Genet., 10*, 704–714.

Erez, A., Shchelochkov, O. A., Plon, S. E., et al. (2011). Insights into the pathogenesis and treatment of cancer from inborn errors of metabolism. *Am. J. Hum. Genet., 88*, 402–421.

Esquela-Kerscher, A., & Slack, F. J. (2006). Oncomirs—microRNAs with a role in cancer. *Nat. Rev. Cancer, 6*, 259–269.

Frank, S. A. (2004). Genetic predisposition to cancer—insights from population genetics. *Nat. Rev. Genet., 5*, 764–772.

Levine, A. J. (1997). p53, the cellular gatekeeper for growth and division. *Cell, 88*, 323.

Lindhurst, M. J., Sapp, J. C., Teer, J. K., et al. (2011). A mosaic activating mutation in AKT1 associated with the proteus syndrome. *N. Engl. J. Med., 365*, 611–619.

Mardis, E. R. (2009). Recurring mutations found by sequencing an acute myeloid leukemia genome. *N. Engl. J. Med., 361*, 1058–1066.

Stratton, M. R. (2011). Exploring the genomes of cancer cells: progress and promise. *Science, 331*, 1553–1558.

Strausberg, R. L. (2009). Cancer genes, genomes, and the environment. In H. F. Willard & G. S. Ginsburg (Eds.), *Genomic and Personalized Medicine*. New York, NY: Elsevier Inc., Ch. 67.

Varmus, H. (2006). The new era in cancer research. *Science, 312*, 1162–1165.

Vogelstein, B., & Kinzler, K. W. (2004). Cancer genes and the pathways they control. *Nat. Med., 10*, 789–799.

Witte, J. S. (2009). Prostate cancer genomics: towards a new understanding. *Nat. Rev. Genet., 10*, 77–82.

Chapter 17

Annas, G. J. (2010). Resurrection of a stem-cell funding barrier—Dickey-Wicker in court. *N. Engl. J. Med., 363*, 1687–1689.

Belmonte, J. C. I., Ellis, J., Hochedlinger, K., et al. (2009). Induced pluripotent stem cells and reprogramming: seeing the science through the hype. *Nat. Rev. Genet., 10*, 878–883.

Copelan, E. A. (2006). Hematopoietic stem-cell transplantation. *N. Engl. J. Med., 354*, 1813–1826.

Dietz, H. C. (2010). New therapeutic approaches to Mendelian disorders. *N. Engl. J. Med., 363*, 852–863.

Goemans, N. M., Tulinius, M. T., van den Akker, J. T., et al. (2011). Systemic administration of PRO051 in Duchenne's muscular dystrophy. *N. Engl. J. Med., 364*, 1513–1522.

González, F., Boué, S., & Belmonte, J. C. I. (2011). Methods for making induced pluripotent stem cells: reprogramming à la carte. *Nat. Rev. Genet., 12*, 231–242.

Hacein-Bey-Abina, S., Hauer, J., Lim, A., et al. (2010). Efficacy of gene therapy for X-linked severe combined immunodeficiency. *N. Engl. J. Med., 363*, 355–364.

Hammond, S. M., & Wood, M. J. A. (2011). Genetic therapies for RNA mis-splicing diseases. *Trends Genet., 27*, 196–205.

Kay, M. A. (2011). State-of-the-art gene-based therapies: the road ahead. *Nat. Rev. Genet., 12*, 316–328.

McCulloch, E. A., & Till, J. E. (2005). Perspectives on the properties of stem cells. *Nat. Med., 11*. v–vii.

Mingozzi, F., & High, K. A. (2011). Therapeutic *in vivo* gene transfer for genetic disease using AAV: progress and challenges. *Nat. Rev. Genet., 12*, 341–355.

Rosenberg, L. E., Lilljeqvist, A., & Hsia, Y. E. (1968). Methylmalonic aciduria: metabolic block localization and vitamin B_{12} dependency. *Science, 162*, 805–807.

Treacy, E. P., Valle, D., & Scriver, C. R. (2001). Treatment of genetic disease. In C. R. Scriver, A. L. Beaudet, W. S. Sly & D. Valle (Eds.), *The Metabolic and Molecular Bases of Inherited Disease* (8th ed.). New York, NY: McGraw-Hill.

Verma, I. M., & Weitzman, M. D. (2005). Gene therapy: twenty-first century medicine. *Annu. Rev. Biochem., 74*, 711–738.

Chapter 18

Foster, M. W., & Sharp, R. R. (2004). Beyond race: towards a whole-genome perspective on human populations and genetic variation. *Nat. Rev. Genet., 5*, 790–796.

Hartl, D. L., & Clark, A. G. (2006). *Principles of Population Genetics* (4th ed.). Sunderland, MA: Sinauer Associates.

Jorde, L. B., Bamshed, M., & Rogers, A. R. (1998). Using mitochondrial and nuclear DNA markers to reconstruct human evolution. *Bioessays, 20,* 126–136.

King, T. E., & Jobling, M. A. (2009). What's in a name? Y chromosomes, surnames and the genetic genealogy revolution. *Trends Genet., 25,* 351–360.

Kuokkhanen, M., Kokkonen, J., Enattah, N. S., et al. (2006). Mutations in the translated region of the lactase gene (LCT) underlie congenital lactase deficiency. *Am. J. Hum. Genet., 78,* 339–344.

Laland, K. N., Odling-Smee, J., & Myles, S. (2010). How culture shaped the human genome: bringing genetics and the human sciences together. *Nat. Rev. Genet., 11,* 137–148.

Rotimi, C. N., & Jorde, L. B. (2010). Ancestry and disease in the age of genomic medicine. *N. Engl. J. Med., 363,* 1551–1558.

Royal, C. D., Novembre, J., Fullerton, S. M., et al. (2010). Inferring genetic ancestry: opportunities, challenges, and implications. *Am. J. Hum. Genet., 86,* 661–673.

Smith, J. J., Baum, D. A., & Moore, A. (2009). The need for molecular genetic perspectives in evolutionary education (and vice versa). *Trends Genet., 25,* 427–429.

Weale, M. E., & Goldstein, D. B. (2009). Concepts of population genomics. In H. F. Willard & G. S. Ginsburg (Eds.), *Genomic and Personalized Medicine.* New York, NY: Elsevier Inc., Ch. 2.

Chapter 19

Altman, R. B., Kroemer, H. K., McCarty, C. A., et al. (2011). Pharmacogenomics: will the promise be fulfilled? *Nat. Rev. Genet., 12,* 69–73.

Annas, G. J. (2009). Protecting privacy and the public —limits on police use of bioidentifiers in Europe. *N. Engl. J. Med., 361,* 196–201.

Bloss, C. S., Schork, N. J., & Topol, E. J. (2011). Effect of direct-to-consumer genome-wide profiling to assess disease risk. *N. Engl. J. Med., 364,* 524–534.

Daly, A. K. (2010). Genome-wide association studies in pharmacogenomics. *Nat. Rev. Genet., 11,* 241–246.

Frueh, F. W., Greely, H. T., Green, R. C., et al. (2011). The future of direct-to-consumer clinical genetic tests. *Nat. Rev. Genet., 12,* 511–515.

Guttmacher, A. E., McGuire, A. L., Ponder, B., et al. (2010). Personalized genomic information: preparing for the future of genetic medicine. *Nat. Rev. Genet., 11,* 161–165.

Grossman, I. (2007). Routine pharmacogenetic testing in clinical practice—dream or reality? *Pharmacogenomics, 8,* 1449–1459.

Jeffreys, A. J. (1993). DNA typing: approaches and applications. *J. Forensic Sci. Soc., 33,* 204–211.

Nadeau, J. H., & Topol, E. J. (2006). The genetics of health. *Nat. Genet., 38,* 1095–1098.

Wagner, J. K. (2010). Understanding FDA regulation of DTC genetic tests within the context of administrative law. *Am. J. Hum. Genet., 87,* 451–456.

Weele, T. V. (2010). Genetic self knowledge and the future of epidemiologic confounding. *Am. J. Hum. Genet., 87,* 168–172.

APPENDIX 1

Answers to Review Questions

CHAPTER 1: FRAMING THE FIELD

1. i. Discovery of structure of DNA served as central organizing principle.
   ii. The field deals with understanding who we are and how we got to be this way.
   iii. Realization that most human disorders result, at least in part, from alteration in DNA structure and function.
   iv. A remarkable series of discoveries has spawned the biotechnology industry, and with it, new kinds of medicines and diagnostic tests.
   v. It has provoked numerous societal collisions which are ongoing.
2. Hopes:
   i. Understanding cause(s) of such common disorders as asthma, diabetes, Alzheimer.
   ii. Answer questions about familial similarities and differences.
   iii. Provide new means to detect and treat disorders in children and adults.

   Fears:

   i. Will lead to genetic determinism (that is, we are nothing more than our genes).
   ii. Will abridge privacy by allowing others access to our individual genetic makeup.
   iii. Will change fundamentally the nature of human reproduction.
3. a. Amniocentesis, because it may lead to abortion, which is unalterably opposed by those who espouse the Right to Life perspective.
   b. Neonatal screening, because it mandates tests which some consider an invasion of privacy.
   c. Gene therapy, because it requires insertion of someone else's DNA, and is, therefore, unnatural.
4. a. For now it is attention-getting because we don't know enough about the human genome to determine the significance of genomic testing. There are those who argue that we already require information about a candidate's health, so why does DNA testing differ in any major way?
   b. More understanding about the human genome; more public education about genetic principles; mitigation of tension between science and religion.

CHAPTER 3: GENETIC VARIABILITY

1. a. Each organism produces its own kind during reproduction—and no other kind.
   b. Aristotle believed that the father and the mother contributed biologic material during reproduction, but he was far from understanding that sperm and egg were the responsible cells.
   c. Because there were no accurate scientific principles with which to explain reproduction before the 19th century.
   d. DNA in the male gamete combines with DNA in the female gamete to produce a zygote specific for each species.

2. a. That genes are inherited as units which segregate independently.
   b. He examined phenotypes (the visual result of gene action) and deduced the behavior of genes (that is, the genotypes).
   c. Archibald Garrod, Frances Crick, James Watson. (Rosalind Franklin is an unsung daughter.)
   d. Genes.
3. Because genes provide the biologic information from which to build an organism, just as an architectural blueprint provides a plan with which to build a house, etc. The biologic structures used as building blocks are proteins, pathways, cells, and organs.

CHAPTER 4: GROWTH, DEVELOPMENT, AND REPRODUCTION

1. a/11; b/3; c/2; d/13; e/8; f/7; g/4; h/9; i/6; j/12; k/1; l/14; m/10; n/5.
2. i. Random chromosome segregation during meiosis I that produces mixing of paternal and maternal chromosomes.
   ii. Crossing over between pairs of homologous chromosomes that increases genetic variability.
3. a. Genome = entire complement of DNA in a cell; chromosome = single linear molecule of DNA and its associated proteins.
   b. Gene = the unit of inheritance; allele = an alternate form of a gene.
   c. Prokaryote = an organism without a nucleus; eukaryote = an organism whose genetic material is contained within a membrane-bound nuclear compartment.
   d. Aneuploidy = any chromosome number not a perfect multiple of the haploid number; polyploidy = having more than two complete sets of chromosomes.
4. i. Male meiosis produces four haploid gametes; female meiosis produces only one gamete.
   ii. Male meiosis is initiated at puberty and continues for remainder of life; female meiosis begins during prenatal development and ends at time of menopause.
   iii. Male meiosis is completed in 64 days; female meiosis is arrested twice and is completed only after many years.
   iv. Males produce 10^{12} sperm during lifetime; females produce only 400 mature eggs.
5. a. 23 chromosomes; 46 chromatids.
   b. 23 chromosomes; 23 chromatids.
   c. Upon fertilization and zygote formation.
6. Identical twins are produced by splitting a single early embryo formed from union of a single sperm and a single egg. Non-identical twins are produced by each of two eggs being fertilized by a single different sperm.
7. To reduce diploid number of chromosomes to haploid number so that diploid number is reconstituted when haploid male gamete fertilizes haploid female gamete.
8. Probability $= 1/2^{23} \times 1/2^{23} = 1/2^{46}$. Could be male or female depending on whether paternal gamete had an X or Y.
9. a. i. Preconceptual screening: to identify genotypes at high risk.
      ii. Pre-implantation genetic diagnosis: to biopsy single cell of early embryo and define its genotype.

iii. *In vitro* fertilization: to obtain sperm and egg from couple and fertilize them in a culture dish with aim of returning zygote to mother.
iv. Prenatal diagnosis: examine fetus directly or by removal of cells through amniocentesis or chorion villus biopsy to diagnose disorders.
v. Artificial insemination: use of sperm donor to fertilize eggs of female partner in situations where male partner is infertile.

b. Technologies i, iii, and v will enable couples to reproduce; thus they should be more ethically defensible to the group described in the question.
c. Each of these technologies enables couples to have healthy children of their own, either by directly managing their infertility, or by removing the couple's anxiety about having a child with a serious disorder.

CHAPTER 5: TRANSMISSION OF GENES

1. a/5; b/8; c/11; d/1; e/3; f/4; g/10; h/2; i/6; j/7; k/9.
2. 1/2; $1/2 \times 1/2 = 1/4$; $1/2 \times 1/2 \times 1/2 = 1/8$; no, because 10 flips is too small a number to expect to obtain precisely equal numbers of heads and tails. (Do the experiment and see.)
3. An example of a "skipped generation" due to incomplete penetrance of the gene responsible for clinodactyly.
4. The Y chromosome contains a very small number of genes—far fewer than any autosome or the X chromosome. A Y-linked trait should be transmitted from an affected father to each of his sons; no females should be affected. A prominent Y-linked characteristic is gender determination controlled by the SRY locus; hairy ears is a Y-linked trait.
5. a. The pedigree:

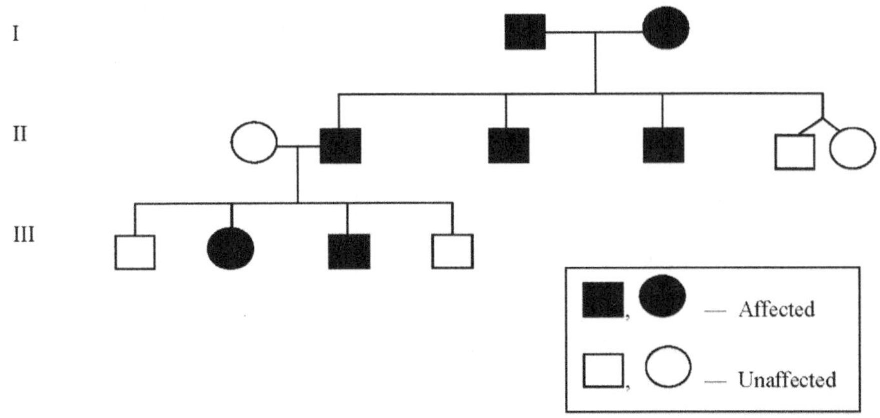

b. Autosomal dominant; vertical transmission, about equal number of affected males and females; each affected child has an affected parent.
c. Male-to-male transmission excludes X-linked condition; mitochondrial inheritance ruled out because some offspring of affected woman are unaffected and because some offspring of affected male are affected.
d. This would indicate that deafness in Amos is caused by a different genetic defect than is deafness in Esther. This is an example of genetic heterogeneity (one phenotype produced by more than one genotype).

6. a. The pedigree:

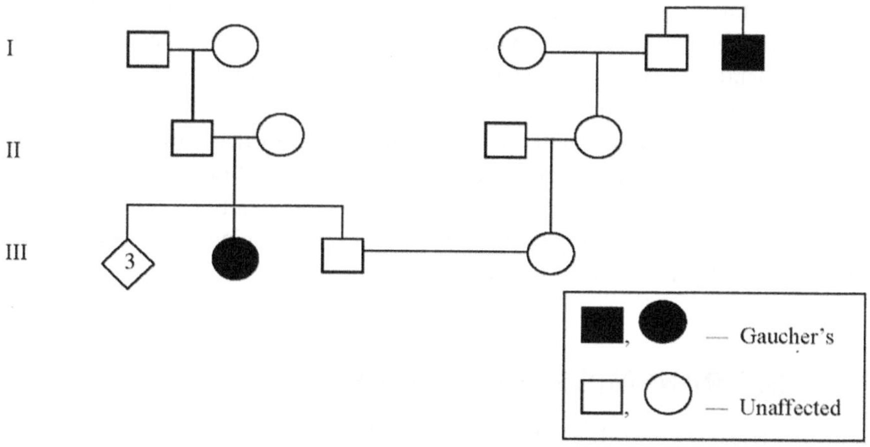

b. Probability that Charles is a carrier: 2/3.
Probability that Rachel is a carrier: $2/3 \times 1/2 \times 1/2 = 1/6$.
Probability that child will be affected: $2/3 \times 1/6 \times 1/4 = 1/36$.
c. $1/36 \times 1/36$.

7. a. The pedigree:

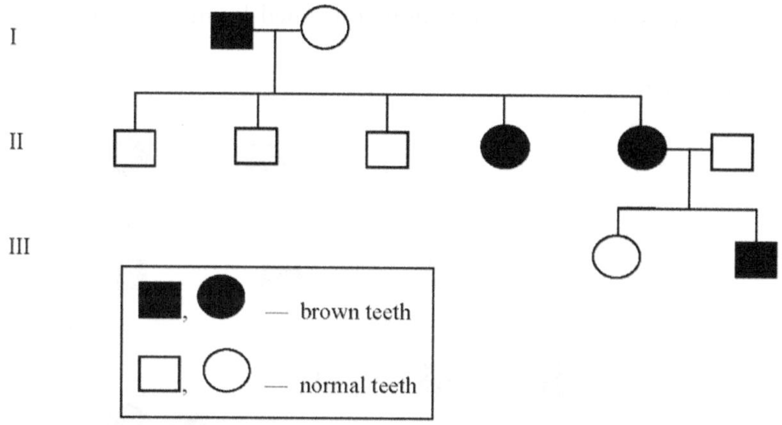

b. X-linked (either recessive or dominant).
c. Autosomal dominant.
d. Mitochondrial.

8. a. Mechanism to account for the observation that the amount of product formed by two copies of a gene is the same as that formed by having only one copy of that gene.
b. X chromosome.
c. Random inactivation of one of the two X chromosomes in each cell early in development.
d. Females heterozygous for X-linked traits may differ from one another phenotypically because of skewing of cells from the expected 50/50 ratio either toward or away from those bearing the normal or mutant X chromosome.

9. a. 1/2.
b. Probability has no memory. Each pregnancy has its own independent probability.
c. Variable expressivity (whose mechanism is unknown).
d. $1/2 \times 1/2 \times 1/2 \times 1/2 = 1/16$.

10. a. The pedigree:

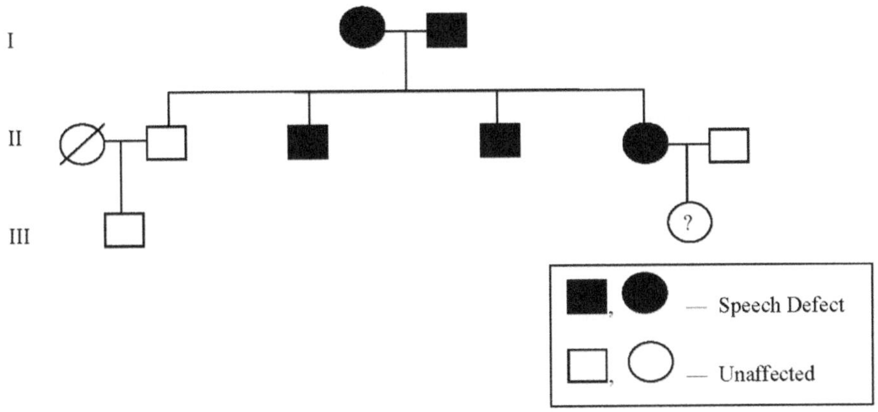

b. Autosomal recessive because Michael is not affected; mitochondrial for the same reason.

c. Autosomal dominant (X-linked dominant also possible).

d. Because George and Lucille are heterozygotes, their affected daughter has a 2/3 probability of being heterozygous and a 1/3 probability of being homozygous affected. If she is a homozygote, her daughter has a 100% chance of being affected; if she is a heterozygote, her daughter has a $2/3 \times 1/2 = 1/3$ probability of being affected.

11. a. The pedigree:

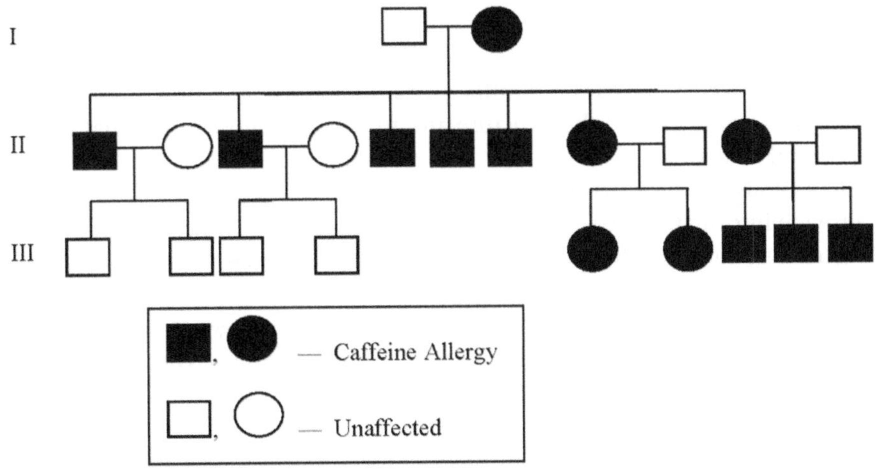

b. Mitochondrial because all offspring of affected female are affected, and no offspring of affected male are affected.

c. No, because Elmo won't transmit the mitochondrial defect to his offspring; mitochondria are maternally inherited.

12. a. i. Short generation time.

ii. Unambiguous discrete phenotypes.

iii. Presence of pure breeding varieties.

iv. Ability to make test crosses at will.

v. Careful control of mating.

b. None of the above characteristics are possible or permissible in humans. Human geneticists make discoveries based on chance matings in other humans.

CHAPTER 6: STRUCTURE OF GENES, CHROMOSOMES, AND GENOMES

1. a/4; b/8; c/6; d/10; e/9; f/13; g/2; h/3; i/1; j/5; k/14; l/11; m/7; n/12.
2. Sulfur.
3. a. Haplotype = genotype of alleles at two or more closely linked loci on a chromosome that are inherited as a unit.
   b. Phenotype = observable manifestations of a genotype.
   c. Karyotype = chromosome constitution of a cell, person or organism usually arrayed in a characteristic fashion to permit identification of individual chromosomes.
   d. Genotype = genetic constitution of an individual or, specifically, the alleles at a specific locus.
4. a. Restriction enzymes: bacterial enzymes that cut DNA at specific sites.
   b. Molecular cloning: the process by which DNA fragments are spliced into bacterial or viral vectors.
   c. Hybridization probes: single-stranded DNA sequences of varying lengths labeled with a radioactive isotope or fluorescent dye.
   d. Polymerase chain reaction: amplification of a piece of DNA up to a million-fold or more using a bacterial polymerase.
   e. DNA sequencing: decodes the order of bases in an isolated DNA molecule.
5. Because it provided a much larger number of linkage markers than had been available using specific disorders or proteins.
6. i. Determining the complete nucleotide sequence of the human genome.
   ii. One dollar per haploid base (3 billion dollars; 3 billion bases).
   iii. That the locus is conserved throughout evolution, and must play a vital function.
   iv. Two nuclear genomes (one inherited from each parent); one mitochondrial genome (inherited from the mother).
   v. Humans have fewer genes per million bases than do other organisms.
   vi. HGP has not identified the structure and function of most genes; that remains a critical next step.
7. i. Identify susceptibility to specific disorders.
   ii. Provide information about response to medicines.
   iii. Determine susceptibility or resistance to specific bacterial or viral pathogens.
8. Because of the distinctive pedigree pattern of X-linked traits or disorders.
9. The critical issue is not the number of genes, but how those genes are "read." Humans have 21,000 genes, but make 200,000 or more protein products. This explains our greater complexity compared to fish or plants.

CHAPTER 7: EXPRESSION OF GENES AND GENOMES

1. a/8; b/7; c/12; d/6; e/14; f/2; g/1; h/5; i/9; j/11; k/3; l/16; m/15; n/13; o/4; p/10.
2. d; b; h; f; c; g; a; e.
3. As named by Crick, the central dogma states that genetic information moves in only one direction in two distinct stages: from DNA to RNA via transcription; and from RNA to protein via translation.
4. i. DNA is double stranded; RNA is (usually) single stranded.
   ii. Deoxyribose is the sugar molecule in DNA; ribose is the sugar molecue in RNA.
   iii. One of the pyrimidine bases in DNA is thymine. Uracil substitutes for thymine in RNA.
5. i. mRNA—RNA that is translated into protein.

- **ii.** tRNA—RNA adapter molecules that place amino acids into their correct position in a growing peptide chain.
- **iii.** rRNA—component of ribosomes, the RNA-protein complexes on which protein synthesis takes place.
- **iv.** miRNA and siRNA—short RNA molecules found in the genome which are transcribed and used to modulate gene expression.

6. a. Translation: because nonsense codon terminates growth of polypeptide chain.
b. Replication: because a cell's DNA content is duplicated during each round of cell division.
c. Transcription: because such a deleterious mutation may interfere with initiation, elongation, or termination of the primary transcript.

7. a. By convention, the left end of a DNA or RNA molecule is called the 5′ end; the right end is called the 3′ end. During transcription the template DNA strand Is read from 5′ to 3′. The primary transcript is formed from 3′ to 5′.
b. Complementary.
c. Identical.

8. Ala—Gly—Cys—Lys—Asn—Phe.

9.

CUU—Leu	UCU—Ser	UUC—Ser
AUU—Ile	UAU—Tyr	UUA—Leu
GUU—Val	UGU—Cys	UUG—Leu

10. The Nobel Prize is never awarded to more than three people for a single discovery. In the case of the discovery of messenger RNA, four scientists (Sydney Brenner, Frances Crick, Francois Jacob, and Jacques Monod) made major contributions— too important to be excluded. The Nobel Committee, therefore, never awarded a prize for this discovery. Each of the four, however, went on to win a Nobel Prize for other contributions to molecular genetics.

11. a. X inactivation silences the great majority of genes on one of the two X chromosomes. Imprinting silences one gene at a time on many different autosomes.
b. X inactivation is random in that either the paternally derived X or the maternally derived X will be affected. In imprinting, either the allele transmitted by the father or that transmitted by the mother will be silenced in an ordered non-random fashion from one generation to the next.

CHAPTER 8: MUTATION

1. a/10; b/5; c/8; d/11; e/13; f/1; g/4; h/16; i/17; j/15; k/7; l/12; m/9; n/2; o/3; p/14; q/6.

2. a. 2.
b. AUU→AAU
AUC→AAC.

3. **i.** Donor site: would lead to portion (or all) of intron being included in the corresponding mRNA.
ii. Acceptor site: would also lead to intron being included in the mRNA.

4. **i.** The substitution could be silent (or synonymous), meaning that it produced no change in the amino acid sequence.
ii. The substitution could produce a change in the amino acid, but this single amino acid change would be neutral (that is, without effect on the function of the protein).
iii. The substitution could occur in an intron and therefore would be removed during splicing of the primary transcript.

5. The average number of generations is $1/5 \times 10-^{4} = 2{,}000$ generations.

6. **a.** No. Substitutions, deletions, insertions, and splicing mutations may all result in premature termination.
   b. Given that there are 3 stop codons out of 64 total codons, the number would be 3/64, or about 1/21.
7. **a.** In intron, expansion could interfere with splicing, and therefore with structure of mRNA.
   b. In exon, expansion could lead either to gain of function or loss of function of corresponding protein. (In fact, little is known about precise mechanism of expansion on phenotype.)
8. Removal of either of two possible phenylalanine codons: UUU or UUC.
9. **a.** Due to the degeneracy of the genetic code, there are several possible sequences. Here is one:
      Non-template strand: AAA AGA CAT CAT TAT CTA
      Template strand: TTT TCT GTA GTA ATA GAT.
   b. Frameshift mutation: insertion of G between first and second codons.
   c. Frameshift mutation: deletion of first nucleotide of third codon.
10. **a.** Would interfere with or eliminate transcription.
    b. Would prevent normal translation.
    c. Would interfere with DNA replication.

CHAPTER 9: BIOLOGICAL EVOLUTION

1. a/4; b/2; c/1; d/6; e/5; f/3.
2. $4 \times 10^6 / 4 \times 10^9 = 1 \times 10{-}^3$, or 1/1,000.
3. Agree, because it is consistent with all of the scientific evidence, including: homology of genes from bacteria to man; common informational macromolecule (DNA) and functional macromolecules (proteins); mutation as driving force for natural selection.
   Disagree, because evolution does not explain how life began, and doesn't leave any room for the profound role of a God as creator.
4. (b) and (c) are true; (a) and (d) are false.
5. One might see a sharp reduction in frequency of sporadic cases (new mutations) of dominant disorders. One might see a reduced frequency of chromosomal disorders—prenatal and postnatal ones.
6. **i.** Molecules that could reproduce themselves had to come together.
   ii. Copies of these molecular assemblies had to show variations.
   iii. Variations had to be heritable.
7. **i.** Paleontology: developed fossil record.
   ii. Geology: showed that Earth is old enough.
   iii. Chemistry: developed dating methods.
   iv. Anthropology: provided insights into human behavior.
   v. Genetics: discovered macromolecular events.

CHAPTER 10: HUMAN INDIVIDUALITY

1. During meiosis when random homologue separation and crossing over between homologues occur. These events produce many more possible genotypes than found in the world's people.

2. The haploid genome consists of 3 billion bases. If any two humans differ at 0.1% of them, or 3 million nucleotides, there is considerable opportunity for individual difference. This, plus all the epigenetic, miRNA- and siRNA-caused, and copy number variation, provides the components for genetic uniqueness.
3. Because identical twins do not have identical environments—prenatally or postnatally.
4. Because it is in the interaction between genes and the environment that uniqueness resides. Accounting for subtle or not-so-subtle differences in environment will surely limit understanding of individual uniqueness.

CHAPTER 11: CHROMOSOME ABNORMALITIES

1. a/11; b/8; c/5; d/10; e/6; f/3; g/12; h/14; i/7; j/1; k/13; l/9; m/4; n/2; o/15.
2. a. Yes, on affected child, to confirm diagnosis, and to look for translocation carrier in either parent, which would increase the risk of recurrence.
   b. Yes, on father and mother, looking for chromosomal error (such as balanced translocation) which might explain miscarriages.
   c. Yes, on fetus, looking for trisomy 21 which is much more frequent in offspring of older mothers.
   d. No, there is no reason to do karyotyping on parents or children. Albinism is a single-gene disorder inherited as an autosomal recessive.
   e. Yes, on son, looking for karyotype associated with Klinefelter syndrome (47, XY, +X).
   f. Yes, on both parents, looking for some kind of chromosomal error.
3. a. The child is a female with a chromosomal deletion of region 9 on chromosome number 13.
   b. The deleted region may contain one or more genes that must be present in two copies for normal development.
   c. If neither of the parents had the deletion, it would suggest that the deletion occurred in the germ line of one of them, thereby being consistent with the idea that the deletion was responsible for the child's phenotype; if one of the healthy parents had the same deletion as the child, it would imply that the deletion was not the cause of the problem.
4. a. Indirect: increased risk with advanced maternal, but not paternal, age. Direct: examination of the maternal and paternal number 21 chromosomes in parents of Down syndrome child.
   b. Meiotic non-disjunction—most often in meiosis I, but occasionally in meiosis II.
   c. Several associated defects: congenital heart abnormalities; increased risk of leukemia; amyloid plaques in brain.
   d. Trisomies for 13,18, X, and Y; monosomy for X only.
   e. Two kinds of sperm—disomic and monosomic for 21; would expect half to be disomic (normal). Two kinds of offspring: those with trisomy 21, and those with normal karyotype.
   f. To exclude the unlikely possibility of Down syndrome due to balanced translocation in one of the parents.
5. a. 46XY + 21/46XY in boy; 45X/46XX in girl.
   b. Early in mitosis.
   c. Patients with mosaic Down syndrom or Turner syndrome are often more mildly affected than those with a single aneuploid cell line.
6. This is purely hypothetical, because neither child with trisomy 18 nor monosomy 18 live long enough to reproduce.
7. a. 45, XX rob (14,21).

b.

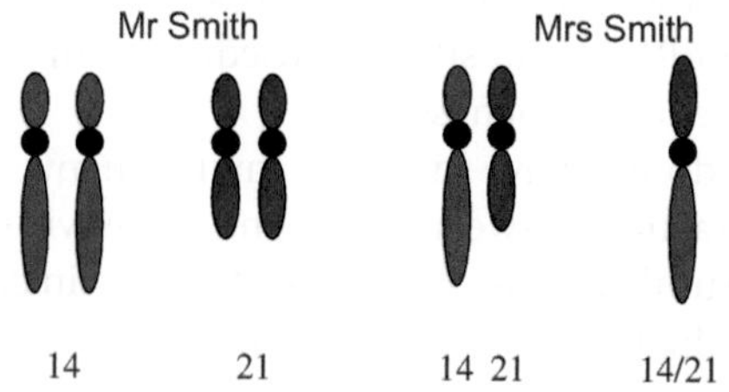

c. Would expect 1/3 of offspring to be normal, 1/3 to have Down syndrome, and 1/3 to be aborted.

8. In genomic imprinting, either the allele from the mother or that from the father is silenced epigenetically; in uniparental disomy, both alleles come from one parent due to meiotic non-disjunction.

9. a. Prada-Willi syndrome: obesity, short stature, developmental delay; Angelman syndrome: autistic behavior, seizures, severe developmental delay.

b. Each comes about as a result of the chromosome deletion and genomic imprinting; in PWS, the deletion occurs on the number 15 from the father, and the imprinting on the 15 from the mother silences the other allele for several genes in the region; in AS, the reciprocal occurs—the deletion affects the maternal homologue, and the imprinting of different gene(s) on the paternal chromosome in the same chromosomal region.

10. a. In the father.

b. Non-disjunction is a biological accident, not the fault of the father in this case, or the mother in most cases; ascribing guilt is misguided and counterproductive.

CHAPTER 12: SINGLE-GENE DEFECTS

1. a/9; b/3; c/6; d/12; e/8; f/14; g/10; h/15; i/1; j/16; k/13; l/4; m/17; n/7; o/2; p/11; q/5.

2. a. Allelic = multiple mutations at a single locus; non-allelic = mutations at two (or more) loci.

b. Non-allelic: children have inherited mutations from different loci—one from each parent—and are double heterozygotes.

c. Allelic: parents presumably have different mutations at the same locus; their children have phenotypic variability depending on which mutations they inherit.

3. Man's risk of being carrier: $1/2 \times 1/2 = 1/4$
Wife's risk of being carrier: $2/3 \times 1/2 = 1/3$
Risk of affected offspring: $1/4 \times 1/2 \times 1/3 \times 1/2 \times 1/4 = 1/192$.

4. a. Because males have only one X (are hemizygous) while females have two (are heterozygous).

b. i. Monosomy X with OTC mutation.

ii. Marked skewing of random X chromosome inactivation favoring cells containing the X with an OTC mutation.

iii. An X-autosome translocation in which an OTC mutation is preferentially expressed.

iv. Mating between a woman heterozygous for an OTC mutation and a man with a mild OTC mutation compatible with long-term survival.

c. Karyotyping the girl would rule in or out possibilities (i) and (iii); measuring OTC activity in her parents would address possibility (iv); if (i), (iii), and (iv) are ruled out, then answer must be (ii).

d. Either because they die before liver transplantation can be performed, or because the paucity of donor livers from suitable children makes transplantation unfeasible.

e. Most affected males are so severely affected that they do not reach reproductive age.

5. a. The pedigree:

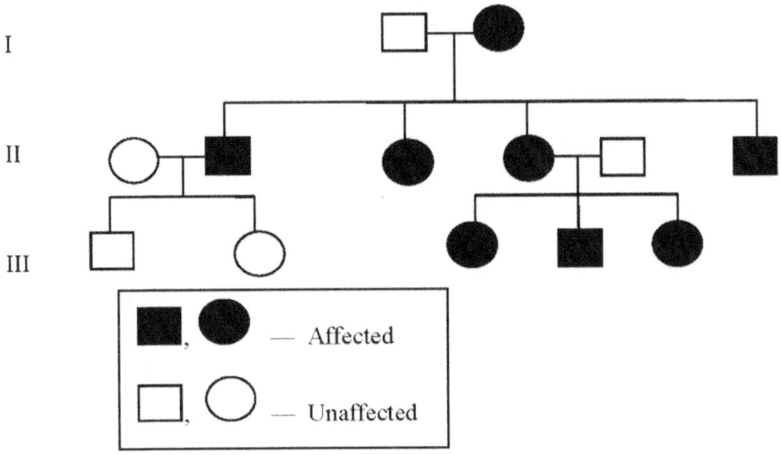

b. Elevated N, reduced I and H.
c. Restrict supply of N in diet; or replace H.
d. Would not be surprised, because some mitochondria are likely not to carry the mutation (called heteroplasmy).

6. a. Point mutation that converts stop codon to one coding for an amino acid; or a splicing mutation that "skips" stop codon.
b. $180 \times 3 = 540$.
c. Each genotype has a different phenotype; heterozygotes are phenotypically discernable from either homozygote.

7. a. i. Allelic heterogeneity at the MCR-1 locus.
ii. Other genes that modify expression of MCR-1.
iii. Mutations of genes that account for 20% of those with red hair who do not have MCR-1 mutations.
b. MCR-1 is expressed in skin cells as well as hair follicles.

8. A nonsense mutation in exon one would lead to a severely truncated protein product with little or no function; a missense mutation would change a single amino acid to another one, or not change the amino acid sequence at all; nonsense will, therefore, affect function to a greater degree than missense.

CHAPTER 13: MULTIFACTORIAL TRAITS

1. a/9; b/6; c/19; d/16; e/3; f/14; g/11; h/15; i/8; j/13; k/1; l/5; m/17; n/10; o/12; p/7; q/20; r/4; s/2; t/18.

2. a.
$$h^2 = \frac{\text{variance observed in relatives}}{\text{genetic relatedness}}$$

$$h^2 = \frac{(\text{variance in DZ twins}) - (\text{variance in MZ twins})}{\text{variance in DZ twins}}$$

b. Because they take do not take into account the environmental component to traits or disorders.
c. The gene pool in the US may differ significantly from that in Australia, thereby leading to differences in gene frequency for loci responsible for the anomaly.

3. a. Apparently so; mean IQ in twins was 95.7, compared to statistical mean in population of 100.
b. Average difference within pairs: 8.2 points.
c. i. Range of differences within pairs goes from 1 to 24, indicating that environment must play an important part in IQ variation.
ii. IQ obeys the rules for a normal distribution.

4. a. Largest role, disorder 3; smallest role, disorder 1.
   b. Disorder 3 is a single-gene disorder: MZ >> DZ = S. There is no way to estimate number of genes in disorder 2: MZ > DZ > S. There is no genetic contribution in disorder 1: MZ = DZ >> S.
   c. If their environments were very different (e.g., if they were separated at birth and reared apart); if their prenatal environments or epigenetic modification differed.
   d. DZ twins share their prenatal environment to a greater degree than do sibs.
5. a. $15/50 = 3/10$ (the variance); degree of relatedness $= 1/2$; $h^2 = 3/10 \div 1/2 = 0.6$.
   b. Sibs may have many environmental factors in common, thereby leading to an overestimation of the genetic components.
6. Fraction of genes shared by great grandparents and their great grandchildren: $1/2 \times 1/2 \times 1/2 = 1/8$ (or 0.125) $h^2 = 0.02/0.125 = 0.16$.
7. 1/2 (you to your parent); 1/2 (your parent to your grandparent); 1/2 (your grandparent to your great uncle or aunt); 1/2 (your great uncle or aunt to their child); thus $1/2 \times 1/2 \times 1/2 \times 1/2 = 1/16$ of your genome.
8. a. GWAS is performed in populations of people with a particular condition and in controls; linkage studies are conducted in families.
   b. Before completion of the human genome project the SNPs used to seek susceptibility alleles were not available, nor were their haplotypes.
   c. GWAS identify susceptibility alleles (necessary but not sufficient to produce phenotype); single-gene disorders are caused by a single mutation (both necessary and sufficient to produce phenotype).
   d. AMD is a reliably detectable phenotype, diabetes mellitus is a syndrome with much variability; it is quite likely that each subtype of diabetes is caused by a different mutation(s) which would not be identifiable when all forms are lumped together.
   e. i. Improved diagnostic criteria.
      ii. Do larger population studies.
      iii. Sequence the entire genome of patients and controls rather than relying on SNPs.
9. a. SNP 3.
   b. i. Study large population of those with baldness and carefully matched controls (age; ethnicity, occupation, etc.).
      ii. Use SNP 3 to seek more SNPs in same region of genome.
      iii. Identify candidate genes in region.
      iv. Sequence gene reasonably expected to be physiologically associated with baldness.
10. Empiric risk in affected families.
    Concordance in twins of affected children.
    Examine offspring of affected; would not expect them to be affected if condition is autosomal recessive.
11. a. Genetic determinism ("it's all in our genes").
    b. They are exaggerated and scientifically unfounded; genes don't "explain" any of the traits described.
    c. No, because the vast majority of the lay public don't understand such fundamental principles as: how genes work; what gene–environment interactions are; the laws of probability; and much more.

CHAPTER 14: DISORDERS OF VARIABLE GENOMIC ARCHITECTURE

1. a/5; b/10; c/13; d/7; e/8; f/3; g/4; h/11; i/6; j/12; k/9; l/1; m/2.
2. (i) ribosomal RNA; (ii) transfer RNAs; (iii) miRNAs and siRNAs; (iv) telomeric DNA; (v) centromeric DNA; (vi) LINES; (vii) SINES; (viii) in/dels; (ix) pseudogenes; (x) segmental duplications; (xi) CNVs.

3. Given that the nucleotide sequence characteristics of telomeric repeats is TTAGGG, the fusion sequence of two chromosome ends would be the following:
   TTAGGGTTAGGG
   AATCCCAATCCC.
4. Understood: centromeric (α-satellite) DNA; telomeric DNA. Not understood: long interspersed nuclear element (LINE); short interspersed nuclear element (SINE).
5. a. Array-based comparative genomic hybridization. Short stretches of DNA from across the genome are prepared from test and reference samples; these are labeled with different fluorochromes, mixed together, and spotted on arrays containing segments of DNA that cover the genome; then the read out compares the fluorescent colors well-by-well.
   b. Complete genomic sequencing in a large population of individuals.
   c. α-Globin gene deletions in α-thalassemia; PMP22 gene duplication in CMT1A; deletion of unknown genes in schizophrenia.
   d. It would suggest that such CNVs either caused some fraction of these affected or were sites of predisposition adjacent to etiologically important genes.
6. It would be convincing evidence that schizophrenia and bipolar disease were "real"—that is that they resulted from brain abnormalities—rather than reflecting imaginary findings or weaknesses; this would go a considerable distance in lessening the stigma still accompanying many common disorders affecting mood and behavior.

CHAPTER 15 BIRTH DEFECTS

1. a/9; b/8; c/3; d/10; e/6; f/13; g/14; h/1; i/5; j/12; k/11; l/7; m/4; n/2; o/15.
2. Because flies and mice can be mutagenized experimentally and humans can't.
3. The word congenital means "present at birth." Congenital (or birth) defects may be inherited, but many (e.g. teratogenic or chromosomal ones) are not inherited.
4. It says that homeobox genes are vital to development of most (perhaps all) multicellular organisms.
5. a. Might produce a larger than normal eye, or a part thereof.
   b. Might produce a small eye or a missing part of the eye.
   c. Gain-of-function would likely be dominant; loss-of-function recessive.
6. This suggests that homeobox genes demonstrate functional redundancy, meaning that more than one gene controls a single function; therefore the need to knock out more than one gene.
7. a. i. Obtain a karyotype in child and parents looking for a chromosomal abnormality.
      ii. Question mother about use of any possible teratogen during early weeks of pregnancy.
      iii. Inquire regarding consanguinity in parents.
   b. Adage about "lightning" is not applicable to birth defects; given that pattern of defects has not been observed before, recurrence risk could range from zero (new dominant mutation) to 25% (recessive disorder) to very likely (chromosomal translocation in one parent, or use of teratogen).
8. Yes, all women of child-bearing age should supplement their diet, because unplanned or unrecognized pregnancy might lead to neural tube defect before supplementation could be initiated; required supplementation is more difficult because it implies governmental regulation of reproduction and medical care.
9. a. Because accutane is very effective for the treatment of severe acne—a condition of great concern to women of child-bearing age; and because accutane's manufacturer has lobbied FDA to leave it on the market.
   b. Because Congress delegates such decision-making to the FDA, an agency with far greater expertise on such a matter than Congress possesses.

CHAPTER 16: THE GENETICS OF CANCER

1. a/19; b/8; c/4; d/7; e/14; f/11; g/18; h/2; i/16; j/9; k/13; l/17; m/5; n/15; o/1; p/12; q/20; r/6; s/21; t/3; u/10.
2. Cancer is genetic because it is caused by mutations; it is rarely inherited, because these mutations occur in somatic cells rather than germ-line cells.
3. Both result from abnormal proliferation of cells; benign tumors remain circumscribed and are not invasive; malignant tumors (synonymous with cancer) invade tissue locally and have the capacity to metastasize. If a benign tumor occurs at a site where it interferes with function (e.g., brain or heart), it may be clinically important—even lethal.
4. About 30 ($2^{30} = \sim 10^9$); given that each mitotic cell cycle in humans takes about 24 hours (1 day), 30 doublings would take about 30 days.
5. a. Dominant because it follows vertical transmission, affects both genders, and is almost completely penetrant; recessive because both Rb1 alleles must be knocked out before tumor results.
   b. i. Inhibit cell proliferation
      ii. Promote apoptosis.
   c. Identification of two kinds of families: those in which retinoblastoma appeared in multiple children, often in both eyes; and those in which retinoblastoma was sporadic and unilateral.
6. Cancer is clonal in that it originates in a single cell; that cell (and its progeny) must suffer several mutations at different loci before it develops the malignant cell phenotype.
7. Mutation of one allele from proto-oncogene to oncogene is sufficient to drive cell proliferation whereas both alleles of a TSG must be mutated before its inhibitory function is lost; apparently mutating one allele of a proto-oncogene to an oncogene disturbs development sufficiently that miscarriage results or the affected live-born infant doesn't survive to reproductive age.
8. a. Because cancer is caused by so many different mutations of hundreds of "cancer genes" that no one form of treatment will cure all types of cancers.
   b. i. Identify protein markers on tumor cells that are of prognostic significance (such as the estrogen receptor in breast cancer cells).
      ii. Do expression profiling looking for genomic signatures of predictive value.
      iii. Sequence entire genome of cancer cells to identify complete repertoire of gene mutations leading to cancer phenotype.
   c. Targeted treatment is aimed at the particular step that makes cancer cells vulnerable; standard chemotherapy aims to destroy cancer cells using one or more cytotoxic drugs whose mechanism of action is generic (e.g., impairs DNA replication or mitotic apparatus), realizing that some normal cells will be killed along with cancer cells.
9. a. They reflect environmental influences; lung cancer prevalence has risen as cigarette smoking has increased, while stomach cancer prevalence has decreased as such carcinogens as nitrite have been eliminated from the diet.
   b. Discourage cigarette smoking by every possible, legal means (taxation, education, public affairs advertising).
   c. Because breast cancer is easier to detect early; because curative surgery is easier to perform; because tumor markers in breast cancer exist and are accessible to treatment.
10. a. Identification of the Philadelphia chromosome in the blood cells of patients with CML; this led ultimately to the discovery that a chromosomal translocation (t9/22) led to activation of the ABL oncogene.
    b. Complete genomic sequencing that identified many genes not previously shown to be mutated in any cancer.

11. It seems likely that some genes (such as *Rb1*) are turned on during prenatal or early neonatal development and are, therefore, prone to deleterious mutations early in life; other genes may be active preferentially in adults, where the series of mutations required to produce cancer occurs over many years.
12. Because genomic stability (DNA replication, chromosome segregation, etc.) is required during each mitotic division; a variety of different disturbances of the cell cycle will, thus, interfere with such genomic stability and result in aneuploidy of many types (monosomy, trisomy, translocations).

CHAPTER 17: DETECTION AND TREATMENT

1. a/5; b/9; c/3; d/13; e/7; f/2; g/12; h/11; i/4; j/18; k/16; l/8; m/1; n/14; o/15; p/10; q/17; r/6.
2. i. To provide effective treatment: phenylketonuria.
   ii. To identify family members at risk: translocation Down syndrome.
   iii. To understand disease mechanism: Alzheimer disease.
3. a. Tay-Sachs is much more prevalent in Ashkenazi Jews than in any other ethnic group; 1/30 Ashkenazi are carriers compared to 1/300 in other whites.
   b. Couples are tested to identify carriers; if both are carriers they have a 1/4 chance of having an affected child with each pregnancy; this information generally leads couples to opt for prenatal diagnosis.
   c. The prevalence of Tay-Sachs in Ashkenazim has decreased by ~90% in the US.
   d. No, because most of the mutant alleles are carried in heterozygotes, not in the homozygous affected.
4. a. For certain recessive disorders that are much more prevalent in one ethnic group than in others (e.g., sickle cell anemia in blacks; cystic fibrosis in whites; β-thalassemia in Italians; Tay-Sachs in Ashkenazi Jews) interracial marriage will decrease the prevalence of affected homozygotes; for disorders that have equal frequency in all ethnic groups, interracial marriage will have no effect.
   b. Brazil and Hawaii.
   c. Because, in a democratic society, marriage and reproduction have always been considered strictly private matters, not in the arena of government.
5. a. Cure means getting rid of a deleterious condition with single course of management (e.g., penicillin for strep throat or surgery for appendicitis); treatment means chronic administration of agent (e.g., insulin for diabetes or statin for elevated serum cholesterol).
   b. i. The role of the majority of human genes is not understood.
      ii. Knowledge of structure and function of any given gene is merely the beginning of developing effective therapies.
      iii. It usually takes 20–30 years from identifying a medical need to developing useful treatment; it's too early to expect clinical "fruits" of genome project.
      iv. Meiotic non-disjunction, the cause of most chromosomal abnormalities, cannot be approached at present because it occurs in gametes rather than in post-zygotic cells, embryos, fetuses, or babies.
6. a. i. Development of technology to clone gene.
      ii. Perfecting vectors to deliver gene.
      iii. Lack of knowledge about where transduced gene will be inserted.
      iv. Failure to achieve sufficient activity of transferred gene.
      v. Need to be sure that gene/vector administration will not produce ill effects.
   b. *In vivo*: gene (and vector) are delivered directly into patient's organ or tissue. *Ex vivo*: cells are removed from patient, gene is inserted into a vector, then propagated in normal cells in culture; then modified cells are returned to the patient.

c. The success was the cure of children with severe combined immunodeficiency disease (SCID); the failure was the first documented death of a patient (Jesse Gelsinger) volunteering to participate in a gene therapy trial; his death provoked a powerful backlash against gene therapy trials that virtually halted such trials in academia, government and industry for several years.

7. a. Fewer than 200,000 affected people in the entire population of the US.
b. **i.** Most clinicians have little or no experience in diagnosing them.
ii. Little information is disseminated through public media.
iii. Little interest by pharmaceutical companies to develop treatments.
c. Because it is as expensive to develop a drug for a rare disease as for a common one, and the return on that investment will be much less for a rare disease.

8. a. That medical treatment can repair damage to vital organ or tissue.
b. The ability to understand how different tissues differentiate, and to use that information clinically.
c. Stem cells are cells that can proliferate and differentiate; thus they hold great promise for development of regenerative technologies.
d. Adult stem cells isolated from bone marrow have been employed in a number of hematologic disorders, including leukemia, immunodeficiency, lymphoma, and aplastic anemia.
e. **i.** Ability to isolate human ESCs was developed only recently (1998).
ii. In response to the "pro life" lobby, President Bush prohibited use of federal funds for isolation of new ESCs (2001).
iii. Effect of above executive decision was to discourage academic research on ESCs.
iv. Reversal of prior executive decision by President Obama led to legal challenges which continue to the present day.
f. No, because it is already clear that iPSCs are not identical to ESCs and may not have full range of potential clinical utility that ESCs do.

9. a. The medical school curriculum does not provide adequate information about genetics; most clinicians cannot spend the large amount of time with consultands to provide accurate and understandable information.
b. **i.** Taking a detailed family history.
ii. Researching the condition for which information is requested.
iii. Sharing information with all members of counseling team.
iv. Delivering information in understandable terms.
c. **i.** Recognition that condition has a genetic basis.
ii. Having some understanding of simple probability (% recurrence risk for example).
iii. Finding an appropriate counseling program.
d. **i.** Increase the number of courses in human genetics in undergraduate colleges and universities.
ii. Expand the number and kind of programs designed to educate genetic counselors.
iii. Subsidize education of genetic counselors, and increase their salaries after graduation.

CHAPTER 18: POPULATION AND EVOLUTIONARY GENETICS

1. a/2; b/12; c/6; d/17; e/19; f/10; g/20; h/9; i/8; j/14; k/13; l/1; m/3; n/7; o/18; p/16; q/5; r/15; s/4; t/11.

2. a. Total alleles = 2,000; number of A alleles = 500 + 500 + 400 = 1,400; number of a alleles = 400 + 100 + 100 = 600; allele frequency for A = 1,400/2,000 = 0.7; allele frequency for a = 600/2,000 = 0.3.
b. $(0.7)^2 + 2(0.7 \times 0.3) + (0.3)^2 = 0.49 + 0.42 + 0.09 = 1.0$; population is in Hardy-Weinberg equilibrium.

c. i. Large population.
   ii. Random mating.
   iii. No selection.
   iv. No mutation.
   v. No assortative mating.
   vi. No inbreeding.

3. a. Total alleles = 10,000; number of M alleles = 2,300 + 2,300 + 2,200 = 6,800; number of N alleles = 2,200 + 500 + 500 = 3,200; allele frequency for M = 6,800/10,000 = 0.68; allele frequency for N = 3,200/10,000 = 0.32.
   b. $(0.68)^2 + 2(0.68 \times 0.32) + (0.32)^2 = 0.46 + 0.22 + 0.10 = 0.78$; this population is not in Hardy-Weinberg equilibrium.
4. a. $q = \sqrt{1/10000} = 1/100$; $2pq = \sim 2/100$ or 1/50.
   b. 1/4 x 1/4 x 1/4 = 1/64.
   c. Mutation; inbreeding; bottleneck, assortative mating.
5. a. $q = \sqrt{1/250{,}000} = 1/500$.
   b. 300,000,000 ÷ 1/500.
   c. 2pq = 1/250; affected = 1/250,000; ratio of carriers to affected = 1/250 over 1/250,000 = 1/100.
   d. Vast majority of mutant alleles are "masked" in heterozygotes where selection doesn't affect them.
6. a. 2pq = 0.04; q = 0.02.
   b. (0.04) × (0.04) = 0.0016 or 1/625.
   c. $(0.02)^2 = 0.0004$.
   d. 1/4.
7. a. Total alleles = 430; frequency of $t = q^2 = 65/215 = 0.3$ therefore q = 0.55; frequency of T = 1 − 0.55 = 0.45.
   b. Genotype frequencies: $TT = p^2 = 0.20$; Tt = 2pq = 0.495; $tt = q^2 = 0.30$; number of heterozygous carriers = 0.495 × 215 = 106.
   c. Frequency of T would rise; frequency of t would fall.
8. a. An evolutionary clock attempts to estimate how long ago particular species appeared by using the frequency of mutations in DNA; mitochondrial DNA could be used because its mutation rate is high and because its mutations are not repaired.
   b. Since mtDNA is maternally inherited, its use can serve as an evolutionary clock for women, but not men.
   c. Studies of the Y chromosome have been used as clocks in males; the limitation here is that the mutation rate is lower than that for mtDNA, and DNA repair mechanisms exist.
9. There is a close inverse correlation between the importance of religion and the belief in evolution. In countries like Japan and Sweden, organized religion is much less a cultural norm than in the United States. Further, if one believes in a literal interpretation of the Bible, there is no way to accept the evidence that life on Earth began more than 4 billion years ago.

CHAPTER 19: PERSONALIZED GENETICS AND GENOMICS

1. a/13; b/6; c/8; d/2; e/12; f/11; g/4; h/1; i/10; j/7; k/9; l/3; m/5.
2. a. Because VNTRs in minisatellites are more polymorphic than STRPs in microsatellites, thereby yielding more observable alleles.
   b. STRPs are 2–10 base pairs long; VNTRs are 10–100 bp long; VNTRs have more alleles than STRPs do.
   c. Some 9/11 victims were so charred that there was no DNA to fingerprint; Copernicus was identified because his skeleton and hair were matched by DNA profiling.
3. a. Because each person is heterozygous at the VNTR probed.

b. No, because some people would be expected to be homozygous at this locus.
c. Cannot exclude paternity in families 1 and 2 because father and child share one allele; in family 3, child shares neither band of "father," thereby demonstrating non-paternity.

4. First, determine fraction of those homozygous for each allele as follows: $0.25 \times 0.25 = 0.0625$; for all four alleles 0.0625×4 or 0.25 or 25% are homozygous; then calculate fraction heterozygous using formula (1 – fraction who are homozygous) or $1 - 0.25 = 0.75$ or 75%.
5. a. i. Determine if other people on drug developed anemia.
   ii. Ascertain that anemia disappears promptly upon stopping drug.
   iii. Determine whether those people with anemia had higher blood levels of drug after dosing than did people who did not develop this adverse reaction.
   b. i. Notify patient to stop taking the drug.
   ii. Notify FDA who maintain drug safety database.
   c. It is most unlikely that such a suit would be successful because drug had been deemed safe and effective by FDA.
6. a. They catalyze the first step of metabolism of > 90% of commonly used pharmaceuticals.
   b. Pharmacokinetic: they affect drug metabolism.
   c. For drug (1), CYP catalyzes first step in drug breakdown; thus, reduced CYP activity would lead to drug accumulation. For drug (2), CYP catalyzes conversion of inactive prodrug to active drug; enhanced CYP activity would yield formation of excess active drug.
   d. It indicates that CYPs developed to respond to molecules in the natural environment which affected the organism—either positively or negatively.
7. a. Treating each person as unique: "the right drug at the right dose for the right indication."
   b. They are correct; we are only at the beginning of personalized medicine for several reasons: the genomic technology has not been perfected or applied; the pharmaceutical industry has not yet used genomics regularly in their research and development programs; the FDA has not mandated GWAS for prospective pharmaceuticals.
   c. DTC genome testing uses SNP-testing in the same way that GWAS do, namely predicting susceptibilities based on informative SNPs.
8. a. DTC genome testing is the commercial approach whereby individuals can subscribe to a service offering them (directly) information about disease susceptibility, drug response, trait presence and ancestry.
   b. i. Identify people with increased risk for many disorders—common or rare, serious or mild.
   ii. Predict those at increased risk of ADR to growing list of medicines.
   iii. Increase knowledge of basic genetic concepts.
   c. i. Invalid testing procedures and/or interpretation.
   ii. Inability of consumer to interpret information and act on it.
   iii. Breach privacy of individuals concerning their health status.
   d. It would not be illegal or unethical to offer DTC genome testing for any of these conditions provided the consumer fully understood quality of information and how it might be used; because there are no good therapies for Alzheimer disease today, there is more debate about identifying those people with increased susceptibility for it than there is for diabetes or Hodgkin's.

Credits and Permissions

CHAPTER 1

Fig. 1.1 ©American Express 1990.
Fig. 1.2 Rights-free image.
Fig. 1.3 ©Leon Rosenberg.
Fig. 1.4 From Richards, J.E. and R.S. Hawley. *The Human Genome*, 3rd Edition. Elsevier Academic Press, 2010; Fig. 6.9, p. 53, courtesy of D.M. Reed.
Fig. 1.5 From Richards, J.E. and R.S. Hawley. *The Human Genome*, 3rd Edition. Elsevier Academic Press, 2010; Fig. 6.10, p. 54, courtesy of D.M. Reed.
Fig. 1.6 Used with permission from Complete Genetics Corporation.

CHAPTER 3

Fig. 3.1 All photographs ©Diane Drobnis Rosenberg.
Fig. 3.2 Used with permission from the Metropolitan Museum of Art, New York.
Fig. 3.4 In public domain.
Table 3.1 ©Leon Rosenberg.

CHAPTER 4

Fig. 4.4 Modified from Nussbaum, R.L., McInnes, R.R., and Willard, H.F. *Thompson & Thompson: Genetics in Medicine*, 7th Edition. Saunders Elsevier, 2007.
Fig. 4.5 Modified from Lewis, R. *Human Genetics: Concepts and Applications*, 9th Edition. McGraw Hill, 2010; Fig. 3.3, p. 47.
Figs. 4.6 and 4.7 Modified from Hartwell, L.H., L. Hood, M. Goldberg *et al. Genetics: From Genes to Genomes*, 3rd Edition. McGraw Hill, 2008.
Fig. 4.8 Modified from Hartwell, L.H., L. Hood, M. Goldberg *et al. Genetics: From Genes to Genomes*, 3rd Edition. McGraw Hill, 2008; Fig. 4.18, p. 104.
Fig. 4.9 Modified from Hartwell, L.H., L. Hood, M. Goldberg *et al. Genetics: From Genes to Genomes*, 3rd Edition. McGraw Hill, 2008; Fig. 4.19, p. 105.
Fig. 4.13 Modified from Lewis, R. *Human Genetics: Concepts and Applications*, 9th Edition. McGraw Hill, 2010; Fig. 3.16, p. 58.
Fig. 4.14 Modified from Lewis, R. *Human Genetics: Concepts and Applications*, 9th Edition. McGraw Hill, 2010; Fig. 3.14.
Fig. 4.15 Used with permission from *Journal of Histochemistry and Cytochemistry*, 2005, Ogilvie *et al.*
Fig. 4.16 (A) Courtesy of NIH Report on Stem Cells, 2004; (B) Used with permission from *Oxford Today*, Oxford University.
Fig. 4.18 Modified from Lewis, R. *Human Genetics: Concepts and Applications*, 9th Edition. McGraw Hill, 2010; Fig. 2.22, p. 37.
Fig. 4.23 ©Leon Rosenberg.

CHAPTER 5

Fig. 5.2 Modified from Hartwell, L.H., L. Hood, M. Goldberg *et al. Genetics: From Genes to Genomes*, 3rd Edition. McGraw Hill, 2008; Fig. 2.76, p. 17.

Fig. 5.4 Modified from Hartwell, L.H., L. Hood, M. Goldberg *et al. Genetics: From Genes to Genomes*, 3rd Edition. McGraw Hill, 2008; Fig. 2.11, p. 21.

Fig. 5.5 Modified from Hartwell, L.H., L. Hood, M. Goldberg *et al. Genetics: From Genes to Genomes*, 3rd Edition. McGraw Hill, 2008; Fig. 2.15, p. 25.

Fig. 5.6 Modified from Nussbaum, R.L., R.R. McInnes and H.F. Willard. *Thompson & Thompson: Genetics in Medicine*, 7th Edition. Saunders Elsevier, 2007; Fig. 7.1, p. 117.

Fig. 5.14 Used with permission from *Thompson & Thompson Genetics in Medicine*, 7th Edition, 2007, Saunders Elsevier, Philadelphia.

CHAPTER 6

Fig. 6.1 Modified from Lewis, R. *Human Genetics: Concepts and Applications*, 9th Edition. McGraw Hill, 2010; Fig. 9.1, p.169.

Fig. 6.9 Modified from Hartwell, L.H., L. Hood, M. Goldberg *et al. Genetics: From Genes to Genomes*, 3rd Edition. McGraw Hill, 2008; Fig. 6.15a.

Fig. 6.11 Modified from Lewis, R. *Human Genetics: Concepts and Applications*, 9th Edition. McGraw Hill, 2010; Fig. 9.13, p. 176.

Fig. 6.12 Modified from Nussbaum, R.L., R.R. McInnes and H.F. Willard. *Thompson & Thompson: Genetics in Medicine*, 7th Edition. Saunders Elsevier, 2007; Fig. 2.8, p. 12.

Fig. 6.13 Used with permission from Donahue, R.P., W.B. Bias, J.H. Renwick and V.A. McKusick. Probable assignment of the Duffy blood group locus to chromosome 1 in man. *Protocols of National Academy of Science USA*: 1968; 61: 949–955.

Fig. 6.15 Modified from Korf, B.R. *Human Genetics and Genomics*, 3rd Edition. Blackwell Publishing, 2007; Fig. 4.7, p. 67.

Fig. 6.19 Modified from Hartwell, L.H., L. Hood, M. Goldberg *et al. Genetics: From Genes to Genomes*, 3rd Edition. McGraw Hill, 2008; Fig. 9.2, p. 304.

Fig. 6.20 Modified from Hartl, D.L. and E.W. Jones. *Genetics: Analysis of Genes and Genomes*, 7th Edition. Jones and Bartlett Publishers, LLC. 2009; Fig. 2.16, p. 55.

Fig. 6.21 Modified from Hartl, D.L. and E.W. Jones. *Genetics: Analysis of Genes and Genomes*, 7th Edition. Jones and Bartlett Publishers, LLC. 2009; Fig. 12.4, p. 435.

Fig. 6.23 Modified from Hartwell, L.H., L. Hood, M. Goldberg *et al. Genetics: From Genes to Genomes*, 3rd Edition, McGraw Hill, 2008.

Fig. 6.24 (B) Modified from Korf, B.R. *Human Genetics and Genomics*, 3rd Edition. Blackwell Publishing, 2007; Fig. 2.15, p. 30.

Fig. 6.25 Modified from Lewis, R. *Human Genetics: Concepts and Applications*, 9th Edition. McGraw Hill, 2010; Fig. 22.1, p. 431.

Fig. 6.26 Modified from Lewis, R. *Human Genetics: Concepts and Applications*, 9th Edition. McGraw Hill, 2010; Fig. 22.3, p. 435.

Table 6.2 Modified from Hartwell, L.H., L. Hood, M. Goldberg *et al. Genetics: From Genes to Genomes*, 3rd Edition. McGraw Hill, 2008; Table 10.1, p. 355.

CHAPTER 7

Fig. 7.3 Modified from Lewis, R. *Human Genetics: Concepts and Applications*, 9th Edition. McGraw Hill, 2010; Fig. 10.8, p. 187.

Fig. 7.5 Modified from Korf, B.R. *Human Genetics and Genomics*, 3rd Edition. Blackwell Publishing, 2007; Fig. 1.10, p. 187.

Fig. 7.8 Modified from Lewis, R. *Human Genetics: Concepts and Applications*, 9th Edition. McGraw Hill, 2010; Fig. 10.15, p. 192.

Fig. 7.9 Modified from Lewis, R. *Human Genetics: Concepts and Applications*, 9th Edition. McGraw Hill, 2010; Fig. 10.16, p. 193.

Fig. 7.13 Modified from Hartwell, L.H., L. Hood, M. Goldberg *et al. Genetics: From Genes to Genomes*, 3rd Edition. McGraw Hill, 2008; Fig. 8, p. 493.

Fig. 7.14 Modified from Hartl, D.L. and E.W. Jones. *Genetics: Analysis of Genes and Genomes*, 7th Edition. Jones and Bartlett Publishers, LLC. 2009; Fig. 11.27, p. 408.

Fig. 7.15 Modified from Hartl, D.L. and E.W. Jones. *Genetics: Analysis of Genes and Genomes*, 7th Edition. Jones and Bartlett Publishers, LLC. 2009, Fig. 11.33, p. 413.

Fig. 7.16 Modified from Hartl, D.L. and E.W. Jones. *Genetics: Analysis of Genes and Genomes*, 7th Edition. Jones and Bartlett Publishers, LLC. 2009; Fig. 11.35, p. 415.

CHAPTER 8

Table 8.1 Modified from Nussbaum, R.L., R.R. McInnes and H.F. Willard. *Thompson & Thompson: Genetics in Medicine*, 7th Edition. Saunders Elsevier, 2007; Table 9.1, p. 176.

Table 8.3 Modified from Hartl, D.L. and E.W. Jones. *Genetics: Analysis of Genes and Genomes*, 7th Edition. Jones and Bartlett Publishers, LLC. 2009; Table 14.6, p. 538.

CHAPTER 9

Fig. 9.3 Modified from National Academy of Sciences and Institute of Medicine. *Science, Evolution, and Creationism*. The National Academies Press, 2008; Figure, p. 3.

Fig. 9.4 Modified from National Academy of Sciences and Institute of Medicine. *Science, Evolution, and Creationism*. The National Academies Press, 2008; Figure, p. 8.

Fig. 9.5 Modified from National Academy of Sciences and Institute of Medicine. *Science, Evolution, and Creationism*. The National Academies Press, 2008; Figure, p. 25.

Fig. 9.6 Modified from National Academy of Sciences and Institute of Medicine. *Science, Evolution, and Creationism*. The National Academies Press, 2008; Figure, p. 30.

Box 1 Figure used with permission from The Field Museum, Chicago, IL.

CHAPTER 11

Fig. 11.1 (A) Used with permission from Gelehrter, T.D., F.S. Collins and D. Ginsburg. *Principles of Medical Genetics*, 2nd Edition. Lippincott, Williams & Wilkins, 1998; Figs 8.1 and 8.6; (B) Courtesy of University of Michigan Cytogenetics Services (PHIPPS' Medical-Surgical Nursing: Health and Illness Perspectives).

Fig. 11.2 Used with permission from Nussbaum, R.L., R.R. McInnes and H.F. Willard. *Thompson & Thompson: Genetics in Medicine*, 7th Edition. Saunders Elsevier, 2007; Fig. 5.1, p. 61.

Fig. 11.5 Modified from Gelehrter, T.D., F.S. Collins, and D. Ginsburg. *Principles of Medical Genetics*, 2nd Edition. Lippincott, Williams & Wilkins, 1998; Fig. 8.11, p. 164.

Fig. 11.6 Modified from Gelehrter, T.D., F.S. Collins and D. Ginsburg. *Principles of Medical Genetics*, 2nd Edition. Lippincott, Williams & Wilkins, 1998; Fig. 8.12, p. 165.

Fig. 11.7 Modified from Gelehrter, T.D., F.S. Collins and D. Ginsburg. *Principles of Medical Genetics*, 2nd Edition. Lippincott, Williams & Wilkins, 1998; Fig. 8.13.

Fig. 11.8 Modified from Gelehrter, T.D., F.S. Collins and D. Ginsburg. *Principles of Medical Genetics*, 2nd Edition. Lippincott, Williams & Wilkins, 1998; Fig. 8.14, p. 167.

Fig. 11.9 Modified from Gelehrter, T.D., F.S. Collins and D. Ginsburg. *Principles of Medical Genetics*, 2nd Edition. Lippincott, Williams & Wilkins, 1998; Fig. 8.15.

Fig. 11.12 Used with permission from Gelehrter, T.D., F.S. Collins and D. Ginsburg. *Principles of Medical Genetics*, 2nd Edition. Lippincott, Williams & Wilkins, 1998; Fig. 8.13b (courtesy of Dr T. Glover).
Fig. 11.14 Modified from Lewis, R. *Human Genetics: Concepts and Applications*, 9th Edition. McGraw Hill, 2010; Fig. 13.5, p. 244.
Fig. 11.18 Used with permission from Nussbaum, R.L., R.R. McInnes, and H.F. Willard. *Thompson & Thompson: Genetics in Medicine*, 7th Edition. Saunders Elsevier, 2007.
Fig. 11.20 Modified from Hartl, D.L. and E.W. Jones. *Genetics: Analysis of Genes and Genomes*, 7th Edition. Jones and Bartlett Publishers, LLC. 2009; Fig. 11.32, p. 412.
Box 1 Quotation from Congressional Record; Human Life Bill, p. 52, 1981.
Table 11.3 Modified from Nussbaum, R.L., R.R. McInnes and H.F. Willard. *Thompson & Thompson: Genetics in Medicine*, 7th Edition. Saunders Elsevier, 2007; Table 5.2, p. 66.
Table 11.4 Modified from Hartl, D.L. and E.W. Jones. *Genetics: Analysis of Genes and Genomes*, 7th Edition. Jones and Bartlett Publishers, LLC. 2009; Table 8.2, p. 264.

CHAPTER 12

Fig. 12.1 Used with permission from http://www.sciencephoto.com/
Fig. 12.5 Modified from Scriver C.R., A.L. Beaudet, S.W. Sly *et al. The Metabolic & Molecular Bases of Inherited Disease*, 8th Edition. McGraw Hill, 2001; Fig. 4.1a, p. 169.
Fig. 12.6 Modified from Scriver C.R., A. L. Beaudet, S.W. Sly *et al. The Metabolic & Molecular Bases of Inherited Disease*, 8th Edition. McGraw Hill, 2001; Fig. 4.7, p. 172.
Fig. 12.10 Modified from Gelehrter, T.D., F.S. Collins and D. Ginsburg. *Principles of Medical Genetics*, 2nd Edition. Lippincott, Williams & Wilkins, 1998; Fig. 6.4, p. 95.
Fig. 12.13 Modified from Gelehrter, T.D., F.S. Collins, and D. Ginsburg. *Principles of Medical Genetics*, 2nd Edition. Lippincott, Williams & Wilkins, 1998; Fig. 6.6, p. 97.
Fig. 12.14 Courtesy of PhotoResearchers.com
Fig. 12.15 Used with permission from Bunn, H.F. and B. Forget. *Hemoglobin: Molecular, Genetic, and Clinical Aspects*. Philadelphia, PA: WB Saunders, 1986.
Fig. 12.17 Modified from Hartl, D.L. and E.W. Jones. *Genetics: Analysis of Genes and Genomes*, 7th Edition. Jones and Bartlett Publishers, LLC. 2009; Fig. 2.26, p. 69.
Fig. 12.18 Modified from Gelehrter, T.D., F.S. Collins and D. Ginsburg. *Principles of Medical Genetics*, 2nd Edition. Lippincott, Williams & Wilkins, 1998; Fig. 9.22, p. 216.
Fig. 12.19 Used with permission from Nicholson, L.V., K. Davison, M.A. Johnson *et al.* Dystrophin in skeletal muscle. II. Immunoreactivity in patients with Xp21 muscular dystrophy. *Journal of Neurological Science*, 1989; 94: 137–146.
Fig. 12.20 Used with permission from Nussbaum, R.L., R.R. McInnes and H.F. Willard. *Thompson & Thompson: Genetics in Medicine*, 7th Edition. Saunders Elsevier, 2007; Fig. 12.17.
Fig. 12.21 Modified from Gelehrter, T.D., F.S. Collins and D. Ginsburg. *Principles of Medical Genetics*, 2nd Edition. Lippincott, Williams & Wilkins, 1998; Fig. 7.8, p. 129.
Fig. 12.22 Modified from Gelehrter, T.D., F.S. Collins and D. Ginsburg. *Principles of Medical Genetics*, 2nd Edition. Lippincott, Williams & Wilkins, 1998; Fig. 7.10.
Fig. 12.23 Used with permission from Lewis, R. *Human Genetics: Concepts and Applications*, 9th Edition. McGraw Hill, 2010; Fig. 5.3.
Fig. 12.24 Used with permission from Nussbaum, R.L., R.R. McInnes and H.F. Willard. *Thompson & Thompson: Genetics in Medicine*, 7th Edition. Saunders Elsevier, 2007; Fig. 7.27, p. 141.
Fig. 12.26 ©Leon Rosenberg.
Box 1 Both figures courtesy of *New York Times*. Photographer: Arian Rosenbloom.
Box 2 Both figures ©Leon Rosenberg.

CHAPTER 13

Fig. 13.3 Used with permission from Hartwell, L.H., L. Hood, M. Goldberg *et al.* *Genetics: From Genes to Genomes*, 3rd Edition. McGraw Hill, 2008; Fig. 3.22, p.71.

Fig. 13.5 Used with permission from Gelehrter, T.D., F.S. Collins and D. Ginsburg. *Principles of Medical Genetics*, 2nd Edition. Lippincott, Williams & Wilkins, 1998; Figure 4.6a, courtesy of R.B. Ross and M.C. Johnson).

Fig. 13.7 Modified from Hartwell, L.H., L. Hood, M. Goldberg *et al. Genetics: From Genes to Genomes*, 3rd Edition. McGraw Hill, 2008; Fig. 11.9, p. 402.

Fig. 13.8 Used with permission from Nussbaum, R.L., R.R. McInnes and H.F. Willard. *Thompson & Thompson: Genetics in Medicine*, 7th Edition. Saunders Elsevier, 2007; Fig. C.3, p. 237.

Fig. 13.9 Modified from Nussbaum, R.L., R.R. McInnes and H.F. Willard. *Thompson & Thompson: Genetics in Medicine*, 7th Edition. Saunders Elsevier, 2007; Fig. 12.27, p. 380.

Fig. 13.10 Modified from Nussbaum, R.L., R.R. McInnes and H.F. Willard. *Thompson & Thompson: Genetics in Medicine*, 7th Edition. Saunders Elsevier, 2007; Fig. 8.7, p. 165.

Fig. 13.13 Used with permission from *Science*, 296:686, American Association for the Advancement of Science, 2002.

Fig. 13.14 Courtesy of Dr. Charles Allen.

Fig. 13.15 Used with permission from *Science*, 308:386, Fig. 1A, American Association for the Advancement of Science, 2005.

Box 13.1 Figure courtesy of *Discover Magazine*, March 2007 issue.

Table 13.3 Modified from Carter C. O. Genetics of Common Disorders. British Medical Bulletin, 1969; 25(1): 52–57.

Table 13.4 Modified from Hartl, D.L. and E.W. Jones. *Genetics: Analysis of Genes and Genomes*, 7th Edition. Jones and Bartlett Publishers, LLC. 2009; Table 18.2, p. 663.

Table 13.6 Modified from Carter C. O. Genetics of Common Disorders. British Medical Bulletin, 1969; 25(1): 52–57.

Table 13.8 Modified from Nussbaum, R.L., R.R. McInnes and H.F. Willard. *Thompson & Thompson: Genetics in Medicine*, 7th Edition. Saunders Elsevier, 2007; Table 8.7, p. 165.

Table 13.10 Modified from Nussbaum, R.L., R.R. McInnes and H.F. Willard. *Thompson & Thompson: Genetics in Medicine*, 7th Edition. Saunders Elsevier, 2007; Table 8.5, p. 164.

CHAPTER 14

Fig. 14.1 Modified from Willard, H.F. and G.S. Ginsburg. *Genomic and Personalized Medicine*. Elsevier, 2009; Fig. 9.3, p. 113.

Fig. 14.2 Modified from Willard, H.F. and G.S. Ginsburg. *Genomic and Personalized Medicine*. Elsevier, 2009; Fig. 1.2, p. 9.

Fig. 14.3 Used with permission from Nussbaum, R.L., R.R. McInnes and H.F. Willard. *Thompson & Thompson: Genetics in Medicine*, 7th Edition. Saunders Elsevier, 2007; Fig. 6.8.

Fig. 14.4 ©Leon Rosenberg.

Fig. 14.5 ©Leon Rosenberg.

Table 14.2 Modified from Willard, H.F. and G.S. Ginsburg. *Genomic and Personalized Medicine*. Elsevier, 2009; Table 1.3, p. 10.

CHAPTER 15

Fig. 15.1 Modified from Mange, E.J. and A.P. Mange. *Basic Human Genetics*, 2nd Edition. Sinauer Associates Inc., 1999; Fig. 15.3, p. 309.
Fig. 15.2 Modified from Mange, E.J. and A.P. Mange. *Basic Human Genetics*, 2nd Edition. Sinauer Associates Inc., 1999; Fig. 15.5, p. 311.
Fig. 15.4 [Image Missing]
Fig. 15.5 Modified from Mange, E.J. and A.P. Mange. *Basic Human Genetics*, 2nd Edition. Sinauer Associates Inc., 1999; Fig. 15.9, p 314.
Fig. 15.6 Used with permission from Nussbaum, R.L., R.R. McInnes, and H.F. Willard. *Thompson & Thompson: Genetics in Medicine*, 7th Edition. Saunders Elsevier, 2007.
Fig. 15.7 Modified from Mange, E.J. and A.P. Mange. *Basic Human Genetics*, 2nd Edition. Sinauer Associates Inc., 1999; Fig. 15.4, p. 310.
Fig. 15.8 Used with permission from Nussbaum, R.L., R.R. McInnes and H.F. Willard. *Thompson & Thompson: Genetics in Medicine*, 7th Edition. Saunders Elsevier, 2007; Fig. 14.4.
Fig. 15.10 Used with permission from Springer Verlag.

CHAPTER 16

Fig. 16.2 Modified from Hartwell, L.H., L. Hood, M. Goldberg *et al. Genetics: From Genes to Genomes*, 3rd Edition. McGraw Hill, 2008; Fig. 19.2, p. 687.
Fig. 16.3 Modified from Hartwell, L.H., L. Hood, M. Goldberg *et al. Genetics: From Genes to Genomes*, 3rd Edition. McGraw Hill, 2008; Fig. 19.7a, p. 690.
Fig. 16.4 Modified from Hartl, D.L. and E.W. Jones. *Genetics: Analysis of Genes and Genomes*, 7th Edition. Jones and Bartlett Publishers, LLC. 2009; Fig. 15.11, p. 560.
Fig. 16.5 Modified from Hartwell, L.H., L. Hood, M. Goldberg *et al. Genetics: From Genes to Genomes*, 3rd Edition. McGraw Hill, 2008; Fig. 19.21, p.703.
Fig. 16.6 ©Leon Rosenberg.
Fig. 16.7 Modified from Gelehrter, T.D., F.S. Collins and D. Ginsburg. *Principles of Medical Genetics*, 2nd Edition. Lippincott, Williams & Wilkins, 1998; Fig. 11.7, p. 253.
Fig. 16.8 Modified from Blattner, W.A., D.B. McGuire, J.J. Mulvihill et al. Genealogy of Cancer in a Family. *J. Am. Med. Assoc.* 1979; 241: 259.
Fig. 16.9 [Image Missing] Used with permission from Mange, E.J. and A.P. Mange. *Basic Human Genetics*, 2nd Edition. Sinauer, 1999. Fig. 17.5.
Fig. 16.10 [Awaiting hi-res image] Used with permission from Nature Publishing Company.
Fig. 16.11 Modified from Nussbaum, R.L., R.R. McInnes and H.F. Willard. *Thompson & Thompson: Genetics in Medicine*, 7th Edition. Saunders Elsevier, 2007; Fig. 16.15B, p. 481.
Fig. 16.12 [Image Missing] Used with permission from Nature Publishing Company.
Fig. 16.14 Used with permission from Gelehrter, T.D., F.S. Collins and D. Ginsburg. *Principles of Medical Genetics*, 2nd Edition. Lippincott, Williams & Wilkins, 1998; Fig. 11.2, courtesy of Dr T. Glover.
Fig. 16.15 Modified from Lewis, R. *Human Genetics: Concepts and Applications*, 9th Edition. McGraw Hill, 2010; Reading 18.1, Fig. 2, p. 367.
Fig. 16.17 Modified from Gelehrter, T.D., F.S. Collins and D. Ginsburg. *Principles of Medical Genetics*, 2nd Edition. Lippincott, Williams & Wilkins, 1998; Fig. 11.13, p. 258.
Fig. 16.19 Modified from Nussbaum, R.L., R.R. McInnes and H.F. Willard. *Thompson & Thompson: Genetics in Medicine*, 7th Edition. Saunders Elsevier, 2007; Fig. 16.10, p. 473.
Table 16.1 Modified from Hartwell, L.H., L. Hood, M. Goldberg *et al. Genetics: From Genes to Genomes*, 3rd Edition. McGraw Hill, 2008; Fig. 19.1, p. 686.

CHAPTER 17

Fig. 17.2 Modified from Nussbaum, R.L., R.R. McInnes and H.F. Willard. *Thompson & Thompson: Genetics in Medicine,* 7th Edition. Saunders Elsevier, 2007; Fig. 13.1, p. 394.

Fig. 17.4 Modified from Gelehrter, T.D., F.S. Collins and D. Ginsburg. *Principles of Medical Genetics,* 2nd Edition. Lippincott, Williams & Wilkins, 1998; Fig. 13.4, p. 320.

Box 1 Both figures ©Leon Rosenberg.

Table 17.3 Modified from Nussbaum, R.L., R.R. McInnes and H.F. Willard. *Thompson & Thompson: Genetics in Medicine,* 7th Edition. Saunders Elsevier, 2007; Box, p. 444.

Table 17.4 Modified from Nussbaum, R.L., R.R. McInnes and H.F. Willard. *Thompson & Thompson: Genetics in Medicine,* 7th Edition. Saunders Elsevier, 2007; Table 19.1, p. 508.

Box 2 Figure1, gotpetsonline, purchase. Figure2, current opinion in Biotechnology

CHAPTER 18

Fig. 18.1 Used with permission from Hartwell, L.H., L. Hood, M. Goldberg *et al. Genetics: From Genes to Genomes,* 3rd Edition. McGraw Hill, 2008; Fig. 21.7.

Fig 18.7 Modified from Hartl, D.L. and E.W. Jones. *Genetics: Analysis of Genes and Genomes,* 7th Edition. Jones and Bartlett Publishers, LLC. 2009; Fig. 17.32, p. 645.

Table 18.2 Modified from Nussbaum, R.L., R.R. McInnes and H.F. Willard. *Thompson & Thompson: Genetics in Medicine,* 7th Edition. Saunders Elsevier, 2007; Table 9.6, p. 193.

Table 18.9 Modified from Lewis, R. *Human Genetics: Concepts and Applications,* 9th Edition. McGraw Hill, 2010; Table 16.3, p. 322.

CHAPTER 19

Fig. 19.2 Used with permission from Robert W. Allen, Oklahoma State University.

Fig. 19.3 Used with permission from Nussbaum, R.L., R.R. McInnes and H.F. Willard. *Thompson & Thompson: Genetics in Medicine,* 7th Edition. Saunders Elsevier, 2007; Fig. 9.6

Fig. 19.4 Modified from Lewis, R. *Human Genetics: Concepts and Applications,* 9th Edition. McGraw Hill, 2010; Reading 14.1, Fig. 1, p. 272.

Fig. 19.7 Modified from Goldstein, D.B., S.K. Tate and S.M. Sisodiya. Pharmacogenetics Goes Genomic. *Nature Reviews Genetics* 2003; 4: 937–947.

Fig. 19.8 Modified from Nussbaum, R.L., R.R. McInnes and H.F. Willard. *Thompson & Thompson: Genetics in Medicine,* 7th Edition. Saunders Elsevier, 2007; Fig. 18.3, p. 499.

Fig. 19.11 Used with permission from Gelehrter, T.D., F.S. Collins and D. Ginsburg. *Principles of Medical Genetics,* 2nd Edition. Lippincott, Williams & Wilkins, 1998; Fig. 7.5, redrawn from Luzzatto, L. and A. Mehta. Glucose-6-phosphate dehydrogenase deficiency. In: Scriver, C.R., A.L. Beaudet, W.S. Sly, eds. *The Metabolic and Molecular Bases of Inherited Disease,* 7th Edition. New York, NY: McGraw-Hill, 1995: 3367–3398.

Fig. 19.12 Modified from Gelehrter, T.D., F.S. Collins, and D. Ginsburg. *Principles of Medical Genetics,* 2nd Edition. Lippincott, Williams & Wilkins, 1998; Fig. 7.3, p. 121.

INDEX

Note: Page numbers followed by *f* indicate figures, *t* indicate tables and *b* indicate boxes.

B

C

D

E

F

H

I

N

O

P

Q

V

W

X

Y

Z

Printed and bound by CPI Group (UK) Ltd, Croydon, CR0 4YY
03/10/2024
01040316-0012